Carbon-Carbon Bond Formation

TECHNIQUES AND APPLICATIONS IN ORGANIC SYNTHESIS

Editor: Robert L. Augustine

Catalytic Hydrogenation
by Robert L. Augustine

Reduction
Robert L. Augustine, Editor

Oxidation, Volume 1
Robert L. Augustine, Editor

Oxidation, Volume 2
Robert L. Augustine and D. J. Trecker, Editors

Introduction to Organic Electrochemistry
by M. R. Rifi and Frank H. Covitz

Carbon-Carbon Bond Formation, Volume 1
Robert L. Augustine, Editor

Other Volumes in Preparation

Carbon-Carbon Bond Formation

VOLUME 1

Edited by
ROBERT L. AUGUSTINE
Department of Chemistry
Seton Hall University
South Orange, New Jersey

MARCEL DEKKER, INC.   New York and Basel

Library of Congress Cataloging in Publication Data
Main entry under title:

Carbon-carbon bond formation.

   (Techniques and applications in organic synthesis)
   Includes bibliographical references and index.
   1. Chemistry, Organic--Synthesis. 2. Chemical bonds.
I. Augustine, Robert L. [Date]
QD262.C29   547'.2   78-31483
ISBN 0-8247-6787-X

Marcel Dekker, Inc.
270 Madison Avenue, New York, New York 10016

Current printing (last digit):
10 9 8 7 6 5 4 3 2 1

Printed in the United States of America

PREFACE

The synthetic organic chemist must, of necessity, be well versed in
the applications and variations of a large number of reactions. As
time goes on and the volume of literature expands, it becomes increas-
ingly difficult for the practicing organic chemist to be aware of
all of the subtleties of a given reaction. It also becomes more
difficult to select the conditions most suitable for a particular
application of a reaction. It is the purpose of this series to
provide the practicing chemist with concise and critical evaluations
of as many reactions of synthetic importance as possible.

The previous volumes of this series have dealt with techniques
used primarily for the modification of functional groups on a particular carbon skeleton. The present volume, and others to follow,
deals with the reactions used in building the carbon skeleton, reactions in which a carbon-carbon bond is formed. These reactions are
not of a nature amenable to extensive generalization as far as
reaction conditions are concerned, and thus, the previous practice
of providing generalized experimental procedures is not applicable
here. Instead, discussions on the effects of the reaction variables
and selected, more specific experimental procedures are given. We
believe that what is presented will be sufficient for the practicing
chemist to utilize without requiring extensive reference to the
original literature.

It is anticipated that subsequent volumes will cover most, if
not all, of the more important reactions used to form carbon-carbon
bonds, and it is hoped that this series will provide the synthetic
organic chemist with a valuable tool to use in the carrying out of

complex syntheses and that through its use more time will be available to devote to the attainment of the final goal.

Robert L. Augustine
South Orange, New Jersey
June 1977

CONTENTS

CONTRIBUTORS

H. J. Bestmann, Institute of Organic Chemistry, The University Erlangen-Nurenberg, Erlangen, Germany

Drury Caine, School of Chemistry, Georgia Institute of Technology, Atlanta, Georgia

Zoltan G. Hajos, Division of Chemical Research, Ortho Pharmaceutical Corporation, Raritan, New Jersey

R. Zimmerman, Institute of Organic Chemistry, The University Erlangen-Nurenberg, Erlangen, Germany

Chapter One

ALDOL AND RELATED REACTIONS

Zoltan G. Hajos

Division of Chemical Research
Ortho Pharmaceutical Corporation
Raritan, New Jersey

I.   INTRODUCTION

The aldol reaction was not invented in a scientific laboratory, but
in the workshop of Nature. It is thus not accidental that, unlike
most other important chemical bond-forming reactions, this one has
not been named after any particular laboratory investigator. The
name of the reaction refers to the reaction product, a compound
with an aldehydic and alcoholic function, i.e., an *aldol*, formed
in the addition reaction of two specific carbonyl compounds.

This, however, amounts to only the prototype of aldol reactions.
One of Nature's original biochemical versions of the reaction produces carbohydrates (e.g., aldol type of aldoses) from two relatively
simple building blocks, the three carbon containing trioses. The
reaction is part of the important life-sustaining process in which
carbon dioxide is converted into carbohydrates.

Emil Fischer's classical synthesis of fructose [1] [cf. Refs.
2(a) and (b)] via an aldol reaction related to the enzyme-catalyzed
natural process has been followed by the discovery of a variety of
laboratory procedures for conducting the aldol reaction. The
directed, intermolecular, mixed aldol reactions have been developed
mainly during the past two decades. They represent the progress
of science in challenging Nature's achievements in this area.

II.   FUNDAMENTALS

A. *Reaction Mechanism and Techniques*

The reaction of an aldehyde or ketone with a carbanion of an active
methylene compound to give an aldehydic or ketonic alcohol (aldol
or ketol) is commonly known as the *aldol condensation reaction*.
Since it involves the addition of an enolate anion to the carbonyl
carbon atom without the formation of a side product, it should

more properly be named an *aldol addition reaction*. Description of
a reaction as a condensation implies that elimination of a smaller
molecular weight side product (e.g., water, alcohol, HCl) is
involved between the two reactants. The distinction is not only
of semantic importance, but it is of great value to the laboratory
experimenter in understanding the reaction mechanism and, through
this, in the proper choice of the experimental techniques.

The aldol or ketol (β-hydroxy carbonyl) reaction product of a
great many aldol type of reactions readily dehydrates under the
applied reaction conditions to yield an α,β-unsaturated carbonyl
compound. Therefore this second step is properly designated as
the condensation step of the aldol reaction. With certain carbonyl
compounds it is possible to find reaction conditions which permit
the isolation of the intermediate aldol or ketol addition product,
if the rate of addition is considerably faster than the rate of
elimination.

Due to the large abundance of aldehydes and ketones as well
as active methylene compounds, the aldol reaction enjoys a high
rating among the available carbon-carbon bond-forming reactions.
The aldehydes or ketones do not react readily without catalysts
because by themselves they are not sufficiently strong electrophiles
or nucleophiles. The catalysts for the aldol addition step are the
same as those used to promote enolization of carbonyl compounds.
Thus, bases as well as acids have been used to catalyze the reaction
by converting the originally unreactive compounds into reactive
donor anions and acceptor cations. A great variety of catalysts
have been used: KOH, CaO, Ba(OH)$_2$, K$_2$CO$_3$, KCN, sodium acetate;
organic bases such as pyrrolidine or piperidine and their salts
with organic acids, basic ion exchange resins; acids, such as HCl
or H$_3$PO$_4$; and Lewis acids, for example, POCl$_3$, BF$_3$, FeCl$_3$, AlCl$_3$,
or ZnCl$_2$. Further details are presented in Secs. III and IV.

In the simplest case of the aldol reaction, the active methylene
donor as well as the carbonyl acceptor is the same compound as
illustrated in the following reactions by acetaldehyde (*1*). Scheme 1

depicts the steps of the base catalyzed reaction consisting of
enolization under the influence of the catalyst (a), aldol addition
(b), and protonation of the aldol product with simultaneous recovery
of catalyst (c).

a. Enolization:

$$CH_3\text{-}CHO + HO^- \rightleftharpoons CH_2\text{=}\underset{\underset{O^-}{|}}{C}\text{-H} + H_2O$$
$$\underline{1}$$

b. Aldol addition reaction:

$$CH_3\text{-}\underset{\underset{O}{\|}}{\overset{\overset{H}{|}}{C}} + CH_2\text{=}\underset{\underset{O^-}{|}}{C} \rightleftharpoons CH_3\text{-}\underset{\underset{O^-}{|}}{CH}\text{-}CH_2\text{-}CHO$$

c. Protonation of aldol and recovery of catalyst:

$$CH_3\text{-}\underset{\underset{O^-}{|}}{CH}\text{-}CH_2\text{-}CHO + H_2O \rightleftharpoons CH_3\text{-}\underset{\underset{OH}{|}}{CH}\text{-}CH_2\text{-}CHO + HO^-$$
$$\underline{2}$$

Scheme 1

If possible, the aldol- or ketol-forming reactions should be carried
out under the *mildest possible reaction conditions,* usually at or
near ambient temperature with as weak a base as will promote the
forward reaction. More drastic conditions may favor the formation
of undesired polycondensation products. Since the reaction products are subject to air oxidation, the reactions are normally
carried out in an *inert atmosphere.*

It will be noted that the formation of the aldol addition
products proceeds through a series of equilibrium reactions. With
most aldehydes of the RCH_2CHO general structure the equilibrium con-
stants are favorable for the forward steps, and thus, for instance;
acetaldol (2) is formed in good yield. In this, as well as in
other cases where the acceptor molecule is sterically unhindered,
step a, the proton removal from the carbon atom alpha to the
carbonyl, constitutes the rate-limiting step. With sterically
more hindered systems, such as with ketones as acceptors, the
equilibrium of the addition reaction (b) becomes less favorable.

On occasion an efficient technique may be applied consisting in the continuous separation of the β-hydroxy ketone reaction product to shift the equilibrium of step b in a forward direction. This will be discussed in more detail in Sec. III.A.2.

The base-catalyzed addition of acetaldehyde (1) to acetaldol (2) is more rapid than the dehydration of the latter to the aldol condensation product, croton aldehyde (3). It is, therefore, easy to obtain 2 in good yield. The aldol addition product 2 may be readily dehydrated either by heating the basic reaction mixture or by a separate acid-catalyzed reaction. The elimination of water from β-hydroxy carbonyl systems is facilitated by the activating effects of the electron-withdrawing carbonyl group on the adjacent hydrogen atom. The acid-catalyzed dehydration of the aldol addition product is depicted in Scheme 2a, and the base-catalyzed dehydration in Scheme 2b.

a. Acid-catalyzed dehydration:

$$CH_3-\underset{\underset{2}{OH}}{\underset{|}{CH}}-CH_2-CHO \overset{H^+}{\rightleftharpoons} CH_3-\underset{\underset{2}{+OH_2}}{\underset{|}{CH}}-CH_2-CHO$$

$$\overset{-H_2O}{\rightleftharpoons} \quad + \quad CH_3-CH-CH_2-CHO \overset{H^+}{\rightleftharpoons} CH_3-CH=CH-CHO$$

$$\underline{3}$$

b. Base-catalyzed dehydration:

$$CH_3-\underset{\underset{2}{OH}}{\underset{|}{CH}}-CH_2-CHO + HO^- \overset{-H_2O}{\rightleftharpoons} CH_3-\underset{\underset{}{OH}}{\underset{|}{CH}}-CH-CHO$$

$$\overset{-HO^-}{\rightleftharpoons} CH_3-CH=CH-CHO$$

$$\underline{3}$$

Scheme 2

Thus, the acid- or base-catalyzed dehydration, that is, the condensation step of the aldol reaction, also proceeds through a series of equilibria. The separate, acid-catalyzed reaction usually permits the isolation of a cleaner dehydration product and is, therefore, the more widely used dehydration technique.

As already mentioned, the aldol addition step may also be catalyzed by acids. This will be discussed in Sec. III.A.2. in connection with the acid-catalyzed aldol (ketol) reaction of acetone.

B. *Classification of Aldol Type of Reactions*

A good understanding of the fundamentals described in the preceding section should enable the investigator to select and to apply the proper reaction techniques to a particular problem in the laboratory. The ancient Greeks assumed that good classification was essential to understand and to solve scientific problems. While this, of course, still holds, it should be emphasized that the various aldol type of reactions should be investigated with a certain amount of flexibility. For example, Knoevenagel-type catalysts have been used in non-Knoevenagel type of aldol reactions, and there is sometimes some overlap within the name reactions related to the aldol reaction. While the importance of differentiation between aldol addition and aldol condensation reactions has already been emphasized (Sec. II.A.), this distinction shall not be used in the classification of the aldol and related reactions. The reason for this is that depending on the experimental techniques and conditions the reaction between the same two reactants may proceed to the stage of aldol addition or further to the stage of aldol condensation; it would thus require an overextension of the classification process.

The aldol type of reactions demonstrate the versatility of the carbonyl group. It is an electrophilic, acceptor type of chemical group, yet at the same time it can support a negative charge on a neighboring carbon atom by enolization, thereby acting as an electron donor in a chemical reaction. The great variety of

reactions may, thus, be classified according to the nature of the
reactants and their function (acceptor or donor) in the particular
reaction in question. Accordingly, an excellent review [3] published
in 1968 lists the self-condensation of aldehydes or ketones; mixed
condensation of aldehydes or ketones; condensations of aldehydes
with acyclic, alicyclic, or aralkyl ketones; intramolecular condensations of dialdehydes, diketones, and keto aldehydes.

It will be noticed from the table of contents of this chapter
that the classification of available material has been adapted to
the needs of understanding the aldol and related reactions as they
are used in organic synthesis. Special consideration has been
given to new techniques developed in recent years to overcome the
problem of ambiguity in the direction of the reaction between nonidentical carbonyl compounds. A review covering certain aspects of
this problem appeared in 1971 [4]. The techniques described in
Sec. V are methods employed to circumvent the problem of the mixed
aldol reaction; Sec. VI, on the other hand, includes methods and
techniques named after their originators, and they constitute an
expansion of the principles of the aldol reaction in its fundamental
role of practical carbon-carbon bond-forming reactions.

III.   INTERMOLECULAR ADDITIONS AND CONDENSATIONS

A.   *Base-, Acid-, or Ion Exchange Resin-*
     Catalyzed Self-additions

     1. Aldehydes
     The aforementioned (Sec. II.A.) aldol reaction of acetaldehyde
(*1*) to give acetaldol (*2*) or croton aldehyde (*3*) has significant
practical applications. The acetaldol (*2*) has been the intermediate
of several industrial 1,3-butadiene syntheses, via hydrogenation
to the intermediate 1,3-butanediol, while *3* has been used in the
industrial production of n-butanol via catalytic hydrogenation and
in the contact catalytic oxidation to maleic anhydride. Thus, it
is not surprising to find an extensive array of data for the preparation of both *2* and *3* in the scientific literature, as well as
in the patent literature.

The preparation of acetaldol (2) constitutes an excellent general example for the preparation of aldols from the lower molecular weight aldehydes. The reaction conditions allow the formation of the aldol addition products and avoid as much as possible the elimination of water to form the unsaturated aldehyde 3. In order to achieve this, relatively low temperatures and not too strongly basic reaction conditions should be used; the conversion to acetaldol (2) should not be forced beyond approximately 60-70% of the addition product 2 in the equilibrium mixture [5].

The reaction may be executed under mildly basic conditions at -12°C in water using a KCN catalyst [6] in diethyl ether at 0°C using a mildly basic saturated aqueous sodium sulfite solution [7], or with 1.25% sodium hydroxide at about 10°C [8]. All these reaction conditions gave approximately a 45% yield of acetaldol (2). A slightly higher yield (50%) can be obtained [9] by placing 1 kg (22.7 mol) of acetaldehyde (1) into a 2 liter flask and adding, dropwise, a 10% sodium hydroxide solution (25 ml) over a period of 20 min while stirring at about 4°-5°C. Stirring is continued for 1 hr at this temperature, and the mixture is then acidified with tartaric acid and filtered, and the filtrate distilled in vacuo to give acetaldehyde (1) and acetaldol (2); redistillation of the aldol fraction gave about 500 g (50%) of acetaldol.

Acetaldol (2) dimerizes readily [10, 11] on standing to paraldol (4). Therefore, distillation of aldols immediately before use is recommended to convert the cyclic hemiacetal 4 to the aldol 2. It is also possible to stabilize acetaldol (2) with a small amount of water or acetaldehyde during storage [5].

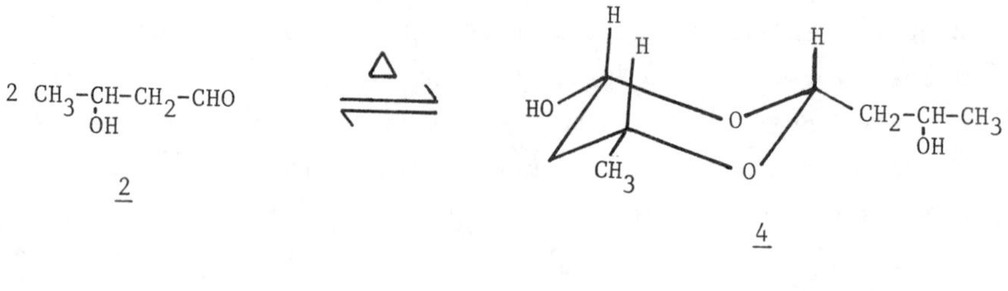

If the α,β-unsaturated aldehyde [e.g., croton aldehyde (3)] is the *desired reaction product*, the preceding procedure can be modified so that a stronger acid (e.g., sulfuric acid, phosphoric acid, or thionyl chloride) is used in the acidification step [12-14].

Alternatively, the α,β-unsaturated aldehyde may also be obtained under basic reaction conditions at higher reaction temperatures than those used for the first step of the reaction. This procedure should not be used with acetaldehyde (1), because of the strong tendency of the intermediate acetaldol (2) to give resinous poly-merization products with alkalies above 25°C.

On the other hand, n-butanal (5) has been converted in good yield to the α,β-unsaturated aldehyde, 2-ethyl-2-hexenal (7). The formation of 7 proceeds via the intermediate aldol 6 in agreement with the mechanism indicated for the formation of croton aldehyde (3) in Scheme 2b. The aldol intermediate 6, however, was not iso-lated in the procedure described in the literature [15].

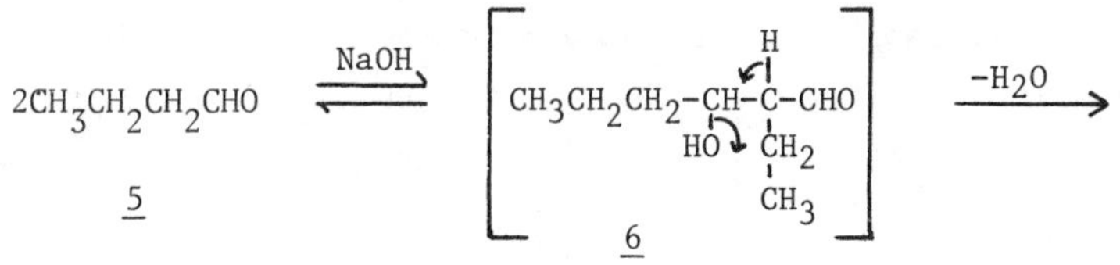

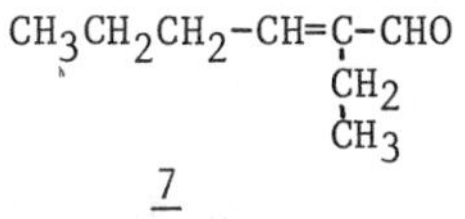

$$2CH_3CH_2CH_2CHO \quad \underset{}{\overset{NaOH}{\rightleftharpoons}} \quad \left[CH_3CH_2CH_2-\overset{\overset{H}{|}}{CH}-\overset{\overset{O}{\parallel}}{C}-CHO \atop HO \searrow \underset{\underset{CH_3}{|}}{CH_2} \right] \quad \overset{-H_2O}{\longrightarrow}$$

$$\underset{\underline{5}}{} \qquad\qquad\qquad \underset{\underline{6}}{}$$

$$CH_3CH_2CH_2-CH=\overset{}{C}-CHO \atop \underset{\underset{CH_3}{|}}{CH_2}$$

$$\underline{7}$$

Freshly distilled n-butanal (2250 g, 35 mol) was added to 1 *N* aqueous NaOH (750 ml) while stirring at 80°C over a 1.5-hr period. After the addition was completed, the mixture was heated to reflux for 1 hr. The organic layer was separated, and distilled through a 150-cm Vigreux column to give 1880 g (86%) of the α,β-unsaturated aldehyde, 7 [15].

In contrast to the preceding aldol reactions of aldehydes, the reaction of ketones is much less likely to proceed in good yield, since the steric and inductive effects of the second alkyl group make nucleophilic self-additions to the carbonyl group more difficult. The equilibrium described in Scheme 1b lies to the left toward the ketone starting material (*retroaldol route*). Acetone (*8*) for instance, furnishes only 2% diacetone alcohol (*9*) at equilibrium in the presence of barium hydroxide. If, however, the same catalyst [Ba(OH)$_2$] is placed into the thimble of a Soxhlet extractor, a 71% yield of *9* the aldol (ketol) addition product is obtained [16]. In the procedure, acetone is heated to boiling and the condensate passes back to the flask via the porous thimble filled with the catalyst. A small amount of the addition product *9* is formed, which drops back into the boiling flask, and since it is much higher boiling than acetone, it will accumulate unchanged in the boiling flask. If *9* is not in contact with the catalyst, it is stable, and the retroaldol reaction to acetone does not take place. The more volatile acetone is continually recycled to the barium hydroxide catalyst, which is insoluble in acetone. With this ingenious technique it is possible to convert 1.5 liters of acetone (*8*) into diacetone alcohol (*9*) in about 4 days.

$$2CH_3\text{-}\overset{\displaystyle O}{\underset{\displaystyle \|}{C}}\text{-}CH_3 \;\underset{\displaystyle \xleftarrow{\hspace{1cm}}}{\overset{\displaystyle Ba(OH)_2}{\longrightarrow}}\; CH_3\text{-}\underset{\displaystyle H\text{-}O}{\overset{\displaystyle CH_3}{\underset{\displaystyle |}{\overset{\displaystyle |}{C}}}}\text{-}CH_2\text{-}\overset{\displaystyle O}{\underset{\displaystyle \|}{C}}\text{-}CH_3$$

<u>8</u>   <u>9</u>

Diacetone alcohol (*9*) may be dehydrated with a small amount of iodine to give the α,β-unsaturated ketone, mesityl oxide (*10*) in 65% yield [17]. The reaction is analogous to the aforementioned conversion of acetaldol (*2*) to croton aldehyde (*3*) in the condensation step of the aldol reaction.

Mesityl oxide (*10*) can also be obtained directly in the acid catalyzed aldol reaction [18] of acetone (*8*). The addition product diacetone alcohol (*9*) has not been isolated under these reaction conditions. The yield of *10* is low, and the substance is accompanied by phorone [*11*], the tri-condensation product of acetone. The conversion of acetone (*8*) to mesityl oxide (*10*) and phorone (*11*) exemplifies the general acid-catalyzed aldol reaction of ketones.

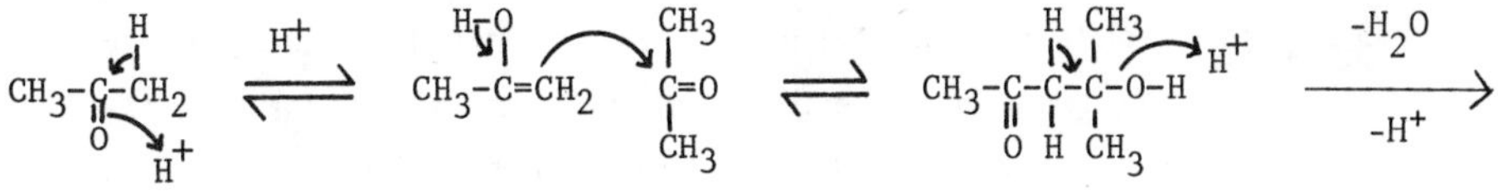

$$8 \qquad\qquad\qquad\qquad 9$$

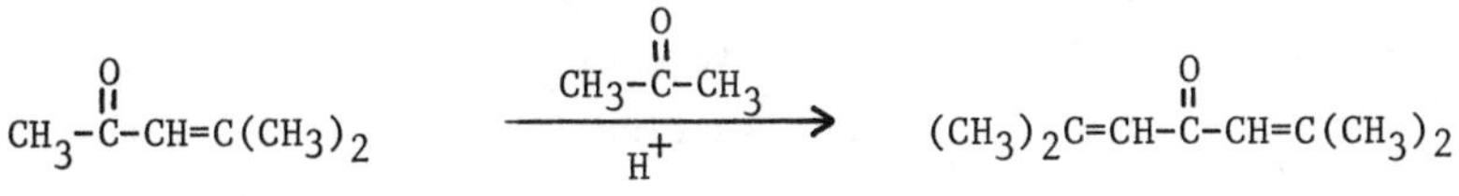

$$\underline{10} \qquad\qquad\qquad\qquad \underline{11}$$

The preferred experimental condition for the preparation of mesityl oxide (*10*) is the treatment of acetone (*8*) with the basic ion exchange resin Dowex 50HH under reflux to give the α,β-unsaturated ketone *10* in 79% yield [19]. The use of barium hydroxide in the heterogeneous, contact catalytic conversion of acetone (*8*) to diace-tone alcohol [9] has already been discussed; a somewhat similar case exists with the ion exchange resins, which are substantially insol-uble catalysts because of their polymeric cross-linked structure. The use of ion exchange resins to promote aldol reactions should be considered whenever the reactants or the reaction product show sensitivity toward conventional acids or bases. Other advantages include their ease of separation by filtration or decantation, the reduction of costs, since the catalyst may be reused repeatedly usually without regeneration, and the increased product yield and efficiency. The major disadvantages of the resins seem to be their thermal and chemical instability under extreme reaction conditions [20, 21].

B. *Mixed Aldol Reactions*

Mixed aldol reactions between nonidentical aldehydes or between
aldehydes and ketones used to be generally considered unsatisfactory
because of the complexity of the resulting product mixture. This
was due to the ambiguity originating from the cross-condensation
and the self-condensation of the reactants. Until a few years ago,
mixed aldol reactions have been useful only in exceptional cases
in which the structural features and/or a difference in the reactivity
of one of the reactants brought about the selectivity of the
direction of aldol formation (Sec. III.B.1.). The more recent
developments in the field of mixed aldol reactions involve preformed
derivatives of the reactant(s) to achieve selectivity in the
direction of the mixed aldol reaction (Sec. III.B.2.).

  1. Selectivity Due to Structural Features
      and/or Different Reactivity of Reactants
      in Base-Catalyzed Mixed Aldol Reactions

     a. *Ketones and Aldehydes (Other Than Formaldehyde and Aromatic*
Aldehydes).   It has already been discussed (Sec. III.A.2) that
ketones in general are less susceptible to nucleophilic self-
additions to the keto-carbonyl group. An early recognition made
use of this fact in the cross-condensations between ketones and
aldehydes, wherein the more reactive aldehydic carbonyl group
served as the acceptor for the nucleophilic attack of the enolized
ketone donor.

     Important industrial use has been made of this phenomenon in
the synthesis of pseudoionone (*14*), the key intermediate in the
manufacturing process for vitamin A. Aldol condensation of 203 g
(1.3 mol) of the terpene aldehyde trans-citral (*12*) with 800 g
(13.8 mol) of acetone (*8*) in the presence of 0.4 mol of sodium
ethoxide in 200 ml absolute ethyl alcohol at $-5^{\circ}C$ gave 120-130 g
(45-49% yield based on *12*) of pseudoionone (*14*), puridied via its
bisulfite addition product and distilled in vacuo. This procedure
[22] made use of the fact that acetone, in the presence of base,
gives only about 2% of the self-addition product, diacetone

alcohol [9]. In the formation of pseudoionone the intermediate
ketol 13 is not isolated; it spontaneously loses water under the
reaction conditions.

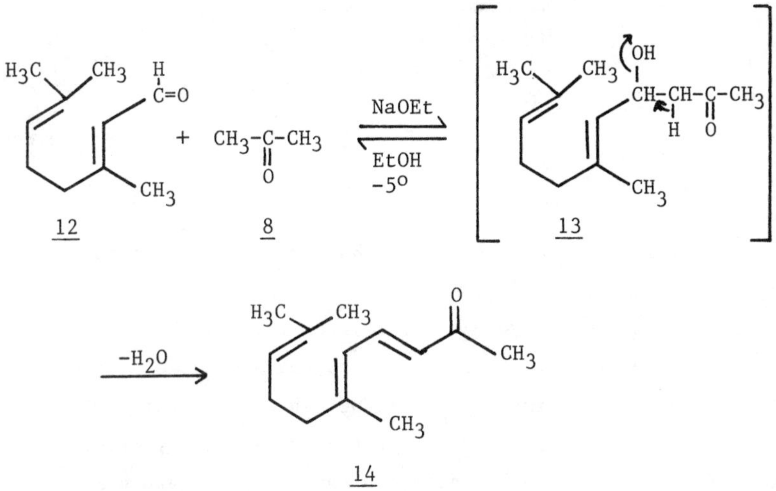

As has already been mentioned in Sec. I, the enzyme-catalyzed
natural process to produce carbohydrates in the mixed aldol reaction
of two relatively simple triose derivatives has also been executed
in the chemical laboratory, as examplified by the nonenzymatic,
simple, base-catalyzed reaction of (+)-glyceraldehyde (15) and
dihydroxyacetone (16). Just as in the aforementioned synthesis of
pseudoionone, this mixed aldol reaction is feasible, since no self-
condensation of the active methylene keto compound 16 occurs in the
presence of the more reactive aldehydic carbonyl acceptor [15].

To 3.15 g of 15 in 100 ml of water was added 3.15 g of freshly
distilled 16 and enough $Ba(OH)_2$ to make the solution 0.01 molar.
After standing for 2 hr at room temperature the mixture was neutralized with dilute sulfuric acid, and the $BaSO_4$ precipitate removed
by centrifugation. The base-catalyzed mixed aldol reaction gave
34.3% of (-)-fructose (17) and 26.4% of (+)-sorbose (18) separated
via their respective phenylosazones [2(a)].

$$
\begin{array}{ccc}
\begin{array}{c}
\text{H-C=O} \\
| \\
\text{H-C-OH} \\
| \\
\text{CH}_2\text{OH}
\end{array}
& + &
\begin{array}{c}
\text{CH}_2\text{-OH} \\
| \\
\text{C=O} \\
| \\
\text{CH}_2\text{-OH}
\end{array}
\end{array}
\quad
\xrightarrow[\text{H}_2\text{O, 20}^\circ\text{, 2 hr}]{\text{0.01M Ba(OH)}_2}
\quad
\begin{array}{c}
\text{CH}_2\text{OH} \\
| \\
\text{C=O} \\
| \\
\text{HO-C-H} \\
| \\
\text{H-C-OH} \\
| \\
\text{H-C-OH} \\
| \\
\text{CH}_2\text{OH}
\end{array}
\quad + \quad
\begin{array}{c}
\text{CH}_2\text{OH} \\
| \\
\text{C=O} \\
| \\
\text{H-C-OH} \\
| \\
\text{HO-C-H} \\
| \\
\text{H-C-OH} \\
| \\
\text{CH}_2\text{OH}
\end{array}
$$

$$\underline{15} \qquad\qquad \underline{16} \qquad\qquad\qquad\qquad\qquad\qquad\qquad \underline{17} \qquad\qquad \underline{18}$$

b. *Formaldehyde and Other Aldehydes or Ketones (Tollens Method) [23(a) and (b)]*. Formaldehyde (19) plays a rather important role in selective mixed aldol reactions with aldehydes or ketones. The selectivity of these base-catalyzed reactions is due to the fact that formaldehyde has no α-hydrogen atom for donation, while its carbonyl group acts as an extremely reactive acceptor for the nucleophilic attack of an enolized carbonyl partner in mixed aldol reaction. The reaction thus has great preparative and industrial importance. One process for the production of acrolein (21) involves the contact catalytic conversion of an equimolecular amount of acetaldehyde (1) and a 30% aqueous formaldehyde (19) solution on silica gel impregnated with 10% sodium silicate at 305°-310°C; 2.2 kg of acetaldehyde and 5 kg of 30% formaldehyde solution rendered 1.15 kg of acrolein, while 1.13 kg of acetaldehyde was recovered [23(c)]. The reaction proceeds via the nonisolated β-hydroxy aldehyde 20; this aldol dehydrates on the surface of the catalyst to the desired α,β-unsaturated aldehyde 21.

$$
\begin{array}{c}
\text{H} \\
| \\
\text{H-C=O}
\end{array}
\quad + \quad
\begin{array}{c}
\text{H} \\
| \\
\text{CH}_3\text{-C=O}
\end{array}
\qquad
\xrightarrow[\text{305-310}^\circ]{\text{Na}_2\text{Si}_2\text{O}_5 \text{ on Silica Gel}}
$$

$$\underline{19} \qquad\qquad\qquad \underline{1}$$

$$
\left[\text{HO-CH}_2\text{-CH}_2\text{-}\overset{\overset{\text{O}}{\|}}{\underset{|}{\text{C}}}\text{=O} \right]
\quad
\xrightarrow{-\text{H}_2\text{O}}
\quad
\begin{array}{c}
\text{H} \\
| \\
\text{CH}_2\text{=CH-C=O}
\end{array}
$$

$$\underline{20} \qquad\qquad\qquad\qquad\qquad\qquad \underline{21}$$

The important synthetic intermediate methyl vinyl ketone has been obtained in 97% yield in a mixed aldol (ketol) reaction of formaldehyde and acetone over a calcium phosphate contact at $450°C$ [24].

An important example of the Tollens reaction is the conversion of formaldehyde (*19*) and acetaldehyde (*1*) into pentaerythritol (*23*) in an aqueous solution of calcium hydroxide [25]. Since a 5:1 ratio of formaldehyde to acetaldehyde is employed in this reaction, the primarily formed aldol *20* reacts further with an additional 2 mol of formaldehyde (*19*) to give the aldo-triol *22*. This then undergoes a crossed-Cannizzaro reaction (an intermolecular oxidation-reduction with an additional mole of formaldehyde) resulting in the reduction of the aldo-triol *22* to pentaerythritol (*23*) and oxidation of formaldehyde to formic acid (*24*).

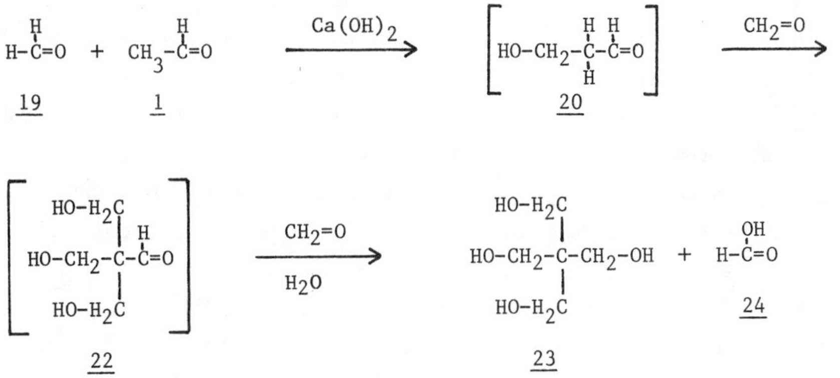

In contrast to this synthesis of pentaerythritol, it is often desired to avoid the secondary Cannizzaro reaction of the aldol with an additional molecule of formaldehyde. Therefore, a 100% excess of the other aldehyde is used. This excess can be separated from the aldol by distillation. In this fashion the aldol intermediate *26* in the synthesis of pantothenic acid can be prepared by the action of potassium carbonate on a mixture of isobutyraldehyde (*25*) and 30% aqueous formaldehyde (*19*) at $20°C$. The aldol 2,2-dimethyl-3-hydroxy-propanal (*26*) can be obtained in 40% yield [26].

$$CH_2O + H-\underset{\underset{CH_3}{|}}{\overset{\overset{CH_3}{|}}{C}}-CHO \xrightarrow{K_2CO_3, \ 20^{\circ}C} HO-CH_2-\underset{\underset{CH_3}{|}}{\overset{\overset{CH_3}{|}}{C}}-CHO$$

<u>19</u>   <u>25</u>   <u>26</u>

Despite the relatively wide industrial application of the Tollens technique, it is often rather difficult to control the reaction under normal laboratory operations. In addition to the often undesired Cannizzaro reaction, the problem of polycondensations with formaldehyde may also arise. A more satisfactory technique for the introduction of the formaldehyde unit is the Mannich reaction (Sec. III.B.2.a.).

    c. Aromatic Aldehydes and Aliphatic (Alicyclic) Aldehydes or Ketones (Claisen-Schmidt Reaction Technique). The reaction technique by which aromatic aldehydes or other aldehydes having no α-hydrogen atom are reacted with an aliphatic or alicyclic aldehyde or ketone in the presence of strong aqueous alkali to form α,β-unsaturated aldehydes or ketones is generally known as the *Claisen-Schmidt method*. It was discovered by Schmidt [27] and developed by Claisen [6, 28(a) and (b)]. One of the reaction's important recent applications has been its use for the selective introduction of angular methyl groups in the total synthesis of steroids and steroid-like compounds. The procedure employed by Johnson and co-workers involves the preparation of the benzylidene derivative of a decalone to protect the α-methylene position from alkylation, methylation at the angular carbon atom, and subsequent removal of the benzylidene group [29]. To 152 g of trans-decalone-1 (*27*) in 1 liter of 95% EtOH was added while stirring 400 ml of 15% NaOH, 113 ml of freshly distilled benzaldehyde (*28*) and 100 ml of water. The pale yellow solution was seeded with trans-2-benzaldecalone-1 (*30*) and allowed to stand at 20°C for 2 days. Work-up of the crystalline precipitate and its mother liquors rendered 178.6 g of *30* in 75% yield, while 23 g of *27* could be recovered. It should be noted that the β-hydroxy ketone (ketol) intermediate *29* has not

been isolated, since the strongly basic reaction conditions of the
Claisen-Schmidt reaction favor dehydration of this intermediate to
yield the α,β-unsaturated carbonyl compound.

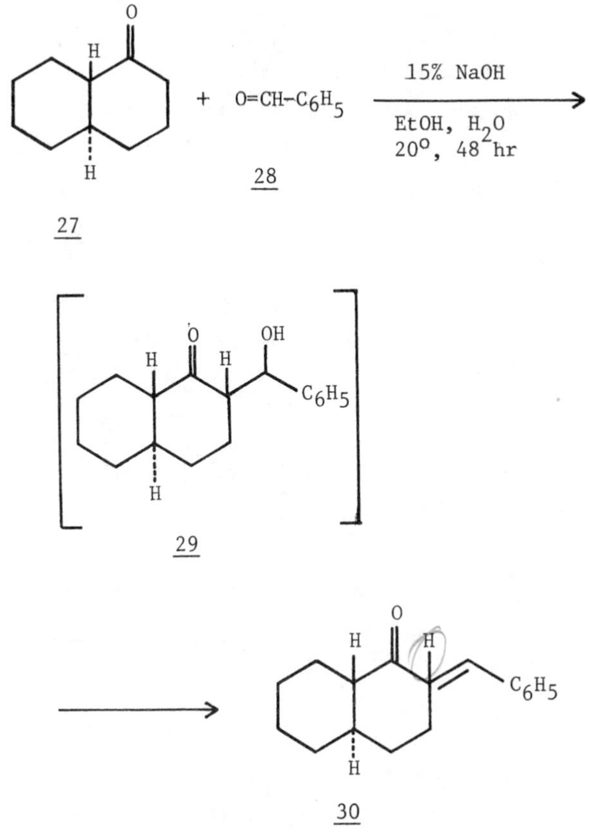

It was subsequently shown that the use of furfural (31) rather than
benzaldehyde (28) is more desirable since the reaction rate of the
Claisen-Schmidt reaction is faster with the former reagent [30(a)
and (b)]. It is also preferential to use methanol rather than ethyl
alcohol as a cosolvent, since the strong base and air may cause the
ethyl alcohol to dehydrogenate to acetaldehyde, and the latter may
be converted to aldol polymerization of polycondensation products,
which result is an undesired yellowing of the reaction mixture.

The Claisen-Schmidt reaction of furfural (*31*) and aliphatic
aldehydes to give α,β-unsaturated aldehydes is preferentially
executed in the presence of basic ion exchange resins [21 and 31].
The advantageous use of the Dowex 50HH resin has already been
discussed (Sec. III.A.2). The Claisen-Schmidt reaction of 120 g
of furfural (*31*) and 110 g of Hexanal (*32*) in the presence of 30
g of the basic ion exchange resin Amberlite IRA-400 gave the α,β-
unsaturated aldehyde 2-furfuryliden-1-hexanal (*34*) after being
stirred and heated at 135°C for 4 hr. The intermediate aldol *33*
was not isolated since it rapidly dehydrated to the aldol condensation product *34* under the applied reaction conditions [31]. The
use of the heterogeneous ion exchange resin catalyst promoted the
aldol addition and condensation reaction, while it avoided the
undesired polymerization of furfural and also of the aliphatic
aldehyde.

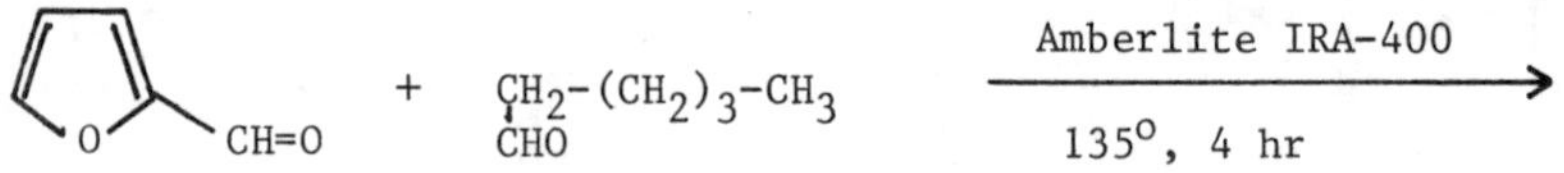

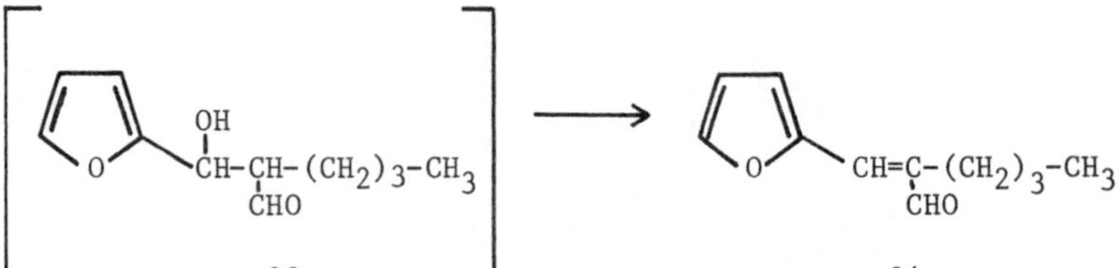

2. Selectivity Through Preformed Derivatives
of the Aldol Reactant(s)

Except for the Mannich reaction technique, the methods to be
described in this section are mainly the result of the scientific
investigations of the last two decades in the field of mixed aldol
reactions. Most of these techniques involve the preformation of
derivatives of only one aldol reactant to achieve selectivity, but
on occasion both reactants may be derivatized before the mixed aldol
reaction is actually executed.

a. *Enolized Ketone Donors with Imminium Salt Acceptors (Mannich Technique)*. It has been pointed out previously, in connection with the Tollens method (Sec. III.B.1.b.), that the Mannich reaction constitutes a more satisfactory technique for the attachment of a single carbon atom to an active methylene compound. In this reaction, a compound containing one or more active hydrogen atoms is treated with formaldehyde and a secondary amine hydrochloride salt to yield a β-amino carbonyl compound. It may therefore be considered the nitrogen analogue of the aldol reaction. As with the β-hydroxy carbonyl aldol products, the β-amino carbonyl analogues may be converted to α,β-unsaturated carbonyl derivatives in a step involving the elimination of the amine rather than water. The reaction has been employed with a wide variety of amines, and it is successful not only with enolizable ketones but also with enolizable aldehydes, esters, and other groups of compounds which are sufficiently reactive electron donors for the imminium salt acceptor system. Due to the nature of this chapter, the discussion of this technique will have to be limited; however, important compendia have been published on this subject in the chemical literature [32].

Possibly one of the most important areas of application of the β-amino carbonyl Mannich-base intermediates has been in the field of the synthesis of polycyclic derivatives of natural product chemistry. An illustration of the preparation of a Mannich base is the synthesis of 4-diethylamino-2-butanone (*37*). In a 3-liter round-bottom flask are placed 176 g (1.6 mol) of diethylamine hydrochloride, 68 g (2.26 mol) of paraformaldehyde, 600 ml (8.2 mol) of acetone, 80 ml of methanol, and 0.2 ml of conc. HCl. The mixture is heated for 12 hr at reflux. Under these reaction conditions it is assumed that the intermediate imminium salt *36* is preformed in the reaction of the "aldol" reactant formaldehyde (*19*) with the diethylamine hydrochloride (*35*). This serves as the acceptor for the nucleophilic attack of the enolized acetone (*8*) donor system. The quaternary HCl salt *37* of the Mannich base is formed and can then be converted to the free base with aqueous sodium hydroxide.

The usual work-up followed by vacuum fractionation gave 4-diethylamino-2-butanone (*37*) in 66-71% yield as a nearly colorless liquid [33].

Steam distillation of the salt *37* affords the α,β-unsaturated ketone methyl vinyl ketone (*38*) via elimination of the amine hydrochloride. Since the yields are usually poor due to polymerization, the preferred pathway consists in converting the Mannich base *37* to its quaternary methiodide salt, and to use this salt as a gradual source of methyl vinyl ketone (*38*) in the Robinson annulation reaction.

$$(C_2H_5)_2NH \cdot HCl \quad + \quad CH_2=O \quad \rightleftharpoons \quad (C_2H_5)_2\overset{+}{N}=CH_2 \quad + \quad CH_2=C-CH_3 \longrightarrow$$

35   19   36   O-H

enol-8

$$(C_2H_5)_2\overset{H^+}{\underset{|}{N}}-CH_2-CH_2-\underset{\underset{O}{\|}}{C}-CH_3 \quad \xrightarrow{-Et_2NH \cdot HCl} \quad CH_2=CH-\underset{\underset{O}{\|}}{C}-CH_3$$

(Cl⁻)

37   38

A useful intermediate, *41*, for the total synthesis of steroids has been recently prepared in excellent yield by use of a modified Mannich reaction [34] in which both reactants have been derivatized. The enolizable ketone donor is introduced in the form of a β-keto acid, *39*, which is allowed to react with the imminium salt of formaldehyde (*19*) and piperidine hydrochloride (*42*) in dimethyl sulfoxide (DMSO) to give directly the α,β-unsaturated ketone *41*. The intermediate salt *40* has not been observed, since it very likely instantly loses *42* in the highly polar, aprotic solvent. Decarboxylation of the β-keto acid *39* takes place under the influence of the DMSO and leads to the regiospecific formation of the desired $\Delta^{4(5)}$-enol donor, which is immediately quenched by the imminium salt.

In protic solvent systems, which were used earlier in related reactions [35], only extensive decarboxylation occurred, and the desired product could not be obtained; stabilization of the enolate by tautomeric proton shift was faster than the desired attack by the imminium salt.

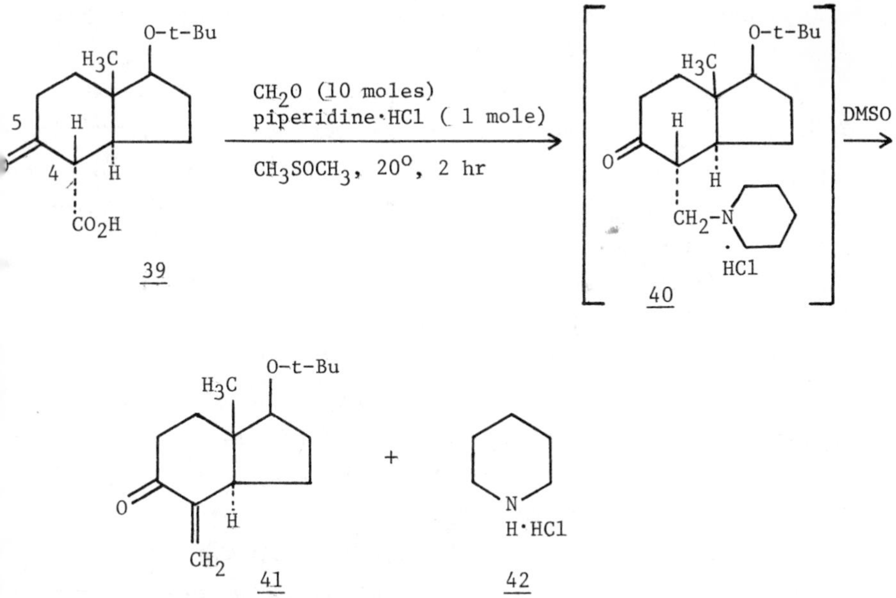

b. *Metalated Schiff-Base Donors with Aldehyde or Ketone Acceptors*. The identification of an unexpected reaction product led Wittig et al. (1962) to the development of a new method for a mixed aldol condensation [36]. Until the discovery of this technique an aldehyde with an active α-hydrogen atom could not be selectively added to the carbonyl group of a ketone, since the preferred reaction was self-addition of the aldehyde. This new method avoids this difficulty by converting the aldehyde into a Schiff base and by metalating the azomethine group of this masked aldehyde derivative. A carbanion donor is thus obtained, which can then be reacted with an aldehyde or ketone acceptor to give the desired α,β-unsaturated carbonyl derivative.

To prepare the Schiff bases, 1 mol of an aldehyde is added dropwise to 1 mol of a primary amine at -20°C, with vigorous stirring. After standing at -20°C for 1 hr sodium sulfate is added, and the organic phase is separated at room temperature. Anhydrous sodium carbonate is then added, and the reaction product (the Schiff base) fractionated through a 30-cm-long packed column, and distilled before use.

In the standard procedure, the Schiff base is metalated at -70°C with lithium diisopropylamide in anhydrous ether or tetrahydrofuran. It is then allowed to react at this temperature (-70°C) with the desired aldehyde or ketone, and the mixture is allowed to come to room temperature.

These directed mixed aldol reactions can be used for the synthesis of many α,β-unsaturated compounds. Extensive reviews have been published in the chemical literature [37 and 4]. They include several examples from natural product syntheses. An illustration is the synthesis of cis- and trans-citral [37].

A solution of 0.1 mol of ethylidenecyclohexylamine (43) in 80 ml of anhydrous ether was added dropwise to 0.1 mol of lithium diisopropylamide in anhydrous ether (approximately 0.6 M solution) at 0°C under argon. After 10 min, it was cooled to -70°C, and 0.1 mol of 6-methyl-5-hepten-2-one (44) was added dropwise to the magnetically stirred solution under argon. The reaction mixture was then allowed to come to room temperature. After standing for 12 hr at 20°C, it was cooled to 0°C and hydrolyzed with 160 ml of 20% aqueous acetic acid. The reaction was completed by stirring for 3 hr at room temperature. The ether phase was neutralized with $NaHCO_3$, dried, and distilled to give a fraction, which according to gas-liquid chromatography (glc) contained 68% of cis-citral (cis-12), 27% of trans-citral (trans-12), and 4% of 6-methyl-5-hepten-2-one (44). This reaction sequence corresponds essentially

$$CH_3CHO + H_2N-C_6H_{11} \xrightarrow{-20^\circ} CH_3-CH=N-C_6H_{11} \xrightarrow[0^\circ, Ar]{LiN(iPr)_2, Et_2O}$$

$\underline{1}$ $\underline{43}$

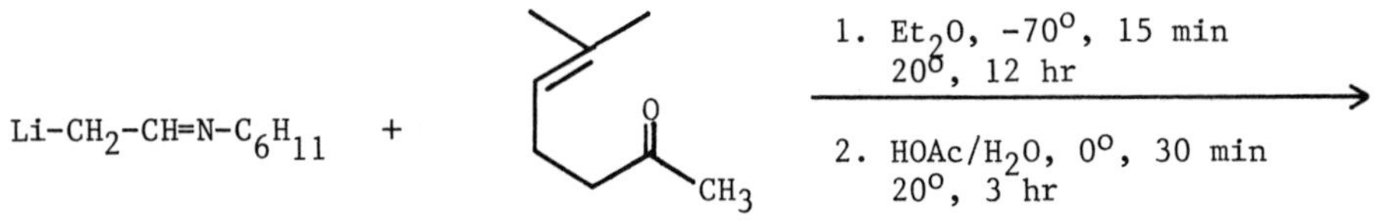

$$Li-CH_2-CH=N-C_6H_{11} \quad + \quad$$

44

1. Et$_2$O, -70°, 15 min
   20°, 12 hr

2. HOAc/H$_2$O, 0°, 30 min
   20°, 3 hr

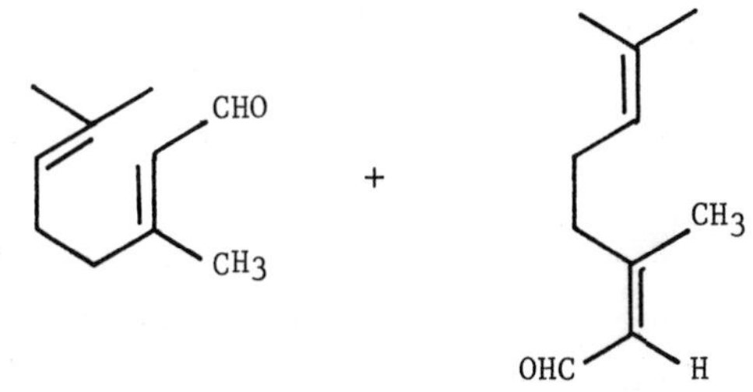

trans-12	+	cis-12
27%		68%

to the mixed aldol reaction between acetaldehyde (*1*) and 6-methyl-5-hepten-2-one (*44*).

The same technique has been used for the highly stereoselective synthesis of the racemic sesquiterpene aldehyde nuciferal (*47*). The mixed aldol reaction of the aldehyde *45* with lithio propylidene-t-butylimine (*46*) gave *47* in 83% yield [38].

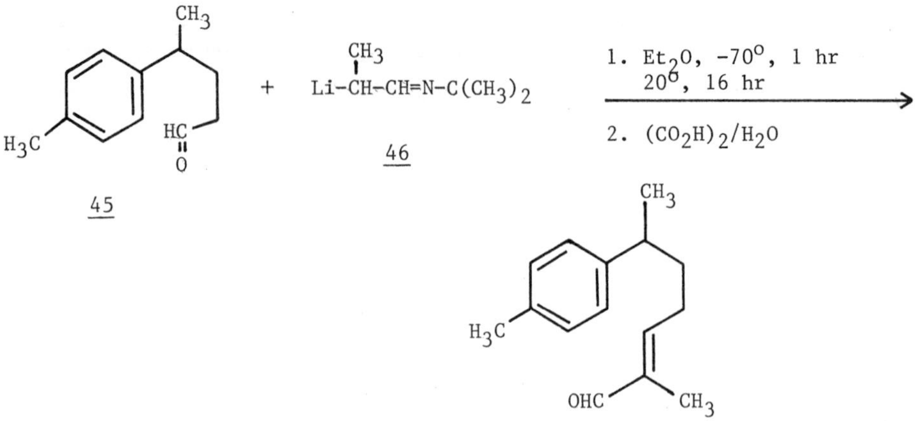

c. *Kinetic Enolate Salts of Ketones with Aldehyde Acceptors.*
Another useful method of avoiding the problem of self-condensation
in mixed aldol reactions consists of the use of preformed kinetic
lithium enolates. The method also provides the regiospecificity
of the reaction by avoiding the rapid equilibration of the initially
formed anion normally encountered.

The kinetic enolate salts may be either directly prepared from
the respective ketones or (preferentially) obtained via an enol
ether or an enol ester derivative thereof. A full discussion of
the formation of these enolate salts is given in Chap. 2. An
earlier technique to obtain the desired regiospecific enolates
consisted of the use of β-keto acids (Sec. III.B.2.a).

The regiospecific mixed aldol reaction of an α,β-unsaturated
ketone with formaldehyde through the use of the kinetic enolate
salt can be illustrated in the following example [39]. The lithium
salt of the α,β-unsaturated ketone *48* has been generated with a 25%
excess of lithium in liquid ammonia in the presence of 0.8 eq of
t-butyl alcohol. The ammonia was completely and carefully evaporated
(finally at $20^{\circ}C$ at 0.5 mm Hg vacuum) from the lithium enolate *49*
followed by the addition of 2.5 eq of trimethylchlorosilane in dry
tetrahydrofuran to trap the kinetic enolate as the trimethyl silyl
ether derivative *50*. This was isolated by work-up with cold
saturated aqueous sodium bicarbonate, and extracted with ether

(90% yield). Regeneration of the kinetic enolate *49* with 1 eq of
methyllithium (1 hr, 20°C), followed by treatment with anhydrous
formaldehyde at -78°C for 20 min, gave the crystalline carbinol
51 in 90% yield [40].

It should be emphasized that in the preceding procedure the
regeneration of the kinetic enolate *49* is performed in THF, free of
t-butyl alcohol to avoid the equilibration of the kinetic enolate.
The alcohol, on the other hand, was the necessary proton source in
the first step of the reaction.

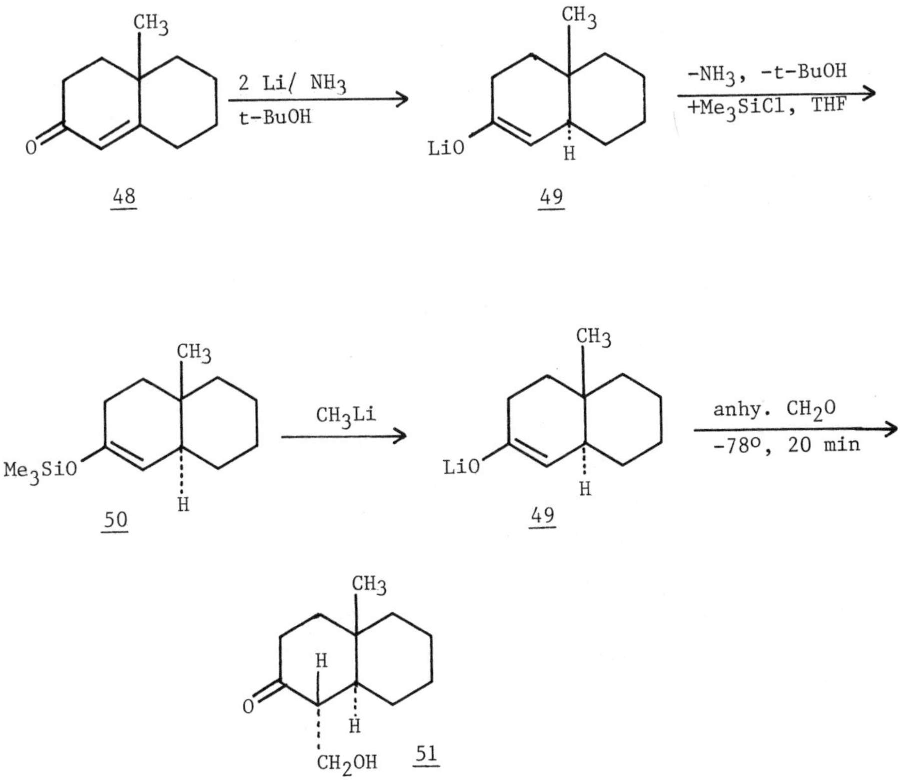

The kinetic enolates of α,β-unsaturated ketones may also be
generated through 1,4-addition of organometallic reagents to the
enones. Addition of methylmagnesium bromide to cyclohexenone in
the presence of 1% CuI·Bu₃P generated its enolate, which was

regiospecifically trapped in ether at -10°C with anhydrous formaldehyde gas (as above). Work-up with ammonium chloride at 0°C followed
by distillation gave 2-hydroxymethyl-3-methylcyclohexanone in 70%
yield. The method has its limitations owing to the introduction
of the β-alkyl group during enolate formation [39].

With saturated methyl ketones or saturated cyclic ketones
the kinetic enolates may be formed directly using 1.1 eq of lithium
diisopropylamide in dry tetrahydrofuran at -78°C, followed by the
careful addition of the solution of the appropriate aldehyde in THF
to the kinetic enolate solution [41, 42].

A related technique involves the conversion of the ketone
reactant of the mixed aldol reaction to its enol acetate and the
conversion of the enol acetate to the kinetic lithium enolate of
the ketone with the help of methyllithium. This preformed lithium
enolate may then be reacted with other carbonyl compounds in a
mixed aldol reaction provided the initially formed product is inter-
cepted as a metal chelate. Although lithium itself may serve as
the chelating metal cation in nonpolar solvents and at low temperatures, it is experimentally preferred to add a divalent metal salt,
such as anhydrous $MgBr_2$ or especially $ZnCl_2$. This technique can
thus displace unfavorable equilibria and avoid common side reactions
such as di- and polycondensations and dehydration followed by undesired Michael addition of an enolate anion. An example of the
technique is as follows [43]: a mixture of 75 g (0.75 mol) of pina-
col (52), 4.5 g (24 mol) of p-TsOH, and 155 g (1.55 mol) of isopropenyl acetate was heated for 24 hr with continuous distillation of
materials boiling below 90°C. The resulting mixture was diluted
with pentane and poured into cold saturated aqueous $NaHCO_3$. Work-
up as usual followed by fractional distillation gave 30.1 g (28%)
of the enol acetate 53. To a cold (-60°C) solution of 53 mmol of
MeLi in 60 ml of ether containing several milligrams of 2,2'-bipyridyl
was added, dropwise and with stirring for 4 min, 3.567 g (25.1 mmol)
of the enol acetate 53. The resulting purple solution of the lithium
enolate 54 was stirred for 15 min at -40°C, warmed to -6°C, and

treated with 3.4 g (25 mmol) of freshly fused $ZnCl_2$ while stirring
for 10 min. To the resulting light purple, heterogeneous mixture
was added, dropwise over 1 min, 2.687 g (31 mmol) of pivaldehyde
(*55*). The resulting cold ($0°C$) mixture was stirred for an additional
4 min. The Zn complex *56* was worked up as usual with aqueous NH_4Cl
to give 3.8 g (82%) of the β-hydroxy ketone *57*. The yield may be
increased to 86% by adding an etheral solution of anhydrous $ZnCl_2$
to the preformed lithium enolate *54* followed by the addition of
the aldehyde *55*.

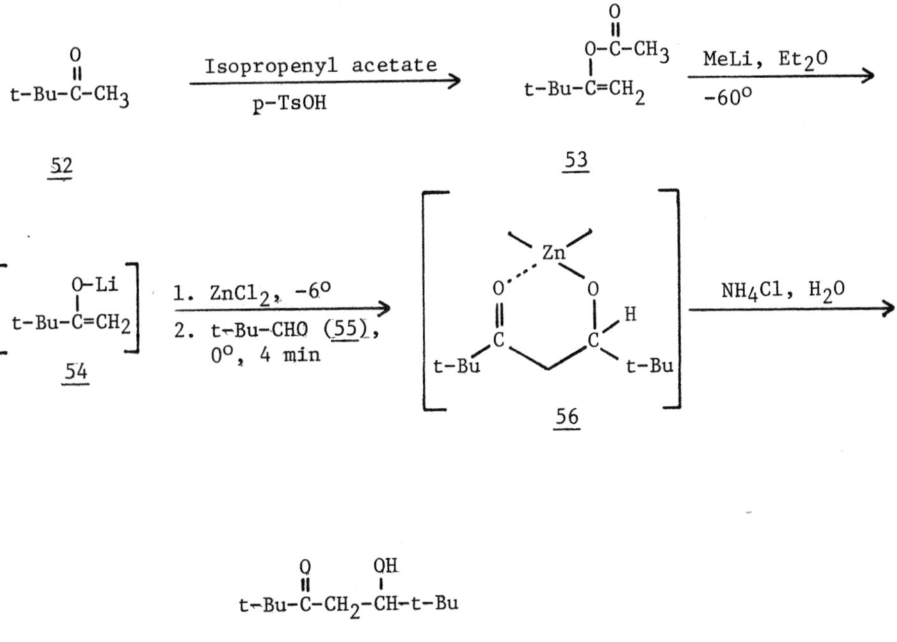

$$t\text{-Bu-C-CH}_3 \quad \xrightarrow[\text{p-TsOH}]{\text{Isopropenyl acetate}} \quad t\text{-Bu-C=CH}_2 \quad \xrightarrow[-60°]{\text{MeLi, Et}_2\text{O}}$$

52   *53*

$$\left[t\text{-Bu-C=CH}_2 \right] \quad \xrightarrow[\substack{2.\ t\text{-Bu-CHO (55),}\\0°,\ 4\ \text{min}}]{1.\ \text{ZnCl}_2,\ -6°} \quad \left[\substack{\text{Zn}\\ \text{O} \quad \text{O} \quad \text{H} \\ \text{C} \quad \text{C} \\ t\text{-Bu} \quad t\text{-Bu}} \right] \quad \xrightarrow{\text{NH}_4\text{Cl, H}_2\text{O}}$$

54   *56*

$$t\text{-Bu-C-CH}_2\text{-CH-t-Bu}$$

57

A further modification of the regiospecific generation of an
enolate is illustrated by the reaction of the bromoketone *58* with
methyl diphenylphosphinite (*59*) in chloroform at $20°C$ for 16 hr
to give the enol phosphinate *60* in about 85% yield. Addition of
1.2 eq of 2 *M* t-butyllithium in pentane to the enol phosphinate *60*
in 2:1 ether-tetrahydrofuran at $-78°C$, under nitrogen over a 2-hr

period generated the lithium enolate *61*. Zinc chloride in tetrahydrofuran (1.2 eq of 0.35 *M*) was added as a chelating agent,
followed by gaseous formaldehyde (*19*) at -78°C. The Zn-chelate
was quenched with acetic acid in the cold to give the hydroxymethyl-
cyclopentanone *62* in 80-90% yield. Conversion of this ketol to
an α,β-unsaturated methylene ketone, *63*, using an excess of methane-
sulfonyl chloride in pyridine at 0°C for 2 hr constitutes a new
approach to prostaglandins [44].

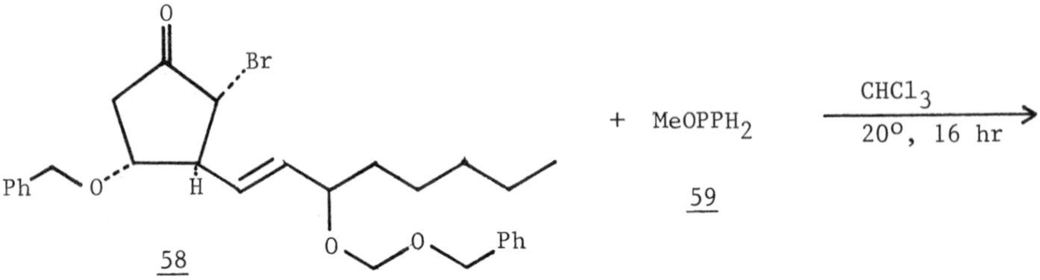

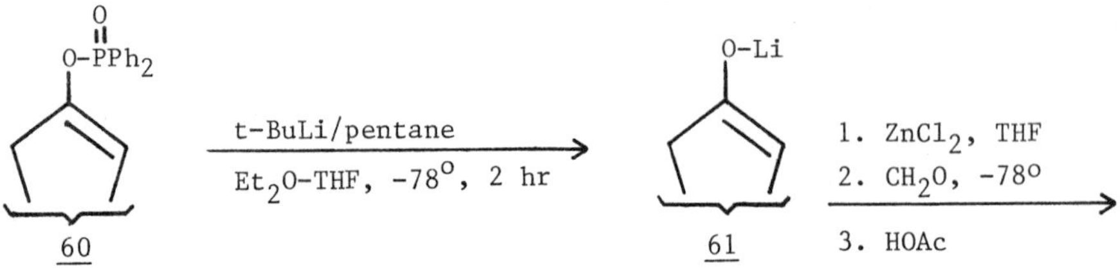

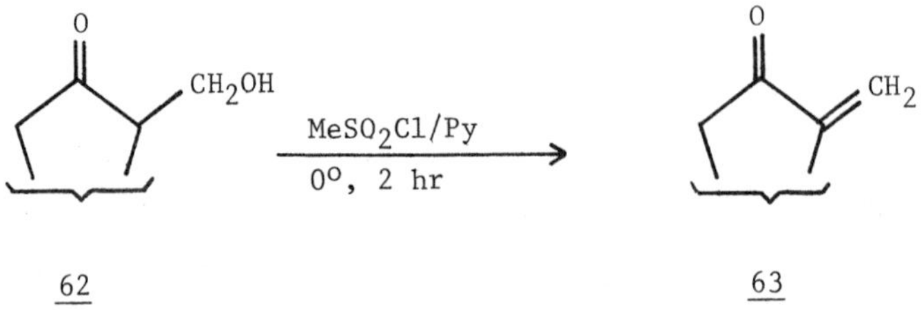

   d. *Bromo Metal Enolate Donors*.   An earlier technique related
to the preceding one consists of the treatment of an α-bromoketone
with zinc in 10:1 benzene-dimethyl sulfoxide to generate an enolate
species, which reacts with electrophiles, such as aldehydes, to form
a new carbon-carbon bond at the original site of the bromine [45].
This procedure, which has its origin in the classical Reformatsky
reaction (Sec. VI.F.), represents a potentially useful method for
mixed aldol reactions [46].   An illustration of the technique is
as follows [45]:   0.1 g mol of zinc powder and 0.04 mol of acetaldehyde (*1*) were added to 0.01 mol of 2α-bromo-3-cholestanone (*64*) in
50 ml of benzene and 5 ml of dimethyl sulfoxide.   The mixture was
stirred under nitrogen at room temperature for 24 hr to give the
β-hydroxyketone *66* in excellent yield.   This crude ketol was
dehydrated with p-toluenesulfonic acid in refluxing benzene to the
α,β-unsaturated ketone, 2-ethylidene-3-cholestanone (*67*) in 90%
overall yield.   Column chromatography on neutral alumina afforded
78% of the pure mixed aldol condensation product *67* after recrystallization.   In agreement with a more recent publication [47] it is
assumed that a bromo zinc enolate species, *65*, is generated.

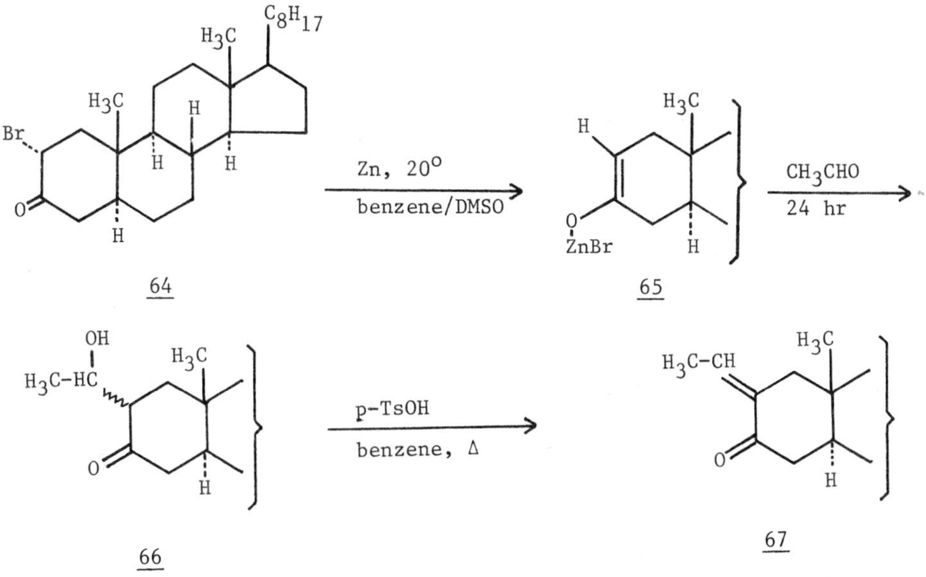

In a few preliminary experiments using ketone acceptors the yields
were much lower; 2-bromocycloheptanone afforded 16% of 2-isopropyl-
idenecycloheptanone with acetone as the electrophilic acceptor
molecule [45].

   e. *Enol Ether and Enol Ester Donors with Acetal, Aldehyde, or
Ketone Acceptors.*   In this mixed aldol reaction technique either
one or both carbonyl reactants are converted to derivatives which
are properly functionalized carbonyl compounds.   In the mixed aldol
reaction of propionaldehyde with acetaldehyde, four products are
to be expected, whereas the reaction of propionaldehyde acetal with
ethyl vinyl ether gave only one aldol product [48].   Selectivity
is due to the fact that the enol ether derivative of the carbonyl
compound acts as the nucleophilic donor, while the acetal derivative
functions as the electrophilic acceptor in the mixed aldol reaction.

   It was first observed in 1939 that enol ethers (71) can add
to acetals (68) in the presence of BF$_3$ [49].   The β-alkoxy-acetal
addition compounds 73 can eliminate alcohol to yield the α,β-
unsaturated aldehydes 74.   A carbonium ion mechanism comparable to
the mechanism of the acid-catalyzed aldol reaction (Sec. III.A.2)
has been proposed [50].

$$R'-CH(OR)_2 + BF_3 \rightleftharpoons R'-\underset{\underset{OR}{|}}{\overset{\overset{BF_3}{\uparrow}}{CH}}-OR \rightleftharpoons R'-\overset{+}{C}H-OR \; ROBF_3^-$$

$$\underline{68} \qquad\qquad\qquad \underline{69} \qquad\qquad\qquad \underline{70}$$

$$\underline{70} + CH_2=CH-OR \rightleftharpoons R'-\underset{\underset{OR}{|}}{CH}-CH_2-\overset{+}{C}H-OR \; ROBF_3^-$$

$$\underline{71} \qquad\qquad\qquad \underline{72}$$

$$\underline{72} \rightleftharpoons R'-\underset{\underset{OR}{|}}{CH}-CH_2-CH(OR)_2 + BF_3$$

$$\underline{73} \longrightarrow R'-CH=CH-CHO$$

$$\underline{74}$$

One industrial synthesis of β-carotene [51] involves the conversion of the α,β-unsaturated aldehyde *75* to its diethyl acetal, *76*, and its reaction with ethyl vinyl ether (*77*) in the presence of $ZnCl_2$

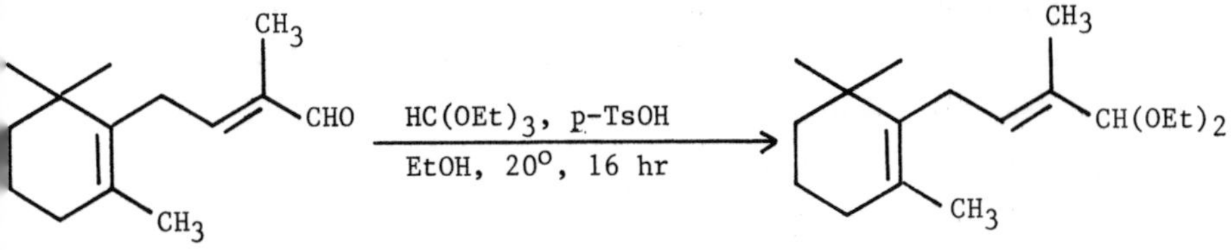
CH3
CHO
CH3
HC(OEt)3, p-TsOH
EtOH, 20°, 16 hr
CH3
CH(OEt)2
CH3

75 76

76 + CH_2=CH-OEt → (10% $ZnCl_2$/EtOAc; 40-45°, 1 hr)

77

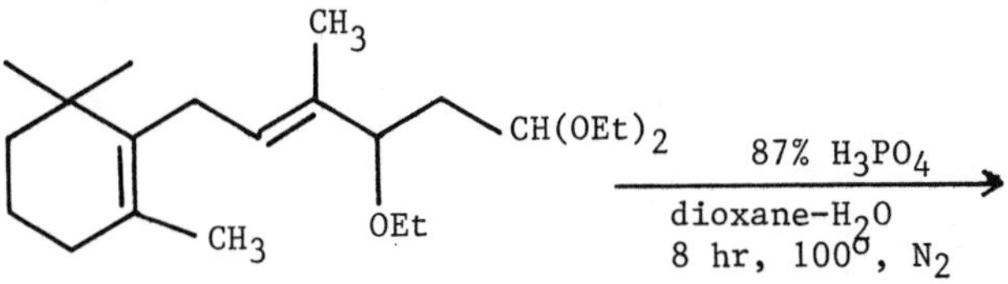
CH3
CH(OEt)2
CH3
OEt
87% H3PO4
dioxane-H2O
8 hr, 100°, N2

78

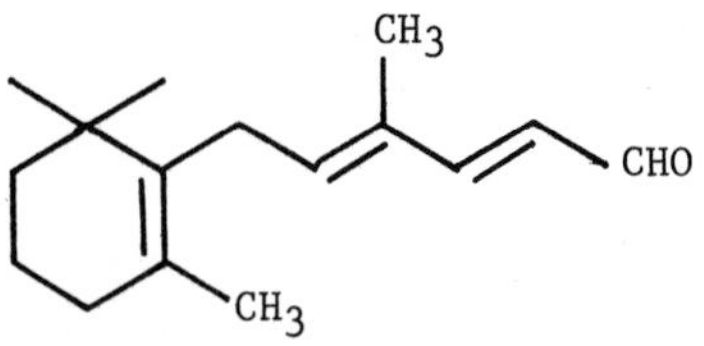
CH3
CHO
CH3

79

to give the β-alkoxy acetal *78*. To avoid further reaction of this
acetal (*78*) with the enol ether *77*, the reaction temperature is
maintained between $40°$ and $45°C$. Conversion of *78* to the unsaturated
aldehyde *79* can be executed with phosphoric acid at $90°-100°C$ through
hydrolysis of the acetal group and elimination of ethyl alcohol [51].
The reaction corresponds to the mixed aldol reaction of the α,β-
unsaturated aldehyde *75* with acetaldehyde (*1*).

In a related reaction it was found that trimethylsilyl enol
ethers (*81*) derived from aldehydes or ketones react smoothly with
acetals (*80*) at $-78°C$ in the presence of $TiCl_4$ to give β-alkoxy
ketones (*82*) in good yield. With trimethyl orthoformate (*80*; R_1 =
H, R_2 = OR', R' = Me) as the electrophilic acceptor, β-keto acetals
have been obtained [52]. In a typical procedure benzaldehyde
diethyl acetal (2.5 mmol) in methylene chloride was added to a solution of $TiCl_4$ (2.6 mmol) in 5 ml of the same solvent at $-78°C$ under
argon. A solution of the trimethylsilyl enol ether of phenyl acetone (2.5 mmol) in methylene chloride was added immediately, and
the reaction mixture was stirred at $-78°C$ for 3 hr. It was then
worked up by hydrolysis and extracted with ether. The reaction
product, 3,4-diphenyl-4-ethoxybutan-2-one (*82*; R' = Et, R_1 = Ph,
R_2 = H, R_3 = CH_3, R_4 = H, and R_5 = Ph) has been isolated in 95%
yield [52].

$$R_1-\underset{\underset{R_2}{|}}{\overset{|}{C}}(OR')_2 + R_3-\overset{\overset{OSiMe_3}{|}}{C} = CR_4R_5 \quad \xrightarrow[\text{2. } H_2O, + 20°C]{\text{1. } TiCl_4, CH_2Cl_2, -73°C} \quad R_1-\underset{\underset{R_2}{|}}{\overset{\overset{OR'}{|}}{C}}-CR_4R_5-\overset{\overset{O}{||}}{C}-R_3$$

   80   81   82

It should be noted that, unlike the related methods previously discussed [48-51], elimination of alcohol from the reaction products
to afford the unsaturated ketones does not take place under these
reaction conditions.

The reaction may also be executed using ketones (*83*) or aldehydes (*83*; R_2 = H) as the electrophilic acceptors. These reactions
may be performed at room temperature to give the aldol-type addition
products *84* in good yield [53, 54].

$$R_1-\overset{\overset{\text{O}}{\|}}{C}-R_2 + \underline{81} \quad \xrightarrow[\text{2. } H_2O]{\text{1. } TiCl_4, \text{ + } 20°C} \quad R_1-\overset{\overset{\text{OH}}{|}}{\underset{\overset{|}{R_2}}{C}}-CR_4R_5-\overset{\overset{\text{O}}{\|}}{C}-R_3$$

$\underline{83}$ $\underline{84}$

The same authors [55] found that an enol acetate rather than an enol ether may be used as the electron donor in the mixed aldol reaction. Isopropenyl acetate (*85*), the enol acetate of acetone, reacts with various acetals (*80*) and with benzaldehyde to afford the corresponding aldol type of addition products *82* in good yield.

$$R_1-CH(OR')_2 + OAc\overset{|}{\underset{\diagdown\!\diagup}{\diagup}} \quad \xrightarrow[\text{2. } H_2O, \text{ } 20°C]{\text{1. } TiCl_4, \text{ } CH_2Cl_2, \text{ low temp.}} \quad R_1-\overset{\overset{\text{OR'}}{|}}{CH}-CH_2-\overset{\overset{\text{O}}{\|}}{C}-CH_3$$

$\underline{80}$ $\underline{85}$ $\underline{82}$

This reaction can also be explained by assuming the initial formation of a $TiCl_4$-acetal complex (see also Sec. III.B.2.c).

IV. INTRAMOLECULAR ALDOL REACTION TECHNIQUES

In the intramolecular aldol reaction cyclic products are formed from dicarbonyl compounds (dialdehydes, diketones, or ketoaldehydes). Similar to the intermolecular aldol reactions the products may be either aldol or ketol addition products or α,β-unsaturated carbonyl condensation products resulting from dehydration of the addition products. Possibly the most important area of application of the technique has been the synthesis of monocyclic, oligocyclic, and polycyclic derivatives in natural product chemistry. Accordingly, examples have been chosen from the monoannular, transannular (intra-annular), interannular, and epiannular mode of the reaction [56a and b]. Basic as well as acidic agents have been used to catalyze the reaction. This may well be considered a special case of the mixed aldol reaction, in which both carbonyl functions happen to be attached to the same molecule. Once again, the problem of regio-selectivity has to be solved either through the construction of

proper structural features into the molecule or through the proper
choice of the reaction conditions, in particular that of the catalyst.
Selectivity may also be enhanced through the use of proper chemical
derivatives of the carbonyl function in the ring closure. The
problem of stereochemistry and that of asymmetric cyclization shall
be discussed in more detail because of its practical implications.

A. Regioselectivity of Cyclization

1. Monoannular Reactions

The conversion of open-chain dicarbonyl compounds to monocyclic
products via intramolecular aldol reactions falls into this category.
The regioselectivity of the cyclization may be governed by structural
features of the reactant or that of the product. In general, ring
formation becomes more favorable as the ring strain of the product
decreases (four → seven → five → six-membered ring systems). Although
the formation of six-membered monocyclic systems has been extensively
investigated, the interest in prostaglandin chemistry has directed
many laboratories to develop new and improved methods of five-membered
ring closures.

Treatment of the keto aldehyde *86* with 0.1 eq of 1,5-diazabicyclo
[4.3.0]-5-nonene (DBN) in methylene chloride at 0°C for 24 hr followed by acetylation and chromatography furnished the crystalline
cyclopentanol derivative *87* in 45% yield. The ketol derivative *87*
is an important intermediate in the total synthesis of the prostaglandins [57].

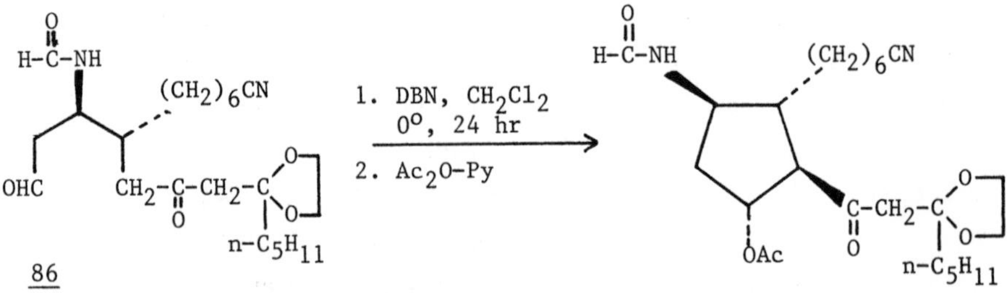

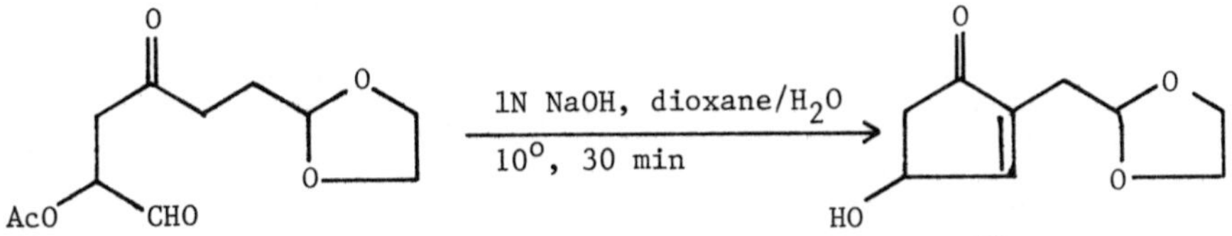

88   89

It should be noted that the regioselectivity of this conversion
(86 → 87) is the result of the higher reactivity of the aldehydic
carbonyl group which served as the acceptor for the nucleophilic
attack by the enolized ketone donor. This same principle has been
utilized in intermolecular mixed aldol reactions (Sec. III.b.1.a).

The use of DBN, the anhydrous reaction medium, the aprotic
solvent of modest polarity, and the relatively low reaction tempera-
ture--all serve to assure successful aldol reactions of base sensitive aldehydes, such as 86.

In a related example, the ketoaldehyde 88 has been cyclized
to the prostaglandin intermediate 89 in 37% yield. This reaction
was executed in a mixture of dioxane and 1 N aqueous sodium hydroxide
(15:0) at 10°C while stirring under nitrogen for 30 min [58]. Here
too, the mild reaction conditions rendered the reaction feasible.

In a similar fashion the 1,4-diketone 90 can be cyclized with
dilute aqueous base in >75% yield to jasmone (91), the important
constituent of the essential oil of jasmine [59].

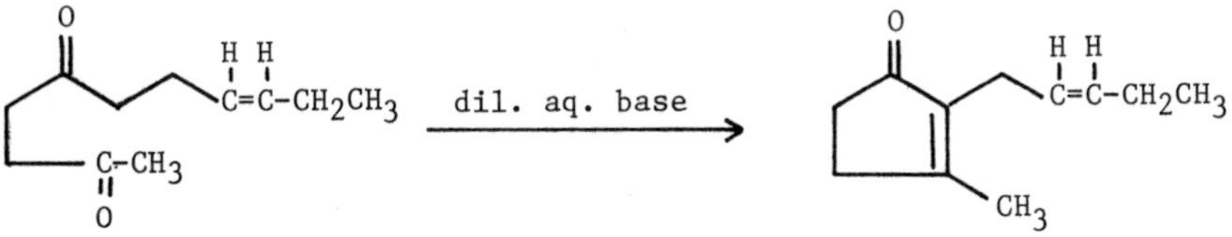

90   91

An important intermediate for the nonenzymic biogenetic-type
olefinic cyclization to steroids has been prepared by treatment of
the 1,4-diketone 92 with 2% sodium hydroxide in aqueous ethyl alcohol

for 6 hr at $105^{\circ}-110^{\circ}$C to give the cyclopentenone derivative *93* in
excellent yield [60].

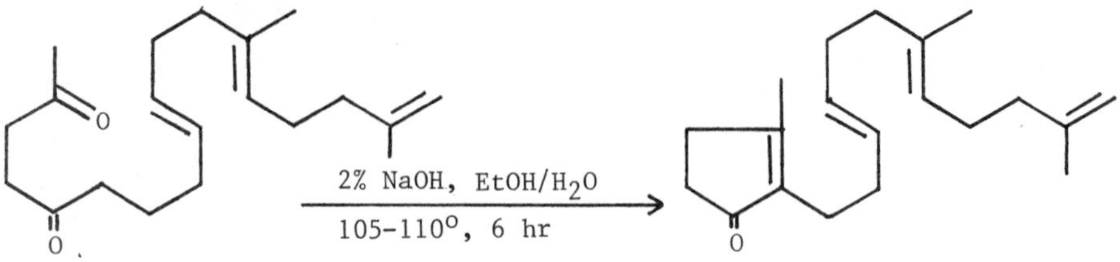

If 1,5-dicarbonyl starting materials are employed in the monoannular reactions, six-membered monocyclic derivatives should be
obtained. The reaction conditions are in general very similar to
those described above for the formation of five-membered ring
systems. For example, the 5-keto aldehyde *94* gave 1-methyl-3-
cyclohexen-2-one (*96*) in 40% yield in the presence of 2 *N* KOH in
methanol [61].

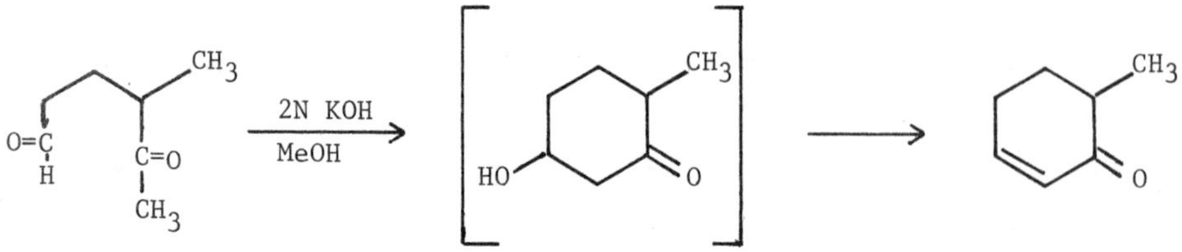

2. Transannular (Intraannular) Reactions

This type of reaction involves an intramolecular aldol reaction
within an already existent dicarbonyl ring compound to form a new
bicyclic ring system. The reaction may be illustrated with the
synthesis of the bicyclic unsaturated ketone *99*, an important intermediate in the improved synthesis of azulene. Cyclodecandione (*97*)
(50 g, 0.3 mol) was refluxed in 500 ml of 5% aqueous sodium carbonate
for 1 hr. The mixture was steam distilled, and the distillate
simultaneously extracted with chloroform. Evaporation of the

chloroform followed by fractional distillation gave 4-keto-
octahydroazulene (*99*) in 96% yield [62].

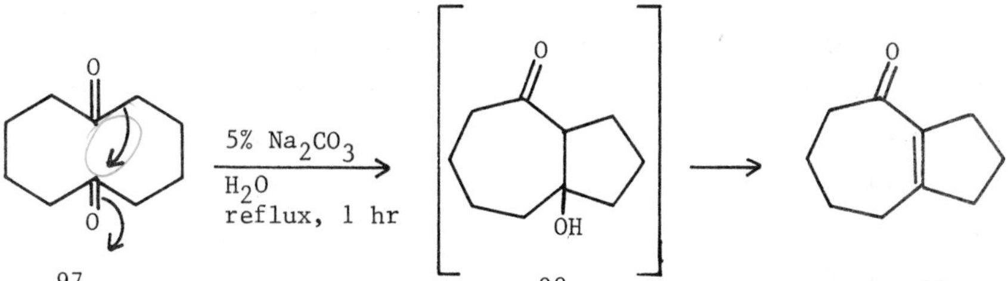

3. Interannular Reactions

An example of this type of intramolecular aldol reaction consists of the conversion of the decalin derivative *100* to a tricyclo[4.4.0.0]decane, a twistane derivative, *101*. The unfavorable
aldol equilibrium was displaced by derivatization of the resulting
aldol hydroxyl function. The methyl ether derivative *101a* of the
twistane aldol can be obtained in modest yield by treating *100a*
with a solution of hydrogen chloride in anhydrous methanol.

The introduction of a second methyl substituent at C10, trans
to the angular methyl group, enhanced the yield of the twistane
ether *101b* to 98% most likely through steric compression [63].

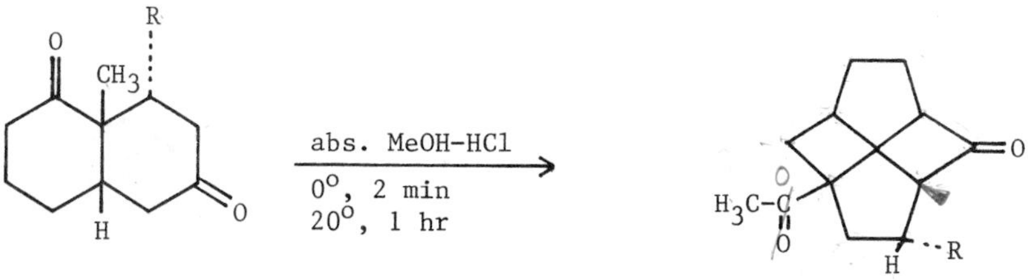

100 a, R = H   101 a, R = H (25%)

    b, R = CH₃   b, R = CH₃(98%)

4. Epiannular Reactions

This type of intramolecular aldol reaction differs from the
previous three categories in that a new ring is being formed on an

already existing ring or ring system. The two carbonyl groups
necessary for the aldol reaction may be located either in two
properly positioned side chains, or in the already existent ring
and a properly positioned side chain. The latter is the case in
the already mentioned Robinson annulation reaction (Sec. III.B.2.a).
The actual ring closing step in the Robinson annulation consists
of an acid- or base-catalyzed aldol cyclization, followed by
dehydration.

Cyclization of the dialdehyde *102* has been achieved in a classical total synthesis of steroids [64] by heating it in dioxane under
nitrogen in a sealed glass tube at $145^{o}C$ for 7.5 hr. The desired
aldehyde *103* has been obtained in 27% yield. The mother liquors
from the crystallization were combined and then chromatographed on
alumina to give lesser amounts of the isomeric aldehyde *104* resulting from attack of the 15-methylene on the 20-aldehyde.

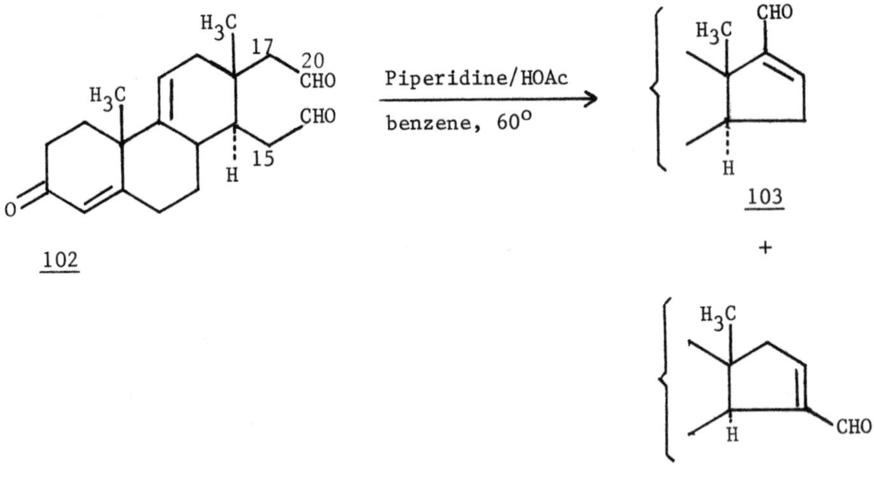

The preferred reaction technique for the ring closure of the
dialdehyde *102* consists of heating it in benzene in the presence of
a catalytic amount of piperidinium acetate at $60^{o}C$ in a slow stream
of nitrogen, with the constant removal of water for 1 hr. The

desired unsaturated aldehyde 103 has been obtained in 66% yield
after crystallization [64]. The reaction is assumed to proceed
via an enamine intermediate.

It seems quite reasonable to assume that the predominance of
103 in the aldol cyclization reaction may be due to the relatively
uncrowded environment of the upper activated methylene group in
102 as compared with the lower (C17 versus C15). This relationship,
which is apparent from inspection of models, permits more ready
access to the upper center (C17), while approach to the lower
center (C15) is hindered by the axial C13 methyl group.

The triketo aldehyde 105, an intermediate in the aforementioned
nonenzymic biogenetic type of olefinic cyclization [60], has been
cyclized to afford racemic 16,17-dehydroprogesterone (106). This
double ring closure has been effected by stirring 105 with 2.5%
aqueous potassium hydroxide for 13 hr at 74°C [60].

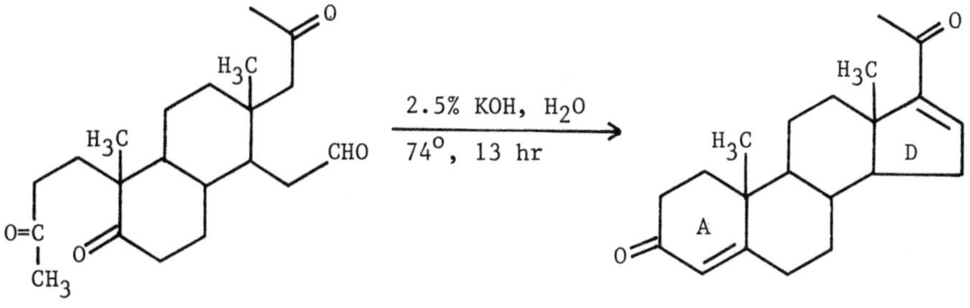

$$\underset{\textbf{105}}{} \xrightarrow[74^\circ,\ 13\ \text{hr}]{2.5\%\ \text{KOH},\ H_2O} \underset{\textbf{106}}{}$$

Regiospecificity of the five-membered D-ring closure is again
governed by the difference in the reactivity of the two carbonyl
groups. The more reactive aldehydic carbonyl group served as the
acceptor toward the nucleophilic attack of the enolized ketone
donor (Sec. IV.A.1. and II.B.1.a).

The regiospecificity of the ring closure of the six-membered
A-ring should also be discussed. It has been shown earlier by
Johnson et al. [65] that a bridged ketol of type 108 is initially
formed in an aldol cyclization reaction involving the α-methylene

group next to the B-ring ketone. The resulting bridged ketol may
then revert to the 1,4-diketo system *107*. Aldol cyclization to the
regular ketol *109* followed by dehydration will then give the desired
steriod *110*.

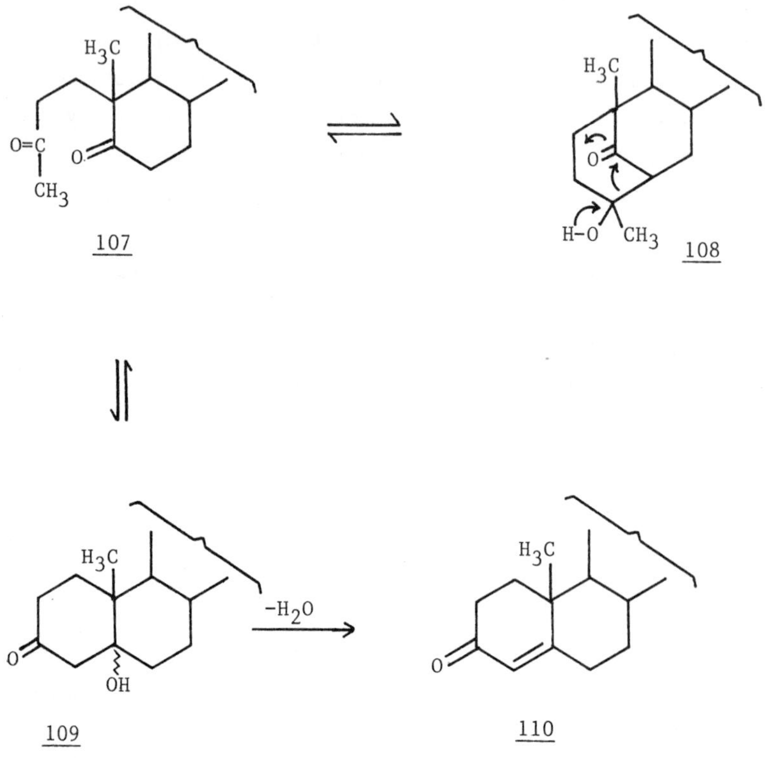

107   108

109   110

The regiospecificity and stereochemistry of the aldol type of
ring closures leading to CD-bicyclic steroidal intermediates have
been carefully investigated with the help of ir and nmr spectrometry
and by chemical correlation [66] in connection with the stereocontrolled total synthesis of 19-norsteroids [34].

Neutral piperidinium acetate in water cyclizes the triketone
111 to the bridged ketol *112* of the [3.2.1]bicyclooctane system in
a regiospecific intramolecular aldol reaction via the enolic intermediate *111a*. This energetically favorable molecular arrangement

allows maximum π-orbital overlap. The six-membered ring thus results
in an almost perpendicular attack by the enolized ring ketone on
the side chain keto group placing the ketol hydroxyl group in an
equatorial arrangement.

Additional piperidine epimerizes the C4 center of the bridged
ketol *112* to give the more stable bridged ketol *113*. The epimeriza-
tion involves retro-aldol ring opening to yield the triketone *111*,
followed by aldol cyclization to a new bridged ketol, *113*, via the
enolic intermediate *111b*. The axial orientation of the newly formed
ketol hydroxyl group in *113* is due to the molecular arrangement of
the enolic intermediate *111b*. By the principle of microscopic
reversibility [67] this bridged ketol, *113*, should be more stable,
since its formation, as well as its possible opening, would have
to proceed through an energetically less favorable intermediate,
111b.

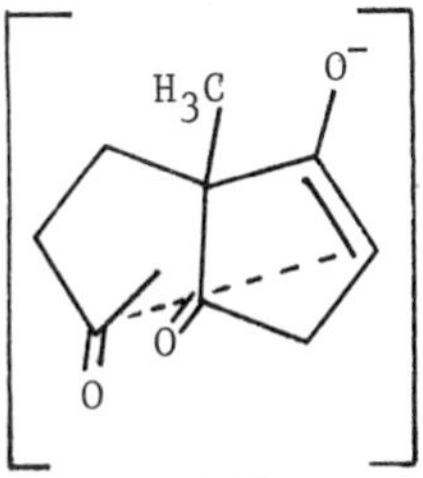

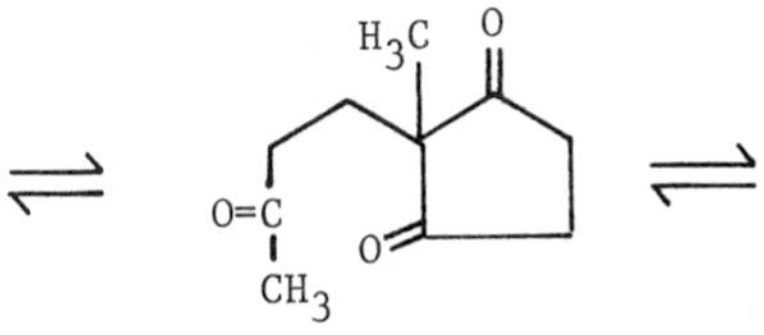

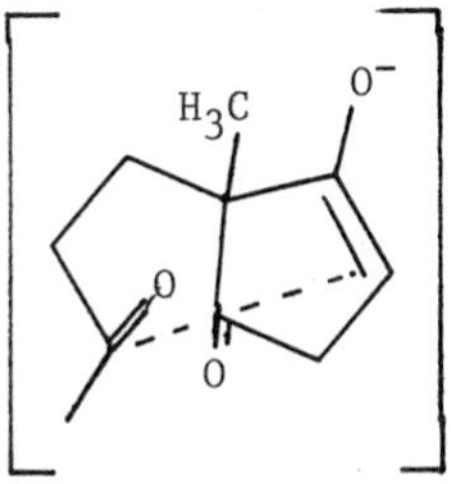

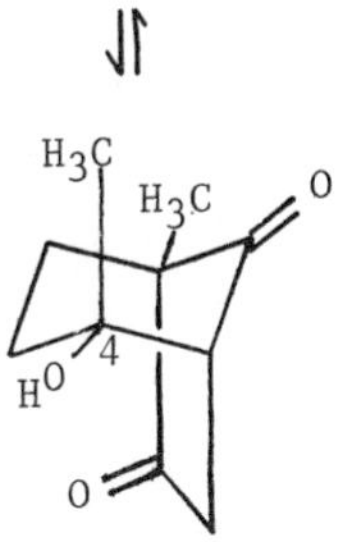

In order to convert the triketone *111* to the bicyclic ketol
114 of the perhydroindane series, the direction of the aldol reaction
has to be changed. The addition of pyrrolidinium acetate to the

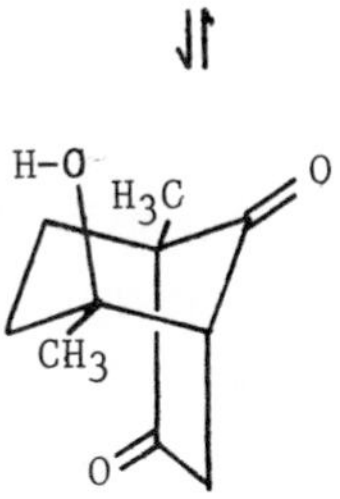

triketone *111* in anhydrous ether at $0^\circ C$ resulted in the formation
of the desired bicyclic ketol *114* in 21% yield and its dehydration
product, the bicyclic enone *115*, in 22% yield [66]. Similar conditions have been used to prepare the homologous ketol cis-9-hydroxy-
10-methyldecalin-2,5-dione [68].

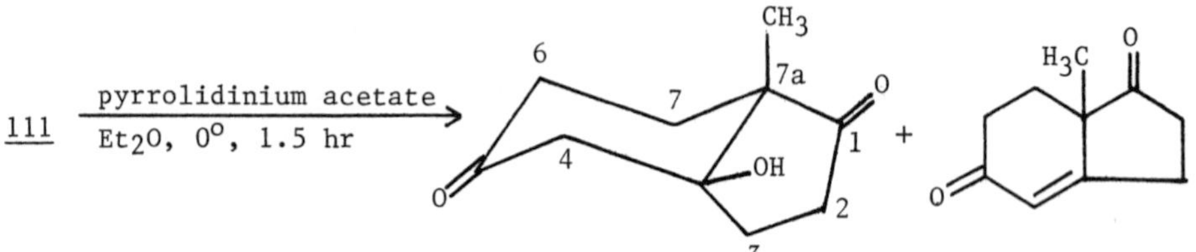

$$\underline{111} \quad \xrightarrow[\text{Et}_2\text{O, } 0^\circ, \text{ 1.5 hr}]{\text{pyrrolidinium acetate}} \quad \underline{114} \quad + \quad \underline{115}$$

The change in the direction of the aldol cyclization reaction
is most likely due to addition of pyrrolidine to the more reactive
cyclic keto group [69] and nucleophilic attack by the enolized
side-chain keto group. On the other hand, the aforementioned
cyclizations with piperidine take a different course due to the
decreased reactivity of piperidine toward ketones [70].

B. *Asymmetric Syntheses through Stereocontrolled*
    Aldol Cyclizations

A well-known phenomenon in natural product chemistry is the finding
that many physiologically active naturally occurring compounds
possess optical activity, and that the biological activity of a
particular compound usually resides almost exclusively in one of
its optically active forms. The scientific and practical importance
of the preparation of specific optical isomers is therefore quite
obvious. A regular chemical laboratory synthesis would normally
yield an equal amount of both enantiomers, which would have to be
separated yielding a theoretical maximum of 50% of the desired
optically active isomer. An asymmetric synthesis, on the other
hand, can theoretically result in a 100% yield of only one
enantiomer [71].

The problems associated with an asymmetric synthesis utilizing a stereocontrolled aldol cyclization are twofold:

1. A new reaction technique has to be found which would promote the aldol reaction in a high chemical yield.
2. The cyclization reaction should occur with maximum transfer of chirality, i.e., with the best possible optical yield.

The solution to the problem is illustrated in the asymmetric conversion of the triketone *111* to the optically active bicyclic ketol (+)-*114* and to its dehydration product, the optically active bicyclic enone (+)-*115*.

It was found [72] that the triketone *111*, a compound of reflective symmetry, could be converted by an asymmetric aldol cyclization to the optically active bicyclic ketol (+)-*114* in 100% chemical and 93.4% optical yield while stirring with a catalytic amount (3% molar eq) of (S)-(-)-proline-(*116*), in anhydrous dimethylformamide under argon at 20°C for 20 hr.

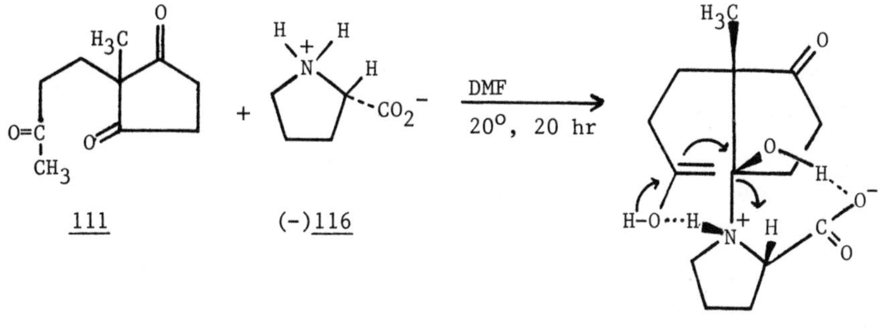

111   (-)116   117

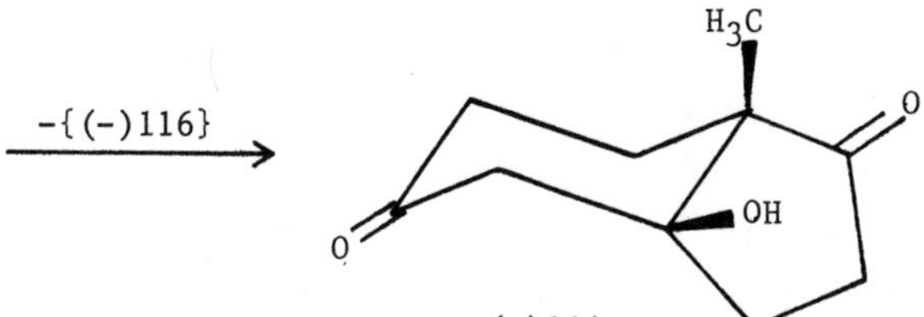

(+)114

It is assumed that the mechanism of this asymmetric aldol cyclization involves the addition of (S)-(-)-proline in its zwitterionic form to one of the carbonyl groups of the cyclopentanedione ring of *111* to form the intermediate *117*, in which two hydrogen bonds provide a 6,6,7-membered conformation and the rigidity necessary to achieve the stereoselectivity of the reaction. The bulky (S)-(-)-proline molecule is thus positioned opposite the β-oriented angular methyl group; C-C bond formation would have to occur from the side opposite the angular methyl group to give a cis-fused product with the concomitant regeneration of (S)-(-)-proline by expulsion from the molecular complex.

The high optical yield can well be explained by invoking the intermediate *117* in which the center of asymmetry of the asymmetric reagent (-)-*116* is three bonds away from the angular methyl group (the prochiral center). The asymmetry of the optically active reagent (-)-*116* is being transferred in the process in which, through the intermediacy of *117*, the center of prochirality is being converted to a center of chirality giving rise to the optically active bicyclic ketol (+)-*114*. The optically active asymmetric reagent is thus able to differentiate between the two identical (*enantiotopic*) carbonyl groups in the five-membered ring of *111*.

Dehydration of the optically active bicyclic ketol (+)-*114* in refluxing benzene with a catalytic amount of p-toluenesulfonic acid gave the optically active enone (+)-*115* in excellent chemical and optical yield [72]. The conditions of the dehydration should not have affected the 7a center, thus the absolute configuration of (+)-*114* at the 7a position should be identical with that of the enone (+)-*115* of known (7aS) absolute configuration [73]. This was confirmed by both the circular dichroism data of (+)-*114* and a single-crystal x-ray diffraction study of the corresponding racemic compound *114*, which clearly indicated the cis conformation with an axial 7a-methyl and an equatorial 3a-hydroxyl group in the six-membered ring of the bicyclic system.

be executed with 0.5 eq of (S)-(-)-proline and 0.25 eq of HClO$_4$ (1 N aqueous solution) in refluxing acetonitrile for 22 hr to give the optically active enone (+)-*115* in 86.6% chemical yield and 84% optical purity [74].

C. β-Keto Derivatives in Aldol Cyclizations

The regiospecificity of intramolecular aldol reactions may also be assured through activation of the donor reaction partner by functionalization beta to its carbonyl group.

1. β-Keto Carboxylic Esters

The reaction of 2-methyl-1,3-cyclohexanedione (*118*) and the methyl vinyl ketone derivative *119* in refluxing tetrahydrofuran in the presence of a catalytic amount of triethylamine gave the α,β-unsaturated β-keto ester *122* and the diastereomeric mixture of the ketol derivative *121* in 31% yield (each) through the intermediacy of the β-keto ester aldol intermediate *120* [75].

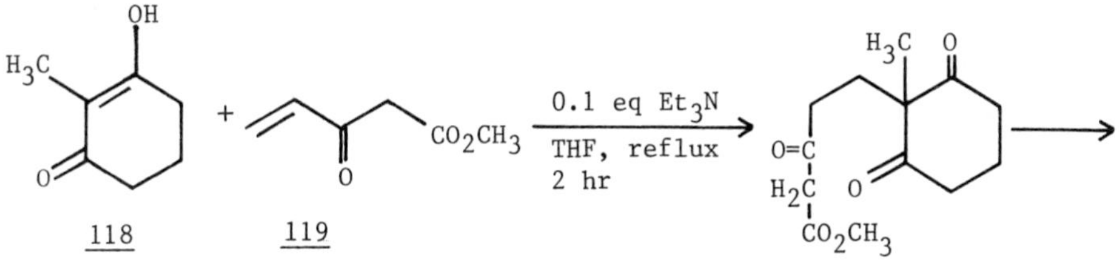

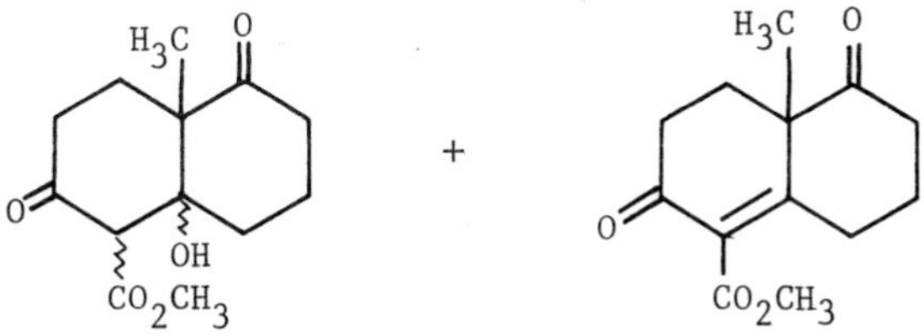

2. β-Keto Phosphonates

An example of this technique is illustrated by an additional
synthesis of cis jasmone (91); an earlier synthesis has already been
discussed (Sec. IV.A.1). Diethyl-cis-3-hexenyl-phosphonate (124)
was prepared from cis 3-hexen-1-ol by conversion to the bromide
(PBr₃, ether) and by subsequent Arbuzov reaction with triethyl
phosphite. Addition of the lithio derivative of 124 (BuLi, THF,
-78°C) to methyl levulinate ethylene ketal (123) gave the β-keto
phosphonate 125, which was hydrolyzed to the diketone 126, in an
overall yield of 91%. Cyclization of 126 with sodium hydride in
THF at room temperature (formation of anion at 0°C) gave cis
jasmone (91) in 68% yield after distillation [76]. The reaction
proceeds via nucleophilic attack of the β-keto phosphonate anion
on the ε-keto carbonyl acceptor group [77].

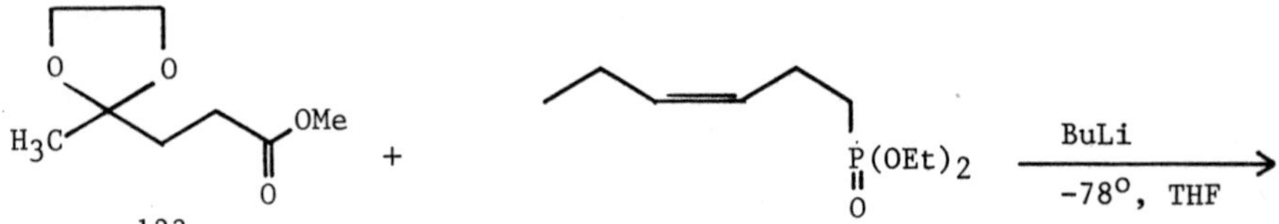

123

124

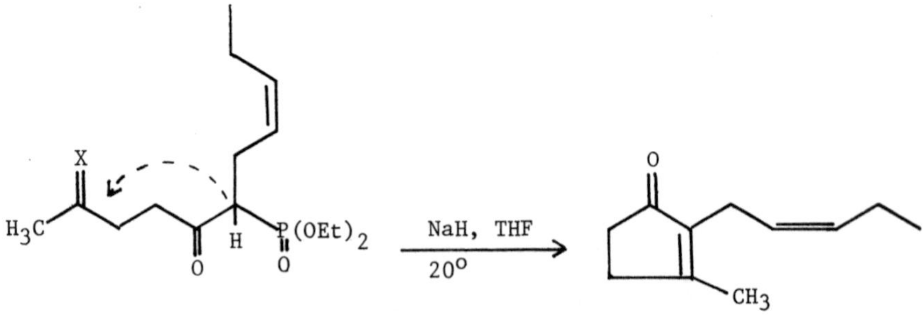

BuLi
-78°, THF

NaH, THF
20°

125, X = O(CH₂)₂O

126, X = O

91

3. β-Keto Sulfones

An intramolecular aldol reaction to produce tetracyclic intermediates has been accomplished by reaction of the diketo sulfone
carboxylic acid *127* with 2 eq of t-butylmagnesium chloride in a
nonpolar reaction medium (DME + THF) at $0^{\circ}C$ for 20 min, followed
by refluxing for 51 hrs. It was then partitioned between chloroform
and aqueous acetic acid. The $CHCl_3$ extract was concentrated and
partitioned between 0.5 *N* aqueous HCl and ethyl acetate to give the
ketol sulfone derivative *129* in 96% yield (after recycling recovered
starting material). The success of this otherwise unfavorable aldol
reaction has been attributed to the formation of a covalent metal
chelate intermediate, *128*, that is stable in the absence of polar
solvents. It is, however, readily reversible to *127* with even very
weak bases such as aqueous sodium bicarbonate, indicating the value
of the metal-complexing procedure [78]. The use of metal salts to
intercept intermolecular mixed aldol products was discussed in Sec.
III.B.2.c [43].

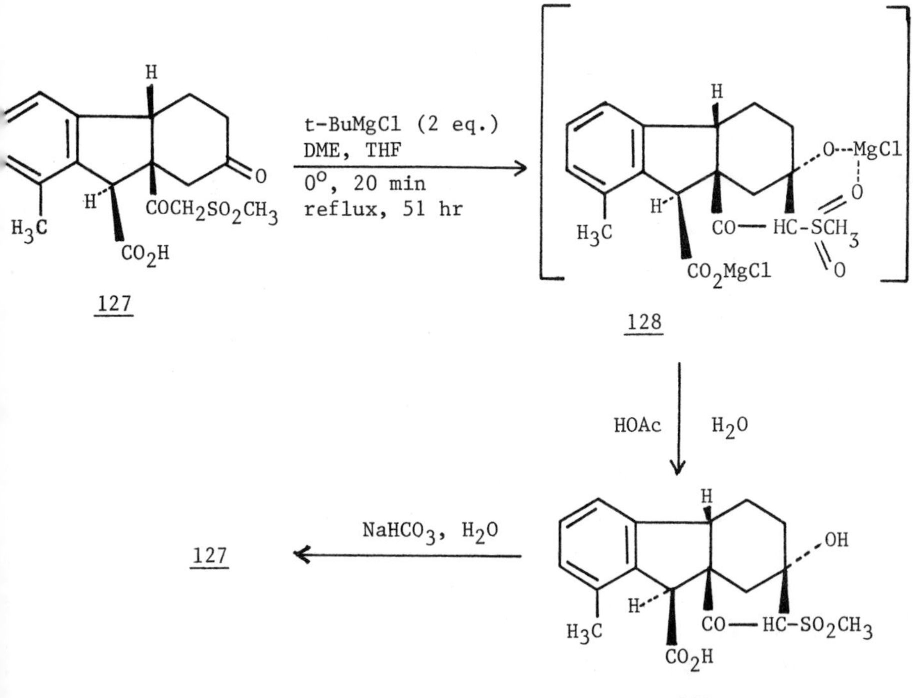

4. Diazo Ketones

This type of β-keto derivatives can readily be prepared by treatment of an acid chloride with diazomethane. Starting with 2-methylcyclohexanone-2-propionic acid (130), the diazoketone 132 can be obtained via the acid chloride 131 in good yield. A methanolic solution (1.0 M) of the crude diazoketone was then treated with aqueous 10% KOH to give the diazo ketol 133 in 65% yield [79].

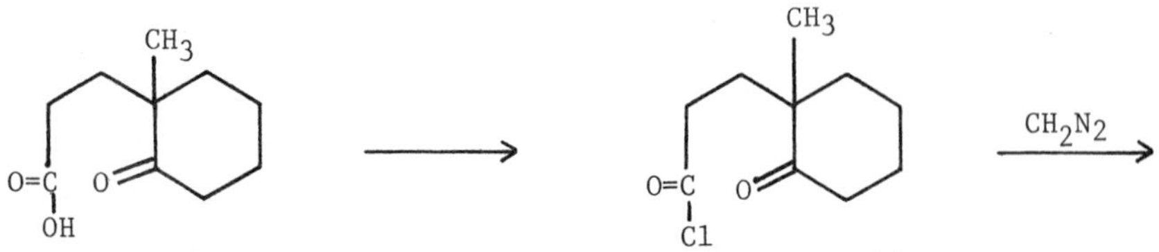

130 → 131 (CH₂N₂)

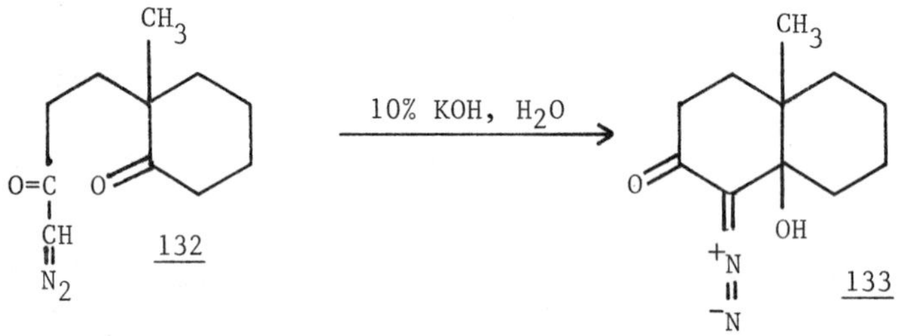

132 (10% KOH, H₂O) → 133

This intramolecular aldol-type reaction of diazo ketones has recently been utilized in the total synthesis of steroids [80].

D. Dicarbonyl Aldol Intermediates through Indirect Methods

1. Vinyl Chlorides

As already discussed (Sec. IV.A.4), the Robinson annulation reaction followed by an intramolecular aldol reaction offers an advantageous pathway towards the synthesis of oligocyclic and/or polycyclic compounds. On occasion, however, undesirable condensations and polymerizations occur between unactivated cyclohexanones with vinyl ketones. In research directed toward resin acid synthesis,

it was found [81] that the reaction of the ketone *134* with methyl
vinyl ketone gave only polymeric tars. To overcome the problem
the scheme devised by Wichterle et al. [82] was applied. The
ketone *134* was found to react with 1,3-dichloro-2-butene (*135*) in
the presence of sodium amide to give *136* in 69% yield. The vinyl
chloride group of this compound was then hydrolyzed by acid treatment (9:1 sulfuric acid and water) for 1.5 hr at 0°C and then for
an additional 1.5 hr at room temperature to afford the dicarbonyl
aldol intermediate *137* in 73% yield. These reaction conditions are
necessary to avoid undesired bridged ketol formation followed by
dehydration [83].

The diketone *137* can be cyclized in an intramolecular aldol
reaction to the bicyclic enone *138* by treatment with potassium t-
butoxide in t-butanol-benzene (100:8) under nitrogen for 16 hr at
room temperature. Treatment with 10% aqueous acetic acid followed
by evaporation in vacuo, extraction with ether, and chromatography
on basic Al_2O_3 afforded the enone *138* in 64% yield. However, a
significant amount of the diketone *137* underwent cleavage in a
reverse Michael reaction to give the cyclohexanone derivative *134*
in 30% yield.

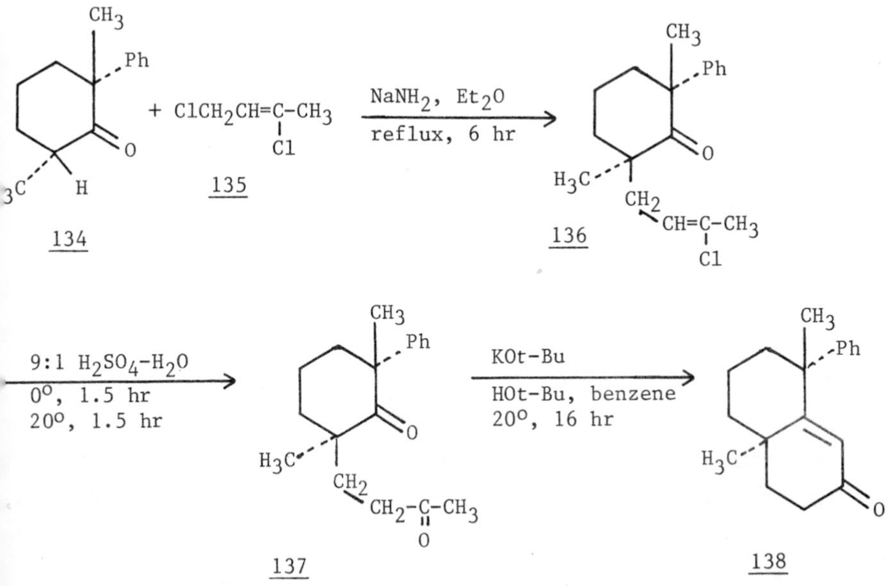

Alternatively, the vinyl chloride intermediate *139* may be dehydrohalogenated to the acetylene derivative *140* in quantitative yield by treatment with 2 eq of sodium amide in liquid ammonia. Further reaction with 3 eq of the same reagent in refluxing toluene for 12 hr isomerized the triple bond to the terminal position to give *141* in 73% yield. Hydration with approximately 2% sulfuric acid in methanol-water containing a catalytic amount of mercuric sulfate for 1.5 hr at room temperature gave the dicarbonyl aldol starting material *142* in 96% yield. Aldol cyclization gave trans 8,10-dimethyl-1(9)-octalone-2 (*143*) in excellent yield [84].

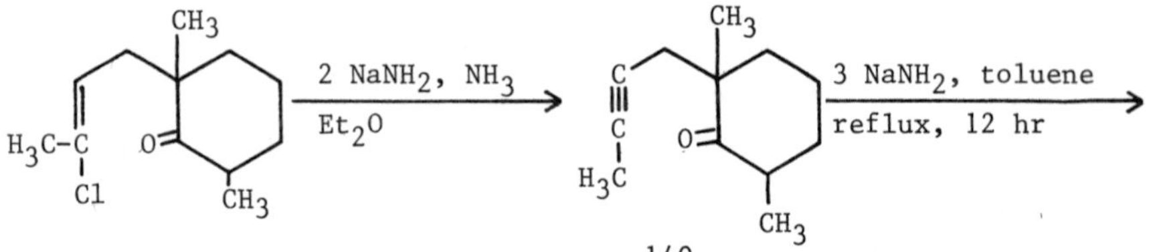

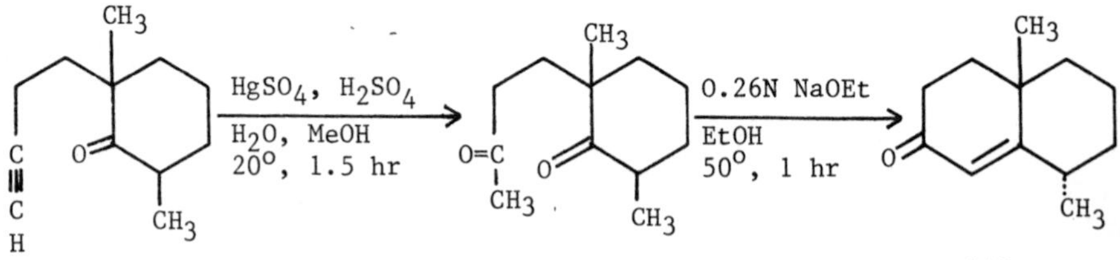

2. Vinyl Silane Intermediates

These intermediates have a structure analogous to the vinyl chloride intermediate *139*. Alkylation of the lithium enolate *144*, kinetically generated from its enol acetate [43], with 1 eq of the iodo vinyl silane *145* in THF at room temperature gave the keto vinyl silane *146* in 91% yield [85]. Epoxidation with a slight excess of

m-chloroperbenzoic acid at $0°C$ for 10 min in methylene chloride
gave largely the epoxysilane *147*. Keeping of the reaction mixture
for an additional 4 hours at room temperature led to the formation
of the dicarbonyl aldol intermediate *148*, which could be cyclized
under the usual basic reaction conditions to the octalone *48*.

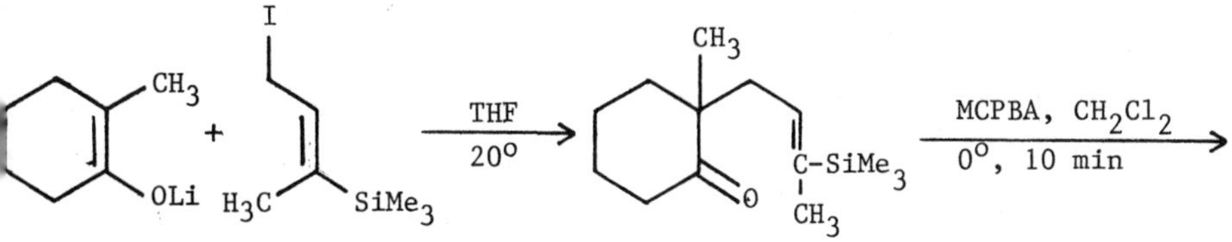

3. Enol Lactone Intermediates

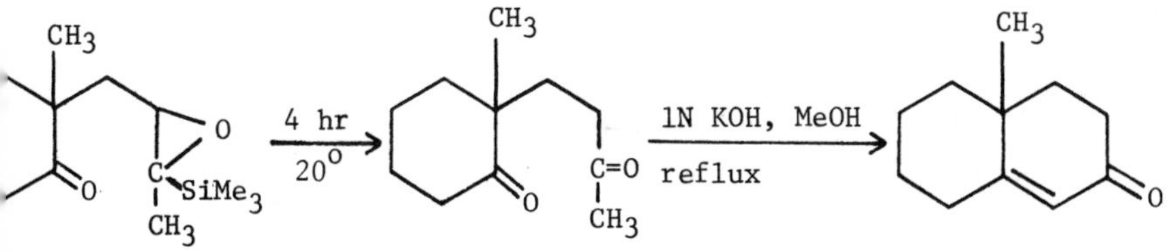

It has been previously mentioned that the synthetic methods
utilized for the preparation of cyclic α,β-unsaturated ketones
often suffer from the disadvantage that alternate condensation
reactions may occur which lead to isomeric unsaturated ketones or
other undesirable products [65, 83]. The regiospecific aldol
cyclization through an enol lactone intermediate offers a useful
general method for the synthesis of cyclic, α,β-unsaturated ketones,
overcoming these difficulties.

As an illustration, the seco acid *149* has been converted to
the enol lactone *150* with acetic anhydride and sodium acetate at
reflux temperature for 2 hr (78% yield). Reaction with methyl

magnesium bromide, followed by aqueous HCl opened the enol lactone
to the dicarbonyl aldol intermediate *151*, which was converted to
$\Delta^{9(11)}$-testosterone (*152*) in 50% yield [86] in an aldol cyclization
reaction with aqueous KOH and methanol at reflux temperature.

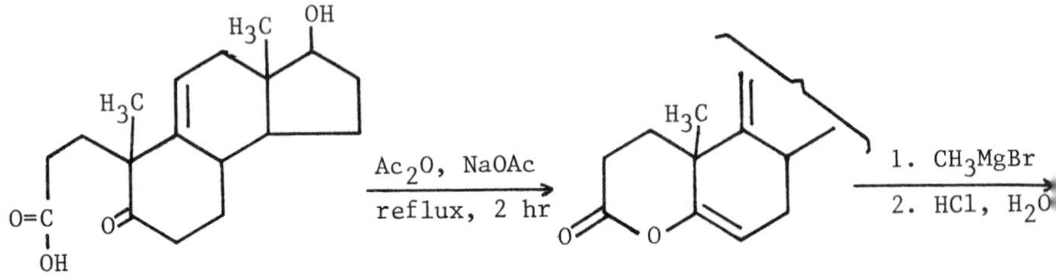

Instead of a Grignard reagent, alkylidenetriphenylphosphoranes
or preferably dialkyl alkylphosphonate anions may be used in the
sequence. The anion of dimethyl methylphosphonate (*154*) reacts with
the enol lactone *153* to give testosterone acetate (*156*) in 50%
yield [87]. The strongly nucleophilic β-ketophosphonate anion
intermediate *155* obviously facilitates ring closure by an intramolecular Wittig reaction (see also Chap. 3).

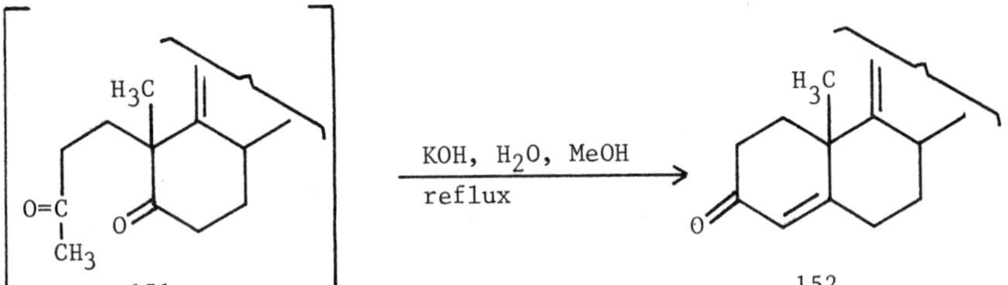

    4. Pyridine Derivatives
    It has been demonstrated [88(a)] that the aldol intermediate
dicarbonyl compound *161* may be prepared in 43% yield by the metal

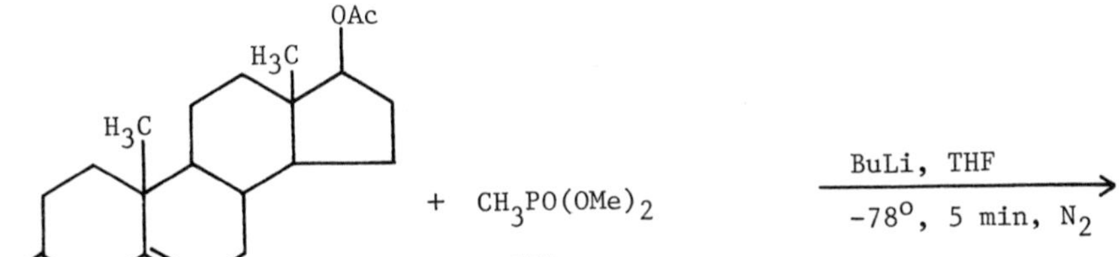

153 + CH$_3$PO(OMe)$_2$ 154

BuLi, THF
$-78°$, 5 min, N$_2$

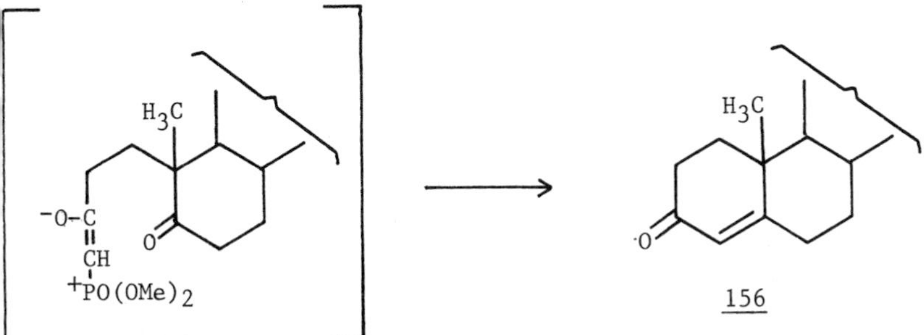

155 → 156

ammonia reduction of the pyridine derivative *160*, followed by treatment with dilute ethanolic sodium hydroxide at room temperature for
5 min. Aldol condensation of *161* with aqueous ethanolic sodium
hydroxide for 17 hr at reflux under nitrogen followed by acidic
deketalization with 10% aqueous HCl for 15 min at room temperature
gave the enedione *162* in quantitative yield. The same sequence
without the isolation of the aldol intermediate *161* gave the enedione
162 in 93% overall yield from *160*. The pyridine derivative *159* has
been prepared from the ketal of the Wieland-Miescher ketone *157* by
Michael addition of *157* to 6-methyl-2-vinyl pyridine (*158*) followed
by hydrolysis and sodium borohydride reduction. The enedione *162*

is an important intermediate in the total synthesis of (±)-D-homoestrone [88(a)]. The pyridine route has also been successfully applied for the synthesis of optically active estrone and 19-norsteroids [88(b)].

V. METHODS OF CIRCUMVENTING MIXED ALDOL REACTIONS

The problem in executing mixed aldol reactions either in intermolecular or in intramolecular systems has already been discussed (Secs.
III.B and IV). Several reaction techniques have been offerred for
the solution of the problem. On occasion, however, the laboratory
investigator shall find it necessary to search for methods that
will allow him to circumvent the problem of mixed aldol reactions
altogether. Some of the more important of these techniques are
listed in this section.

A. *Formylation and Reduction*

The attachment of a single hydroxymethyl group adjacent to a carbonyl
function is accomplished most often by the base-catalyzed aldol
reaction of an active methylene compound and formaldehyde (Sec.
III.B.1.b). As already discussed, it is often difficult to control
the process at the monoalkylation stage, resulting in complex mixtures of polycondensation products as well as Cannizzaro-type
products.

Alternative processes include the Mannich technique (Sec.
III.B.2.a) and the generation of kinetic enolate salts in conjunction
with the use of gaseous, anhydrous formaldehyde (Sec. III.B.2.c).

However, to circumvent the mixed aldol reaction altogether, a
formylation and reduction sequence has been introduced. The 3-mono-
ethylene ketal of progesterone, *163*, has been formylated with iso-
amyl formate in the presence of a threefold excess of sodium hydride
in benzene under a nitrogen atmosphere at 15°C. The mixture was
then refluxed for 4 hr to give the sodium salt of the 21-
hydroxymethylene derivative *164*. This was reduced with an excess
of sodium borohydride in methanol at 0°C for 1 hr and at room
temperature for 16 hr to give the diol *165* in 97% yield. Compound
165 was then selectively tritylated at the primary alcohol group
with trityl chloride in pyridine for 40 hr at room temperature to
give *166* in 88% yield. Oxidation with chromium trioxide in pyridine
for 16 hr at room temperature gave *167* in 91% yield. Mild acid

hydrolysis with 80% aqueous acetic acid for 17 hr at room temperature removed both the trityl and the ethylene ketal group to yield
21-hydroxymethyl-progesterone (*168*) in a 78% yield [89].

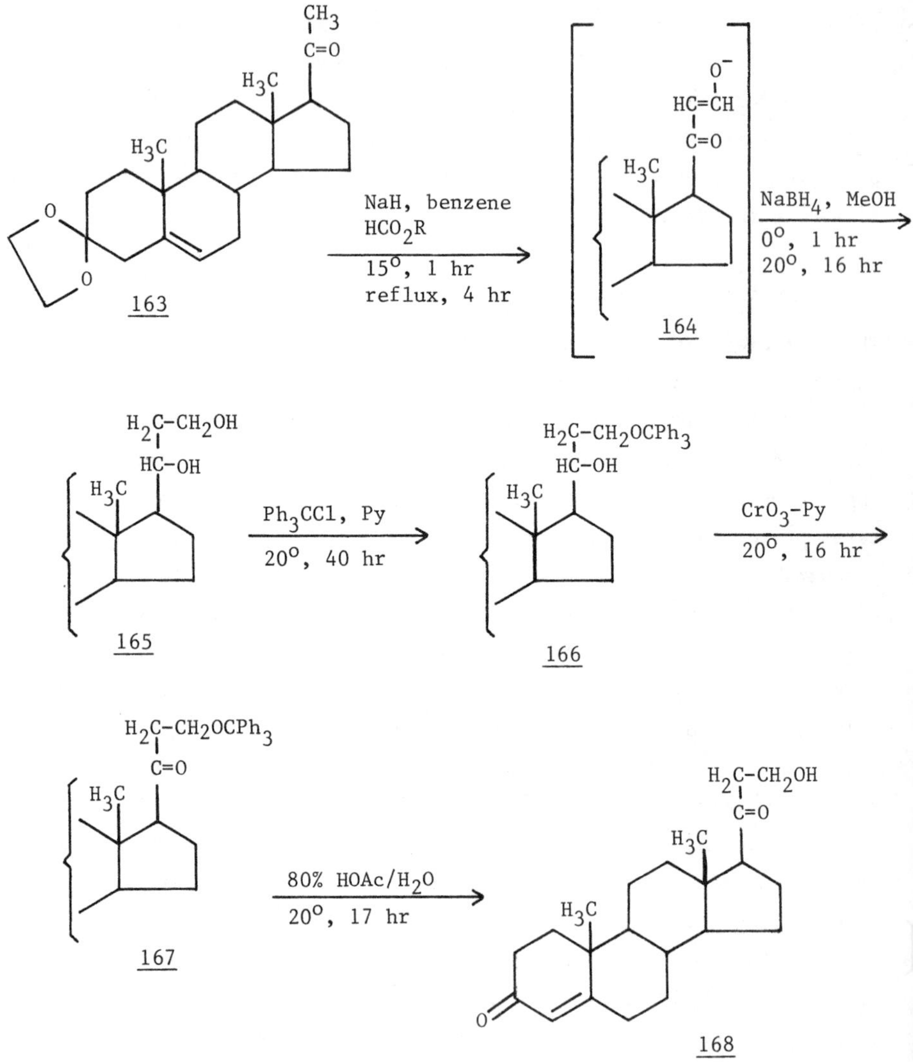

A simplification of this method consists in the selective
reduction of the sodium enolate of the hydroxymethylene derivative
170 with aluminum hydride in tetrahydrofuran for 1.25 hr at room
temperature to give the 2-hydroxymethyl ketone *171* in 90% yield [90].

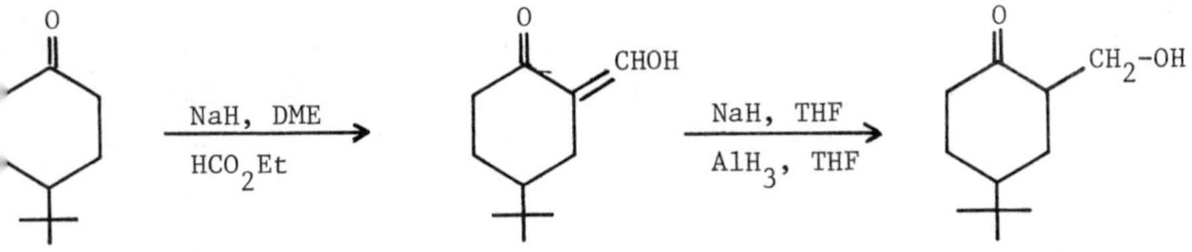

169   170   171

B. *Acylation of Olefinic Compounds*

To avoid the synthesis of the keto aldehyde *172* but still form its
intramolecular mixed aldol condensation product, tetrahydroacetophenone (*173*), cyclohexene (*174*) may be acylated with acetyl chloride (*175*) in the presence of aluminum chloride in carbon disulfide
at $0^{\circ}C$, and the resulting β-chloro ketone *176* isolated by pouring
the mixture into ice water. On treatment with a tertiary base,
such as diethylaniline, hydrogen chloride is eliminated, and tetrahydroacetophenone (*173*) is obtained [91].

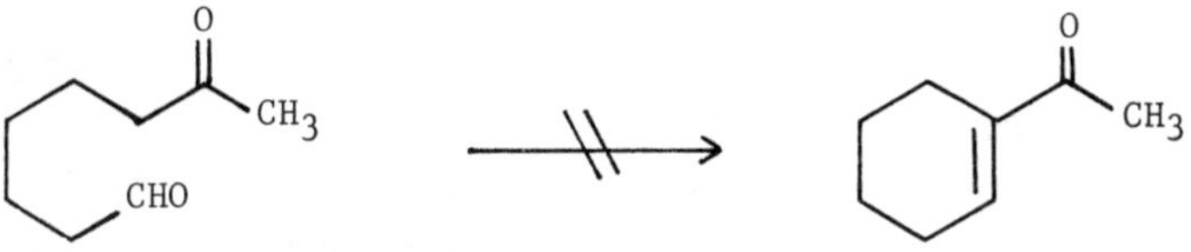

172   173

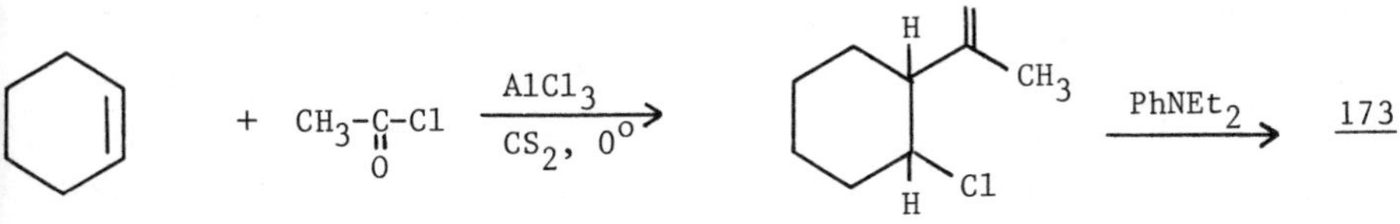

174   175   176

The β-halo ketones may be considered the halo analogues of the
β-hydroxy carbonyl (aldol) compounds. An illustration from the

aliphatic series is the preparation of 1-bromo-3-pentanone (*177*) through the Friedel-Crafts type of acylation of ethylene with propionyl bromide at -10°C in methylene chloride in the presence of aluminum bromide [92].

$$CH_3-CH_2-\underset{O}{\overset{O}{\underset{||}{C}}}-Br \quad \xrightarrow[\substack{AlBr_3/CH_2Cl_2 \\ -10° \text{ to } 0°, \ 2 \text{ hr}}]{CH_2=CH_2} \quad CH_3-CH_2-\underset{O}{\overset{O}{\underset{||}{C}}}-CH_2-CH_2-Br$$

177

C. Vinyl Ether Lithium

An elegant way of circumventing the mixed aldol reaction has become feasible through the discovery of useful synthetic equivalents of the acyl anion [93]. Thus, α-methoxyvinyllithium (MVL), (*178*) and related species are readily prepared acyl anion equivalents of considerable synthetic value. In a typical procedure [94] t-butyllithium (100 mmol, 62.5 ml, 1.6 *M* in n-pentane) was added dropwise to a solution of methyl vinyl ether (9.2 g, 160 mmol) in dry tetrahydrofuran at -65°C under nitrogen. The cooling bath was removed, and the solution became colorless between -5° and 0°C. This solution, which contains a quantitative amount of MVL (*178*), was cooled to -65°C for most reactions. Addition of an electrophilic species such as cyclohexanone (*179*) in tetrahydrofuran at -65° gave the vinyl ether addition product *180* in 90% yield after quenching with a 20% aqueous ammonium chloride solution. Hydrolysis with aqueous methanolic 0.02 *N* HCl at 20°C for 30 min gave the α-hydroxy carbonyl compound *181* in excellent yield. It should be remembered that aldol reactions give β-hydroxy carbonyl intermediates. Although not described [94], dehydration of the α-ketol *181* should give tetrahydroacetophenone (*173*). The preparation of this compound by another technique has already been described (Sec. V.B).

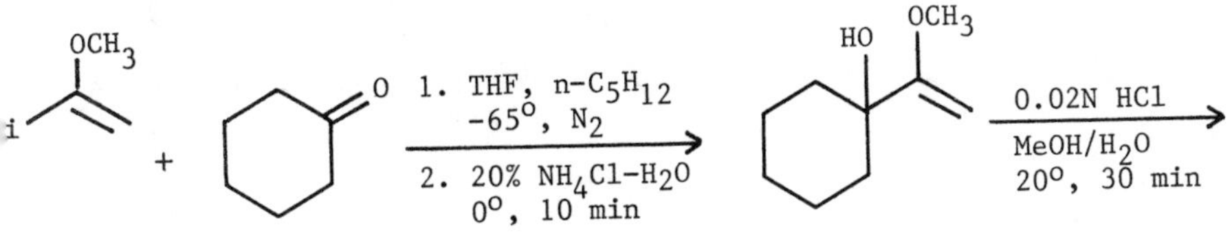

D. Lithio Dithianes

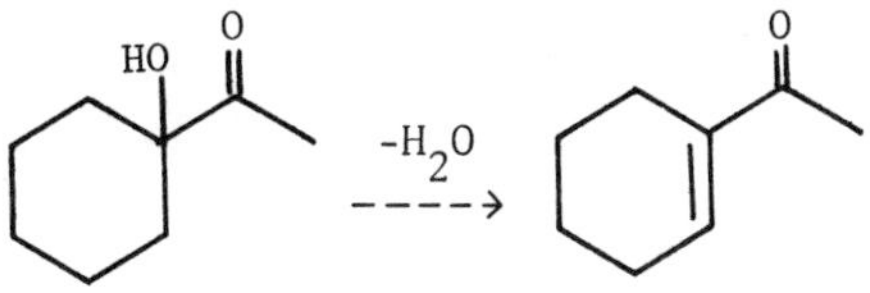

Another synthetic equivalent of the acyl anion can be obtained from dithianes by the action of n-butyl lithium [95]. A 0.2-0.5 M in THF solution of the dithiane *183* at -30°C under nitrogen is stirred and treated with 1 eq of 1.5 M n-butyllithium in hexane at the rate of 3-5 ml/min. The mixture is stirred at -30° to -20°C for 1.5 hr to give a clear, colorless solution of the lithio dithiane *184*. The dithiane *183* in turn may be obtained from acetaldehyde (*1*) and 1,3-dimercaptopropane (*182*) in chloroform in the presence of BF_3-etherate at 5°C for 15 hr [96].

The reaction of the lithio dithiane *184* with styrene oxide (*185*) in tetrahydrofuran at -78°C for 1 hr, then at 0°C for 24 hr, followed by 20°C for 24 hr [96] gave the β-hydroxy mercaptal *186* in 70% yield [95]. This aldol type of compound yields benzylideneacetone (*187*) on hydrolysis and dehydration. Hydrolytic cleavage of dithianes may readily be executed with the help of mercuri-salts [96]. This particular example is an alternate for the mixed aldol reaction of benzaldehyde (*28*) and acetone (*8*).

1 + HS(CH$_2$)$_3$SH $\xrightarrow[\text{CHCl}_3]{\text{BF}_3\cdot\text{Et}_2\text{O}}$ [183 dithiane structure with CH$_3$, S, S, H] $\xrightarrow[\text{THF, hexane}]{\text{n-BuLi}} \xrightarrow{-25°, 1.5 \text{ hr}}$ [184 dithiane structure with CH$_3$, S, S, Li]

182   183   184

184 + CH$_2$—CH-Ph (epoxide with O) $\xrightarrow{\text{THF}}$ [186 dithiane structure with CH$_3$, S, S, CH$_2$-CH-Ph, OH] $\xrightarrow[\text{reflux}]{\substack{\text{HgCl}_2\text{-HgO} \\ 80\% \text{ MeOH}}}$ CH$_3$-C-CH=CH-Ph (with O)

                                           186   187

E. Dihydro-1,3-oxazines

In 1969 a series of brief reports appeared which outlined a technique for the preparation of aldehydes based upon the dihydro-1,3-oxazine (DHO) ring system [97]. One of its several applications permits one to circumvent certain mixed aldol reactions leading to α,β-unsaturated aldehydes. The scheme may, in effect, be considered a two-carbon homologation of electrophilic carbonyl compounds; i.e., the aldehyde equivalent of the malonic ester synthesis. A somewhat similar concept utilizing the versatile lithiodithiane system has already been discussed (Sec. V.D).

The dihydro-1,3-oxazine starting materials are readily available through the reaction of simple nitriles with 2-methyl-2,4-pentanediol (*188*) in concentrated sulfuric acid at 0°C [98]. Thus, 90.2 g (2.2 mol) of acetonitrile (*189*) was added at 0°-5°C to 400 ml of cold concentrated sulfuric acid dropwise while stirring. After the addition was complete, 236 g (2.0 mol) of the diol *188* was added at a rate that maintained the same temperature (0°-5°C). The mixture was stirred for an additional 1 hr at this temperature and then poured onto 1500 g of crushed ice. Work-up of the reaction mixture gave 183.2 g (65%) of 2,4,4,6-tetramethyldihydro-1,3-oxazine [*190*].

The anion of the DHO *191* can be generated at -78°C in anhydrous tetrahydrofuran using 1.1 eq of n-butyllithium in hexane. The n-butyllithium was added dropwise over a period of 1 hr while stirring at -78°C. Approximately 1 hr after the addition was complete, a yellow precipitate formed, which was indicative of complete anion formation.

The reaction of this lithio DHO, *191*, with a variety of electro-philic carbonyl compounds at -78°C produced β-hydroxy DHO deriva-tives analogous to aldol type of products. Thus, 0.11 mol of acetone (*8*) in 25 ml of anhydrous tetrahydrofuran was added slowly to 0.1 mol of *191* while stirring at -78°C over a period of 30 min. The reaction mixture was allowed to warm slowly to room temperature, at which time the yellow precipitate disappeared. The mixture was then poured into 100 ml icewater and acidified with 9 *N* HCl to pH 2-3. Extraction with pentane removed impurities. The aqueous system was cooled with ice and made basic by the careful addition of a 40% NaOH solution. Work-up with ether gave the crude β-hydroxy DHO derivative *192*, which was used without further purification for the reduction to the tetrahydro derivative *193*.

The crude β-hydroxy DHO derivative *192* was reduced in a mixture of 100 ml of THF and 100 ml of 95% ethyl alcohol. The mixture was cooled to about -40°C, and 9 *N* HCl was added to the magnetically stirred solution until approximately pH 7 was obtained. A sodium borohydride solution (3.78 g, 0.1 mol in 4-5 ml of water and 1 drop of 40% NaOH) and a 9 *N* HCl solution were introduced alternately so that pH 6-8 was maintained, and temperature was controlled to about -40°C with cooling. After the borohydride addition was com-plete, the solution was stirred for an additional 1 hr with the occasional addition of 9 *N* HCl to maintain pH 7. It was then poured into 100 ml of water and made basic with 40% NaOH. Extraction with ether gave the crude tetrahydrooxazine (THO) *193*, which was used without further purification for the hydrolysis.

The crude THO derivative (0.1 mol) was added to 50.4 g (0.4 mol) of oxalic acid dihydrate in 150 ml of water and heated to

reflux for 2 hours. Work-up by extraction gave 3-methyl-
crotonaldehyde (194) in 50% overall yield [99]. This example of
DHO technique circumvents the mixed aldol reaction of acetone with
acetaldehyde.

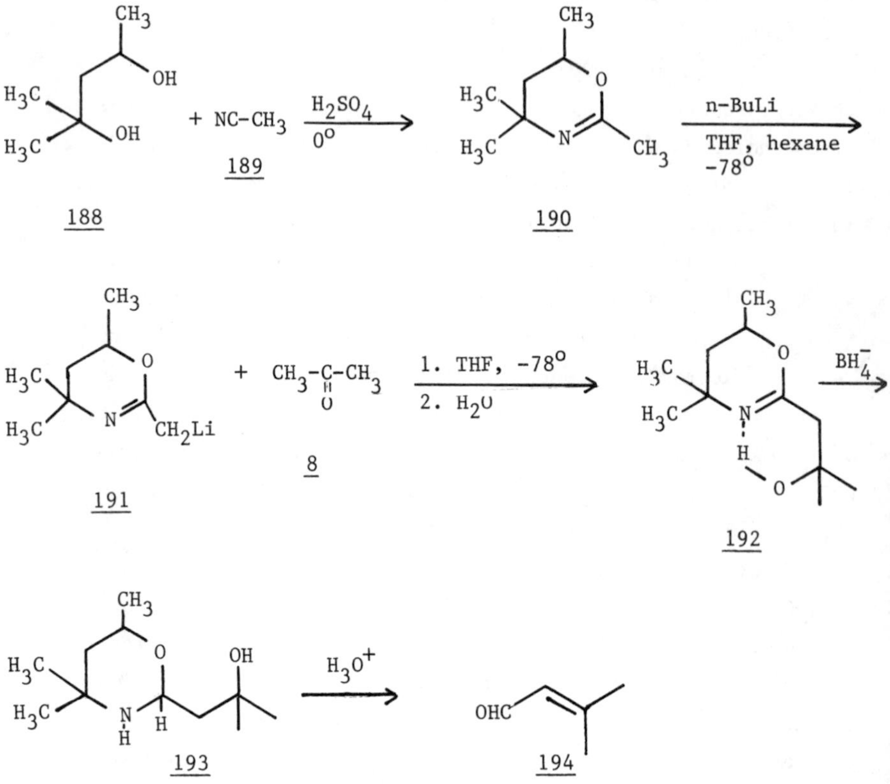

F. Ethynyl Lithium

Besides the vinyl ether lithium reagent (Sec. V.C), the addition of
ethynyl lithium to carbonyl compounds (e.g., ketones) to effect a
two-carbon homologation can be used as an alternative to the mixed
aldol reaction. The initially formed acetylenic carbinol addition
products may either be rearranged to α,β-unsaturated aldehydes or to

α,β-unsaturated ketones [100]. These reactions have been reinvesti-
gated [101] in connection with the synthesis of the unsaturated
ketone *173* and the unsaturated aldehyde *197*.

Metallic sodium (11.5 g) was added slowly with stirring to 500
ml of liquid ammonia which had been perfused with acetylene. Following the addition of sodium, 49 g of cyclohexanone (*179*) was added
slowly through a dropping funnel. The liquid ammonia was allowed
to evaporate and the resulting solid was hydrolyzed with water.
Work-up with ether gave 25 g (40%) of the α-ethynyl carbinol *195*.
This was refluxed with an excess of formic acid for periods of 0.25
to 8 hr. The three major components of the reaction mixture were
detected and isolated by gas-liquid partition chromatography.

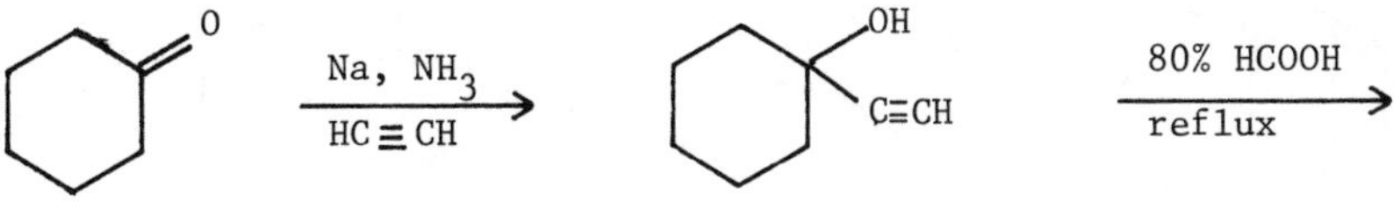

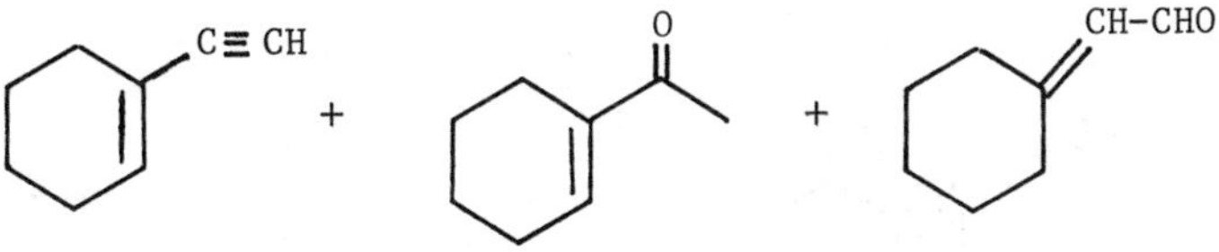

After 0.25 hr of reflux (14% conversion), the major reaction
product is the α,β-unsaturated aldehyde *197*. This was formed in
75% yield via the Meyer-Schuster type of mechanism [100]. The
amount of *196* and *173* has been determined to be 17% and 9%,
respectively, at the same time.

If the refluxing were continued, the Rupe mechanism [100]
took over, and at 8 hr (95% conversion) the product ratio was 1%
of *196*, 96% of *173*, and 1% of *197*. The Rupe mechanism, thus,

clearly favors the formation of the α,β-unsaturated ketone *173*, while the Meyer-Schuster mechanism produces the α,β-unsaturated aldehyde *197*.

As an alternative to the problematic mixed aldol reaction of the keto aldehyde *198* to the α,β-unsaturated methyl ketone *199*, the α-ethynyl carbinol *201* has been prepared from the tetralone *200* with lithium acetylide in liquid ammonia. This was converted in a modified Rupe rearrangement in refluxing methanol-water with a mercurated cation exchange resin to give the α,β-unsaturated methyl ketone *199* in 40% yield [102].

198 199

200 201

$$\underline{201} \xrightarrow[\text{MeOH, } H_2O \text{ reflux, 72 hr}]{\text{Dowex 50W-X8/}Hg^{\cdot\cdot}} \underline{199}$$

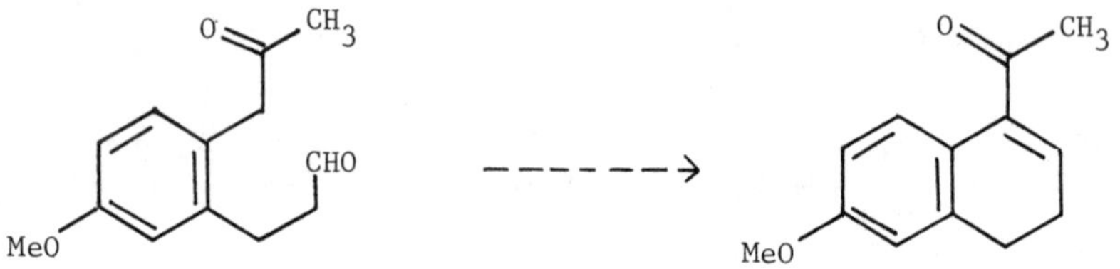

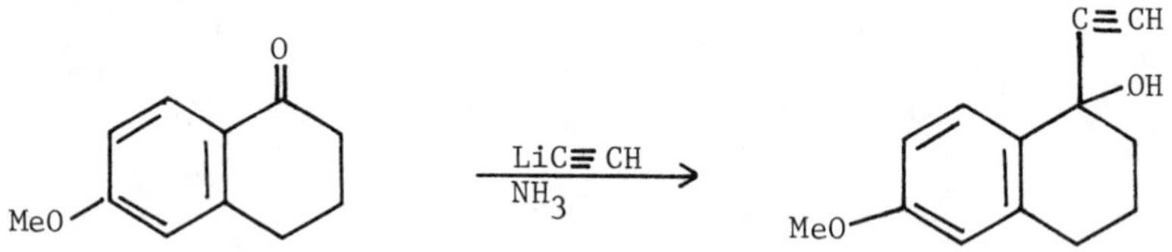

G. *Sulfenylation-[2,3]-Sigmatropic Rearrangement*

Reaction of carbonyl compounds with vinyllithium reagents or addition of organolithium reagents to enones produces allylic alcohols. Treatment with benzenesulfenyl chloride produces allylic sulfoxides via a [2,3]-sigmatropic rearrangement. Addition of the allylic sulfoxide to 2 eq of lithium diethylamide in THF at -78°C generates the anion which is inversely quenched by addition to 1 eq of diphenyl disulfide in THF at 0°C. The initial product of sulfenylation undergoes in situ rearrangement and desulfenylation to yield the hydroxy enol thioether directly. Hydrolysis of this with 1-3 eq of mercuric chloride liberates the α,β-unsaturated carbonyl compound in 44-80% overall yields [103].

This technique has been used as an alternative to the mixed aldol condensation of estrone methyl ether (*202*) with propionaldehyde (*203*) to give the α,β-unsaturated aldehyde *204*. The allylic sulfoxide *208* has been obtained in 80% overall yield from estrone methyl ether (*202*); sulfenylation followed by rearrangement and desulfenylation gave the vinyl sulfide *211* in 81% yield and high isomer purity [nuclear magnetic resonance (nmr)]. Nevertheless, hydrolysis led to a 1:1.2 E-Z mixture of the unsaturated aldehyde *204*. The major isomer could be assigned the E stereochemistry based upon the higher field position of its angular methyl signal in the proton nmr spectrum.

VI. RELATED REACTIONS

A number of addition-elimination reactions mechanistically related to the aldol reaction are known. These constitute an important group of organic chemical reaction techniques and involve the reaction of aldehyde or ketone acceptors with enolate donors of active methylene compounds other than aldehydes or ketones. Unlike the aldol reaction itself, these have been designated by the names of the investigators who discovered or developed them (Sec. VI.A-F).

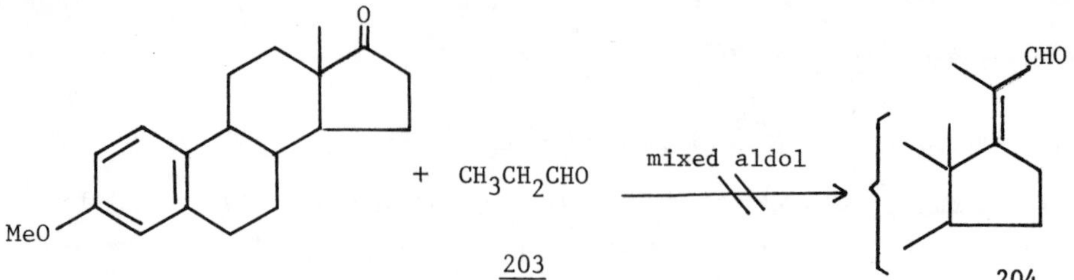

202 + CH₃CH₂CHO $\xrightarrow[\text{mixed aldol}]{//}$ 204

202 + CH₂=C-Li (205) with CH₃ $\xrightarrow[\text{1. THF, -78°, 4 hr}\ \text{2. 20°, 1 hr}\ \text{3. 2N HOAc-H}_2\text{O}]{}$ 206 $\xrightarrow[\text{1. n-BuLi THF, hexane, -78°}\ \text{2. PhSCl}]{}$

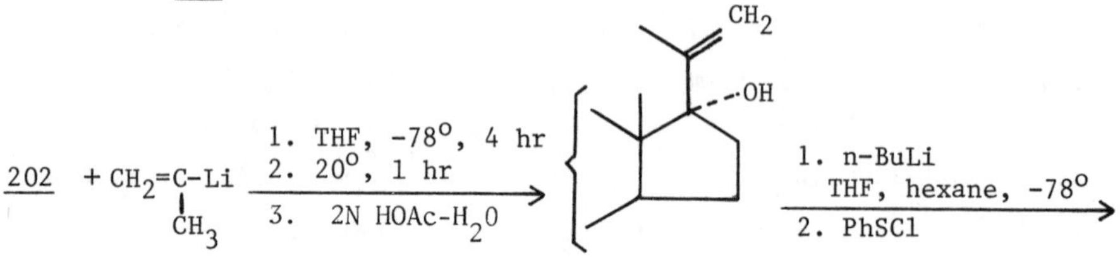

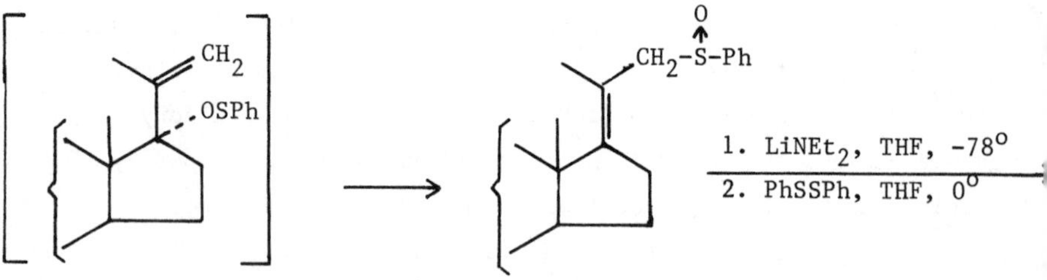

207 $\longrightarrow$ 208 $\xrightarrow[\text{1. LiNEt}_2\text{, THF, -78°}\ \text{2. PhSSPh, THF, 0°}]{}$

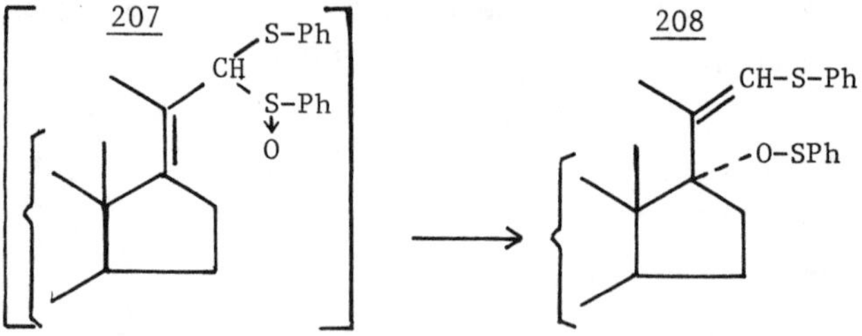

209 $\longrightarrow$ 210

211 $\xrightarrow[\text{HgCl}_2\ \text{CH}_3\text{CN-H}_2\text{O}]{}$ 204, E:Z = 1:1.2

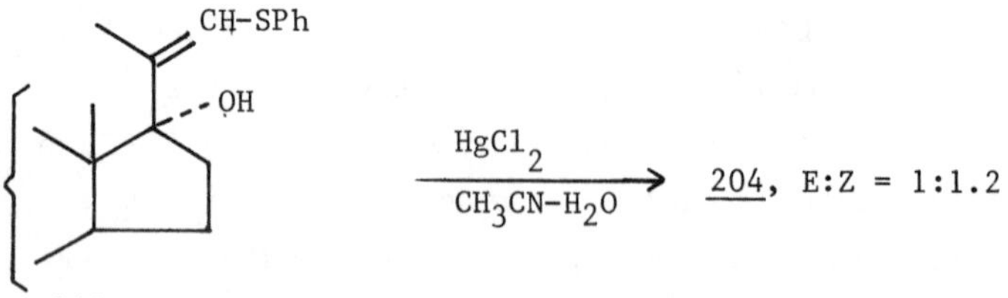

Since, however, almost all of them had been the subject of extensive monographs, only a limited discussion appears justified at this point.

A. Perkin Reaction

In 1868 W. H. Perkin described a synthesis of coumarin (the lactone of o-hydroxy-cis-cinnamic acid) by heating the sodium salt of salicyclaldehyde with acetic anhydride [104]. Further study of the reaction furnished a new general method for the preparation of cinnamic acid (216) and its analogues. It consists of the aldol type of reaction of an aromatic or α,β-unsaturated aldehyde, which cannot undergo self-condensation, with an aliphatic acid anhydride as the active methylene donor compound in the presence of a basic catalyst, e.g., the alkali salt of the corresponding acid, or triethylamine.

Anhydrides are relatively unreactive methylene compounds and require long treatment at relatively high temperatures (150°-200°C) even with the highly electrophilic aromatic aldehydes as reaction partners. Cinnamic acid (216) may be obtained in 55-60% yield by heating a mixture of 21 g (0.2 mol) of benzaldehyde (28), 30 g (0.3 mol) of 95% acetic anhydride (212), and 12 g (0.12 mol) of freshly fused potassium acetate for 5 hr at 170°-175°C using an air-cooled condenser [104]. This permits the acetic acid to escape; the equilibrium is thereby favorably shifted in the direction of cinnamic acid (216) formation. The reaction mechanism is essentially identical to other aldol-type reactions. It proceeds through the β-hydroxy-carbonyl addition product 214, which readily hydrolyzes to 215, and then dehydrates to the α,β-unsaturated carbonyl condensation product 216.

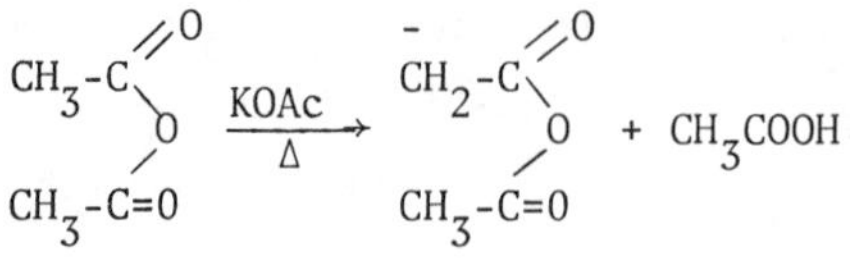

$$CH_3-C\overset{O}{\underset{O}{\diagdown}} \xrightarrow[\Delta]{KOAc} \overset{-}{CH_2}-C\overset{O}{\underset{O}{\diagdown}} + CH_3COOH$$
$$CH_3-C=O \qquad\qquad CH_3-C=O$$

212   212-anion   213

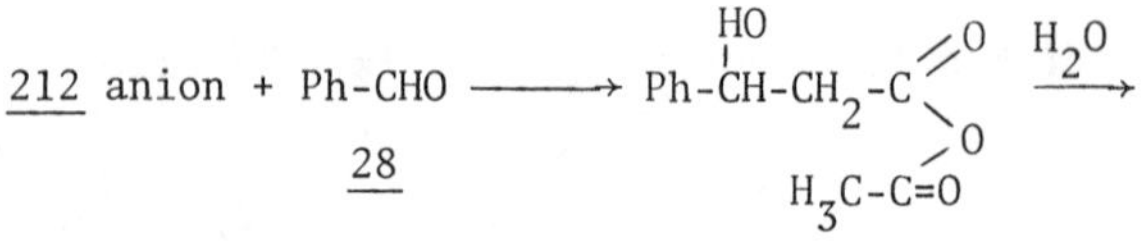

$$\underline{212}\text{ anion } + \text{ Ph-CHO} \longrightarrow \text{Ph-}\overset{HO}{\underset{|}{CH}}\text{-CH}_2\text{-C}\overset{O}{\underset{O}{\diagdown}} \xrightarrow{H_2O}$$
$$\underline{28} \qquad\qquad\qquad\qquad\qquad H_3C-C=O$$

214

$$\underset{215}{\overset{OH}{\underset{|}{\text{Ph-CH-CH}_2\text{-COOH}}}} + \underline{213} \xrightarrow{-H_2O} \underset{216}{\text{Ph-CH=CH-COOH}}$$

B. Claisen Reaction

This is a modification of the Perkin reaction for the preparation
of cinnamic acid derivatives. In it an aldehyde that bears no α-
hydrogen atom is reacted with a ketone or an ester in the presence
of metallic sodium and a trace of ethyl alcohol. An illustration
is the synthesis of cinnamic acid ethyl ester (218) from benzaldehyde
(28) and ethyl acetate (217) [104]. An improved yield of 68-74%
reflects the advantage of this modification over the original Perkin
reaction.

$$\text{Ph-CHO} + CH_3\text{-COOEt} \xrightarrow[2\text{ hr}]{NaOEt,\ 0^O C} \text{PhCH=CHCOOEt} + H_2O$$

$\underline{28}$ $\underline{217}$ $\underline{218}$

It should be pointed out that in this case the ester 217
serves as the active methylene donor compound, while in the Claisen
condensation a properly chosen aldehyde plays the same role in the
reaction.

C. Knoevenagel Reaction and Doebner Modification

The Knoevenagel reaction is also an alternative to the Perkin reaction. It consists of the reaction of an active methylene compound other than an aldehyde or ketone with electrophilic carbonyl compounds (aldehydes or ketones) in the presence of primary, secondary, and tertiary amines (as the free bases or their salts) or weakly basic amine types of ion exchange resins. Piperidine and diethylamine are particularly effective catalysts. The Doebner modification involves the use of piperidine as the catalyst in pyridine as the reaction medium. Besides amine salts, Lewis acids such as BF_3, $AlCl_3$ and $TiCl_4$ have been used as catalysts. Potassium fluoride has also been used as a catalyst in the Knoevenagel reaction [105].

The nucleophilic donor compounds of the reaction usually have relatively acidic methylene groups, activated by two electron-withdrawing substituents, e.g., malonic, cyanoacetic, or acetoacetic esters. The use of these activated species permits the reaction to proceed under relatively mild conditions. Aliphatic aldehyde acceptors can, therefore, be used, since the reaction conditions avoid their self-condensation.

The Knoevenagel reaction is similar in principle to the aldol reaction, having the same two-step addition-elimination mechanism. The initially formed β-hydroxy carbonyl adducts are usually dehydrated to the α,β-unsaturated carbonyl derivatives under the applied reaction conditions. Although the reactions may be carried out at room temperature, they are normally performed in refluxing benzene with the continuous separation of water as it is formed. Benzylidene malonate (223) has been prepared in 89-91% yield utilizing this technique [106] in the piperidinium benzoate-catalyzed Knoevenagel reaction of benzaldehyde (28) with the anion of diethyl malonate (221). It is probable, that the initial stage of the reaction involves the formation of an imine or an imminium salt, 220, from the aldehyde and the amine. Subsequent reaction of 220 with 221

gave the β-amino carbonyl derivative *222*, which in turn furnished
the α,β-unsaturated carbonyl product *223* by elimination of piperidine
(*219*).

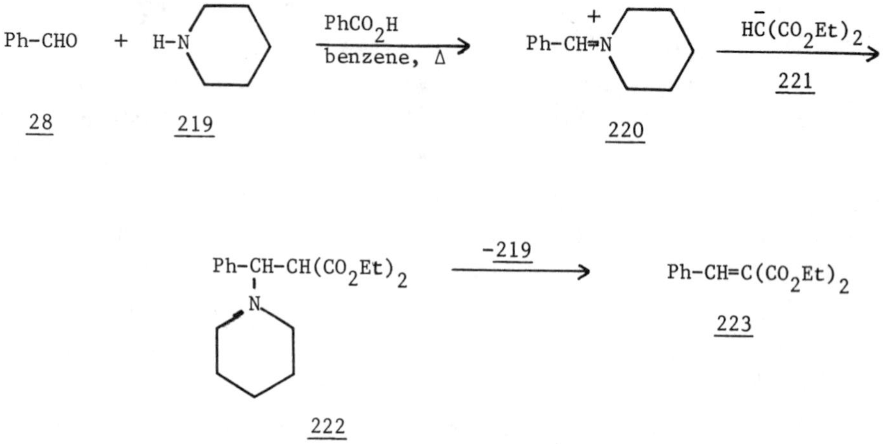

$$Ph-CHO \quad + \quad H-N\diagup\bigcirc \quad \xrightarrow[\text{benzene, } \Delta]{PhCO_2H} \quad Ph-CH{=}\overset{+}{N}\diagup\bigcirc \quad \xrightarrow{\overset{-}{HC}(CO_2Et)_2}$$

<u>28</u>   <u>219</u>   <u>220</u>   <u>221</u>

$$Ph-CH-CH(CO_2Et)_2 \quad \xrightarrow{-219} \quad Ph-CH{=}C(CO_2Et)_2$$

<u>222</u>   <u>223</u>

D. Stobbe Reaction

In this reaction [107] the anion of a succinic ester, *225*, acting
as an active methylene donor is added to an aldehyde or to a ketone,
224, in the presence of strong base, such as sodium hydride or
potassium t-butoxide in an aldol type of reaction. The first adduct
226 cyclizes to a five-membered γ-lactone, *227*, involving the
alkoxide generated in the addition reaction and the more remote
ester carbonyl group. Base-catalyzed opening of the γ-lactone *227*
yields the Stobbe reaction product, or arylidene (or alkylidene)
succinic acid half-ester, *228*. This involves the formation of a
carboxylate anion, *228*, sodium salt, and it is therefore an
essentially irreversible step allowing the reaction to be successfully applied to even relatively hindered ketones.

A solution of 8.56 g (0.04 mol) of 2-propionyl-6-methoxy-
napthalene (*224*) and 17.52 g (0.12 mol) of dimethyl succinate (*225*)
was heated at $50°-58°C$ in 60 ml of benzene with 2.4 g (0.10 mol)
of sodium hydride and a few drops of methanol. After the gas

evolution had stopped, the mixture was acidified with acetic acid
and diluted with ether. The half-ester *228* was extracted from the
organic solution with 5% sodium bicarbonate solution. This was then
acidified and extracted with ether. Work-up as usual gave a quantitative yield of the crude, oily half-ester *228*[108].

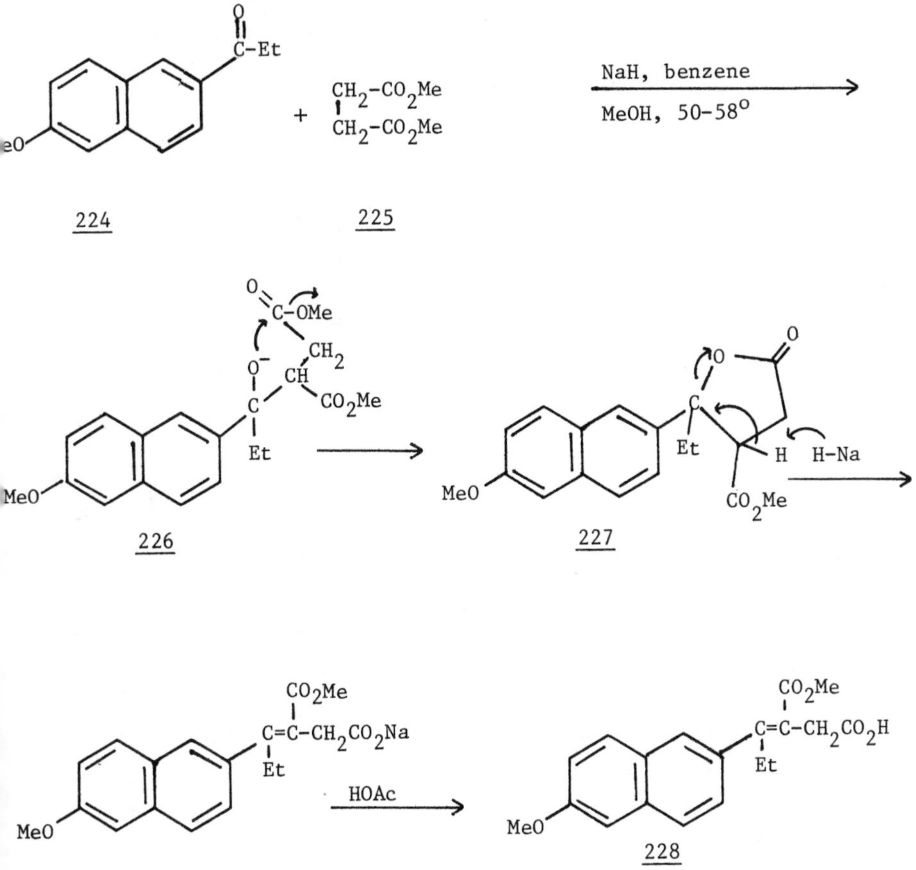

224   225

226   227

228

The half-ester type of Stobbe products *228* are important synthetic intermediates, which may be decarboxylated to propionic acid
derivatives when heated with a mixture of 48% hydrobromic acid and
glacial acetic acid [109]. Either these propionic acid derivatives
or the half-esters themselves may be used for the preparation of
cyclic ketones.

Hydrogenation of the crude oily mixture of the half-esters *228* in methanol in the presence of platinum oxide at room temperature and an initial pressure of 41 psi gave the saturated crystalline half-ester, *229* in 52% overall yield based on *224*. When a benzene solution of *229* was treated for 45 min in the cold with phosphorous pentachloride and then for 10 min at $0^{\circ}C$ with stannic chloride, cyclization occurred affording the crystalline tricyclic keto ester *230* in 67% yield [108].

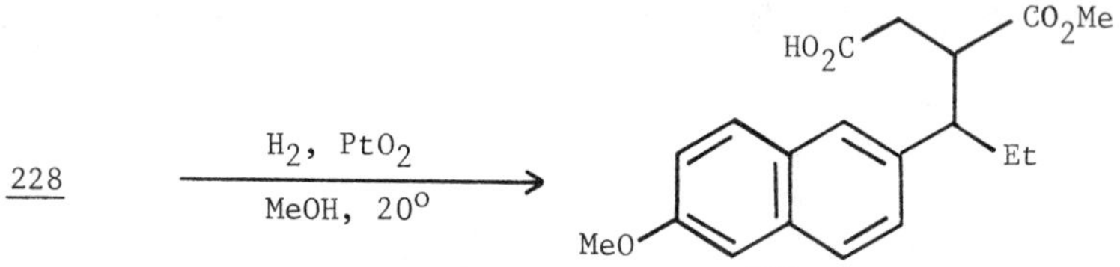

228 $\quad\xrightarrow[\text{MeOH, }20^{\circ}]{\text{H}_2,\ \text{PtO}_2}\quad$ 229

229 $\quad\xrightarrow[\substack{2.\ \text{SnCl}_4 \\ 0^{\circ},\ 10\ \text{min}}]{\substack{1.\ \text{PCl}_5,\ \text{benzene} \\ 0^{\circ},\ 45\ \text{min}}}\quad$ 230

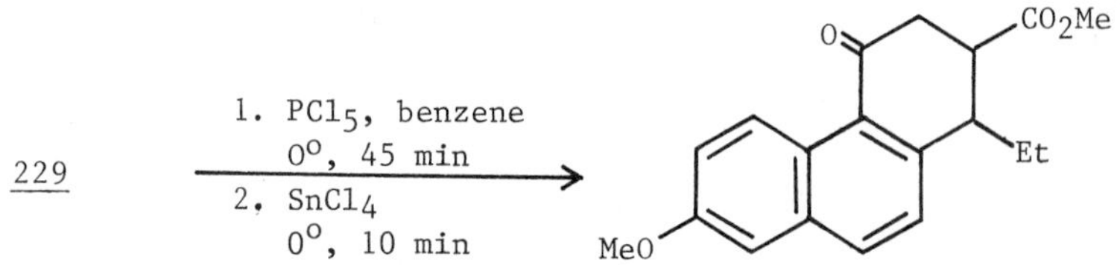

The Stobbe reaction has been extended to glutaric esters. Excellent yields of hydroxy diesters have been obtained using a wide variety of ketone acceptors with the anion of di-t-butyl glutarate formed with lithium amide in liquid ammonia and ether. The success of the reaction was dependent upon the cation employed, as only low yields of desired products resulted when sodium or potassium salts were used. The method failed altogether when attempts were made to use the diethyl rather than the di-t-butyl ester of glutaric acid, or to extend it to include di-t-butyl adipate [110].

Stobbe reactions have been run successfully on γ- and δ-keto esters to give the expected products, while with β-keto esters only cleavage products of the β-keto ester were formed [111]. Attempted condensation of α-keto esters with di-t-butyl succinate in the presence of potassium t-butoxide or sodium hydride gave only the self-condensation products of the α-keto ester, while almost the entire amount of the succinate was recovered [112].

E. Darzens Glycidic Ester Synthesis

Although the first synthesis of a glycidic ester was performed by Erlenmeyer in 1892, it remained for Darzens to develop and generalize the reaction named after him [113]. It involves the aldol type of addition of the anion of an α-halo ester, 232 to a ketone, 231, or to an aldehyde (perferably aromatic) in the presence of a strong base such as potassium t-butoxide or sodium amide. An intermediate aldol type of product, the half alkoxide 233, undergoes intramolecular ring closure to the α,β-epoxy ester 234, called a glycidic ester. The predominant geometrical isomer produced is frequently the one in which the ester function is trans to the larger group at the beta carbon atom. The reaction has been utilized in a synthesis of (+)-yohimbane [114]. To a mixture of 630 g (3.42 mol) of 2-(carboethoxymethyl)cyclohexanone (231) and 420 g (3.43 mol) of ethyl chloroacetate (232) was added a solution of 384 g (3.43 mol) of potassium t-butoxide in 3.2 l of t-butyl alcohol over a period of 2 hr while at 0°-15°C, and then at 20°C for 15 hr. Work-up as usual with ether followed by distillation of the extract gave 572 g (61%) of the glycidic ester 234.

The glycidic esters are of interest primarily because they can be converted into aldehydes and ketones having more carbon atoms than those employed as starting materials. Careful saponification of the glycidic ester 234 with 2.6 N sodium ethoxide in ethyl alcohol containing 6% of water at 0°-15°C for 1 hr, and at 20°C for 15 hr gave the glycidic acid sodio salt 235 in 80-85% yield.

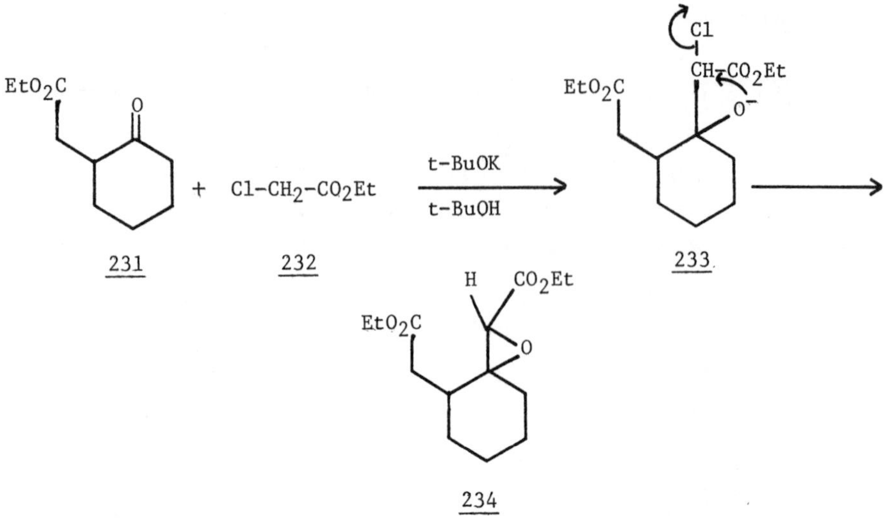

$$\underset{231}{\text{EtO}_2\text{C}\text{—cyclohexanone}} \quad + \quad \underset{232}{\text{Cl–CH}_2\text{–CO}_2\text{Et}} \quad \xrightarrow[\text{t–BuOH}]{\text{t–BuOK}} \quad \underset{233}{\text{EtO}_2\text{C}} \quad \longrightarrow$$

234

The salt was dissolved in water; the solution was cooled to $10°-15°$C and carefully acidified with hydrochloric acid. Extraction with ether gave the thermally unstable glycidic acid *236* in 65% yield. To this was added copper powder (3% by weight), and the mixture was pyrolyzed at $250°$C (100 mm) for 5 hr and the distillate was

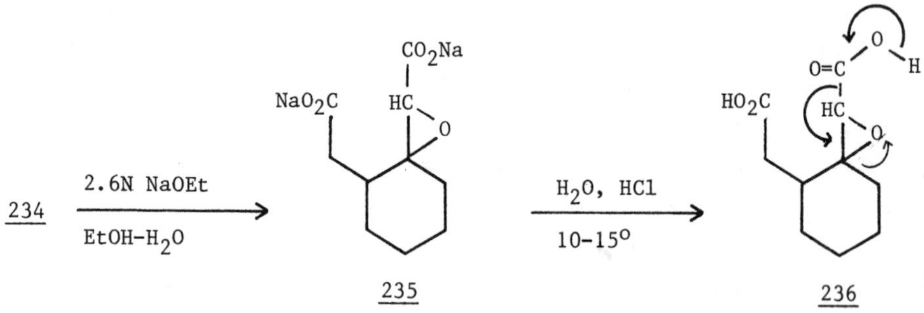

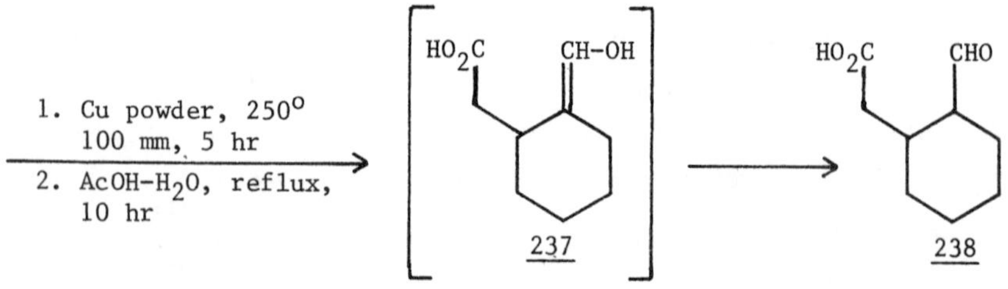

$$\underset{234}{} \xrightarrow[\text{EtOH–H}_2\text{O}]{\text{2.6N NaOEt}} \quad \underset{235}{} \xrightarrow[\text{10–15}°]{\text{H}_2\text{O, HCl}} \quad \underset{236}{}$$

$$\xrightarrow[\substack{\text{2. AcOH–H}_2\text{O, reflux,} \\ \text{10 hr}}]{\substack{\text{1. Cu powder, 250}° \\ \text{100 mm, 5 hr}}} \quad \left[\underset{237}{}\right] \quad \longrightarrow \quad \underset{238}{}$$

refluxed vigorously with a 5% acetic acid solution in water for 10
hr to give the aldehyde *238* in 65% yield via decarboxylation of
236. This was then used for the synthesis of (+)-yohimbane [114].

F. Reformatsky Reaction

In 1887 Reformatsky described an aldol type of reaction, which has
been used extensively in synthetic schemes and was later named for
him [115]. The method involves the generation of an ester enolate
donor by reaction of an α-halo ester, *239*, or an α,β-unsaturated
γ-halo ester with zinc metal in an inert solvent such as benzene
or ether in the presence of a carbonyl acceptor compound such as
an aldehyde or a ketone, *83*. This is a convenient method for the
extension of a carbon skeleton by two carbon atoms giving a β-
hydroxy ester, *240*, after hydrolysis. This is a unique aldol type
of reaction since it permits the isolation of the initial aldol
type of addition product, even when relatively hindered ketones
are employed (Scheme 3). The β-hydroxy ester *240* may, of course,
be dehydrated to the α,β-unsaturated ester *241*. A related reaction
yielding α,β-unsaturated ketones has already been discussed (Sec.
III.B.2.d).

$$\underset{83}{\overset{R_1}{\underset{R_2}{>}}C=O} + \underset{239}{Hal-CH_2-CO_2R} \quad \overset{1.\ Zn}{\underset{2.\ H_3O^+}{\longrightarrow}} \quad \underset{240}{\overset{R_1\ OH}{\underset{R_2}{>}}C-CH_2-CO_2R} \longrightarrow \underset{241}{\overset{R_1\ H}{\underset{R_2}{>}}C=C-CO_2R}$$

Scheme 3

The reported yields of β-hydroxy esters, *240*, from the original
version of the Reformatsky technique vary from less than 10% to over
90% with an average of a little under 60%. It appears that lower
yields are commonly due to production of β-keto ester side products
by attack of the Reformatsky reagent either on itself or on the
α-halo ester starting material *239*.

To avoid this and other side reactions α-halo t-butyl esters have been prepared and subjected to Reformatsky conditions in the presence of iodine activated zinc [116]. These esters are far less susceptible to nucleophilic attack than the more commonly used ethyl esters. [The advantageous use of t-butyl glutarate has been previously discussed in connection with the modified Stobbe reaction (Sec. VI.D).]

Another technique which decreased the possibility of side reactions consists of conducting the Reformatsky reaction in the presence of trimethyl borate, which provides a mildly acidic medium, yet allows the reaction to proceed in an undisturbed fashion. Since it neutralizes the basic alkoxide products, it minimizes the base-catalyzed side reactions responsible for the low yields. It also allows the reaction to be carried out at room temperature rather than at reflux temperatures in benzene or benzene-ether solvents [117]. Reaction of acetaldehyde (1) with 1 eq of zinc and ethyl bromoacetate (239) in a 1:1 mixture of trimethyl borate and tetrahydrofuran at 25°C for 2 hr gave ethyl β-hydroxy butyrate (240) in 95% yield (83 = 1, Hal = Br, R = Et, R_1 = CH_3, R_2 = H in Scheme 3).

It should be pointed out for comparison that when the same reaction was conducted without trimethyl borate in refluxing benzene, only a 22% yield of 240 was obtained, while at 25°C for 4 hr a 65% yield of the same compound was realized [117].

Another procedure [118(a) and (b)] superior to the original Reformatsky method consists of the preparation of lithio ethyl acetate (217-Li) by the addition of ethyl acetate (217) to a solution of lithium bis(trimethylsilyl) amide (242-Li) in tetrahydrofuran at -78°C. At the same low temperature 217-Li reacts almost instantly with aldehydes or ketones, 83, to give, after hydrolysis, excellent yields (80-93%) of the corresponding β-hydroxy esters 240[118(a)].

$$H_3C-CO_2Et + LiN\begin{array}{c} SiMe_3 \\ \diagup \\ \diagdown \\ SiMe_3 \end{array} \xrightarrow{THF, -78°C} LiCH_2-CO_2Et + \underline{242}$$

$$\underline{217} \qquad \underline{242}\text{-Li} \qquad \underline{217}\text{-Li}$$

$$\underline{217}\text{-Li} + \underline{83} \xrightarrow{THF, -78°C} R_1-\overset{\overset{\displaystyle OLi}{|}}{\underset{\underset{\displaystyle R_2}{|}}{C}}-CH_2-CO_2Et \xrightarrow{H_2O} R_1-\overset{\overset{\displaystyle OH}{|}}{\underset{\underset{\displaystyle R_2}{|}}{C}}-CH_2-CO_2Et$$

$$\underline{240}$$

VII. RETROALDOL REACTIONS

A discussion of the aldol and related reactions should include the important retrograde technique, the retroaldol reaction. This is feasible, because of the reversible nature of the aldol reaction at the stage of aldol addition as well as condensation through a series of equilibrium reactions (Sec. II.A). Retroaldol reactions are, thus, common with β-hydroxy carbonyl compounds and with α,β-unsaturated carbonyl derivatives.

A previously discussed (Sec. III.A.2) example is the formation of acetone (8) from the β-hydroxy ketone 9, which furnishes 98% of the ketone 8, at equilibrium in the presence of barium hydroxide catalyst.

With α,β-unsaturated carbonyl compounds it is assumed that an alkoxide of the corresponding β-hydroxy carbonyl compound is the intermediate for the retroaldol reaction. For example, the α,β-unsaturated aldehyde, citral (12) undergoes Michael addition to give the intermediate alkoxide, 242b, of the β-hydroxy aldehyde, followed by retroaldol reaction to furnish the ketone 44 in 96% yield. Removal of the volatile acetaldehyde (1) shifts the equilibrium in the direction of the retroaldol reaction [119]. The reaction is indeed the reversal of the already discussed (Sec. III.B.2.b) directed aldol reaction of acetaldol (1) to citral (12).

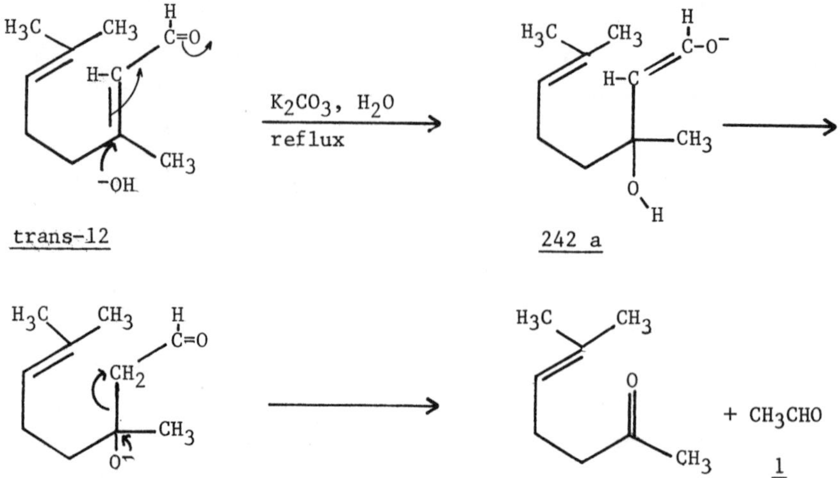

trans-12   242 a

242 b   44   + CH$_3$CHO
                                                             1

This example demonstrates the use of the retroaldol reaction
in the systematic degradation of natural products. The reaction,
however, has also gained prominence as a synthetic pathway enabling
less stable systems to rearrange to more stable ones (Sec. IV.A.4).
A synthesis of cyclohexenones with a β-propionic ester group at
the 4-position, *245*, has been achieved via the dropwise addition
of methyl 4-hydroxy-1-methylbicyclo [2.2.2]oct-2-ene-5-carboxylate
(*243*) to a stirred suspension of sodium hydride in dry ether at
room temperature to give the retroaldol product *245* in 82% yield
[120]. The ring opening is assumed to proceed via *244* by the retroaldol reaction mechanism.

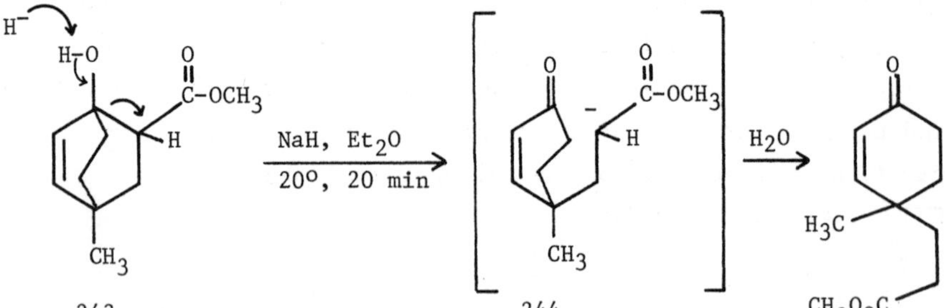

243   244   245

It should be pointed out that the retroaldol reaction may also be executed thermally [121]. The pyrolysis of β-hydroxy ketones gave a mixture of aldehydes and/or ketones in a decomposition that is the reverse of the aldol reaction.

9    $\xrightarrow{\Delta}$

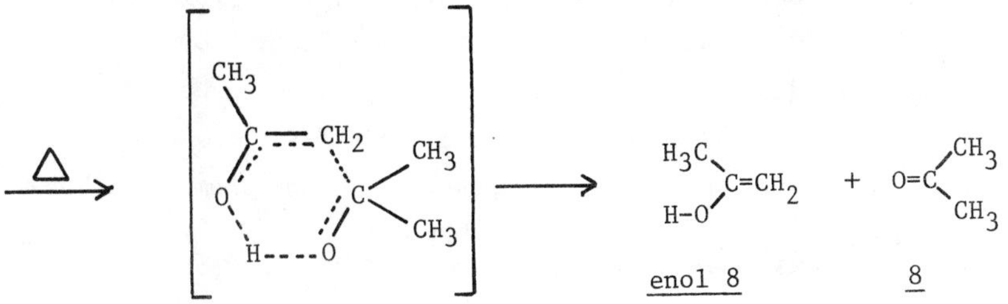

$$\left[\begin{array}{c} \underset{\text{H}----\text{O}}{\overset{\text{CH}_3}{\underset{\text{O}}{\text{C}}}} \cdots \text{CH}_2 \underset{\text{CH}_3}{\overset{\text{CH}_3}{\text{C}}} \end{array} \right] \longrightarrow \underset{\text{H}-\text{O}}{\overset{\text{H}_3\text{C}}{}} C=CH_2 \; + \; O=C\underset{\text{CH}_3}{\overset{\text{CH}_3}{}}$$

enol 8   8

ACKNOWLEDGMENT

Special thanks are due Dr. Seymour D. Levine of Ortho Pharmaceutical Corporation for many helpful suggestions and corrections concerning the manuscript and to Mrs. Mary Westbrook for her meticulous typing of the same.

REFERENCES

1. E. Fischer and J. Tafel, Ber., 20, 1088 (1887).

2. (a) H. O. L. Fischer and E. Baer, Helv. Chim. Acta, 19, 519 (1936); (b) W. D. Walters and K. F. Bonhoeffer, Z. Phys. Chem., A182, 265 (1938).

3. A. T. Nielsen and W. J. Houlihan, Org. React., 16, 1 (1968).

4. H. Reiff, Newer Methods Prep. Org. Chem., 6, 48 (1971).

5. Q. Bayer, in Houben-Weyl's Methoden der Organischen Chemie (E. Müller, ed.), Vol. VII, Part 1, G. Thieme, Stuttgart, 1954, p. 75.

6. L. Claisen, Ann., 306, 323 (1899).

7. V. Grignard and J. Reif, Bull. Soc. Chim. Fr. 1(4), 114 (1907).

8. H. Hammarsten, Ann., 420, 273 (1920).

9. L. P. Kyriakides, J. Amer. Chem. Soc., 36, 530 (1914).

10. E. Späth and H. Schmid, Ber., 74, 859 (1941).

11. M. Vogel and D. Rhum, J. Org. Chem., 31, 1775 (1966).

12. M. Backès, *Bull. Soc. Chim. Fr.*, *9*(5), 60 (1942).

13. M. Delépine, *Compt. Rend.*, *147*, 1316 (1908).

14. A. Kekulé, *Ann.*, *162*, 77 (1872).

15. M. Häusermann, *Helv. Chim. Acta*, *34*, 1482 (1951).

16. J. B. Conant and N. Tuttle, *Org. Syn.*, *Coll. Vol.*, *1*, 199 (1941).

17. J. B. Conant and N. Tuttle, *Org. Syn.*, *Coll. Vol.*, *1*, 345 (1941).

18. For leading references see Ref. 1 as cited in Ref. 17.

19. N. B. Lorette, *J. Org. Chem.*, *22*, 346 (1957).

20. M. J. Astle and J. A. Zaslowsky, *Ind. Eng. Chem.*, *44*, 2869 (1952).

21. G. V. Austerweil and R. Pallaud, *Bull. Chim. Soc. Fr.*, *20*(5), 678 (1953).

22. A. Russell and R. L. Kenyon, *Org. Syn.*, *Coll. Vol.*, *3*, 747 (1955).

23. (a) B. Tollens and P. Wigand, *Ann.*, *265*, 316 (1891);
    (b) P. Rave and B. Tollens, *Ann.*, *276*, 58 (1893);
    (c) H. Schulz and H. Wagner, *Agnew. Chem.*, *62*, 105 (1950).

24. S. Malinowski, S. Basinski and B. Polenska, *Rocz. Chem.*, *38*, 23 (1964).

25. H. B. J. Schurink, *Org. Syn.*, *Coll. Vol.*, *1*, 425 (1944).

26. L. Wessely, *Monatsh.*, *21*, 216 (1900).

27. J. G. Schmidt, *Ber.*, *13*, 2342 (1880); *14*, 1459 (1881).

28. (a) L. Claisen and A. Claparède, *Ber.*, *14*, 349 (1881);
    (b) L. Claisen, *Ber.*, *20*, 655 (1887).

29. W. S. Johnson, *J. Amer. Chem. Soc.*, *65*, 1317 (1943).

30. (a) W. S. Johnson, B. Bannister, and R. Pappo, *J. Amer. Chem. Soc.*, *78*, 6331 (1956); (b) W. S. Johnson, D. S. Allen, Jr., R. R. Hindersinn, G. N. Sausen, and R. Pappo, *J. Amer. Chem. Soc.*, *84*, 2181 (1962).

31. P. Mastagli, Z. Zafiriadis, G. Durr, A. Floch, and G. Lagrange, *Bull. Chim. Soc. Fr.*, *20*(5), 693 (1953).

32. (a) F. F. Blicke, *Org. React.*, *1*, 303 (1942); (b) B. Reichert, Die Mannich Reaktion, Springer Verlag, Berlin, 1959.

33. A. L. Wilds, R. M. Nowak, and K. E. McCaleb, *Org. Syn.*, *Coll. Vol.*, *4*, 281 (1963).

34. Z. G. Hajos and D. R. Parrish, *J. Org. Chem.*, *38*, 3244 (1973).

35. (a) C. Mannich and M. Bauroth, *Chem. Ber.*, *57*, 1108 (1924);
    (b) R. Robinson, *J. Chem. Soc.*, *111*, 766 (1917); (c) C. Schöpf and G. Lehmann, *Justus Liebigs Ann. Chem.*, *518*, 1 (1935).

36. G. Wittig, H. J. Schmidt, and H. Renner, *Chem. Ber.*, *95*, 2377 (1962).

37. G. Wittig and H. Reiff, *Agnew. Chem. (Int. Ed.)*, *7*, 7 (1968).

38. G. Büchi and H. Wüest, *J. Org. Chem.*, *34*, 1122 (1969).

39. G. Stork and J. D'Angelo, *J. Amer. Chem. Soc.*, *96*, 7114 (1974).

40. For other methods for the kinetic generation of enolates see Refs. 2 and 5 cited in Ref. 39 as well as Chap. 2 in this volume.

41. G. Stork, G. A. Kraus, and G. A. Garcia, *J. Org. Chem.*, *39*, 3459 (1974).

42. M. Gaudemar, *Compt. Rend.*, *Ser. C*, *279*, 961 (1974).

43. H. O. House, D. S. Crumrine, A. Y. Teranishi, and H. D. Olmstead, *J. Amer. Chem. Soc.*, *95*, 3310 (1973).

44. G. Stork and M. Isobe, *J. Amer. Chem. Soc.*, *97*, 4745 (1975).

45. T. A. Spencer, R. W. Britton, and D. S. Watt, *J. Amer. Chem. Soc.*, *89*, 5727 (1967).

46. Halomagnesium enolates have been used in an analogous way: see Refs. 2, 7, 8(b) and (g) cited in Ref. 43.

47. C. Chassin, E. A. Schmidt, and H. M. R. Hoffmann, *J. Amer. Chem. Soc.*, *96*, 606 (1974).

48. F. Effenberger, *Angew. Chem. (Int. Ed.)*, *8*, 295 (1969).

49. M. Müller-Cunradi and K. Pieroh, U. S. Patent 2165962 (1939); *Chem. Abstr.*, *33*, 8210 (1939).

50. R. I. Hoaglin and D. H. Hirsh, *J. Amer. Chem. Soc.*, *71*, 3468 (1949).

51. O. Isler, H. Lindlar, M. Montavon, R. Rüegg, and P. Zeller, *Helv. Chim. Acta*, *39*, 249 (1956).

52. T. Mukaiyama and M. Hayashi, *Chem. Lett.*, 15 (1974).

53. T. Mukaiyama, K. Narasaka, and K. Banno, *Chem. Lett.*, 1011 (1973).

54. T. Mukaiyama, K. Banno, and K. Narasaka, *J. Amer. Chem. Soc.*, *96*, 7503 (1974).

55. T. Mukaiyama, T. Izawa, and K. Saigo, *Chem. Lett.*, 323 (1974).

56. (a). The word annular means ring forming. Transannular is a well accepted designation in the chemical literature and the others will be discussed along with the specific mode of reaction. (b). J. E. Baldwin, *Chem. Commun.*, 734 (1976).

57. E. J. Corey, N. H. Andersen, R. M. Carlson, J. Paust, E. Vedejs, I. Vlattas, and R. E. K. Winter, *J. Amer. Chem. Soc.*, *90*, 3245 (1968).

58. R. A. Ellison, E. R. Luckenbach, and C. Chiu, *Tetrahedron Lett.*, 499 (1975).

59. G. Stork and R. Borch, *J. Amer. Chem. Soc.*, *86*, 936 (1964).

60. W. S. Johnson, M. F. Semmelhack, M. U. S. Sultanbawa, and L. A. Dolak, *J. Amer. Chem. Soc.*, *90*, 2994 (1968).

61. J. Colonge, J. Dreux, and M. Thiers, *Compt. Rend.*, *243*, 1425 (1956).

62. A. G. Anderson and J. A. Nelson, *J. Amer. Chem. Soc.*, *73*, 232 (1951).

63. J. D. Yordi and W. Reusch, *J. Org. Chem.*, *40*, 2086 (1975).

64. R. B. Woodward, F. Sondheimer, D. Taub, K. Heusler, and W. M. McLamore, *J. Amer. Chem. Soc.*, *74*, 4223 (1952).

65. W. S. Johnson, J. J. Korst, R. A. Clement, and J. Dutta, *J. Amer. Chem. Soc.*, *82*, 614 (1960).

66. Z. G. Hajos and D. R. Parrish, *J. Org. Chem.*, *39*, 1612 (1974).

67. E. S. Gould, *Mechanism and Structure in Organic Chemistry*, Holt, Rinehart, and Winston, New York, 1965, p. 319.

68. T. A. Spencer, H. S. Noel, D. C. Ward, and K. L. Williamson, *J. Org. Chem.*, *31*, 434 (1966).

69. Reference 67, p. 379.

70. J. A. West, *J. Chem. Educ.*, *40*, 194 (1963).

71. J. D. Morrison and H. S. Mosher, *Asymmetric Organic Reactions*, Prentice-Hall, Englewood Cliffs, N.J., 1971.

72. Z. G. Hajos and D. R. Parrish, *J. Org. Chem.*, *39*, 1615 (1974).

73. W. Acklin, V. Prelog, and A. P. Prieto, *Helv. Chim. Acta*, *41*, 1416 (1958).

74. U. Eder, G. Sauer, and R. Wiechert, *Angew. Chem. (Int. Ed.)*, *10*, 496 (1971).

75. K. Hiroi and S. Yamada, *Chem. Pharm. Bull.*, *23*, 1103 (1975).

76. R. D. Clark, L. G. Kozar, and C. H. Heathcock, *Syn. Commun.*, *5*, 1 (1975).

77. For cyclization of β,ξ-diketo phosphonates see P. A. Grieco and C. S. Pogonowski, *Synthesis*, 425 (1973).

78. H. O. House, D. G. Melillo, and F. J. Sauter, *J. Org. Chem.*, *38*, 741 (1973).

79. T. L. Burkoth, *Tetrahedron Lett.*, 5049 (1969).

80. A. R. Daniewski, *J. Org. Chem.*, *31*, 3135 (1975).

81. R. E. Ireland and R. C. Kierstead, *J. Org. Chem.*, *31*, 2543 (1966).

82. O. Wichterle, J. Prochazka, and J. Hofmann, *Collect. Czech. Chem. Commun.*, *13*, 300 (1948).

83. J. A. Marshall and D. J. Schaeffer, *J. Org. Chem.*, *30*, 3642 (1965).

84. D. Caine and F. N. Tuller, *J. Org. Chem.*, *34*, 222 (1969).

85. G. Stork and M. E. Jung, *J. Amer. Chem. Soc.*, *96*, 3682 (1974).

86. J. A. Vida and M. Gut, *J. Org. Chem.*, *30*, 1244 (1965).

87. C. A. Henrick, E. Böhme, J. A. Edwards, and J. H. Fried, *J. Amer. Chem. Soc.*, *90*, 5926 (1968).

88. (a) S. Danishefsky, P. Cain, and A. Nagel, *J. Amer. Chem. Soc.*, *97*, 380 (1975); (b) S. Danishefsky and P. Cain, *J. Amer. Chem. Soc.*, *97*, 5282 (1975).

89. A. F. Hirsch and G. I. Fujimoto, *J. Org. Chem.*, *35*, 495 (1970).

90. E. J. Corey and D. E. Cane, *J. Org. Chem.*, *36*, 3070 (1971).

91. G. Darzens, *Compt. Rend.*, *150*, 707 (1910).

92. Z. G. Hajos, R. A. Micheli, D. R. Parrish, and E. P. Oliveto, *J. Org. Chem.*, *32*, 3008 (1967).

93. See Refs. 1-11 cited in Ref. 94.

94. J. A. Baldwin, G. A. Höfle, and O. W. Lever, Jr., *J. Amer. Chem. Soc.*, *96*, 7125 (1974).

95. E. J. Corey and D. Seebach, *Angew. Chem. (Int. Ed.)*, *4*, 1075 (1965).

96. D. Seebach, Synthesis, 17 (1969).

97. See Refs. 2-4 cited in Ref. 99.

98. E. J. Tillmanns and J. J. Ritter, *J. Org. Chem.*, *22*, 839 (1957).

99. A. I. Meyers, A. Nabeya, H. W. Adickes, I. R. Politzer, G. R. Malone, A. C. Kovelesky, R. L. Nolen, and R. C. Portnoy, *J. Org. Chem.*, *38*, 36 (1973).

100. P. B. D. de la Mare, in *Molecular Rearrangements* (P. de Mayo, ed.), Part 1, Interscience, New York, 1963, p. 87-88.

101. R. W. Hasbrouck and D. A. Kiessling, *J. Org. Chem.*, *38*, 2103 (1973).

102. Z. G. Hajos, K. J. Doebel, and M. W. Goldberg, *J. Org. Chem.*, *29*, 2527 (1964).

103. B. M. Trost and J. L. Stanton, *J. Amer. Chem. Soc.*, *97*, 4018 (1975).

104. J. R. Johnson, *Org. React.*, *1*, 210 (1942).

105. G. Jones, *Org. React.*, *15*, 204 (1967).

106. C. F. H. Allen and F. W. Spangler, *Org. Syn.*, *Coll. Vol.*, *3*, 377 (1955).

107. W. S. Johnson and G. H. Daub, *Org. React.*, *6*, 1 (1951).

108. D. L. Turner, B. K. Bhattacharyya, R. P. Graber, and W. S. Johnson, *J. Amer. Chem. Soc.*, *72*, 5654 (1950).

109. W. S. Johnson, J. W. Petersen, and W. P. Schneider, *J. Amer. Chem. Soc.*, *69*, 74 (1947).

110. W. H. Puterbaugh, *J. Org. Chem.*, *27*, 4010 (1962).

111. See Refs. 3-5 cited in Ref. 112.

112. R. L. Augustine and L. P. Calbo, Jr., *J. Org. Chem.*, *33*, 838 (1968).

113. M. S. Newman and B. J. Magerlain, *Org. React.*, *5*, 413 (1949).

114. G. C. Morrison, W. A. Cetenko, and J. Shavel, Jr., *J. Org. Chem.*, *31*, 2695 (1966).

115. M. W. Rathke, *Org. React.*, *22*, 423 (1975).

116. D. A. Cornforth, A. E. Opara, and G. Read, *J. Chem. Soc.*, *C*, 2799 (1969).

117. M. W. Rathke and A. Lindert, *J. Org. Chem.*, *35*, 3966 (1970).

118. (a) M. W. Rathke, *J. Amer. Chem. Soc.*, *92*, 3222 (1970);
      (b) D. Ivanov, G. Vassilev, and I. Panayotov, *Synthesis*, 83 (1975).

119. F. Tiemann, *Ber.*, *32*, 830 (1899).

120. A. J. Birch and J. S. Hill, *J. Chem. Soc.*, *C*, 125 (1967).

121. B. L. Yates and J. Quijano, *J. Org. Chem.*, *34*, 2506 (1969).

Chapter Two

ALKYLATION AND RELATED REACTIONS OF KETONES
AND ALDEHYDES VIA METAL ENOLATES

Drury Caine

School of Chemistry
Georgia Institute of Technology
Atlanta, Georgia

I. INTRODUCTION

Protons bonded to carbon atoms α to a number of unsaturated
functional groups are relatively acidic because such groups withdraw
electrons inductively and stabilize the negative charge by electron
delocalization once the proton is lost. Thus, monocarbonyl compounds,
such as ketones and aldehydes (1) (and also various carboxylic acid
derivatives, including nitriles), which contain α protons are converted to the corresponding resonance-stabilized anions 2, generally
termed *enolate anions*, in the presence of bases.

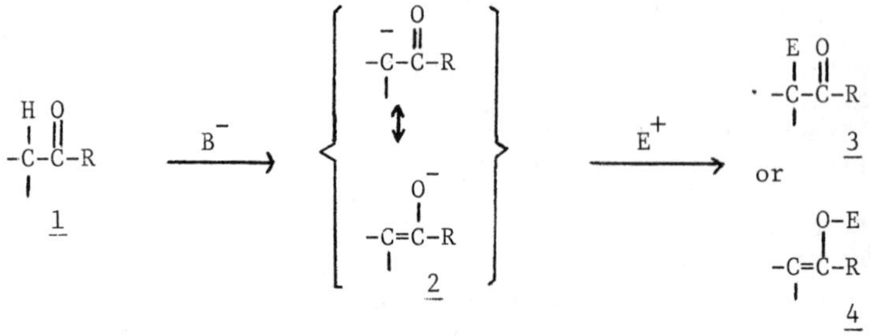

Metal enolates of carbonyl compounds undergo reactions with a
variety of hetero- and carbon-atom electrophiles. In view of the cen-
tral role of the carbonyl group in organic chemistry, these reactions
are of vast synthetic importance. In fact, reactions of metal enolate
with carbon electrophiles rank alongside addition reactions of
carbanionic species to carbonyl groups as the most important methods
of carbon-carbon bond formation for synthesis of complex systems.
Metal enolates normally react with carbon electrophiles at the α
position to form products of the type 3. However, the C and O modes o

reaction may compete; with certain highly reactive electrophiles,
enol derivatives of the type 4 may be the exclusive reaction products.

The most common reactions of metal enolates which lead to
carbon-carbon bond formation are (a) reactions with alkyl halides
and related reagents (normally referred to simply as alkylation
reactions) [Eq. (1)] [1-5], (b) reactions with various acylating
agents such as esters, acid anhydrides including carbon dioxide,
and acid chlorides (the Claisen and related reactions [Eq. (2)] [6-8],
(c) reactions involving 1,2-additions to carbonyl groups of aldehydes
and ketones (the aldol and related reactions) [Eq. (3)] [9, 10], and
(d) 1,4-additions to α,β-unsaturated carbonyl compounds (the Michael
and related reactions) [Eq. (4)] [11, 12].

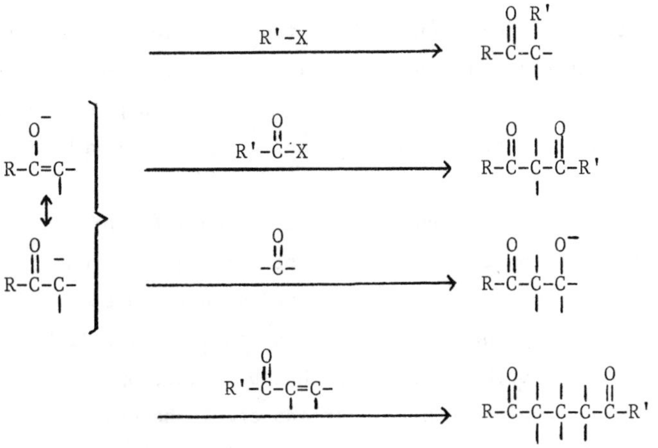

$$
\underset{R'-X}{\xrightarrow{\hspace{3cm}}} \quad R-\overset{\overset{O}{\|}}{C}-\overset{\overset{R'}{|}}{\underset{|}{C}}- \qquad (1)
$$

$$
\underset{R'-\overset{\overset{O}{\|}}{C}-X}{\xrightarrow{\hspace{3cm}}} \quad R-\overset{\overset{O}{\|}}{C}-\underset{|}{C}-\overset{\overset{O}{\|}}{C}-R' \qquad (2)
$$

$$
\underset{-\overset{\overset{O}{\|}}{C}-}{\xrightarrow{\hspace{3cm}}} \quad R-\overset{\overset{O}{\|}}{C}-\underset{|}{C}-\overset{\overset{O^{-}}{|}}{\underset{|}{C}}- \qquad (3)
$$

$$
\underset{R'-\overset{\overset{O}{\|}}{C}-C=C-}{\xrightarrow{\hspace{3cm}}} \quad R-\overset{\overset{O}{\|}}{C}-\underset{|}{C}-\underset{|}{C}-\underset{|}{C}-\overset{\overset{O}{\|}}{C}-R' \qquad (4)
$$

(Equilibrium enolate structures shown at left:

$$
R-\overset{\overset{O^{-}}{|}}{C}=\underset{|}{C}- \;\rightleftharpoons\; R-\overset{\overset{O}{\|}}{C}-\underset{|}{C}-
$$
)

The major portion of this chapter will be devoted to the treatment of reactions of the above types in which metal enolates of
saturated and unsaturated ketones are involved. The emphasis will
be on procedures in which the metal enolate is formed quantitatively,
either by treatment of a ketone with a strong base or by various
indirect methods, prior to the introduction of the electrophilic
reagent. Therefore, no attempt will be made to cover in detail
conventional methods for the Claisen, aldol, and Michael reactions
in which the enolate is generated only in a low equilibrium

concentration in the presence of the electrophilic species. The aldol condensation is discussed in Chap. 1, while the other reactions will be treated in other volumes in this series.

In recent years, there has been a great deal of interest in the reaction of ketone metal enolates with various hetero-atom electrophiles. The products of such reactions, α-substituted carbonyl compounds of the type *1* or enol derivatives of the type *2*, have been widely used as intermediates for a variety of synthetically useful transformations. The more important reactions of this type as well as reactions involving the synthesis of 1,4-dicarbonyl compounds by enolate coupling reactions will be discussed. Again, only those procedures involving reactions of *preformed* enolates will be considered in detail.

The preparation and reactions of enolates of various α-substituted ketones will be described. However, reactions of β-dicarbonyl compounds and related relatively acidic active methylene compounds will not be covered. Such reactions will be treated in another volume in this series.

The synthetic utility of reactions of metal enolates of aldehydes is not nearly as important as the corresponding reactions of ketone enolates. However, problems associated with the use of aldehyde enolates and some of the solutions to these problems will be described. Much of the methodology which has been developed in connection with the preparation and reactions of ketone enolates is also applicable to the formation of metal enolates of carboxylic acid derivatives and nitriles and to the reactions of these species with electrophiles. However, the literature concerning the preparation of α-substituted acid derivatives and nitriles via metal enolates has become too voluminous to permit coverage here. A thorough review of the work in this area prior to 1957 has been published [2].

There are many excellent procedures for the synthesis of α-substituted ketones which involve reactions of derivatives of carbonyl compounds or metalated carbonyl derivatives with electrophilic

reagents, followed by unmasking of the carbonyl group. These reactions serve as alternative procedures to the use of enolate anions
for the synthesis of α-substituted carbonyl compounds. Among the
best known of these are the Stork enamine alkylation and acylation
procedures [13-16], the reactions of metal salts of imines with
electrophilic carbon atoms discovered by Stork, Dowd, and others [17-
22], and the reactions of carbanionic derivatives of 2-alkyl-dihydro-
1,3-oxazines, 2-alkyl oxazolines, and related compounds with electrophiles discovered and developed by Meyers and co-workers [23]. A
new procedure of this type which was recently introduced by Corey
and co-workers [24-26] and which involves the reaction of metalated
N,N-dimethylhydrazones with electrophilic reagents also appears to
have a great deal of promise. Reactions of ketone derivatives and
metalated ketone derivatives will not be described in detail in this
chapter, but in some cases comparisons of the synthetic value of
these reactions with the corresponding enolate reactions will be
made. However, reactions of aldehyde derivatives provide the best
methods for the conversion of aldehydes into the corresponding α-
substituted compounds; therefore, procedures for these transformations
will be described. Full coverage of the enamine and related reactions
will appear in another volume in this Series.

II.   ALKYLATION AND RELATED REACTIONS OF SATURATED
      KETONES VIA METAL ENOLATES

A. *Introduction*

Ketones are the most important class of monocarbonyl compounds.
Alkylations and related reactions of these systems have attracted
the interest of organic chemists for many years. One of the most
important synthetic problems in this field has been the regiospecific
introduction of alkyl groups or other substituents at α positions
of unsymmetrical ketones. Reactions of this type very often constitute key steps in the synthesis of steroids, terpenoids, and other
complex compounds. The simplest and most fundamental method of
accomplishing such a reaction involves the regiospecific formation

and reaction of the desired metal enolate of an unsymmetrical ketone.
While simple in theory, there are a number of problems associated
with this procedure. A great deal of research has been required for
the synthetic chemist to gain a proper appreciation of these problems
and to find practical methods for avoiding them. Fortunately, this
area of research has reached a relatively mature stage, and the
synthetic chemist may often choose from among several excellent
methods to accomplish a particular transformation.

The pK_a values for the ionization of the α protons of simple
ketones lie within the 20-27 range depending on the exact structure
and the nature of the solvent [27]. Therefore, treatment of these
compounds with weak bases (or catalytic quantities of strong bases)
leads to the production of a low equilibrium concentration of the
enolate anion, and aldol condensation of this species with the unionized parent ketone can occur [Eq. (3)]. Because of this, it was
soon recognized that the best conditions for direct alkylation of
ketones involved the quantitative conversion of the ketone to the
metal enolate with a strong base prior to the addition of the
alkylating agent. Under these conditions an unactivated, unsymmetrical
ketone, such as 2-methylcyclohexanone (5), can lose a proton to give
either of two possible structurally isomeric enolate anions, 5A or
5B. In 1963 Conia [3] surveyed the results of alkylations of sodium
and potassium enolates of a variety of unsymmetrical saturated
cyclic and acyclic ketones; he showed that normally mixtures of
alkylation products were obtained and that in most cases the monoalkylation product derived from reaction at the more highly substituted position predominated. Cardwell [28] had earlier provided an
explanation for these results. He noted that on treatment of an
unsymmetrical ketone with a strong base in the absence of a proton
donor enolate formation should be essentially irreversible and that
the composition of the kinetic mixture of enolates should depend
only on the relative rates of proton removal from the alternate α
positions of the ketone. However, if a proton donor such as an
alcohol or unionized ketone were available, equilibration among the

structurally isomeric enolates 5A and 5B by a pathway such as that
shown in Eq. (5) would be possible.

$$5A \quad\underset{MB}{\overset{HB}{\rightleftharpoons}}\quad 5 \quad\underset{HB}{\overset{MB}{\rightleftharpoons}}\quad 5B \tag{5}$$

If equilibrium was completely established, the composition of
the enolate mixture would depend on the relative thermodynamic stabilities of 5A and 5B. The results of studies on the kinetic and
thermodynamic acidities of nitroalkanes suggested to Cardwell [28]
that, in general, proton transfer should be more rapid from the less-
substituted position of an unsymmetrical ketone, but that the more
highly substituted enolate anion should be the more stable species.
As a consequence of this, the composition of the enolate mixture
would depend on whether kinetic or thermodynamic conditions were
employed in its formation. The additional point was made that,
although mixtures of kinetic enolates should be obtained by adding
the ketone slowly to a strong base in an aprotic solvent, on addition
of the alkylating agent (e.g., a methyl halide), the initially formed
alkylated ketone could cause equilibration of the enolate mixture
to occur as shown in Eq. (6).

When sodium or potassium enolates are involved, enolate equilibrium is normally faster than alkylation. This accounts for the
fact that the major monoalkylation product of the sodium or potassium
enolates of 5 is 2,2-dimethylcyclohexanone (7). Rapid proton transfer reactions would also account for the formation of di- and tri-
alkylation products which are usually produced in significant amounts
during the alkylation of sodium or potassium enolates.

The first important solutions to the problem of controlling the
position of alkylation of unsymmetrical ketones involved the introduction of a blocking group at the less highly substituted α carbon

(6)

to allow exclusive alkylation at the more highly substituted α position [29-35], or the addition of an activating group at the less-substituted position so as to enhance the acidity of protons there and stabilize the desired enolate anion [36, 37]. Removal of the blocking group or activating group then yielded the regiospecifically alkylated product. While a number of ingenious procedures involving blocking and activating groups have been developed, the large number of steps required for such transformations has limited the practicality of these approaches.

What might be termed the "modern era of enolate reactions" began in 1961 when Stork, et al. [38, 39] showed that the enolate 8A formed by lithium-liquid ammonia reduction of 10-methyl-1(9)-octalin-2-one (8), could be alkylated (with relatively reactive alkylating agents in liquid ammonia) to produce trans-1-alkyl-10-methyl-2-decalones (9). The enolate 8A is the less stable one of the trans-10-methyl-2-decalone system (10), and direct alkylation of 10 using strong bases which give sodium or potassium enolates leads mainly to 3-alkyl-substituted products 11 via the more stable 2-enolate 10A[37]. The integrity of the 1-enolate 8A was not maintained if sodium or potassium was

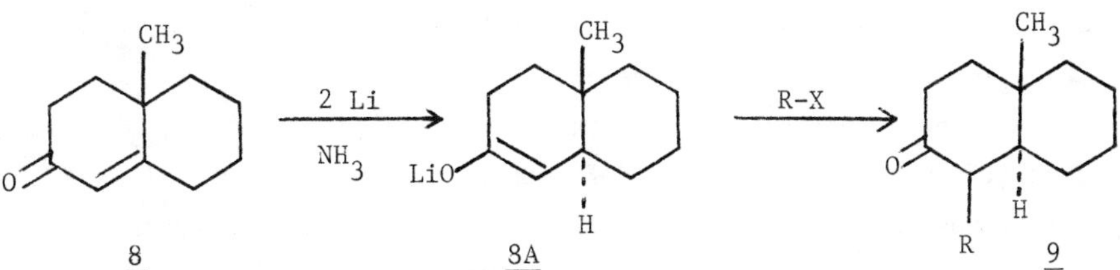

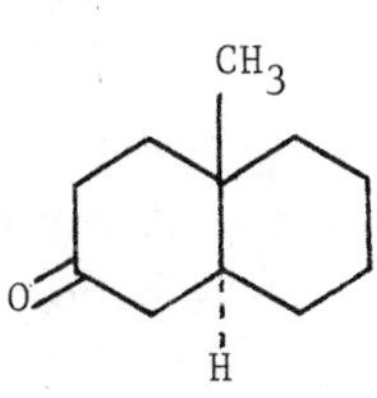

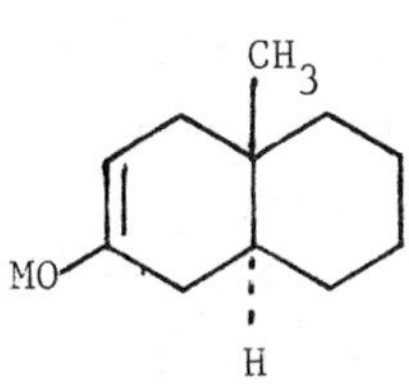

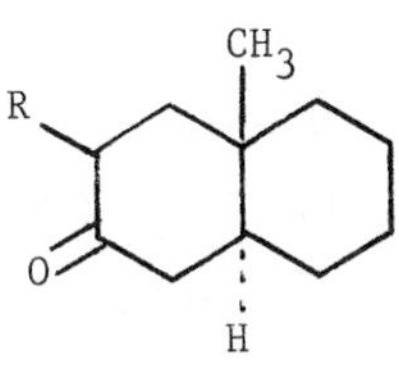

employed as the reducing metal or if the lithium enolate was transferred to dimethylsulfoxide (DMSO) before addition of the alkylating
agent; instead, the 3-alkyl-substituted ketone *11* was the major product. Stork pointed out that the success of the reduction-alkylation
of *8* depended on the fact that in liquid ammonia C-alkylation of the
lithium enolate *8A* is more rapid than equilibration to the more stable
enolate *10A* by proton transfer between *8A* and the initially produced alkylated product. As will be discussed in detail later, it
was quickly demonstrated in a number of laboratories that specific
lithium enolates of a variety of unsymmetrical ketones could be trapped with relatively reactive alkylating and other electrophilic
reagents without equilibration. This unique behavior of lithium
enolates compared with those of sodium and potassium suggested that
other metals which, like lithium, may form relatively covalent metal-
oxygen bonds could be utilized in regiospecific enolate reactions.
This has proved to be the case. In addition to the reduction of
enones, a number of other highly useful methods for generating
specific lithium and other metal enolates, as well as the conditions
for reacting them with electrophilic reagents, have been developed.
Likewise, new procedures for the base-promoted generation of kinetic
and equilibrium enolate mixtures have been reported. A review on
these subjects has appeared recently [40].

Another tremendously important advance in the field of enolate
chemistry was the development of methods for the determination of
the composition of mixtures of metal enolates of unsymmetrical
ketones. This work was pioneered by House and co-workers [4, 41-45],
who determined the composition of kinetic and equilibrium mixtures
of a variety of unsymmetrical acyclic, monocyclic, and bicyclic
ketones. Similar studies were performed on several monocyclic
ketones by Stork and Hudrlik [46, 47], and by Caine and co-workers
[48-51]. These studies demonstrated that in addition to the structure
of the ketone, factors such as the structure of the base, the nature
of the metal cation, and the solvent can play an important role in
determining the composition of kinetic and equilibrium mixtures of

enolate anions and the rates of equilibration of these species. They
have provided information which will generally allow a determination
to be made as to whether base-catalyzed enolization of the parent
ketone or some indirect method will be required to obtain a particular
enolate in an unsymmetrical system.

B. *Base-Promoted Enolization of Saturated Ketones*

   1. Bases and solvents
   Because of the relatively low acidity of simple ketones, bases
stronger than ethoxide and solvents more weakly acidic than ethanol
are required for their conversion to the corresponding anions in significant concentrations. However, there are many strong bases and
solvent systems which bring about complete enolization and provide
solutions of metal enolates suitable for reaction with electrophilic
reagents. The selection of a base-solvent combination required to
achieve a particular result depends on a number of factors. Among
these are (a) the structure of the ketone and its propensity for
undergoing aldol condensation, (b) the degree of regioselectivity
required for the formation and subsequent reaction of the metal
enolate(s) in the case of unsymmetrical ketones, and (c) the type
of electrophilic reagent to be reacted with the metal enolate. Of
course, the expense, the ease of preparation and/or handling and
purification of the base and solvent, and the ease of separation of
the conjugate acid of the base and the solvent from the reaction
products are also important considerations.

   Potassium t-butoxide in t-butyl alcohol or in other solvents
has been widely used to generate potassium enolates of carbonyl
compounds [52-55]. The strength of this case can be significantly
influenced by the solvent. It is strongly basic in DMSO, existing
essentially as a monomer. The base strength decreases significantly
in solvents such as benzene, tetrahydrofuran (THF), and 1,2-
dimethyloxyethane (DME) in which it is aggregated (probably averaging
a tetramer), and it is most weakly basic in t-butyl alcohol. Unless
ketones bear activating groups at the α position, metal alkoxides

generally do not bring about quantitative enolate formation [53-55].
House, et al. [53] have found that treatment of 2-methylcyclohexanone
(5) with 1 eq of potassium t-butoxide in t-butyl alcohol gave
approximately 15% of a mixture of the potassium enolates 5A (M = K)
and 5B (M = K). In DME this base was more effective and gave
approximately 60% of the same enolate mixture. A relatively high
concentration of the enolate anion was produced when methyl t-butyl
ketone was treated with potassium t-butoxide in THF [55]. As a
result of slow and/or incomplete enolate formation, aldol condensation
between enolate and free ketone may be an important side reaction
when potassium t-butoxide is employed as the base, particularly when
reactive ketones such as cyclopentanone are involved [3]. Improved
yields have been achieved in alkylations of unstable ketones through
the use of sodium t-amylate in ether or hydrocarbon solvents [3].
The increased molecular weight of this base compared with potassium
t-butoxide apparently leads to increased solubility. Brown [55]
has found that increasing the size of the alkyl group markedly
increases the basicity of potassium alkoxides in THF. Increased
steric hindrance, which presumably interferes with aggregation and
solvation of the cation in the ion pair, probably accounts for this.

Lithium [56], sodium [57] and potassium amide [58] are much
stronger bases than the alkali metal alkoxides are. In liquid
ammonia, ethereal, or aromatic hydrocarbon solvents, these bases
have been used to prepare the corresponding metal enolates of
ketones for use in alkylation and related reactions [3, 57, 59-61].
However, because of the low solubility of alkali metal amides,
especially in ethereal and aromatic hydrocarbon solvents, enolization
of the ketone is usually slow, so that proton transfer reactions
leading to equilibration of the enolates [59] and/or in the case of
reactive ketones, aldol condensation [3] cannot be avoided. Potassium
and sodium amide are relatively soluble in liquid ammonia. It appears
that kinetic formation of the less-substituted enolate of ketones
such as 2-methylcyclohexanone (5) is possible using sodium amide in
liquid ammonia for deprotonation [28, 60].

Sodium hydride in diethyl ether, THF, DME, aromatic hydrocarbons, or dipolar aprotic solvents has been used widely to produce sodium enolates of cyclic and acyclic ketones [41, 42, 46, 62-65]. Lithium hydride has been used somewhat rarely [62] because the reactivity of this metal hydride is too low for it to be of practical value in synthesis. Potassium hydride has been introduced recently by Brown [66] as a reagent for the metalation of carbonyl and other active hydrogen compounds. Sodium hydride has very limited solubility in organic solvents. Again, this means that enolate formation will be slow and that only thermodynamic mixtures of sodium enolates of unsymmetrical ketones will be formed. Also, for reactive ketones such as cyclopentanone and acetone, aldol condensation will be a significant problem. Both sodium [67] and potassium hydride [68] apparently do not react directly with ketones to form enolates but first react with traces of alcohols present to form metal alkoxides which are the proton abstracting species. Sodium hydride has been reported to reduce certain strained ketones [63].

House and Kramar [41] have found that the enolate equilibration problem can be partially avoided by treating a ketone with sodium hydride in the presence of an alkylating agent. In this way, the enolate concentration remains low, and the kinetically formed enolate mixture is trapped.

Potassium hydride in THF produces high yields of potassium enolates [66]. This holds true even for ketones which are very reactive in aldol condensations. The fact that enolization is very rapid and the potassium cation is relatively poor at providing stabilization of aldol condensation products which are in equilibrium with the potassium enolate and the free ketone [Eq. (3)] provides a reasonable explanation for these results. However, reaction of unsymmetrical ketones with potassium hydride is apparently too slow to prevent equilibration of enolates, because thermodynamic mixtures of enolates are obtained [66].

Tryllithium [42, 70-72], tritylsodium [73], and tritylpotassium [74, 75] have been used as bases to prepare lithium [3, 41, 42, 48,

49, 50], sodium [76, 77], and potassium [41, 42, 46] enolates of cyclic and acyclic unsymmetrical ketones. With the exception of trityllithium, which is not dissociated in diethyl ether, all of these bases give approximately 1 M solutions of the red-colored trityl anion and the metal cation in diethyl ether, THF, DME, or liquid ammonia. DME solutions of these bases have been most frequently employed for deprotonation reactions [42]. These solutions may be titrated with the ketone until the red color is just discharged; thus rapid and quantitative formation of enolates in the absence of excess ketone is possible. These conditions permit the formation of kinetic mixtures of enolates of unsymmetrical ketones which are stable in the absence of proton donors. Of course, the conversion of kinetic enolate mixtures to thermodynamic mixtures can be carried out by adding an excess of the ketone and allowing time for equilibrium to be established. The unique behavior of lithium enolates has promoted a great deal of interest in the use of trityllithium to enolize unsymmetrical ketones. The major difficulty associated with the use of trityllithium and other alkali metal derivatives of triphenylmethane is the removal of the hydrocarbon from the reaction products. Separations are normally effected by absorption chromatography or distillation if the products are relatively volatile.

As will be discussed later, trityllithium cannot be used as the base for enolization of α' protons of α,β-unsaturated ketones because electron transfer to the enone system competes with the deprotonation reaction [78, 79].

Interest in methods of preparation of lithium enolates has led House and co-workers to investigate the use of lithium dialkylamides such as lithium diethyl- [42] and lithium diisopropylamide (LDA) [43, 44] for this purpose. LDA has proved to be an especially effective base which offers the advantages of (a) being very effective for the formation of less substituted enolates of unsymmetrical ketones; (b) having a relatively high solubility in ether, THF, DME, and hexane-pentane mixtures; and (c) giving a deep orange to red color in the presence of bipyridyl indicator.

Approximately 1 M solutions of LDA in DME can be prepared by
removing the ether from a solution of methyllithium, adding an
appropriate quantity of DME, cooling to -20° to 50°C, adding 1 eq
of diisopropylamine dropwise with stirring at -20°C, and stirring for
2-3 min at -20°C. (Methyllithium or LDA attacks DME at temperatures
above 0°C.) Approximately 0.5 M solutions of LDA in hexane-pentane,
which are stable for weeks at 25°C, can be prepared by the dropwise
addition of a slight excess of diisopropylamine with stirring to an
approximately 0.5 M solution of n-butyllithium in approximately 2:3
hexane-pentane [69]. It is also common practice to prepare 0.2-0.5
M solutions of LDA in THF-hexane by adding dropwise 1 eq of n-butyllithium (1.4-2.0 M) in hexane with stirring to 3-7 volumes of THF
containing 1 eq of diisopropylamine at -78°C. After stirring for
15 min at -78°C, the solution is ready for use.

Lithium diethylamide may be prepared by the reaction of lithium
metal with diethylamine in benzene containing 1 eq of HMPA [80].
This base in HMPA(hexamethylphosphoramide)-benzene-ether appears to
be useful for the kinetic enolization of acyclic ketones [81].
Lithium isoproplycyclohexylamide (LICA) [82] and lithium 2,2,6,6-
tetramethylpiperidide (LTMP) [83] are other hindered dialkylamides
that have been used for deprotonation of carbonyl compounds. However, except for use for the α' deprotonation of α,β-unsaturated
ketones [79], these bases have been employed largely for deprotonation
of esters and lactones [80, 83]. The 1:1 complex of LDA and HMPA
is an especially basic nonnucleophilic reagent, which has been used
to prepare ester enolates [84]. The complex is prepared and used
for deprotonation at low temperatures. It seems possible that a
reaction between LDA and HMPA occurs at higher temperatures.

Alkali metal derivatives of hexamethyldisilazane, e.g., lithium
hexamethyldisilazide (LHDS), which are prepared by reaction in benzene
of this amine with the corresponding alkali metal amides [85, 86],
are soluble in hydrocarbon solvents and are effective bases for
preparing ketone enolates [66, 87-91].

A large number of other strong bases have been used less frequently to generate metal enolates of aliphatic ketones. Dimsyl
sodium in DMSO [92] has been used in intramolecular enolate alkylations of expoxy ketones [93-95] and keto tosylates [95-99] in
sesquiterpene synthesis. However, this base-solvent combination has
disadvantages: dimsyl sodium may undergo nucleophilic addition to
carbonyl groups and DMSO itself reacts with certain alkylating
agents [100].

Radical anions of aromatic hydrocarbons may function as reducing
agents [101-103] or bases [102-105]. For example, sodium anthracene
has been used to prepare sodium enolates of acyclic ketones and aldehydes [104]. However, these bases have found more general use in the
formation of dianions of saturated [106-107] and unsaturated [108,
109] carboxylic acids.

Enolates containing metals other than alkali metals are normally
prepared by treating solutions of alkali metal enolates with the
required metal halide or by reacting various enol derivatives of
ketones. However, it has been shown that solutions of Grignard
reagents such as n-butyl or isopropylmagnesium chloride in HMPA are
strongly basic and give solutions of halomagnesium enolates of hindered and unhindered ketones in high yield [110, 111]. Isopropyl
magnesium halides yield halomagnesium enolates on reaction with t-
butyl alkyl ketones in ether [112, 113]. N-Methylanilinomagnesium
bromide has also been used as a base for preparing halomagnesium
enolates of ketones, particularly in aldol condensations [10].

The choice of solvent in base-promoted enolizations of ketones
depends upon the structure of the ketone and the nature of the base.
Since the solvent may have a profound effect on the reaction of
enolates with various electrophilic reagents, subsequent reactions
of the enolate should be borne in mind. However, sometimes it is
necessary to change the reactivity of the metal enolate through the
use of solvent additives or to change the solvent entirely before
addition of the electrophilic reagent. There have been many
examples in which good results were achieved by the preparation and

(subsequent) reaction of metal enolates under heterogenous reaction conditions. However, as noted above, under these conditions the metal enolate will be formed in the presence of free ketone; therefore, yields may be reduced by aldol condensation, and equilibrium mixtures of enolates of unsymmetrical ketones will be formed. Consequently, best results are usually achieved by choosing a solvent in which the ketone, the base, and the metal enolate are soluble.

Metal enolate solutions may consist of molecular aggregates, 12, such as dimers, trimers, and tetramers [44, 48, 62, 114-118] in equilibrium with monomeric covalently bonded species, contact ion pairs, and/or solvent-separated ion pairs of the type 13-15, as shown in Eq. (7). While certain metal cations, such as Hg^{2+}, form α-metalated ketones, 16 (M = HgX), [117-119], the common alkali metal as well as magnesium and zinc enolates appear to exist exclusively as structures such as 12-15 having the metal cation associated with oxygen [47, 117, 118]. Simple ketone metal enolates do not appear to exist as "free" anions and cations even in highly ionizing solvents [47, 117, 118]. This also appears to be true for stabilized metal enolates, such as those derived from malonic ester derivatives [120, 121].

$$\left(\underset{\underset{12}{}}{\text{OM}}\right)_n \rightleftharpoons \left(\underset{}{\text{OM}}\right)_{n-1} + \underset{\underset{13}{}}{\text{OM}} \rightleftharpoons \underset{\underset{14}{}}{\text{O}^-\text{M}^+} \rightleftharpoons \underset{\underset{15}{}}{\text{O}^-} \text{M}^+ \qquad (7)$$

$$\underset{\underset{16}{}}{\overset{\text{O}}{\text{M}}}$$

The nature of the solvent may have a profound influence on the degree of aggregation and the nature of the association between the enolate anion and the metal. This, in turn, will influence the reactivity of the enolate toward electrophilic reagents and its ability to function as a base, i.e., participate in proton transfer reactions. Of course, the structure of the metal enolate and the metal cation will influence reactivity as well; these factors will be discussed later.

In nonpolar solvents such as aromatic hydrocarbons and diethyl ether, the solubility of metal enolates is low, and in solution the metal enolate exists primarily in the aggregated form *12*. The use of dipolar aprotic solvents increases solubility and shifts the equilibrium toward various monomeric species. Solvents such as THF and DME specifically solvate cations by an n-donor mechanism [114]. This forces the equilibrium toward monomeric species and greatly enhances enolate reactivity toward electrophilic reagents. Because of the ability of DME to interact with metal cations via a didentate donor mechanism, it is a significantly better solvent for metal eno-lates than THF [114, 122].

The highly ionizing solvents, such as DMF, DMSO [123], and HMPA [124, 125], specifically solvate metal cations via π-donor mechanisms. In these solvents the equilibrium shown in Eq. (7) is strongly shifted toward solvent-separated ion pairs. Such species are highly nucleophilic, so they will undergo reactions with alkylating agents and other electrophilic species at very rapid rates [114, 126-128]. However, enolate anions in the form of solvent-separated ion pairs are also extremely basic, so the rates of proton transfer are also quite rapid. Thus, in dipolar aprotic solvents equilibration of structurally isomeric enolates and extensive polyalkylation are usually observed. Also, the reactive solvent-separated ion pairs present in these solvents are prone to O-alkylation, particularly with reactive, "hard" alkylating agents [128-131]. These factors limit the value of dipolar aprotic solvents for use in metal enolate preparation and trapping.

A useful technique is to prepare solutions of metal enolates in solvents such as diethyl ether or THF and to adjust the reactivity by adding an appropriate quantity (1 eq or more) of a dipolar aprotic solvent (usually HMPA) prior to addition of the electrophilic reagent. In this way the desired level of nucleophilicity, vis-à-vis,

basicity can be achieved. The addition of 4 eq of a dipolar aprotic
solvent such as HMPA converts lithium enolates into highly reactive
solvent-separated ion pairs. If additional quantities of such solvents are employed, enolate reactivity is not influenced significantly, but work-up of the alkylation reaction mixtures is made more
complicated [69]. Crown ethers, which interact with metal cations
by a polydentate donor mechanism, also appear to be very useful
additives for increasing the reactivity of certain metal enolates
[121]. However, these additives have not been used very widely in
preparative reactions [121].

Because of the relatively high basicity of metal enolates of
ketones, t-butyl alcohol and liquid ammonia are the only protic solvents which have found wide use in reactions involving these species.
As stated earlier, most simple ketones are not quantitatively converted to metal enolates in the presence of t-butyl alcohol. Liquid
ammonia has been rather widely used as a solvent for alkylations
of enolate anions, particularly those generated by metal-ammonia
reduction of α,β-unsaturated ketones [132]. The dipolar hydrogen
bonding character of ammonia provides stabilization of both cations
and anions, but it is not sufficiently acidic to protonate alkali
metal enolates of simple ketones. However, there is evidence that
calcium-ammonia and magnesium-ammonia solutions are sufficiently
acidic to bring about protonation of the corresponding metal enolates
[132].

Solvents for enolate reactions should be anhydrous and free of
all protic impurities. Protic solvents such as t-butyl alcohol and
liquid ammonia are easily dried by refluxing them with a small quantity of freshly cut sodium metal and distilling them just prior to
use. Ethereal solvents such as diethyl ether, THF, and DME may be
dried by refluxing them over lithium aluminum hydride and distilling
them just prior to use.

Caution: Anhydrous reagent grade ether and reagent grade THF
and DME may be treated directly with lithium aluminum hydride. Lower

grades of these solvents should be predried before adding the metal hydride. Under no circumstances should distillations from the metal hydride be conducted to dryness, as it may undergo explosive decomposition at temperatures above $120^{\circ}C$. Distillation of these ethereal solvents from sodium benzophenone ketyl also provides an effective means of drying.

DMF may be dried by allowing it to stand over Linde AW-500 molecular sieves for 48 hr and then distilling it under vacuum from phosphorous pentoxide through a 15- to 20-cm Vigreaux column [133] or by distilling it from calcium hydride [43]. HMPA may be dried by vacuum distillation over Linde 13X molecular sieves (calcinated under nitrogen at $350^{\circ}C$ for 4 hr) [134] or by vacuum distillation from sodium [135]. DMSO is dried by vacuum distillation over molecular sieves followed by vacuum distillation from freshly prepared potassium amide [136] or by vacuum distillation from calcium hydride [92]. For more thorough drying of solvents the sieves should be removed before distillation, since they may liberate water upon heating.

Because of the ease with which enolate anions react with molecular oxygen [137], all reactions involving the formation or reaction of this species should be conducted in a dry, oxygen-free nitrogen or an argon atmosphere. For careful work, solvents should be distilled under nitrogen.

2. Effects of Ketone Structure, the Metal
Cation, and the Reaction Conditions

For stereoelectronic reasons, the preferred transition state for the removal of a proton α to a carbonyl group by a base is one in which the C-H bond is perpendicular to the plane of the α carbon, the carbonyl group, and the α' carbon, as in *17* [138-145]. This conformation allows maximum orbital overlap of the σ C-H bond which is being cleaved with the π bond of the carbonyl group. It seems likely that protons adjacent to carbonyl groups are not acidic unless they occupy or can attain a conformation in which proton removal is assisted by σ-π orbital overlap. The nature of the transition state involved in the enolization of ketones by weak bases is open to question [144]. However, when strong bases are employed for deprotonation, there is considerable evidence to indicate that the geometry of the transition state resembles the starting ketone much more closely than the metal enolate product, although some carbanionic character has built up at the α carbon [145].

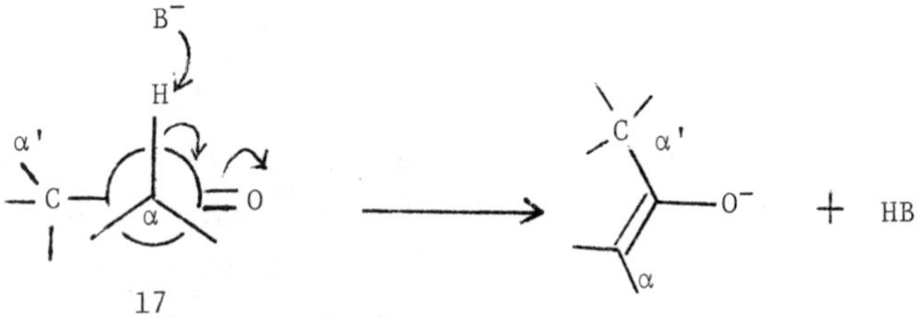

17

The above considerations suggest that substituents at the α position should retard the rates of removal of protons sterically (this seems to be true for alkyl substituents at either α position) and, depending on their electron-releasing or electron-withdrawing ability, should retard or enhance the rates of proton removal electronically.

Ketones of the type 18 in which Y is a strong activating group, such as -NO$_2$, COR, CO$_2$R, SO$_2$R, SOR, or CN, yield exclusively the α-substituted enolate 19 on treatment with 1 eq of a strong base or even a relatively weak base, such as sodium ethoxide. As noted earlier, a chapter dealing with alkylation and related reactions of enolates such as 19 is planned for a later volume in this series.

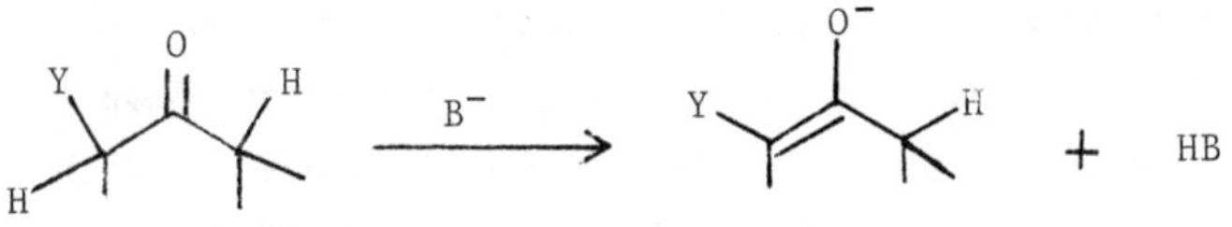

18   19

The structures of many simple ketones allow for the formation of only a single enolate anion or a mixture of geometric isomers. Systems of this type include symmetrical cyclic (20) and acyclic (21) ketones, aryl alkyl ketones (22), and alkyl ketones (23) having one α position fully substituted.

$$\underset{\underline{20}}{\overset{\displaystyle O}{\underset{(CH_2)_n}{\text{H–}\overset{\quad}{\underset{R}{}}\text{–H, R}}}} \qquad \underset{\underline{21}}{R_2HC\overset{\displaystyle O}{-C-}CHR_2} \qquad \underset{\underline{22}}{Ar\overset{\displaystyle O}{-C-}CHR_2} \qquad \underset{\underline{23}}{R_3C\overset{\displaystyle O}{-C-}CHR_2}$$

R = H, alkyl, aryl, etc.

In preparing and reacting metal enolates of such compounds, the major problems involve the prevention of aldol condensation during enolization and the prevention of the formation of polysubstitution products when the enolates are treated with electrophilic reagents. As has been noted previously, simple symmetrical ketones such as cyclopentanone and acetone are highly reactive in aldol condensations and especially prone to proton transfer reactions which lead to polysubstitution products on reaction with electrophilic reagents [3]. Thus, successful generation of metal enolates of these compounds requires the slow addition of the ketone to a strong base (preferably in solution) so that the enolate is generated rapidly. Examples of successful conversions of cyclopentanone to its potassium enolate with potassium hydride in THF [66] and its lithium enolate with LDA in THF at low temperature [146-148] and of formation of the potassium enolate of acetone with potassium amide or potassium t-butoxide in liquid ammonia [149] indicate that under proper conditions deprotonation of these ketones can be accomplished without difficulty.

Cyclohexanone is much less prone to aldol condensation than cyclopentanone is, and metal enolates of the former ketone have been prepared with a number of base-solvent combinations including sodium amide in ether [57], trityllithium in DME [50, 150], and LDA in DME [151]. Studies on the kinetics of ethylation of the lithium enolate of cyclohexanone in DME suggest that it exists primarily in the form of molecular aggregates (tetramers, trimers, etc.) even at low concentration [50]. It seems to be generally true that metal enolates having small substituents on the double bond are prone to aggregation, especially when metal cations that can form relatively covalent M-O bonds are involved [44, 115, 116, 152, 153]. Bulky substituents on the double bond of the enolate would be expected to sterically impede the formation of aggregates.

Acyclic ketones, including symmetrical ketones such as diethyl ketone [*24a*], can yield mixtures of geometric isomers upon deprotonation. Studies on the deprotonation of unsymmetrical acyclic ketones [41-43] have revealed that Z-enolates such as *24Z* are normally more stable than the E-enolates *24E*. Therefore, the former species will normally be the major components of enolate mixtures prepared under thermodynamic conditions. Under kinetic conditions mixtures of E- and Z-enolates are usually observed [41-43]. Bases which may yield relatively free carbanions and metal cations in solution normally yield Z-enolates preferentially. However, enolization of various acyclic ketones [41-43] and esters [154] with the highly associated base LDA leads to mixtures containing relatively large amounts of the less stable E-enolates. House [1] has suggested that when associated bases are used for deprotonation, steric interactions at both the C-H bond undergoing cleavage and the metal-oxygen bond which is forming may be important. Application of this idea to the enolization of ketones such as *24* would suggest that the transition state *25* which leads to the E-enolate *24E* might be lower in energy than the transition state *26*, which yields the Z-enolate *24Z*. However, it has been shown that ethyl t-butyl ketone *24B* yields exclusively the Z-enolate upon treatment with LDA in THF at -70°C [155]. Apparently, when a bulky R group is present, transition states such as *26*, in which steric interactions of this group with the methyl group are minimized, are more favorable.

Nuclear magnetic resonance (nmr) studies on metal enolate solutions suggest that Z isomers have a greater tendency to form solvent-separated ion pairs (cf. *15*) than the corresponding E isomers do because of the cis interaction of the substituent with the metal-oxygen solvent shell in the former [117]. It has been shown that C-/O-acylation ratios [117] and the stereochemistry of the products

$$\underset{\underline{24}}{\overset{O}{\underset{\parallel}{R-C-CH_2CH_3}}}$$

a; R = CH$_2$CH$_3$

b; R = t-Bu

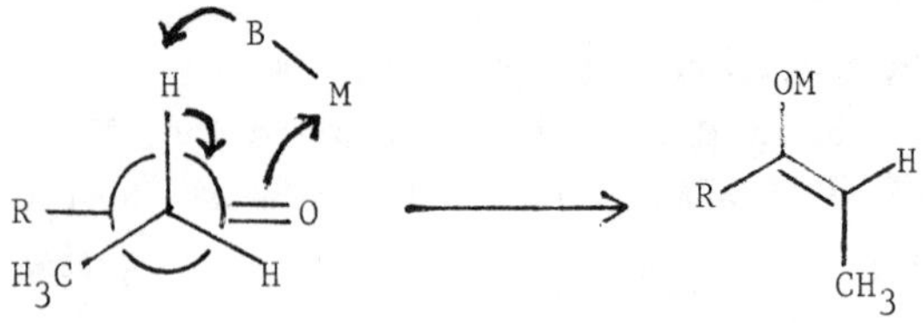

of aldol condensations of metal enolates [155, 156] are dependent
on which geometric isomer of the E-/Z-enolate pair is employed.
As noted, unsymmetrical ketones generally yield enolate mixtures of different compositions, depending on whether kinetic or
thermodynamic conditions are employed in the enolization process.
Kinetic conditions imply that the ketone is added slowly to a solution
containing an excess of a strong base. Equilibrium conditions imply
that an excess of the ketone is present after enolate formation and
that sufficient time is allowed for complete equilibration of structurally isomeric enolates to occur. The methods which have generally
been utilized for the determination of the compositions of enolate
mixtures are (a) quenching with deuterio acids, such as deuterio-
acetic acid, under conditions in which carbon-bound hydrogen and
carbon-bound deuterium are not exchanged and determining the position
of incorporation of deuterium, usually by mass spectrometry [4, 41,
42]; and (b) inverse addition of the enolate mixture to excess acetic
anhydride [4, 41, 42, 48, 49, 142, 157, 158] or direct addition of
excess trimethylsilyl chloride (or other trialkylsilyl chlorides)
to the enolate mixture [43, 46, 87] and determination of the composition of the resulting mixtures of enol acetates or enol silyl ethers
by gas-liquid chromatography (glc) or by spectroscopic methods.

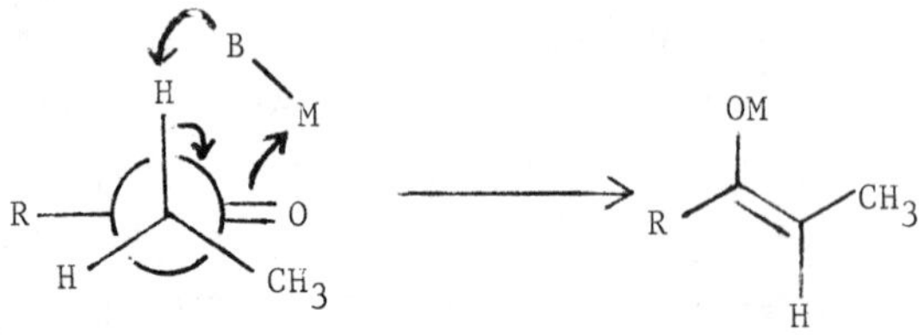

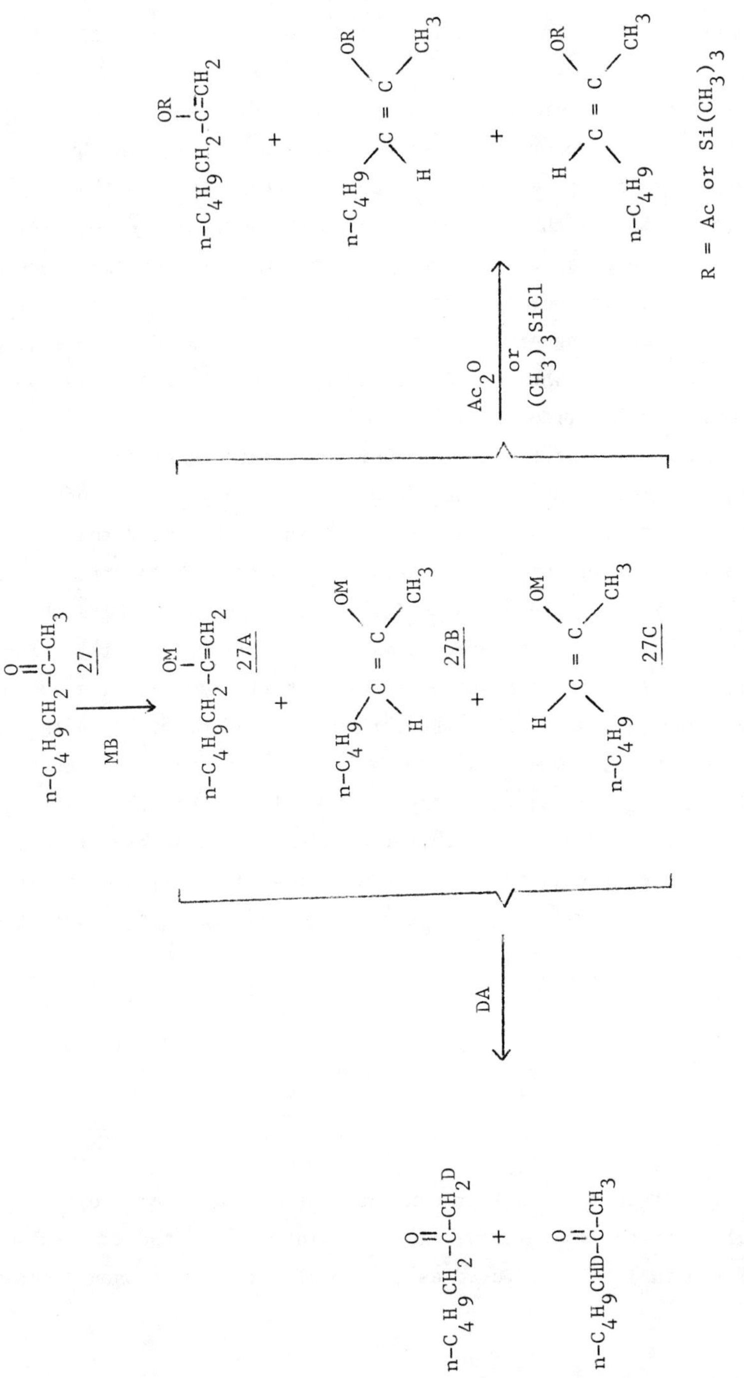

$$n\text{-}C_4H_9CH_2\text{-}\overset{\overset{\displaystyle O}{\|}}{C}\text{-}CH_3 \quad \underline{27}$$

$$\xrightarrow{\text{MB}}$$

$$n\text{-}C_4H_9CH_2\text{-}\overset{\overset{\displaystyle OM}{|}}{C}{=}CH_2 \quad \underline{27A}$$

$$+ \quad \overset{n\text{-}C_4H_9}{\underset{H}{>}}C{=}C\overset{OM}{\underset{CH_3}{<}} \quad \underline{27B}$$

$$+ \quad \overset{H}{\underset{n\text{-}C_4H_9}{>}}C{=}C\overset{OM}{\underset{CH_3}{<}} \quad \underline{27C}$$

$$\xrightarrow[\text{or}\;(CH_3)_3SiCl]{Ac_2O}$$

$$n\text{-}C_4H_9CH_2\text{-}\overset{\overset{\displaystyle OR}{|}}{C}{=}CH_2$$

$$+ \quad \overset{n\text{-}C_4H_9}{\underset{H}{>}}C{=}C\overset{OR}{\underset{CH_3}{<}}$$

$$+ \quad \overset{H}{\underset{n\text{-}C_4H_9}{>}}C{=}C\overset{OR}{\underset{CH_3}{<}}$$

$$R = Ac \text{ or } Si(CH_3)_3 \tag{8}$$

$$\xrightarrow{\text{DA}}$$

$$n\text{-}C_4H_9CH_2\text{-}\overset{\overset{\displaystyle O}{\|}}{C}\text{-}CH_2D$$

$$+ \quad n\text{-}C_4H_9CHD\text{-}\overset{\overset{\displaystyle O}{\|}}{C}\text{-}CH_3$$

The application of these procedures to the determination of the composition of the metal enolates *27A-C* of 2-heptanone (*27*) is shown in Eq. (8).

These different procedures provide results which usually agree within ± 5%. However, the acetic anhydride or the trimethylsilyl chloride quenching methods are much more commonly used than the deuterio acid quenching method because they are simpler experimentally and they also allow the determination of the ratio of the geometrical isomers of a particular structural isomer. The use of the acetic anhydride quenching procedure may lead to C-acylated as well as O-acylated products when halomagnesium enolates in relatively nonpolar solvents are involved [117].

The results of the analysis of kinetic and thermodynamic enolate mixtures produced by treating several simple acyclic ketones with strong bases are shown in Table I. It should be noted that in most cases mixtures containing significant amounts of both structurally isomeric enolates are produced. The thermodynamic results show that enolate stability is related to alkene stability; that is, trisubstituted enolates are more stable than disubstituted enolates except when there is branching at the carbon attached directly to the double bond (cf. methylisobutyl ketone, *28*). In cases where tetrasubstituted enolates are involved (cf. ethyl isopropyl ketone, *29*), steric interactions of the four groups on the double bond cause the trisubstituted enolate to be more stable. In general (again paralleling alkene stabilities), Z-enolates are more stable than the corresponding E isomers.

$$CH_3-\overset{\overset{O}{\|}}{C}-CH_2CH(CH_3)_2 \qquad\qquad CH_3CH_2-\overset{\overset{O}{\|}}{C}-CH(CH_3)_2$$

$$\underline{28} \qquad\qquad\qquad\qquad\qquad \underline{29}$$

Except in systems that can form tetrasubstituted enolates, the position of the equilibrium lies more in favor of the more-substituted isomer when lithium is used as the metal cation than when potassium

Ketone	Conditions (Base)[a,b]	Enolate Composition, %			Ref.

Kinetic

$CH_3CH_2\overset{\overset{\displaystyle O}{\|}}{C}CH_3$

LDA, 0°

Enolate structures:

$\overset{O^-}{\underset{CH_3CH_2}{C}}{=}CH_2$ (with CH_3) — 71

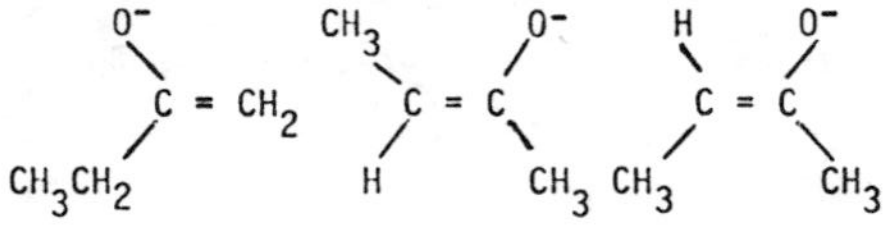

$\overset{O^-}{C}{=}\overset{CH_3}{C}\overset{H}{}$ (CH₃) — 13

$\overset{O^-}{C}{=}\overset{H}{C}\overset{CH_3}{}$ (CH₃) — 16

Ref. 43

Equilibrium

$(CH_3)_2CH\overset{\overset{\displaystyle O}{\|}}{C}CH_3$

KH, THF, 20°

$(CH_3)_2\overset{\overset{\displaystyle O^-}{|}}{C}HC{=}CH_2$ — 88

$(CH_3)_2C{=}\overset{\overset{\displaystyle O^-}{|}}{C}-CH_3$ — 12

Ref. 66

Kinetic

$((CH_3)_3Si)_2NK$, THF, −78°

>99   <1

Ref. 66

Equilibrium

$n-C_4H_9CH_2\overset{\overset{\displaystyle O}{\|}}{C}CH_3$

$\overset{O^-}{\underset{n-C_4H_9CH_2}{C}}{=}CH_2$

$\overset{O^-}{C}{=}\overset{n-C_4H_9}{\underset{H}{C}}$ (CH₃)

$\overset{O^-}{C}{=}\overset{H}{\underset{CH_3\ n-C_4H_9}{C}}$ (CH₃)

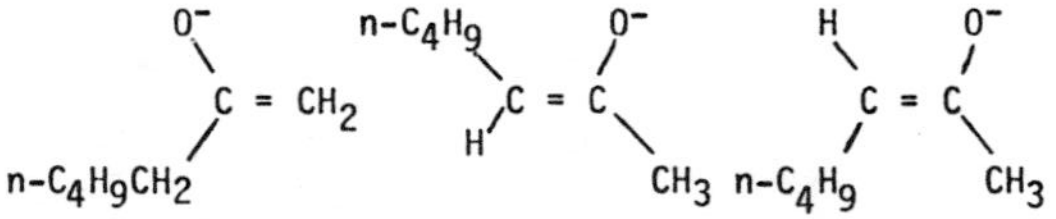

	$(C_6H_5)_3CK$	42	46	12	42
	$(C_6H_5)_3CLi$	13	65	22	42
	KH, THF, 20°	46	54		66

(Continued)

Ketone	Conditions (Base)[a,b]	Enolate Composition, %			Ref.
$C_6H_5CH_2\overset{\overset{\displaystyle O}{\|}}{C}CH_3$	**Kinetic**	$\underset{C_6H_5CH_2}{\overset{\displaystyle {}^-O}{}}C=CH_2$	$\underset{\underset{H}{\|}}{C_6H_5}C=\underset{CH_3}{\overset{O^-}{}}$	$\underset{\underset{C_6H_5}{\|}}{\overset{H}{}}C=\underset{CH_3}{\overset{O^-}{}}$	
	(C₆H₅)₃CK (App. kinetic)	54	37	9	42
	(C₆H₅)₃CK	50	50		46
	(C₆H₅)₃CLi	~88	~12		42
	LDA,0°	84	7	9	43
	LDA,THF,-78°	100	-	-	162
	Equilibrium				
	NaH	2	98		41
$(CH_3)_2CH-\overset{\overset{\displaystyle O}{\|}}{C}-CH_2CH_3$	**Kinetic**	$\underset{CH_3}{\overset{CH_3}{}}C=\underset{CH_2CH_3}{\overset{O^-}{}}$	$\underset{CH(CH_3)_2}{\overset{O^-}{}}C=\underset{H}{\overset{CH_3}{}}$	$\underset{CH(CH_3)_2}{\overset{O^-}{}}C=\underset{\underset{CH_3}{\|}}{\overset{H}{}}$	
	LDA,0°	14	86		43
	Equilibrium				
	(C₆H₅)₃CK	12	74	14	42
	(C₆H₅)₃CLi	<1	>98	<1	42

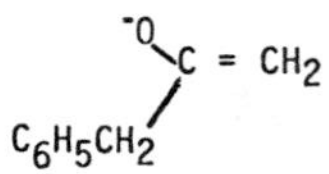

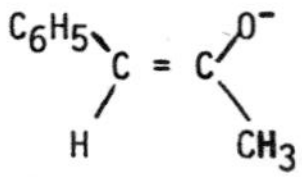

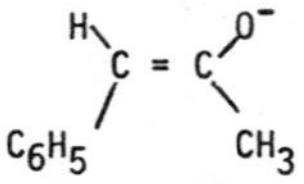

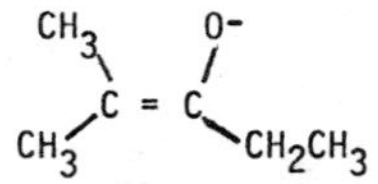

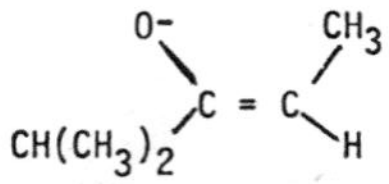

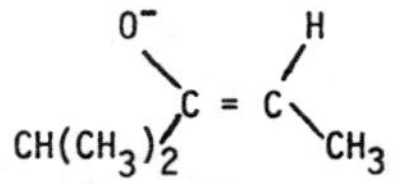

Kinetic

LDA, 0°

$(CH_3)_2CHCH_2\overset{\overset{O}{\|}}{C}CH_3$ 5   42   53   43

Equilibrium

$(C_6H_5)_3CK$

$(CH_3)_2CH\underset{H}{\overset{O^-}{\underset{|}{C}}}=C\overset{CH_3}{\underset{}{}}$ 18 $(CH_3)_2CH\underset{}{\overset{O^-}{C}}=C\underset{H}{\overset{CH_3}{}}$ 7 $(CH_3)_2CH-CH_2-C\overset{O^-}{}=CH_2$ 75   41

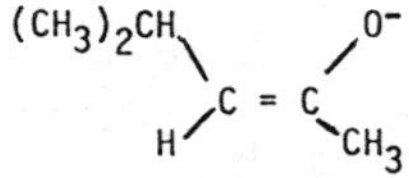

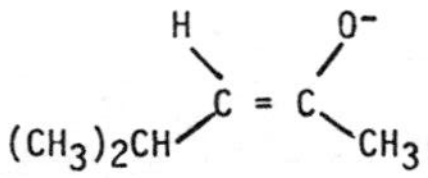

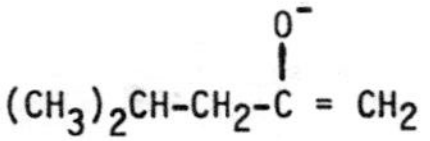

$(C_2H_5)_2CH-\overset{\overset{O}{\|}}{C}CH_2C_2H_5$

$\underset{C_2H_5}{\overset{C_2H_5}{C}}=C\underset{CH_2C_2H_5}{\overset{O^-}{}}$ 3 $(C_2H_5)_2CH-C\underset{C_2H_5}{\overset{O^-}{}}=C\underset{(C_2H_5)_2CH}{\overset{H}{}}$ 49 $\underset{}{\overset{O^-}{C}}=C\underset{H}{\overset{C_2H_5}{}}$ 48   41

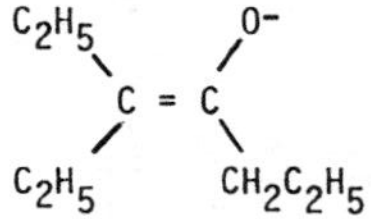

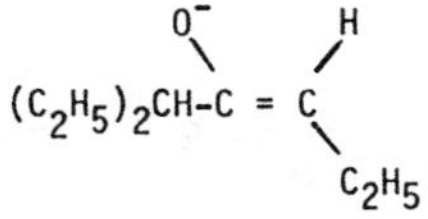

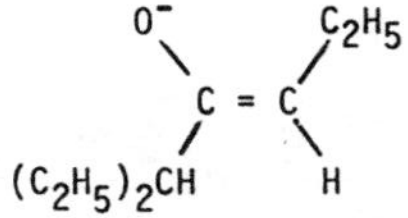

$(C_6H_5)_3CK$

a. The solvent was DME unless specified otherwise.

b. Unless otherwise specified the temperature is assumed to be 25°.

is employed. The explanation for this result seems to lie in the
fact that the oxygen-lithium bond is more covalent than the oxygen-
potassium bond; therefore, more negative charge would reside on the
α-carbon atom in the potassium than in the lithium case [117, 118].
With all simple aliphatic ketones, kinetically controlled
deprotonation favors the production of the less highly substituted
metal enolate. The nature of the base can have a significant effect
on the kinetic composition of enolate mixtures. In DME, because
of its lower reactivity, trityllithium exhibits more regioselectivity
than tritylpotassium. In the same solvent, LDA is considerably
more regioselective than trityllithium is. For this reason it is
clearly the base of choice if the formation of the less substituted
enolate of an unsymmetrical ketone is desired [43].

The rates of equilibration of kinetic mixtures of enolates are
dependent on the nature of the metal cation and the solvent. In DME
at room-temperature potassium enolates undergo equilibration within
a few minutes in the presence of a slight excess of free ketone.
On the other hand, equilibration of lithium enolates requires a
significant excess of the ketone and a time period of 30 min or more
under these conditions. Rates of equilibration of lithium enolates
increase markedly in the presence of dipolar aprotic solvents, such
as DMSO and HMPA, in which solvent-separated ion pairs (cf. *15*) are
favored. Also, in these solvents the position of the equilibrium
among the structurally isomeric enolates is shifted toward the less-
substituted isomer compared with solvents such as DME [41]. As the
degree of dissociation of the oxygen-metal bond increases, the
charge density on the α carbon of the enolate anion increases, and
the less-substituted enolate becomes more favored.

The compositions of enolate mixtures derived from methyl cyclo-
alkyl ketones have apparently not been measured directly, but the
results of alkylation studies indicate that for methylcyclobutyl
ketone (*30*) and methylcyclopentyl ketone (*31*) the more stable enolate
has the double bond exocyclic to the ring, but for methylcyclopropyl
(*32*) and methylcyclohexyl ketone (*33*) the more stable enolate is

derived from enolization toward the methyl group [3]. For ketones
30 and 31 the formation of the more-substituted enolate leads to
relief of eclipsing interactions of groups attached to the ring [161].
However, for ketone 32 enolization toward the three-membered ring
would lead to a significant increase in strain energy [159, 160], and
for 33 enolization toward the ring is unfavorable because of A(1,3)-
strain [143, 161].

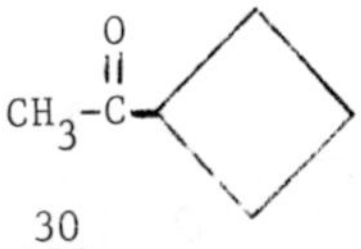

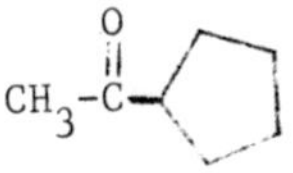

Enolate compositions for several unsymmetrical cyclic ketones
are shown in Table II. For 2-methylcyclopentanone (34) and 2-methyl-
cyclohexanone (5) the more-substituted enolate is favored at equilibrium for all alkali metals. The stability here again parallels
alkene stability, with the enolate having the greater number of substituents on the double bond being preferred. As with acyclic
ketones, the more-substituted enolate is favored to a greater extent
with lithium than with potassium. For both 2-methylcyclopentanone
(34) and 2-methylcyclohexanone (5) the less-substituted enolates are
favored kinetically. Again LDA is a highly selective base for production of the less-substituted enolate. The effect of temperature
on the compositions of kinetic lithium and potassium enolates of 5
is as expected, with the ratio of the less- to the more-substituted
enolate increasing as the temperature is lowered [50, 66].

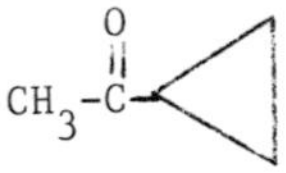

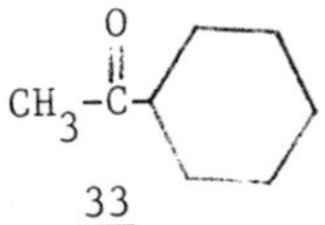

For a given base, the kinetic 6-/2 enolate ratio increases for
2-alkylcyclohexanones as the size of the 2-substituent is increased.
The rate of proton removal from C2 is retarded by the steric effect
of the bulky group [51].

Table II. Compositions of Kinetic and Equilibrium Mixtures of Metal Enolates of Cyclic Ketones in 1,2-Dimethoxyethane

Ketone	Conditions[a,b]	Enolate Mixture Composition; %		Ref.
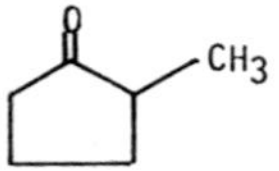		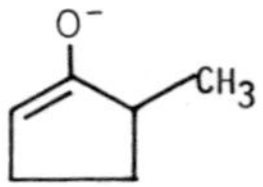	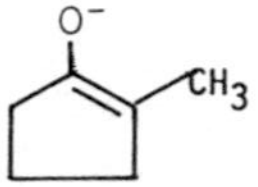	
	Equilibrium			
	$(C_6H_5)_3CK$	22	78	42
	$(C_6H_5)_3CK$	30	70	46
	NaH	29	71	46
	$(C_6H_5)_3CLi$	6	94	42
	Kinetic			
	$(C_6H_5)_3CK$	45	55	42
	$(C_6H_5)_3CLi$	72	28	42
	$(C_6H_5)_3CLi$,THF	78	22	51
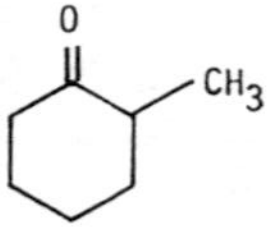	Equilibrium	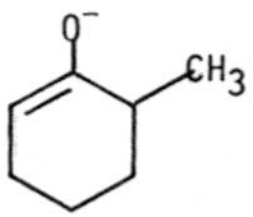	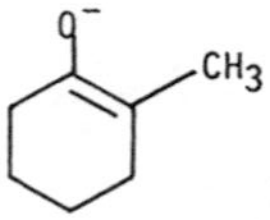	
	$(C_6H_5)_3CK$	35	65	41
	$(C_6H_5)_3CK$	38	62	46
	$(C_6H_5)_3CK$	30	70	50

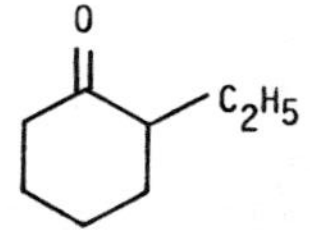

	O⁻ / C₂H₅	O⁻ / C₂H₅	
NaH	26	74	46
(C₆H₅)₃CLi	10	90	50
iPrMgCl, HMPA	9	91	111
Kinetic			
(C₆H₅)₃CK	67	33	46
(C₆H₅)₃CK	68	32	50
(C₆H₅)₃CLi,84°	78	22	50
(C₆H₅)₃CLi	86	14	48
(C₆H₅)₃CLi,-78°	90	10	50
(C₆H₅)₃CLi,THF,-78°	92	8	50
((CH₃)₃Si)₂NK,THF,-78°	95	5	66
LDA, DME, 0°	99	1	43
i-PrMgCl,HMPA	89	11	111
Equilibrium	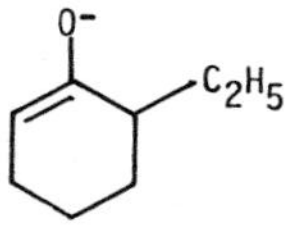	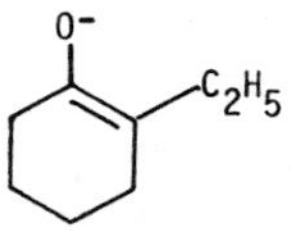	
(C₆H₅)₃CK	48	52	51
(C₆H₅)₃CLi	20	80	51

Ketone	Conditions	Enolate Mixture Composition; %		Ref.
		![enolate with CH(CH₃)₂, non-conjugated]	![enolate with CH(CH₃)₂, conjugated]	
2-isopropylcyclohexanone (O, CH(CH$_3$)$_2$)	**Kinetic**			
	(C$_6$H$_5$)$_3$CK	67	33	51
	(C$_6$H$_5$)$_3$CLi	96	4	51
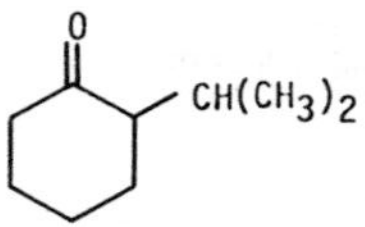	**Equilibrium**	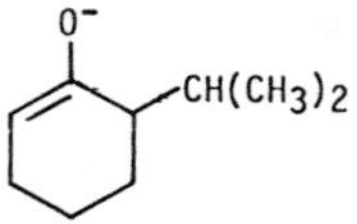	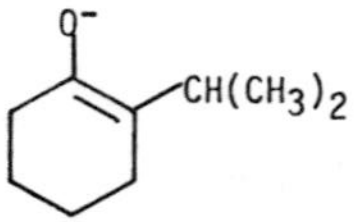	
	(C$_6$H$_5$)$_3$CK	35	65	51
	(C$_6$H$_5$)$_3$CLi	19	81	51
		![enolate CH₃, t-Bu non-conjugated]	![enolate CH₃, t-Bu conjugated]	
2-methyl-4-t-butylcyclohexanone (O, CH$_3$, t-Bu)	**Kinetic**			
	(C$_6$H$_5$)$_3$CK	79	21	51
	(C$_6$H$_5$)$_3$CLi	100	-	51
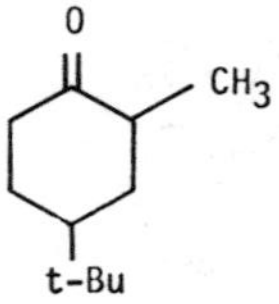	**Equilibrium**	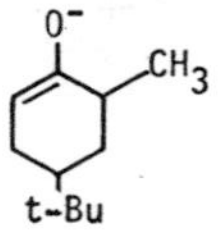	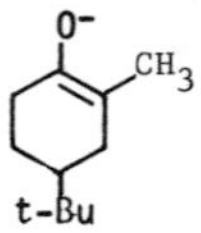	
	(C$_6$H$_5$)$_3$CK	31	69	51
	(C$_6$H$_5$)$_3$CLi	11	89	51
	Kinetic			
	(C$_6$H$_5$)$_3$CK	68	32	51

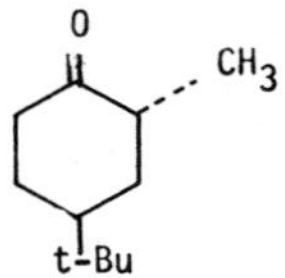

Substrate	Condition	Reagent	Enolate A	Enolate B	Ref.
O=cyclohexanone with CH₃ and t-Bu	Kinetic	$(C_6H_5)_3CK$	100	–	51
		$(C_6H_5)_3CLi$	100	–	51
O=cyclohexanone with CH₃ (3-methyl)	Equilibrium	$(C_6H_5)_3CK$	52	48	163
		$(C_6H_5)_3CNa$	57	43	163
		$(C_6H_5)_3CLi$	54	46	163
		$(C_6H_5)_3CLi,THF$	58	42	163
O=cyclohexanone with C₆H₅ (2-phenyl)	Kinetic	$(C_6H_5)_3CLi$	82	18	163
	Kinetic	LDA	99	1	171

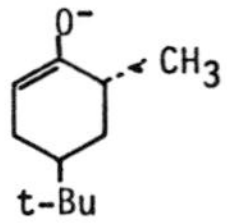

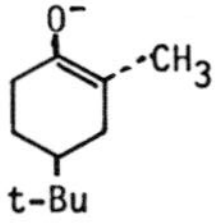

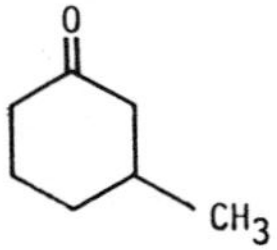

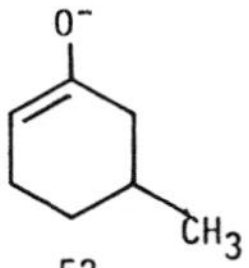

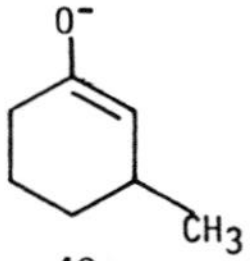

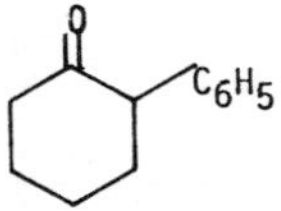

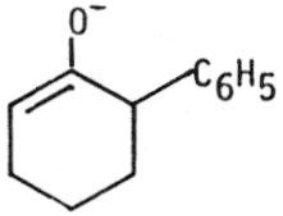

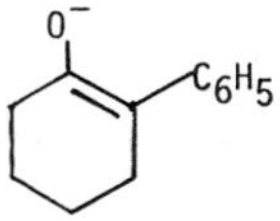

119

Ketone	Conditions	Enolate Mixture Composition; %			Ref.
	Equilibrium				
	NaH	50	50		46
	$(C_6H_5)_3CLi$	66	21	13	42
	Kinetic				
	$(C_6H_5)_3CK$(app.Kinetic)	27	58	13	42
	$(C_6H_5)_3CLi$	10	68	22	42
	LDA	2	71	27	43
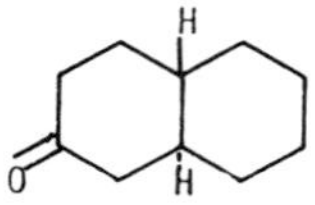	Equilibrium	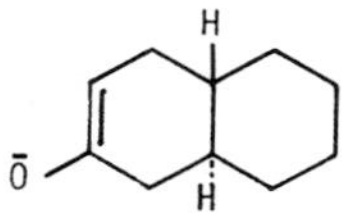	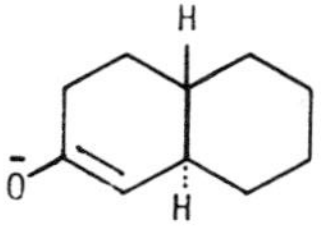		
	$(C_6H_5)_3CLi$	47	53		42

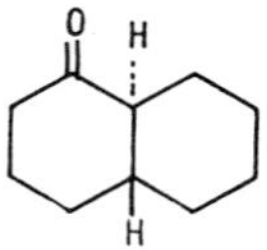

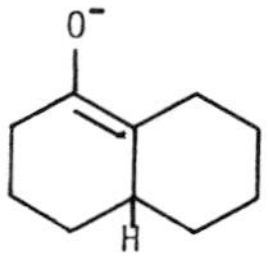

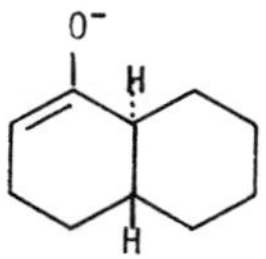

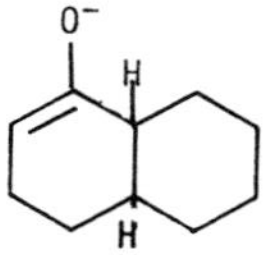

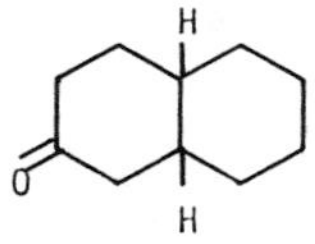

Kinetic				
$(C_6H_5)_3CLi$	87	13		42
$(C_6H_5)_3CK$,(app.Kinetic)	48	52		42
Equilibrium				
$(C_6H_5)_3CLi$	32	68		42
Kinetic				
$(C_6H_5)_3CLi$	32	68		42
$(C_6H_5)_3CK$,(app.Kinetic)	41	59		42

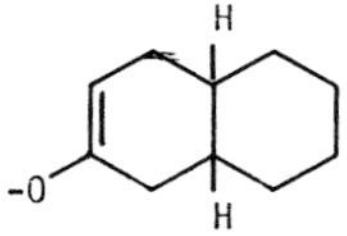

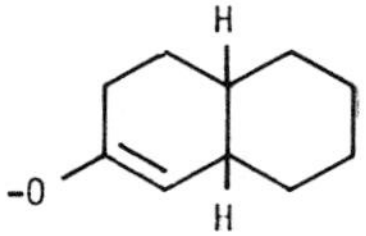

[a] The solvent was DME unless specified otherwise.

[b] Unless otherwise specified the temperature is assumed to be 25°.

It is of interest to note that the 6-enolate (*35A*) of trans-2-
methyl-4-t-butylcyclohexanone (*35*) is formed exclusively with trityllithium or tritylpotassium in DME under kinetically controlled conditions [49]. The usual kinetic preference for formation of the less-
substituted enolate coupled with the importance of stereoelectronic
control in the enolization process, i.e., axial proton removal, provides an explanation for this result.

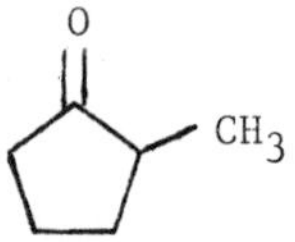

34 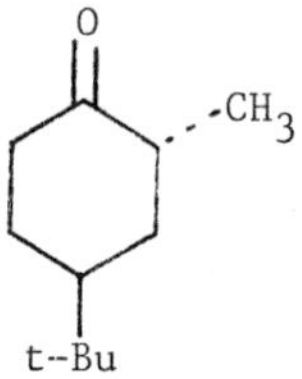   35 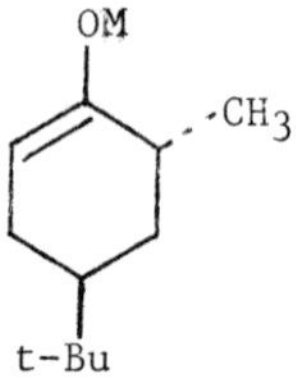   35A

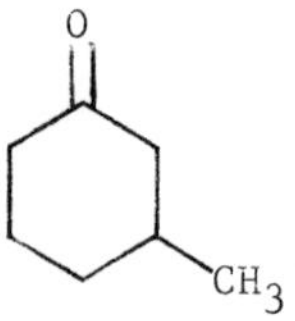

36 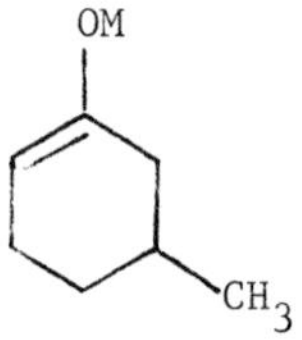   36A 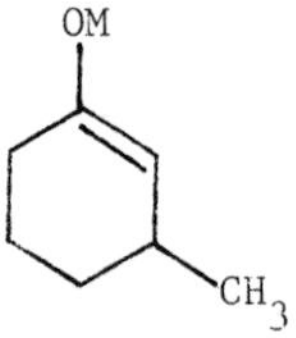   36B

In the case of 3-methylcyclohexanone (*36*) there is a kinetic
preference for the formation of the 6-enolate *36A*, because the 3-
methyl group hinders the approach of the base to C2 [163]. The 6-
enolate is slightly favored thermodynamically as well [163]. A(1,2)-
strain [143] involving the 3-methyl group and the hydrogen at C2 of
the 2-enolate *36B* could account for this result. Interestingly, this
factor does not seem to be as important in determining the position
of the equilibrium in the enolate system as it is in the corresponding
pyrrolidine enamines of *36* in which the thermodynamic 6-/2-enamine
ratio is about 70:30 [143].

The compositions of the kinetic (using trityllithium as the base)
and thermodynamic mixtures of the lithium enolates of 2-
methylcyclohexanone (*5*) in DME compare closely with the ratios

observed for the halomagnesium enolates (using isopropylmagnesium chloride as the base) in HMPA [111]. The difference in the solvent polarities probably accounts for these results. In the same solvent the magnesium-oxygen bond should have more covalent character than the lithium-oxygen bond, and an increased preference for the 2-enolate would therefore be expected when the metal involved is magnesium.

With 1-decalone (*37*) the cis-trans mixture of the 1(2)-enolates *37A* is favored over the 1(9)-enolate *37B* kinetically by a 98/2 ratio when LDA in DME is employed for deprotonation [43]. At equilibrium, enolate *37B* is slightly favored when the cation is lithium, but a 50/ 50 mixture of *37A* and *37B* is observed when the cation is potassium. The 2-enolate of trans-2-decalone (*38*) is preferred kinetically when trityllithium is used as the base, but when tritylpotassium is used, the 2-enolate/1-enolate ratio is about 50/50. It is of interest that the equilibrium mixture of enolates of *38* contains much less of the 2-enolate than would have been expected on the basis of the fact that trans-10-methyl-2-decalone (*10*) and related 5α-3-keto steroids (*39*) are alkylated under equilibrium conditions largely at C3 and C2, respectively. The steric interaction between the 10-methyl group and the 8β proton in the 2-enolate of *10*, that is, *8A*, causes this species to be less stable than in the unsubstituted system *38* [164]. Similar reasoning would account for greater stability of the 2-enolate relative to the 3-enolate in the steroid series [164, 165]. cis-2 Decalone provides an interesting case in which the structurally isomeric enolate ratios are the same under both kinetic and thermo- dynamic control. Studies on the base-promoted alkylation of steroidal ketones indicate that in such systems metal enolate stability rela- tionships closely parellel the corresponding alkene stabilities. The influence of a double bond in the B ring of the cholestan-3-one system upon the position of base-promoted alkylation is of particular interest. 5α-Cholest-6-en-3-one (*40*) undergoes methylation at C2, as does the parent saturated system [166], whereas the presence of the 7,8-double bond in 5α-cholest-7-en-3-one (*41*) causes methylation to occur at C4 [167-168]. In the 5β series, where methylation of the saturated

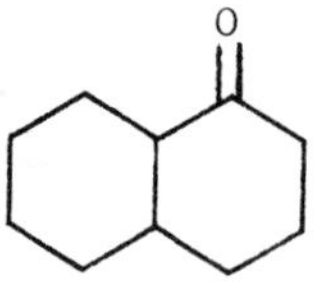

ketone occurs at C4 [37], the introduction of a 7,8-double bond causes
methylation to occur largely at C2 [169]. The influence of the
remote double bonds upon the stabilities of the 2,3- vs the 3,4-
enolates in these systems may be explained in terms of Bucourt's
rules for torsional angle changes [170].

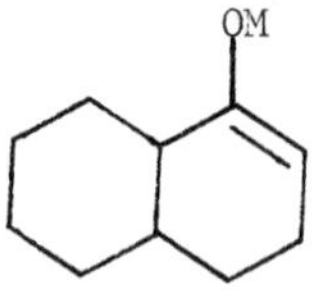

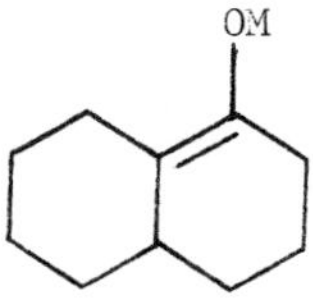

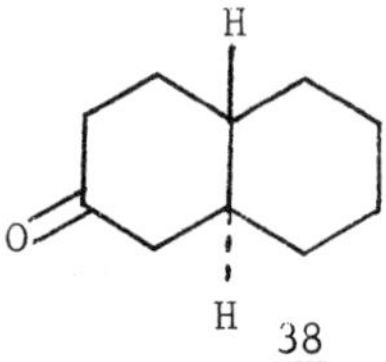

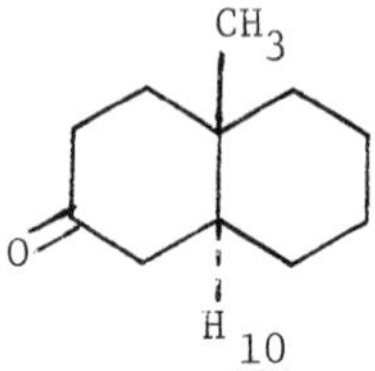

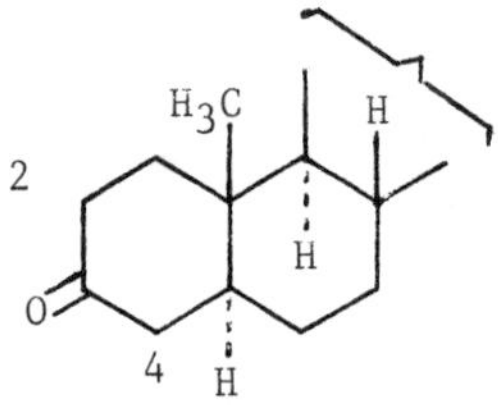

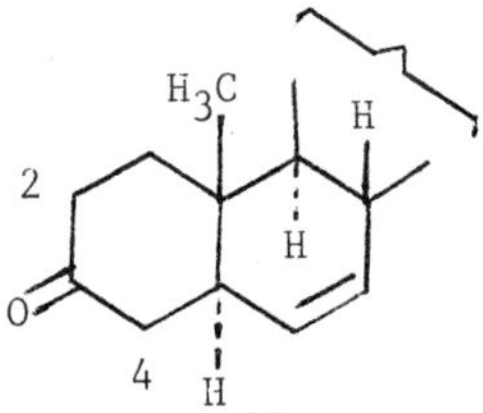

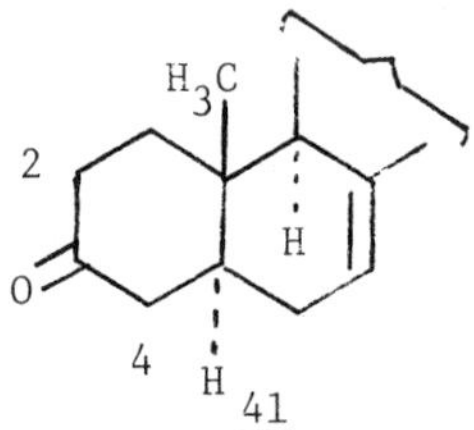

α-Phenyl substituents strongly stabilize metal enolates of
unsymmetrical ketones. Comparison of the results of enolization of
phenylacetone (42) and 2-phenylcyclohexanone (43) is of interest.
In the case of 42, both kinetically and thermodynamically controlled
enolizations lead almost exclusively to mixtures of the E- and Z-
trisubstituted enolates 42A and 42B. The E-enolate 42A (M = Li), is
significantly favored kinetically with LDA. However, with 43 the 6-
enolate 43A (M = Li), is favored over the 2-enolate 43B (M = Li), by
a ratio of 99:1 [171, 172] when LDA in DME is used for kinetic
deprotonation. The results of the alkylation of 43 with methyl iodide

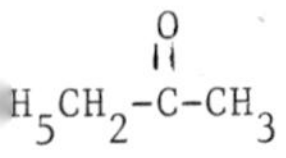

$$H_5CH_2-\overset{\overset{\displaystyle O}{\|}}{C}-CH_3$$

42

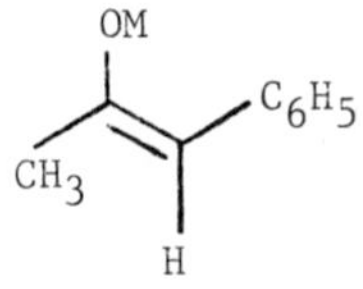

OM
C_6H_5
CH_3
H

42B

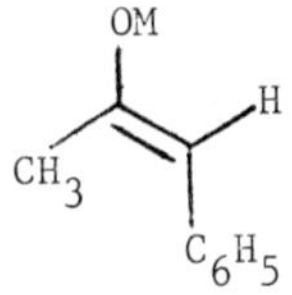

OM
H
CH_3
C_6H_5

42A

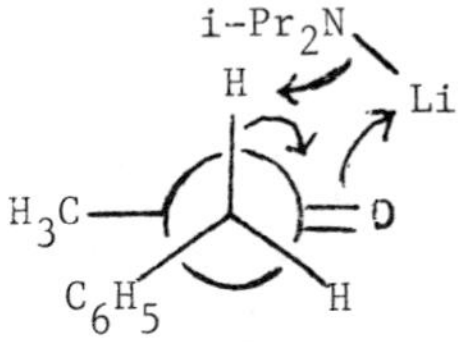

i-Pr$_2$N
H — Li
H_3C
C_6H_5 — O
H

42C

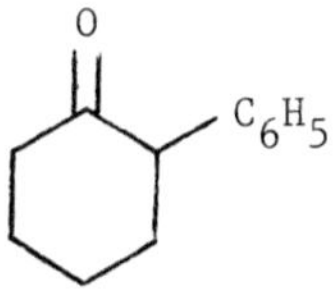

O
C_6H_5

43

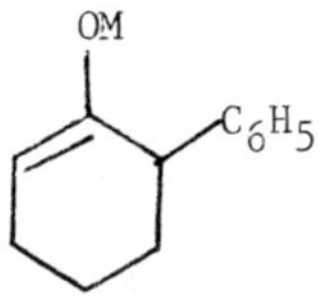

OM
C_6H_5

43A

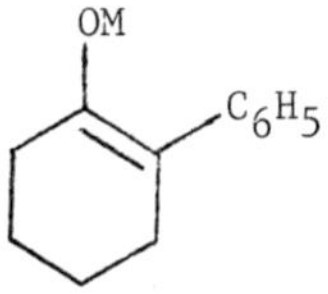

OM
C_6H_5

43B

i-Pr$_2$N
H — Li
O
C_6H_5

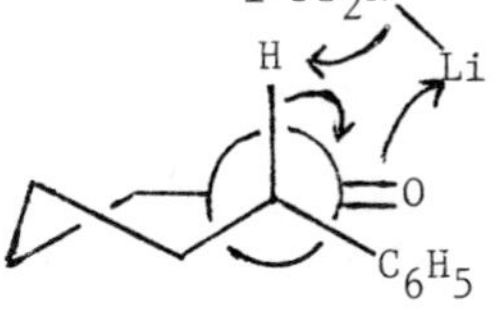

43C

using sodium amide as the base [173] clearly show that the more-
substituted enolate 43B is strongly favored thermodynamically.
Apparently, kinetic deprotonation by LDA of the phenyl-substituted
carbon in 42 is possible because the transition state leading to 42A,
that is, 42C, is sterically favorable and the inductive electron
withdrawal by the phenyl ring enhances the acidity of the methylene
compared with the methyl protons. However, for 43 the acid-
strengthening electron withdrawal by the phenyl group is strongly
overshadowed by the steric effect of this group, which would be
involved in the transition state 43C leading to the 2-enolate 43B.

α-Halogen atoms increase the acidity of α protons of ketones.
α-Haloketones such as 2-chlorocyclohexanone (44) yield exclusively
the more-substituted enolate 44A on treatment with bases such as
sodium methoxide or potassium t-butoxide [174].

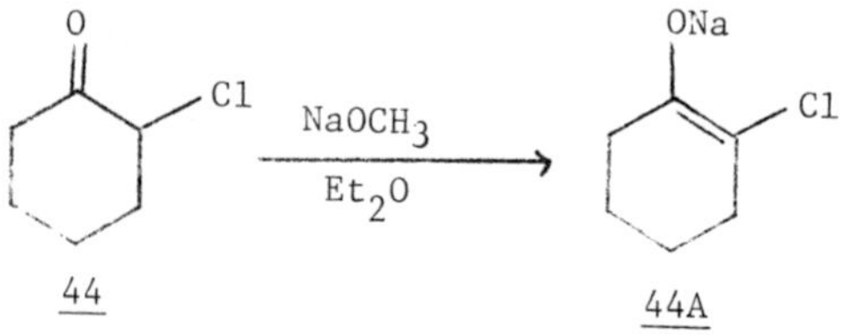

In addition to halogens, a number of other simple hetero-atom
substituents stabilize enolate anions of monocarbonyl compounds.
These include hydroxyl [175], α-thiophenyl [176-180] or thioalkyl
[181-184], α-phenylseleno [185, 186], and α-trimethylsilyl groups
[187, 188]. Under equilibrium conditions, α-thiophenylcyclohexanone
(45) and other cyclic α-thiophenylketones undergo base-promoted
enolization followed by alkylation to yield exclusively products
such as 46, derived from reaction of the hetero-atom-substituted
enolate 45A [176, 180]. Similar results are obtained with α-t-
butylthio acylic ketones such as 47 [181].

However, the α-t-butylthio group apparently does not increase
the kinetic acidity of adjacent protons markedly. For example,
kinetically controlled enolization of 3-t-butylthio-2-octanone (47)

or 4-t-butylthio-3-nonanone with LDA or LHDS in THF followed by addition of C-alkylating agents yields products (*48*) derived from reaction of the unsubstituted enolate *47A*[181].

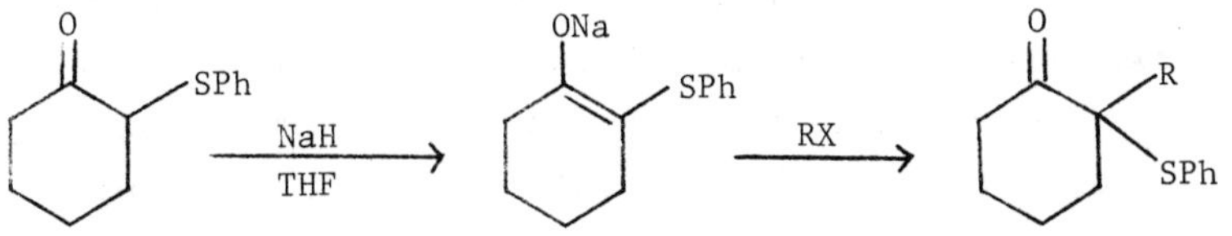

or 4-t-butylthio-3-nonanone with LDA or LHDS in THF followed by addition of C-alkylating agents yields products (*48*) derived from reaction of the unsubstituted enolate *47A*[181].

(scheme 45 → 45A → 46)

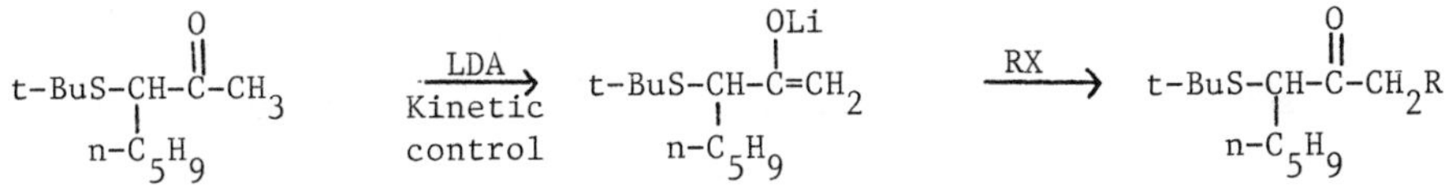

(scheme 47 → 47A → 48)

3. Experimental Procedures for the Deprotonation of Ketones and the Trapping of Metal Enolates

Caution: All of the following procedures should be performed under an anhydrous, oxygen-free nitrogen atmosphere. All reagents and solvents should be anhydrous and all transfers should be carried out with a hypodermic syringe or with a flask-to-flask cannular arrangement. While the enolate solutions prepared as described in this section may be stable under an inert atmosphere at low temperatures for extended time periods, it is recommended that they be prepared immediately before use whenever possible.

a. *Generation of a Kinetic Mixture of Lithium Enolates of an Unsymmetrical Ketone with LDA in DME [43, 45].* The ether is removed under reduced pressure from a solution of 100 mmol of methyllithium containing about 20 mg of 2,2-dipyridyl as an indicator. Anhydrous DME (100 ml) is added, the resulting purple solution is cooled to -50°C, and 10.10 g (100 mmol) of diisopropylamine (freshly distilled from calcium hydride) are added dropwise with stirring over 2-3 min while the temperature is kept below -20°C. The resulting solution is stirred for an additional 3 min, and the temperature is adjusted to a desired point in the 0° to -78^oC range. Methyl ketones and other ketones prone to aldol condensation should be enolized at -78^oC. (LDA slowly cleaves DME at temperatures above 0^oC.) The ketone (99 mmol) in 15 ml of DME is added dropwise with stirring over a 10-min period until the red color of the indicator is almost completely discharged. The kinetic enolate mixture is then ready to be quenched with acetic anhydride or trimethylsilyl chloride to determine its composition.

   b. *Generation of a thermodynamic mixture of lithium enolates of an unsymmetrical ketone [50].* The lithium enolate mixture prepared as described in Sec. II.B.3.a is treated with an additional 20 mmol of the ketone to provide a 20% excess, and the mixture is allowed to equilibrate for a 1- to 18-hr period with stirring at room temperature. Ketones prone to aldol condensation are likely to undergo extensive polymerization during the equilibration period.

   c. *Generation of a kinetic mixture of potassium enolates of an unsymmetrical ketone with tritylpotassium in DME [41, 50].* A solution of tritylpotassium is prepared from 23.4 g (92 mmol) of triphenyl-methane, 3.28 g (85 mg-atm) of freshly cut potassium and 1.0g (19 mmol) of butadiene in 35 ml of DME. The mixture is stirred for 30 min at room temperature, and an additional 300-ml portion of anhy-drous DME is added. The ketone (80 mmol) is then added dropwise with stirring at room temperature over a 15-min period. The mixture is stirred for 15-30 min, and the kinetic enolate mixture is ready to be quenched with an electrophilic reagent.

   d. *Generation of a thermodynamic mixture of potassium enolates of an unsymmetrical ketone in DME.* The potassium enolate mixture prepared as described in Sec. II.B.3.c is treated with an additional 8 mmol of the ketone to provide an approximately 10% excess, and the mixture is stirred for 1 hr at room temperature. The solution of the thermodynamic mixture of potassium enolates is then ready to be quenched with an electrophilic reagent.

e. *Quenching of an alkali metal enolate mixture with trimethyl-silylchloride [43].* A quenching solution prepared from 18.4 g (170 mmol) of trimethylsilyl chloride and 4.4 g (44 mmol) of triethylamine in 50 ml of anhydrous DME and filtered prior to use to remove any triethylammonium chloride is added rapidly with stirring to a solution containing about 100 mmol of a mixture of alkali metal enolates in DME at 0°C. The resulting mixture is allowed to warm and then is partitioned between cold pentane and cold aqueous sodium bicarbonate. The organic layer is separated, dried, and concentrated to give a mixture of trimethylsilyl enol ether derivatives of the ketone.

f. *Quenching of metal enolate mixtures with acetic anhydride [41].* A solution containing about 100 mmol of a mixture of metal enolates in DME is added dropwise with stirring over 30 min to freshly distilled acetic anhydride (0.5-1 mol) at room temperature. The mixture is stirred for 30 min at room temperature and then poured into a mixture of 500 ml of pentane and 500 ml of a saturated solution of aqueous sodium bicarbonate, cooled to 0°C, and stirred for 30 min while solid sodium bicarbonate is added until the evolution of carbon dioxide ceases. The pentane layer is separated, dried, and concentrated to give the crude mixture of enol acetates.

C. *Regiospecific Formation of Metal Enolates*
   of Saturated Ketones

1. Introduction

It should be clear from the preceding discussion that under either kinetic or thermodynamic conditions the base-promoted enolization of most unsymmetrical ketones yields mixtures of metal enolates. Therefore, in order to prepare a particular α-substituted ketone or enol derivative, it is often necessary to regiospecifically form the desired metal enolate by an indirect method and trap it with an appropriate electrophilic reagent. In this section some of the more important methods of generating specific metal enolates will be described.

2. Metal-Ammonia Reduction of α,β-Unsaturated Ketones

Since the original investigations of Stork and co-workers [38, 39], the lithium-ammonia reduction of α,β-unsaturated ketones has been widely used to generate specific lithium enolates of decalones, hydrindanones, cyclohexanones, acyclic unsymmetrical ketones, and steroidal and related ketones which may be trapped with relatively reactive alkylating agents and other electrophilic species [132]. The general type of transformation involved is illustrated in Eq. (9).

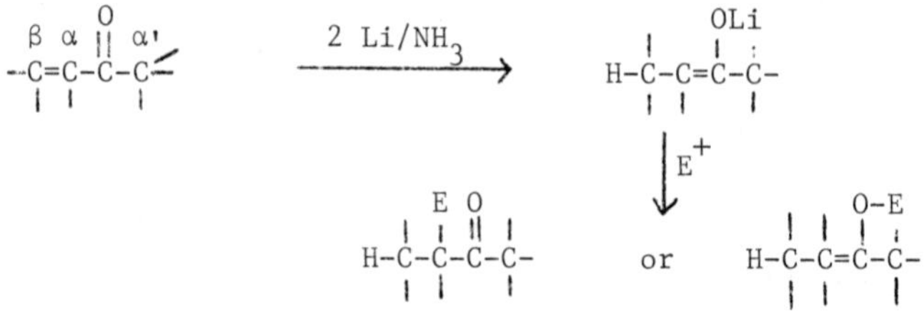

$$(9)$$

Sodium and potassium enolates are formed on reduction of α,β-unsaturated ketones with the corresponding metals, but the value of these enolates is quite limited because their equilibration via proton transfer reactions is generally too rapid to allow trapping with alkylating agents. Calcium and magnesium enolates are also obtained by enone reductions with the corresponding metals in liquid ammonia. However, reductions with these metals often lead to the formation of saturated alcohols even in the absence of proton donors [189, 190]. This suggests that calcium-ammonia and magnesium-ammonia solutions are sufficiently acidic to bring about protonation of the corresponding metal enolates [132]. To the extent that these enolates are destroyed in liquid ammonia, the product yields in trapping reactions would be lowered.

The generation of a specific lithium enolate is usually carried out by adding the appropriate enone to a solution of 2 eq of lithium in liquid ammonia usually containing a cosolvent such as diethyl ether or THF. If no proton donor is added, 1 eq of lithium amide will be formed along with the metal enolate. Better yields of reduction products are often achieved in lithium-ammonia reductions,

particularly of β-unsubstituted cyclohexenones [191] and acyclic
ketones [192], if 1 eq of a proton donor is added along with the
enone. This leads to the formation of 1 eq of the conjugate base of
the proton donor. Lithium amide and other relatively strong bases
such as lithium t-butoxide are able to convert neutral reaction products such as monoalkylated ketones into the corresponding metal
enolates. This leads to the formation of significant amounts of
dialkylated materials. Enolates to be used directly in alkylation
reactions should be prepared by adding 1 eq of water along with the
enone to the lithium-ammonia solution [191-194]. The relatively
weak base lithium hydroxide, which is formed when water is the proton
donor, is not effective in promoting enolization of alkylation products, and water does not appear to cause a significant amount of
over-reduction of the enone to the saturated alcohol [191, 194].

Various electrophilic reagents may be added directly to solutions
or suspensions of metal enolates in liquid ammonia containing cosol-
vents. However, if the enolate trapping reaction is slow or the
electrophilic reagent is highly reactive toward ammonia, the formation
of acidic ammonium salts may occur [192], which will bring about protonation of the metal enolate to produce the simple reduction product of
the unsaturated ketone [Eq. (10)]. If this is the case, the ammonia may

$$RX \ + \ NH_3 \ \longrightarrow \ RNH_3^+ \ Cl^-$$

$$RNH_3^+ \ + \ H-\overset{|}{\underset{|}{C}}-\overset{OLi}{\underset{|}{C}}=\overset{|}{C}-\overset{|}{\underset{|}{C}}- \ \longrightarrow \ H-\overset{|}{\underset{|}{C}}-\overset{H}{\underset{|}{C}}-\overset{O}{\underset{}{\overset{||}{C}}}-\overset{|}{\underset{|}{C}}- \tag{10}$$

be removed completely and the desired solvent added prior to addition
of the electrophilic reagent. The reduction of 3-methylcyclohexenone
(*49*) to the lithium enolate *36B* followed by addition of allyl bromide
to produce 2-allyl-3-methylcyclohexanone (*50*) provides an example of
this sequence [194].

An interesting method of selective geminal alkylation of methylene
ketones, which is based on the enone reduction-enolate trapping

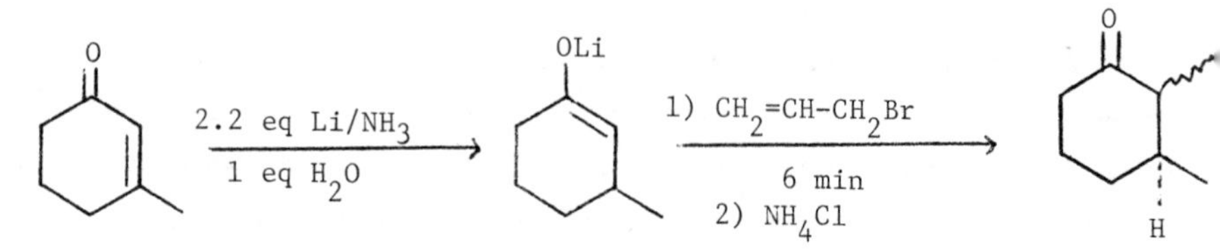

49   36B   47%   50

                                                              20:1   trans:c

sequence, has been developed recently by Coates and Sowerby [193].
These investigators found that n-butylthiomethylene derivatives of
ketones undergo double reduction with lithium in liquid ammonia to
form α-methyl-substituted lithium enolates that can be trapped with
methyl or other alkyl halides, as shown in Eq. (11). This sequence,

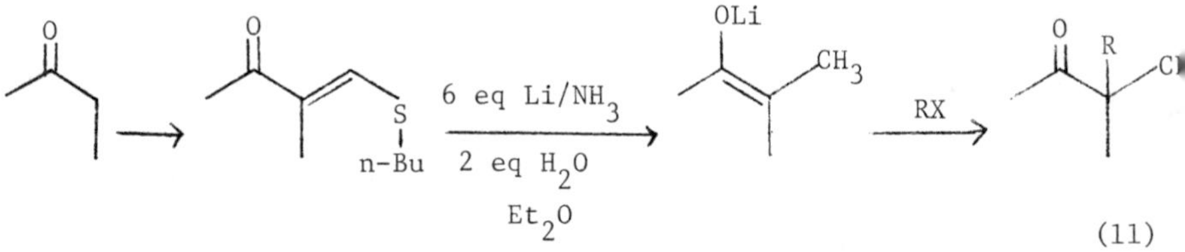

(11)

like the reduction of 9-methyl-6-octalin-1-one to 2,9-dimethyl-6-
octalin-1-ol reported by Ireland and Marshall [34], presumably proceeds via enolate elimination of the n-butylthio anion [193].

    3. Metal-Ammonia Reductions of Ketones
        with Leaving Groups at the α-Position

    Chemical reductions of ketones which have leaving groups at the
α-position may be used to generate specific metal enolates [Eq. (12)]
The lithium-ammonia reductions of α-chloromercuri- [59], α-acetoxy-
[195-197], α-thioalkyl- [181], and α-thiophenyl ketones [176] provide
examples of such reactions. Lithium enolates have also been prepared
by reduction of α-bromoketones with lithium in HMPA-ether [198].

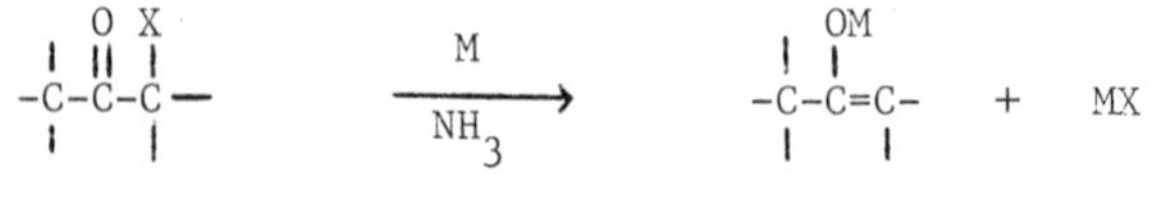

(12)

Sodium, calcium, and barium in liquid ammonia have been employed to
reduce α-acetoxy steroidal ketones to the corresponding metal enolates,
but rather poor yields of products were obtained after trapping
reactions [196]; overreduction of the enolate to the saturated
alcohol is a possible explanation for these results. Lithium-ammonia
reduction of the tetrahydropyranyl ether of 17α-acetoxypregn-5-en-
3-ol-20-one provides an excellent route to pregn-20-one-17-enolate
anions, which have been methylated in high yield [195].

One of the major problems associated with the chemical reduction
of α-halo ketones is that aldol condensation of the enolate with the
unreduced ketone is often a competing side reaction [Eq. (13)] [199].
The γ-halo-β-alkoxy ketones formed in this way can yield furans on
acidification, when structural features permit [199].

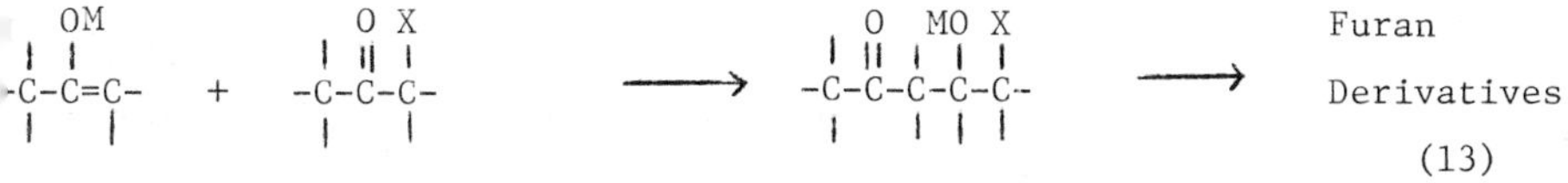

$$
\begin{array}{ccc}
\underset{\substack{|\;\;\;|\\-C-C=C-\\|\;\;\;\;\;\;|}}{OM} & + & \underset{\substack{|\;\;|\;\;|\\-C-C-C-\\|\;\;\;\;\;\;|}}{O\;\;X} \end{array}
\longrightarrow
\underset{\substack{|\;\;\;||\;|\;\;|\;\;|\\-C-C-C-C-C-\\|\;\;\;\;\;|\;\;|\;\;|}}{O\;\;\;MO\;X}
\longrightarrow
\begin{array}{c}\text{Furan}\\\text{Derivatives}\end{array}
$$

$$\tag{13}$$

2-Chloro-2-methylcyclohexanone (*51*) is converted to the lithium
enolate *5B* on reduction with lithium in liquid ammonia at high dilution, because the addition of excess methyl iodide gives 2,2-
dimethylcyclohexanone in 85% yield [50]. However, attempted reduction
of 2-bromocyclopentanone to the corresponding enolate failed, apparently because of the high propensity of cyclopentanone enolates to
undergo aldol condensation [200].

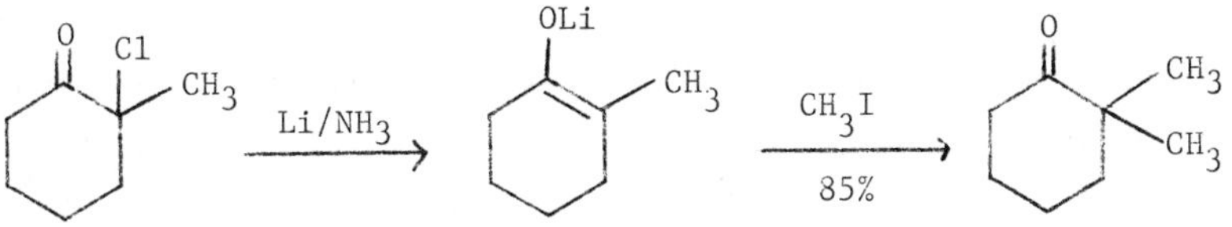

The lithium-ammonia reductions of α-thiophenyl ketones have been
used by Coates and co-workers [176] to prepare lithium enolates of

cyclic ketones, and Nagata and co-workers [181] have prepared enolates
of acyclic ketones by similar reductions of t-butylthio ketones.
The excellent yields of enolate alkylation products obtained in these
reactions indicate that α-thioaryl or α-thioalkyl groups are the
best leaving groups in reductions of this type.
The reduction of α-bromo derivatives of large-ring cyclic ketones
or hindered ketones with zinc [199] in benzene-DMSO or magnesium
[201] in the presence of alkylating agents has been used to prepare
the corresponding α-alkyl ketones. These reactions apparently involve
metal enolate intermediates, but in these cases the enolates were
generated in low concentration and trapped immediately. Bromomagnesium enolates have been preformed from α-bromoketones for use in
studies of aldol condensations [202].

Reductive opening of conjugated cyclopropyl ketones with lithium
in liquid ammonia also provides a means of regiospecific generation
of lithium enolates which can be trapped with electrophilic reagents
without loss of structural integrity [203-205]. The reductive
cleavage of the cyclopropyl ketone 52 followed by alkylation of the
enolate 52A with an allyl bromide provides a recent example of this
procedure [205].

$$\underset{52}{} \xrightarrow[\text{DME}]{\text{Li/NH}_3} \underset{52A}{} \xrightarrow[70\%]{\text{CH}_2=\text{CH-CH}_2\text{Br}} $$

52   52A

Reductive cleavage of α,β-epoxy ketones with lithium in liquid
ammonia yields β-alkoxy lithium enolates which may be alkylated and
dehydrated to produce α-alkyl-α,β-unsaturated ketones [206] [Eq. (14)]

4. Conjugate Additions of Organometallic
   Reagents to α,β-Unsaturated Ketones

In recent years there have been many reports of the cuprous-ion-
catalyzed conjugate addition of Grignard reagents [46, 158, 207-214]

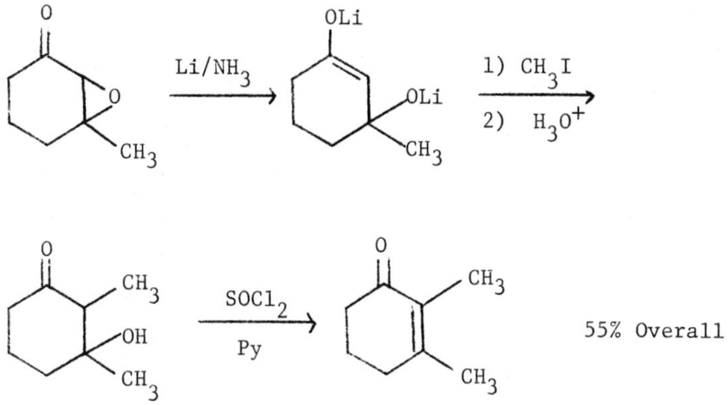

(14)

55% Overall

or of lithium dialkylcuprates [97, 171, 215-238] to α,β-unsaturated ketones to produce metal enolates, which can be trapped under non-equilibrating conditions with alkylating agents or other electrophiles. Reactions of this type permit the introduction of substituents at both the α and β positions of α,β-unsaturated ketones in a "one-pot" operation [Eq. (15)] [207].

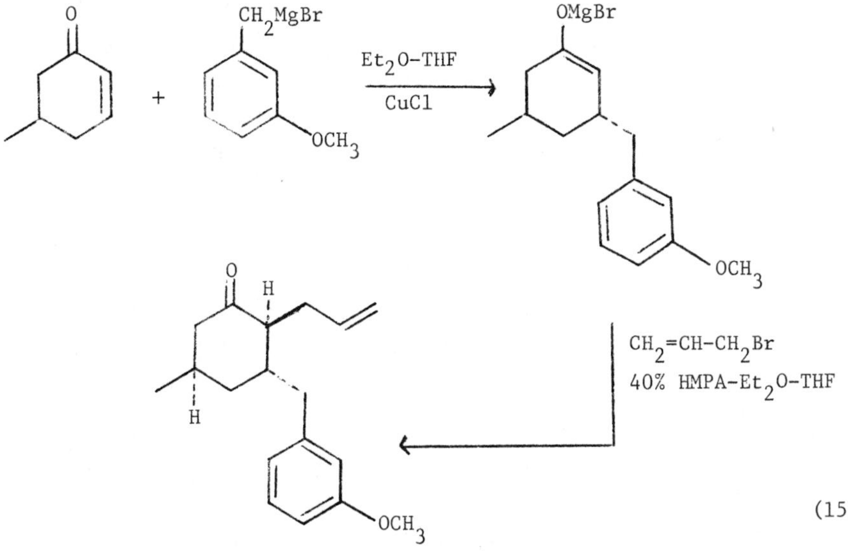

(15)

The nature of the metal enolates obtained upon addition of lithium dialkylcuprates to enones has been a controversial subject.

These species have been formulated as lithium enolates [238-240], copper (I) enolates [220-223, 225], or species having copper bonded to the α carbon or to the Π system of the enolate [236]. Early studies involving reactions of such enolates indicated that they might have properties different from those of the corresponding pure lithium enolates [193, 220, 221]. However, the observed differences may be attributed to the fact that alkylation reactions conducted in diethyl ether (a good solvent for conjugate addition of the cuprate to the enone) were being compared with lithium enolate alkylations conducted in DME. Coates and Sandefur [225] have demonstrated that in pure DME, "copper-lithium" enolates gave alkylation rates and product yields comparable with those of lithium enolates. Boeckman [222] has made similar observations regarding the Michael addition of the enolate 36B to α-silylated vinyl ketones.

House and Wilkens [238] have shown quite recently that the ether solution of the enolate, produced by addition of lithium dimethyl-cuprate to 3-methylcyclohexenone (49) at 25°C followed by removal of insoluble methyl copper, contained a negligible amount of copper and exhibited ^{13}C nmr absorptions which corresponded to those obtained for an ether solution of the pure lithium enolate 52A. These obser-vations were in line with earlier work of House and Fischer [239] which showed that added soluble copper(I) compounds seemed to have no effect on the rates of Michael and alkylation reactions of lithium enolates. Thus, it seems clear that lithium enolates rather than copper(I) species are the reactive intermediates formed on the addi-tion of lithium dialkylcuprates to enones at 25°C. However, Katzenellenbogen and Crumrine [241] have found that the metal enolates [presumably copper(I) dienolates] of α,β-unsaturated carboxylic acids, which are produced by addition of copper(I) salts to dilithium dienolates, exhibit regioselectivity in alkylation reactions remark-ably different from the corresponding dilithium species themselves.

1,4-Additions of lithiated bis(methylthio)silyl and bis(methyl-thio)stannyl methanes [242], trimethylsilyllithium [243], and lithiated protected cyanohydrins [244] to cyclic α,β-unsaturated

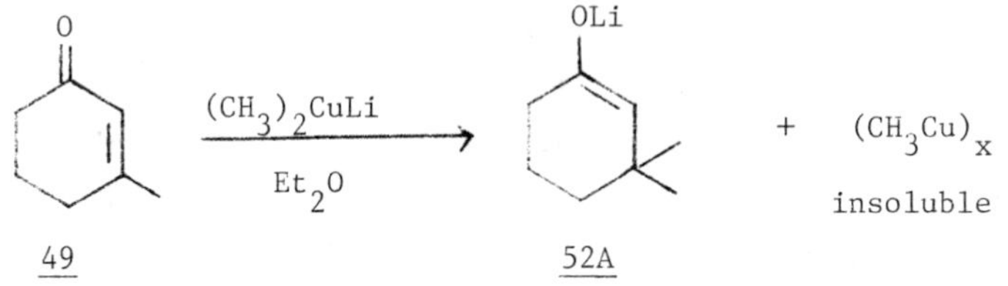

ketones provide interesting examples of other reactions which yield
specific lithium enolates. The latter reaction applied to cyclopentenones followed by enolate alkylation provides a convenient route
to 13-oxoprostanoids [244].

Conjugate addition of trimethylaluminum catalyzed by nickel
acetylacetonate [245] and free-radical conjugate addition of trialkyltin hydrides [246-248] provide routes to the corresponding dimethylaluminum enolates or trialkyltin enol ethers which may be reacted
directly with electrophilic reagents. Simple potassium or lithium
enolates or potassium or lithium trialkylborate enolates which may be
alkylated are produced on reduction of β-unsubstituted cyclohexenones
with potassium or lithium tri-sec-butylborohydride in THF [249, 250].

5. Cleavage of Enol Derivatives of Ketones

The reaction of a number of enol derivatives (including enol
acetates [4, 119, 251], trialkylsilyl enol ethers [43, 44, 46, 47,
223, 228], vinyloxyboranes [252], enol phosphorylated species [150,
253], and trialkyltin enol ethers [254]) with organolithium compounds
[43, 44, 46, 47] or Grignard reagents [46, 47] (or lithium amide
[223, 228], in the case of silyl enol ethers) has been used to produce
specific metal enolates which may be reacted with electrophilic
reagents under nonequilibrating conditions.

The procedure involving cleavage of enol acetates with 2 eq of
methyllithium in solvents such as DME was developed by House and
Trost [119] and is illustrated in Eq. (16). The enolates required

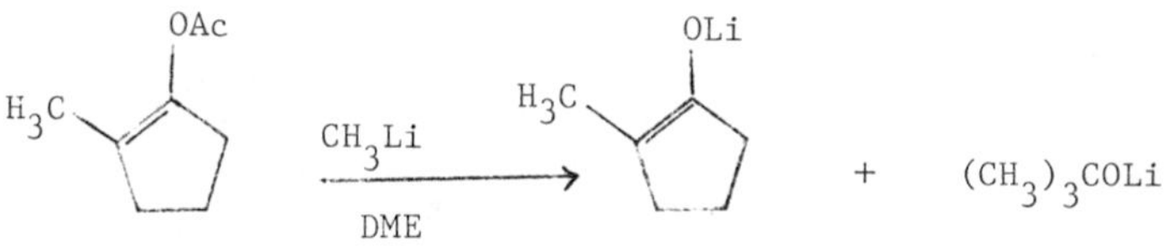

for such transformations may be obtained by acetylation of metal
enolates with excess acetic anhydride or acetyl chloride. They may
also be prepared by reaction of ketones with isopropenyl acetate or
acetic anhydride in the presence of a catalytic quantity of an acid
such as p-toluenesulfonic acid (PTSA) [119]. A single enol acetate
is obtained by these reactions from symmetrical ketones and ketones
that can enolize in only one direction. However, with unsymmetrical
ketones, mixtures of enol acetates are usually formed. When isopropenyl acetate with a trace of PTSA is employed for the acetylation,
the less-substituted enol acetate of an unsymmetrical ketone usually
forms faster, and the reaction may be halted while it is the predominant product. In many cases the less highly substituted enol acetate
can then be separated from the mixture by chemical or chromatographic
methods. Acid-catalyzed equilibration of enol acetate mixtures or
direct reaction of unsymmetrical ketones with acetic anhydride in
the presence of PTSA or perchloric acid gives thermodynamic mixtures
of enol acetates in which the more highly substituted isomer greatly
predominates. For example, enol acetylation of 2-methylcyclohexanone
(5) and 1-decalone (37) under thermodynamic conditions gives mixtures
containing >95% and ∿94%, respectively, of the more substituted isomers [45, 119]. Cleavage of thermodynamic enol acetate mixtures is
often the best method available for producing more highly substituted
metal enolates of unsymmetrical ketones [4].

The limitations of this method are

1. The desired enol acetate is often not readily available
2. One equivalent of the strong base, lithium t-butoxide, is
   produced along with the enolate.

This base may cause fairly rapid enolization of the products of monoalkylation of the enolate and therefore lead to the formation of
significant amounts of polyalkylated materials.

The cleavage of silyl enol ethers with organolithium reagents
in diethyl ether or DME or Grignard reagents in DME to produce specifi◄
lithium or halomagnesium enolates, introduced initially by Stork and
Hudrlik [46, 47], has been widely used as a method of producing

specific metal enolates. The cleavage of the trimethylsilyl enol
ether of cyclohexanone *53* to the lithium enolate *54* (M = Li), with
methyllithium provides an example of this reaction. Trimethylsilyl
enol ethers are cleaved more rapidly than the hydrolytically more
stable t-butyldimethylsilyl compounds are, but even cleavage of the

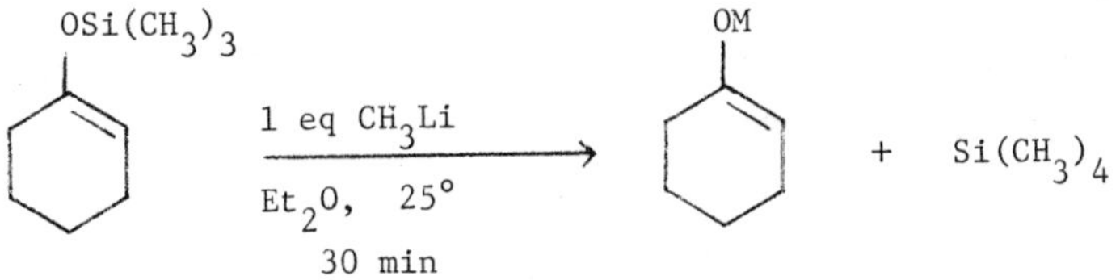

former species with Grignard reagents is relatively slow; approximately
24 hr at reflux in DME is the required reaction time. It has recently
been shown that trialkylsilyl enol ethers are cleaved at convenient
rates to the corresponding lithium enolates and trialkylsilylamines
with lithium amide in liquid ammonia [223, 228].

The availability of the desired silyl enol ether is of course
the major limitation of this procedure. These compounds may be
obtained by the trapping of metal enolates produced by base-promoted
enolization of ketones or by various indirect methods, such as metal-
ammonia reduction of α,β-unsaturated ketones and 1,4-additions of
dialkylcuprates to enones. House and co-workers [43] have shown that
treatment of ketones with trimethylsilyl chloride in the presence of
triethylamine or other tertiary amines in refluxing DMF yields silyl
enol ethers. Symmetrical ketones and those which may enolize in only
one direction yield a single product. However, with unsymmetrical
ketones, equilibration of structurally isomeric species occurs under
the reaction conditions, and after extended reaction times equilibrium
mixtures of products are obtained. At equilibrium the more highly
substituted silyl enol ether predominates, but usually not as greatly
as in the case of the corresponding enol acetate [43]. An improved
procedure for the synthesis of silyl enol ethers which involves the
treatment of ketones with ethyl trimethylsilylacetate using

tetra-n-butylammonium fluoride as a catalyst has been reported [255].
It is sometimes feasible to separate a desired enol silyl ether from
a mixture of isomers by chromatographic methods.

Thermal rearrangements of trimethylsilyloxyvinylcyclopropanes
(55 → 56), reported by Trost and Bogdanowicz [256], and silatropic
rearrangements of trimethylsilyl β-keto esters (57 → 58), reported
by Coates et al. [257], provide two interesting new methods for the
synthesis of specific silyl enol ethers.

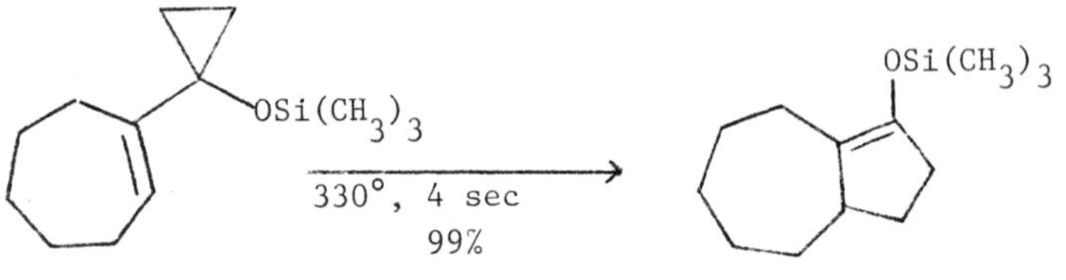

In cleavages of silyl enol ethers with organolithium or Grignard
reagents only the metal enolate and 1 eq of the volatile, relatively
inert tetraalkylsilane are formed [46, 47]. Thus, minimal amounts of
polyalkylation products are obtained when enolates produced in this
way are treated with alkylating agents. For this reason it is often
desirable to trap specific enolates generated by various indirect
methods with trimethylsilyl chloride or t-butyldimethylsilyl chloride,
to isolate, purify, and characterize the enol silyl ether product,
and to subject it to cleavage in a subsequent step [188, 212]. In
this way a relatively "clean" solution of the enolate is obtained.

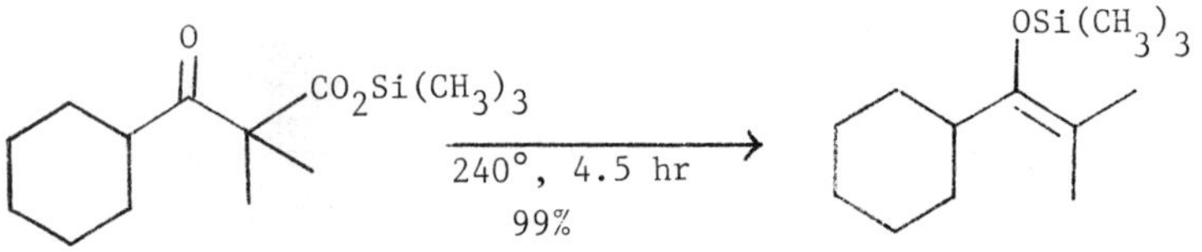

Kuwajima and Nakamura [258] have recently reported an interesting method of regiospecific alkylation of ketones in which a trimethylsilyl enol ether in THF is treated with benzyltrimethylammonium

fluoride (BTAF) in the presence of an alkylating agent [Eq. (17)].
These authors have suggested that this reaction proceeds via a highly
reactive quaternary ammonium enolate, which is trapped without
equilibration. This is difficult to understand in view of the fact
that potassium and sodium enolates undergo rapid equilibration (vide
infra). It seems possible that ionic pentavalent silicon species,
which could be sufficiently electrophilic to react with alkyl halides,
may be involved.

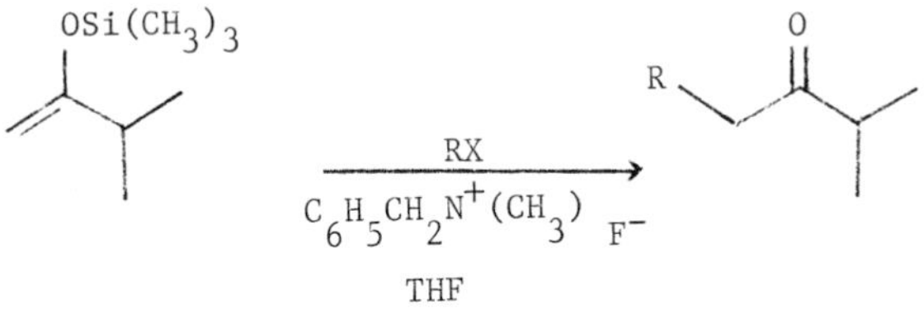

$$OSi(CH_3)_3 \xrightarrow[\substack{C_6H_5CH_2N^+(CH_3) \ F^- \\ THF}]{RX} R\text{-C(=O)-} \tag{17}$$

Vinyloxyboranes, which may be formed by the reaction of diazo
ketones with trialkylboranes or by radical addition of trialkylboranes
to vinyl ketones, may be reacted with 2 eq of methyl- or n-
butyllithium to form the corresponding lithium enolates and lithium
tetraalkylboronates [252]. The formation of the more substituted
enolate of 2-n-butylcyclohexanone from diazocyclohexanone illustrates
this procedure [Eq. (18)]. A single vinyloxyborane is produced in
the initial step so that separation of enol derivatives is not
required.

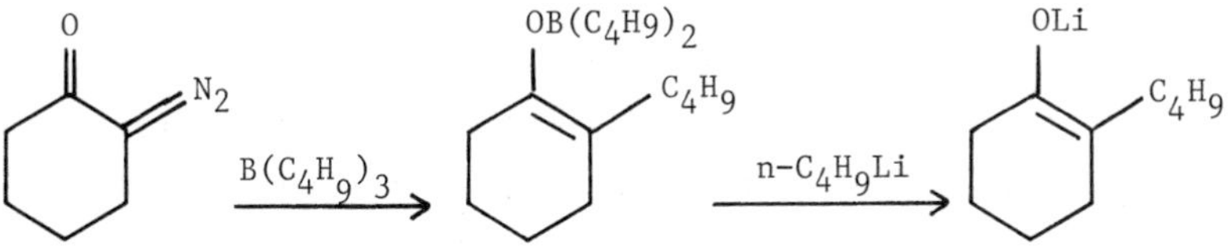

$$\tag{18}$$

Borowitz and co-workers [150] have prepared specific enolates
by the cleavage of enol phosphinates, phosphonates, or phosphates
with organolithium reagents in DME. The enol phosphorylated species
may be prepared by reaction of α-bromo ketones with the appropriate
phorphorous reagents. The phosphorus-containing by-products of the

cleavage reaction presumably do not interfere with subsequent reactions of the lithium enolates. The conversion of 2-bromo-6-
methylcyclohexanone (*59*) to the less-substituted lithium enolate of
2-methylcyclohexanone (*5A*) using a triethylphosphate intermediate
provides an illustration of this reaction.

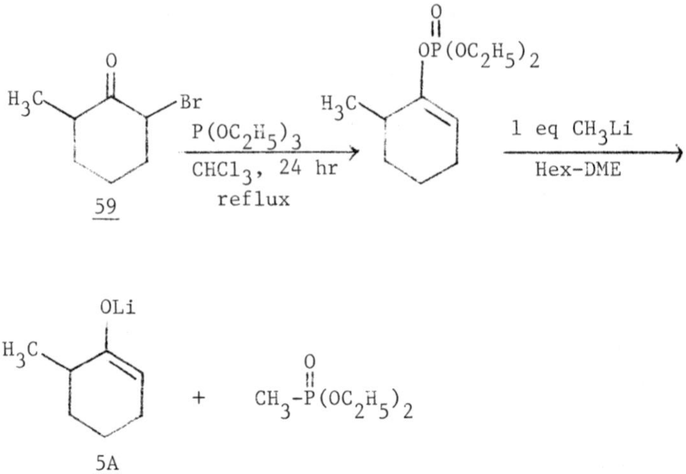

The cleavage of trialkyltin enol ethers with methyl- or phenyllithium and with methylmagnesium iodide has also been used to produce
specific lithium or halomagnesium enolates of cyclic and acyclic
ketones [254].

6. Reaction of α-Halo- or α,α'-Dihaloketones
   with Organometallic Reagents

The reaction of lithium dialkyl or lithium t-butoxyalkylcuprates
with cyclic α,α'-dibromoketones such as *60* leads to the formation of

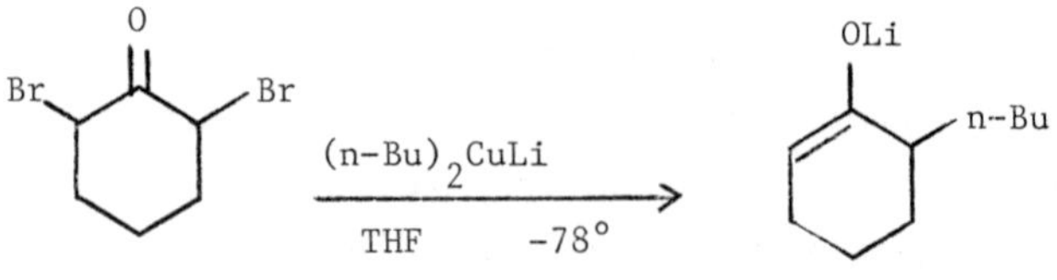

specific lithium enolates, such as 61, of alkylated ketones, which
may be trapped with electrophiles [227]. Likewise, acyclic α,α'-
dibromoketones undergo similar reactions [259].

A number of other reactions in which α-bromo ketones have been
converted to α-alkyl ketones by treatment with organometallic reagents
presumably involve halogen-metal exchange, perhaps via an electron
transfer pathway [215], and proceed by way of metal enolate intermediates. These include the formation of 9α-methyl-11-keto steroids
by reaction of 9α-bromo-11-keto steroids with methylmagnesium iodide
in the presence of excess methyl iodide [251, 260, 261] as well as
the reaction of acyclic α-bromo ketones with primary and secondary
lithium dialkylcuprates [262]. Treatment of α-bromoketones having
a hydrogen atom at the β position with lithium di-t-butylcuprate
leads to elimination of hydrogen bromide followed by conjugate addition of the cuprate to the enone [263]. α-Chlorocyclohexanone primarily undergoes aldol condensation in the presence of lithium dimethylcuprate [264].

7. Experimental Procedures for Generating
   Specific Metal Enolates

Caution: All of the following procedures should be performed
under an anhydrous, oxygen-free nitrogen atmosphere. All reagents
and solvents should be anhydrous and all transfers should be carried
out with a hydrodermic syringe or with a flask-to-flask cannular
arrangement. While the enolate solutions prepared as described below
may be stable under an inert atmosphere at low temperatures for
extended time periods, it is recommended that they be prepared
immediately before use, whenever possible.

a. Lithium-ammonia reduction of an α,β-unsaturated ketone [194].
Freshly cut lithium wire (2.77 g, 0.04 g-atom) is introduced into 1
liter of anhydrous liquid ammonia at reflux temperature, and the mixture is stirred for 20 min. A solution of 0.182 mol of the enone
and 0.182 mol of water (as a proton donor) in 400 ml of anhydrous
ether is added dropwise with stirring over 60 min. When the addition

is complete, the mixture is stirred for 10 min. The enolate solution
is then ready to be reacted with an alkylating agent in liquid ammonia
ether. When an electrophilic reagent incompatible with liquid ammonia
is to be used, the ammonia is removed and replaced by an appropriate
solvent. This is accomplished by allowing the ammonia to evaporate,
adding an appropriate quantity of the new solvent, and removing this
along with residual ammonia at reduced pressure (finally at about
1 mm Hg). A 0.5 to 1.0 M solution of the lithium enolate is then
prepared by the addition of an appropriate quantity of the new solvent. The solution is then ready to be treated with the desired
electrophilic reagent.

     b. *Reaction of an α,β-unsaturated ketone with lithium dimethyl-*
cuprate [239]. A solution of lithium dimethylcuprate is prepared
by the dropwise addition of 27.5 ml of an ether solution containing
48.7 mmol of halide-free methyllithium to 5.0 g (24.3 mmol) of Me_2SCuB
while the temperature is maingained at $18°$-$20°C$. An enone (23.2 mmol)
(having an appropriate reduction potential [240]) is added dropwise
to the solution with stirring over 15 min. The reaction mixture is
then stirred at $20°$-$25°C$ for 20 min. [The solid $(MeCu)_n$ can then be
removed by centrifugation, if desired.] The ether solution of the
lithium enolate is then ready for reaction with an electrophilic
reagent, or the ether is removed under reduced pressure and replaced
by an appropriate amount of a suitable solvent to obtain a 0.5-1.0 M
solution before the addition of the electrophilic reagent. With
appropriate modification of the experimental conditions, various
lithium dialkyl as well as diphenyl and divinylcuprates may be substituted for lithium dimethylcuprate in the above procedure.

     c. *Cleavage of an enol acetate with methyllithium [45].* The
ether is removed under reduced pressure from an ether solution containing 0.40 mol of methyllithium and about 20 mg of 2,2-bipyridyl
as an indicator. Anhydrous DME (400 ml) is added and the resulting
purple solution is cooled to $0°$-$10°C$. The enol acetate (0.190 mol)
is added dropwise with stirring over 35-45 min while the temperature
is maintained at $0°$-$10°C$ with an ice bath. (The mixture should retain

a light red-orange color indicating the presence of a trace of excess methyllithium.) The enolate solution is then ready for reaction with the desired electrophilic reagent.

d. Cleavage of a trimetlylsilyl enol ether

i. Methyllithium [44]. A mixture of 90 mmol of a trimethylsilyl enol ether in 65 ml of an ether solution containing 91 mmol of methyllithium is stirred at $25^{\circ}C$ for 30 min. The ether is removed under reduced pressure from the suspension of the lithium enolate. The residue is dissolved in 50-100 ml of DME or THF. The enolate solution is then ready to be reacted with the electrophilic reagent.

ii. Methylmagnesium bromide [46, 47]. The procedure is similar to that employed in Sec. II.C.7.d.i except that the trimethylsilyl enol ether is refluxed with methylmagnesium bromide in DME for 24 hr. The bromomagnesium enolate is then ready for reaction with the appropriate electrophilic reagent.

iii. Lithium amide in liquid ammonia [228]. A slurry of lithium amide in liquid ammonia is prepared from 0.0175 g (2.5 mg-atom) of lithium and a trace of ferric nitrate in 13 ml of anhydrous liquid ammonia at $-33^{\circ}C$. A solution of 2.5 mmol of the trimethylsilyl enol ether in 10 ml of dry THF is added slowly with stirring. After stirring for a 30-min period at $-33^{\circ}C$, the THF-ammonia solution is ready for reaction with alkylating agents. If other electrophilic reagents are to be used, the ammonia is replaced by an unreactive solvent as described in Sec. II.C.7.a.

D. Alkylation of Metal Enolates of Saturated Ketones

1. Introduction

A large body of evidence indicates that the alkylation of metal enolates with alphatic halides, sulfates, tosylates, etc., generally proceeds by an S_N 2-type mechanism. In the C-alkylation process, approach of the alkylating agent perpendicular to the plane of the enolate system, as shown in 62, is favorable for stereoelectronic

reasons [137-143, 265-269]. This allows maximum orbital overlap
between the developing C-C bond and the π orbital of the carbonyl
group to be maintained in the transition state. In connection with
studies on the rates of C-alkylation of β-dicarbonyl compounds [270]
and on the influence of metal cations on the rates of alkylation of
enolates of simple ketones [62, 115, 128], a six-membered cyclic
transition state of the type 63 has been proposed. (A transition
state related to 63 has been invoked for the alkylation of chiral N-
lithioenamines of cyclohexanone, allowing the asymmetric synthesis
of 2-alkylcyclohexanones [271].) In this type of transition state
the coordination of the leaving group, e.g., the halide ion, with
the metal cation would be expected to facilitate the reaction [62].
However, "endocyclic" transition states in which the S_N2 reaction
is occurring at a tetrahedal carbon atom are unfavorable, because
the nucleophilic carbon atom, the carbon atom at which displacement
is occurring, and the leaving group cannot attain a collinear arrangement [272]. It seems possible that in aggregated systems, transition
states related to 62 may be involved in which the cation of a second
metal enolate ion pair assists the reaction by coordination with the
leaving group. The transition state for O-alkylation of metal enolates may be pictured as in 64. The results of intramolecular alkylation studies (Sec. II.D.10) indicate that O-alkylation involves
attack of the alkylating agent on the nonbonding electron pair in
the plane of the enolate system.

    62   63   64

2. Factors Influencing Reactivity of
   Saturated Ketone Enolates

As noted earlier, metal enolates may exist in the form of high-
molecular-weight aggregates or as monomeric species with various

degrees of association of the oxygen-metal bond. In general, factors which favor a loose association of the metal cation with the enolate anion, i.e., solvent-separated ion pairs of the type 15, lead to increased nucleophilicity toward alkylating agents. The nature of the solvent has a tremendous influence on the reactivity of metal enolates [114, 115, 120, 121, 128]. Enolate reactivity is particularly great in dipolar aprotic solvents which favor the formation of solvent-separated ion pairs.

The nature of the metal cation may also have a significant effect on rates of enolate alkylation. For example, the potassium enolate of butyrophenone (65, M = K) reacts with ethyl bromide in ether 6000 times faster than the lithium enolate (65, M = Li) [62]. Likewise, potassium enolates of cyclohexanone derivatives are significantly more reactive with alkylating agents in DME than the corresponding lithium enolates [48]. It has also been shown that halomagnesium [46, 47, 110, 207], tin [273], and aluminum enolates [273] are much less reactive than lithium enolates. It is clear that association between cations and enolate anions is stronger, i.e., the oxygen-metal bond has more covalent character, when small metal cations or less electropositive metals are employed.

The structure of the metal enolate also has a small but observable effect on its reactivity [1, 48, 152]. In general more highly substituted enolates appear to be more reactive than less highly substituted enolates in alkylation reactions. The electron-releasing influence of alkyl groups may enhance the nucleophilicity of metal enolates, but the steric bulk of these groups would also be expected to retard alkylation rates. These two effects are apparently quite small and essentially balance each other [48]. The lower reactivity of less-substituted enolates suggests that these species may be more highly associated than more-substituted enolates in which the large substituents interfere with association [1, 48]. Substituents located at positions other than the α position may influence the extent of enolate aggregation in a particular solvent [49] or may sterically hinder the approach of the alkylating agent to the enolate [48, 49,

143]. These effects would be expected to influence the rates and
the stereochemistry (see Sec. II.D.9) of enolate alkylations.

3. Alkylating Agents: C- vs O-Alkylation

Activated allylic, benzylic, propargylic halides, and α-haloesters
as well as methyl and primary halides, are excellent reagents for C-
alkylation of metal enolates. Unless strongly polar aprotic solvents
(DMSO, HMPA, etc.) are used, reactions of metal enolates with alkyl
halides normally lead to little, if any, O-alkylation. The order of
reactivity of the various halogens is as expected for an S_N2 process,
that is, I > Br > Cl [1]. The order of reactivity for a particular
halide is usually benzyl > allyl > 1°, and branching at the β carbon
reduces reactivity [152]. Branching at the α carbon reduces reactivity significantly, so that isopropyl and related halides react with
metal enolates slowly and alkylation products are produced in poor
yields [152, 193, 196]. Tertiary halides undergo elimination almost
exclusively. In reactions of sodium enolates with alkyl halides in
DME or diglyme, it has been shown that the extent of elimination
increases as the basicity of the enolate is increased [153].

β-Vinyl and β-phenyl halides are also prone to elimination.
The alkylations of the sodium enolates of norbornanone and 3-
methylnorbornanone with 2-methyl-5-iodo-2-pentene in dioxane which
were employed in the synthesis of (+)-β-santalene and its epimer
provide examples of the successful use of β-vinyl halides as alkylating agents [274]. However, 4-halo-1-butenes are extensively
dehydrohalogenated under the basic conditions of alkylation reactions
[176, 275]. β-Bromopropionitrile, another alkylating agent prone to
elimination, alkylates the enolate 54 (M = Li) in liquid ammonia in
moderate yield [228], the β-vinyl halide, 2,5-dichloro-2-pentene, has
also been used to alkylate the conjugate enolate of 1(9)-octalin-2-
one [276]. Likewise, β-phenyl halides are useful for alkylations of
conjugate enolates of tetrahydroindanone and octalone derivatives
[277]. Such reactions have been employed as key steps in the total
synthesis of steroids [277]. β-Haloketones undergo elimination to
vinyl ketones on base treatment. These reagents have been used in

place of Mannich-base methiodides in annulation reactions [278, 279].
4-Halo-1-butynes also undergo elimination rather than substitution
on reaction with metal enolates [275].

Alkyl tosylates and benzenesulfonates are also useful alkylating
agents. They normally exhibit reactivity similar to the corresponding
alkyl iodides [3, 119, 280] and, like alkyl iodides, give largely C-
alkylation products when reacted with metal enolates in solvents of
relatively low polarity.

The same factors which tend to increase rates of alkylation of
metal enolates tend to increase the extent of O-alkylation. As the
degree of dissociation between the enolate anion and the metal cation
is increased, the anion becomes more capable of exercising its ambident character [128-131]. The competition between C- and O-alkylation
can be explained in terms of the Pearson-Klopman principle, which
states that hard reagents attack the hard site of an ambident anion
and soft reagents attack the soft site [281, 282]. In an enolate
anion the relatively small electronegative oxygen atom acts as the
hard site, while the larger more polarizable carbon atom acts as the
soft site.

In highly polar solvents reactions of metal enolates with alkyl
halides can lead to appreciable quantities of O-alkylation products.
The O-alkylation/C-alkylation ratios which have been determined from
the reaction of the sodium enolate of butyrophenone (65, M = Na) with
n-pentyl halides in DMSO [128] are shown in Eq. (19). The amount of
O-alkylation increased as the leaving group was varied from iodide to
bromide to chloride. This of course, parallels the increase in the
hardness of the alkyl halide. Interestingly, the O-alkylation/C-
alkylation ratios for alkylation of enolates of 65 in DMSO were
essentially unchanged as the metal was varied from lithium to sodium
to potassium [128]. This suggests that in DMSO all of these metal
enolates reacted primarily as solvent-separated ion pairs. However,
rates of alkylation were still shown to be dependent upon the nature
of the metal cation [128].

$$
\begin{array}{c}
\underset{\underset{65}{}}{\overset{\overset{OM}{|}}{C_6H_5C{=}CHCH_2CH_3}}
\quad
\xrightarrow[\text{DMSO}]{n\text{-}C_5H_{11}X}
\quad
\overset{\overset{O-C_5H_{11}}{|}}{C_6H_5C{=}CHCH_2CH_3}
\;+\;
\overset{\overset{O}{\|}\;\overset{C_5H_{11}}{|}}{C_6H_5C{-}CHCH_2CH_3}
\end{array}
$$

$$
\begin{array}{ccc}
X = Cl & 1.2 & 1 \\
Br & 0.64 & 1 \\
I & 0.23 & 1
\end{array}
$$

(19)

Dialkyl sulfates and trialkyl oxonium salts are much more reactive than are alkyl halides, tosylates, etc. [3, 119, 283], and these hard alkylating agents may yield substantial amounts of O-alkylation products when reacted with metal enolates, even in ethereal solvents such as DME [283]. For example, alkylation of the potassium enolate of cyclohexanone with dimethyl sulfate in DME at room temperature gave a 24:44 ratio of O-/C-alkylation products. When DMSO was employe as the solvent the O/C ratio increased to 71:18 [283]. Chloromethyl alkyl [284] and chloromethyl benzyl ethers [285] are also hard alkylating agents which may give rise to significant amounts of O-alkylation products even in relatively nonpolar solvents. In contrast, bromomethylbenzyl sulfide has been found to C-alkylate lithium enolates in reasonable yields in DME [286].

The structure of the metal enolate and the alkylating agent may also influence O-alkylation/C-alkylation ratios [115]. Normally secondary alkyl halides give more O-alkylation than primary halides, and increasing the bulk of substituents at the α position of the enolate also leads to an increase in the amount of O-alkylation [91, 115, 129-131]. For example, the enol ether was found to be the sole product of ethylation of the sodium enolate of diphenyl acetophenone in diglyme [114].

The bicyclic diketone 66 has been converted to an 8:1 mixture of O- and C-alkylated products, 67 and 68, in nearly quantitative yield by treatment with sodium hexamethyldisilazide in HMPA followed by addition of isopropyl bromide [91]. The selective protection of the enolizable carbonyl function in 66 was utilized by Piers and co-workers [91] as an important step in their total synthesis of the ylango sesquiterpenes. The use of HMPA as a solvent provided a substantially greater amount of the enol ether 67 than did DMSO or sulfolane.

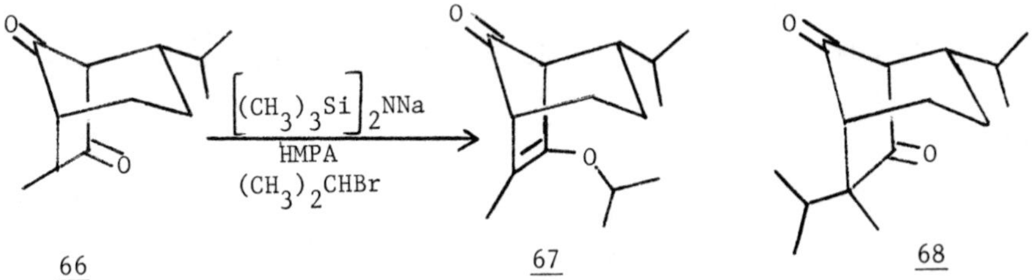

66   67   68

3 : 1

Magnesium or halomagnesium cations are particularly effective in suppressing the O-alkylation reaction. For example, alkylation of halomagnesium enolates of diisopropyl ketone and cyclohexanone with dimethyl sulfate in HMPA yielded little, if any, O-alkylated materials [110].

Although nucleophilic ring openings of epoxides with a variety of carbanionic species are well known, the first examples of the intermolecular reaction of a simple ketone enolate with an epoxide were reported recently [287]. The lithium enolate 54 and the bromomagnesium enolates of cyclohexanone were found to give low yields of the alkylation product 69 on reaction with propylene oxide in DME [287]. A significantly better yield of 69 was obtained when the

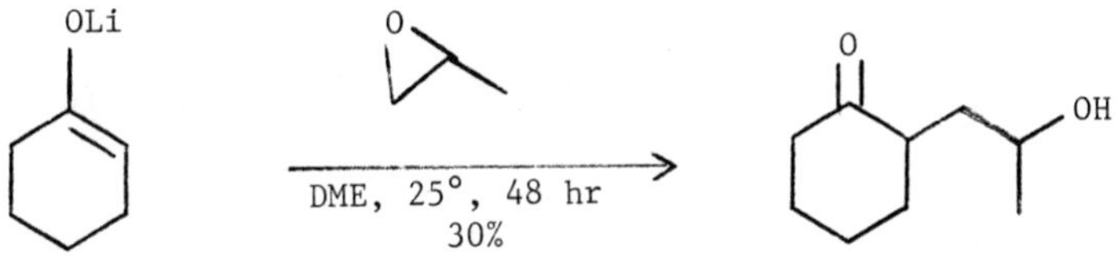

54   69

chloromagnesium salt of the cyclohexylimine of cyclohexanone in THF was employed as the nucleophile. The imine salt also reacted with oxetane to give a ring-opened product [287].

4. Arylation and Vinylation of Metal Enolates

Vinyl halides and aromatic halides (except for those activated by the presence of electron-withdrawing groups at the ortho and/or para positions) are normally unreactive toward bimolecular displacement reactions and thus fail to react with metal enolates under the usual conditions. However, various methods of α-vinylation and α-

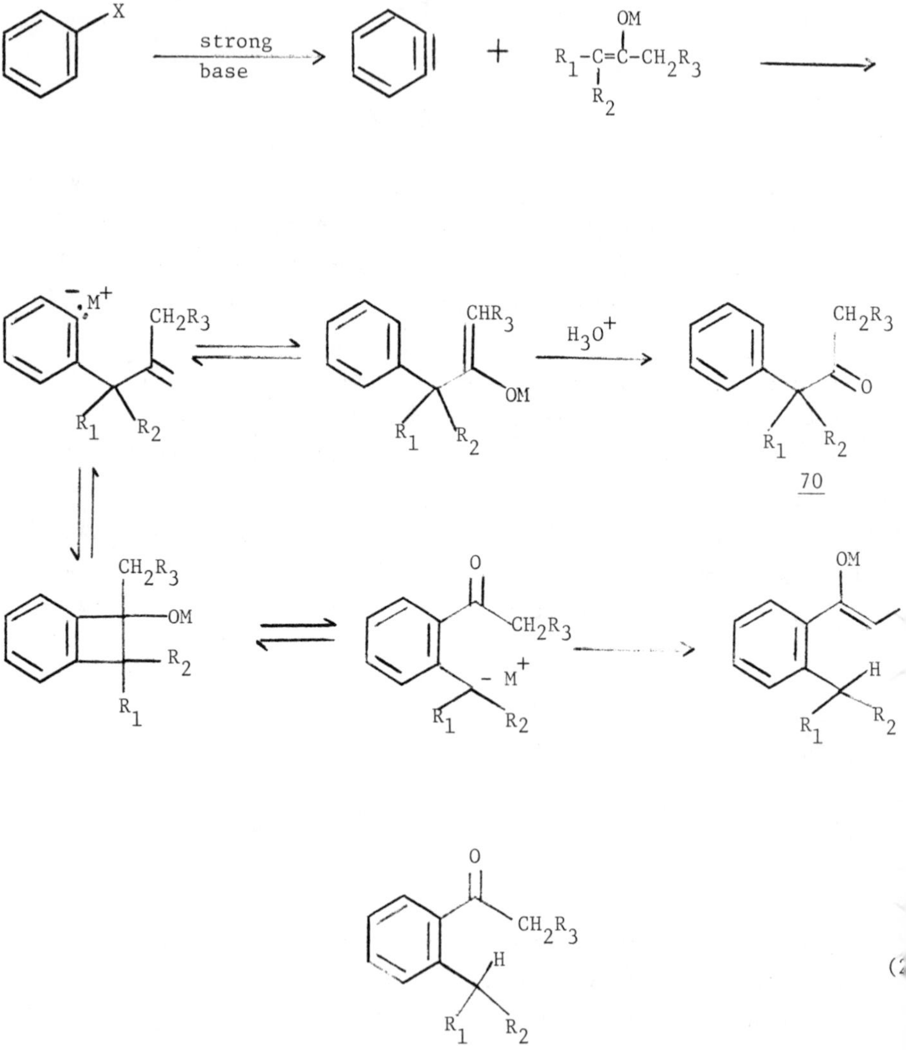

$$(2$$

phenylation of metal enolates are now available. For example, treatment of a solution of a metal enolate with an aryl halide or a 1-halo-1-cyclohexene or a 1-halo-1-cycloheptene in the presence of a strong base such as the 1:1 complex of sodium amide and sodium t-butoxide leads to such products [288]. In the case of aryl halides, the strong base causes dehydrohalogenation to form a benzyne intermediate which undergoes nucleophilic attack by the metal enolate. As shown in Eq. (20), α-arylated ketones of the type 70 and/or aromatic ketones of the type 71 may be obtained, depending on the structure of the metal enolate and the reaction conditions.

Strained alkynes or 1,2-dienes are formed when 1-halocycloalkenes of the type 72 (n = 2 or 3) are treated with strong bases. Reactions of these species with metal enolates leads to mixtures of anionic

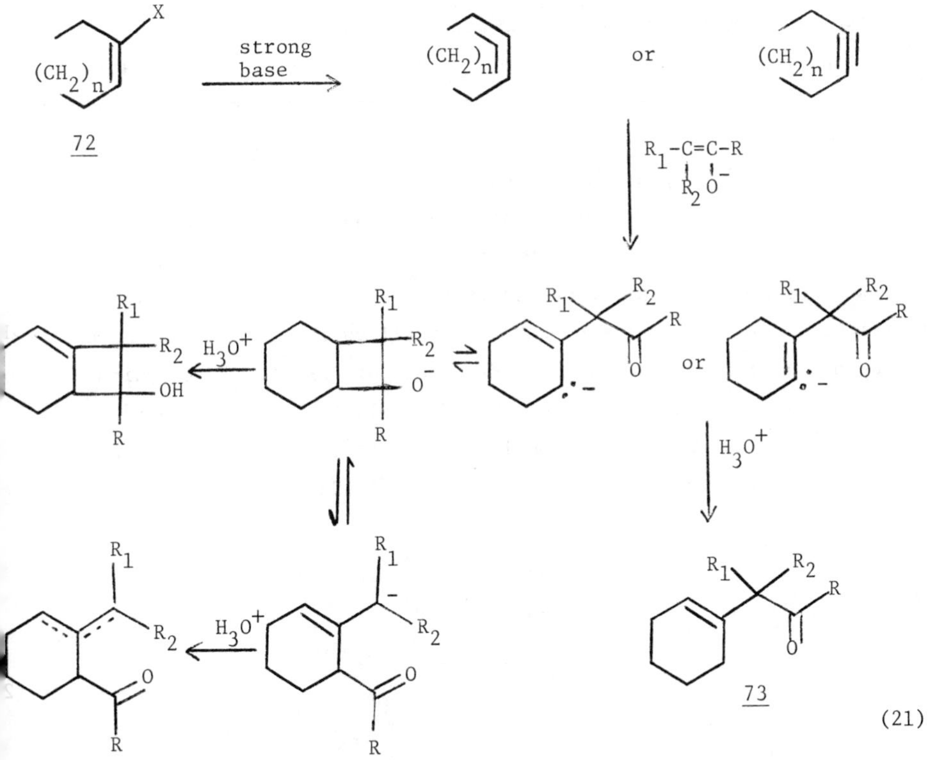

(21)

intermediates which yield mixtures of ketones, including the α-
vinylated product *73*, on hydrolysis [Eq. (21)]. α-Vinylated products
are generally favored in more polar solvents such as HMPA.

Diphenyliodonium chloride has been found to be a useful reagent
for α-phenylation of certain ketones [289]. For example, reaction
of this reagent with the sodium enolate of isobutyrophenone in t-amyl
alcohol at $0^{o}C$ gave an 81% yield of the α-phenylated ketone [Eq. (22)].

$$
\underset{\underset{CH_3}{|}}{C_6H_5-\overset{\overset{O}{||}}{C}-CH-CH_3} \quad \xrightarrow[\substack{NaO-t-amyl \\ t-amyl-OH, \; 0^{\circ} \\ 81\%}]{(C_6H_5)_2\overset{+}{I}Cl^{-}} \quad \underset{\underset{CH_3}{|}}{C_6H_5-\overset{\overset{O}{||}}{C}-\overset{\overset{CH_3}{|}}{C}-C_6H_5} \tag{22}
$$

However, a much lower yield was obtained when the sodium enolate of
isovalerophenone was treated in the same way. An electron transfer,
radical-pair mechanism has been proposed for these arylation reactions
[290].

Synthetically useful procedures for α-phenylation [149, 291-294]
and α-vinylation [295] of ketone enolates have recently been developed
by Bunnett and co-workers. Electrophilic phenyl radicals may be produced by alkali metal reduction or ultraviolet irradiation of iodo-
or bromobenzenes in the presence of metal enolates in liquid ammonia.
α-Phenylation occurs via a radical-chain process ($S_{RN}1$ mechanism),
as outlined in Eq. (23) [149, 291-293]. In the photosimulated process,
electron transfer from the enolate anion to the aryl halide apparently
serves to initiate the reaction [149, 292, 293].

$$
\begin{aligned}
e^{-} \; + \; ArX \quad &\longrightarrow \quad (ArX)^{\cdot-} \\
(ArX)^{\cdot-} \quad &\longrightarrow \quad Ar\cdot \; + \; X^{-} \\
Ar\cdot \; + \; \underset{|}{-\overset{\overset{O^{-}}{|}}{C}=C-} \quad &\longrightarrow \quad (Ar-\overset{\overset{O}{||}}{\underset{|}{C}}-\overset{}{C}-)^{\cdot-} \\
(Ar-\overset{\overset{O}{||}}{\underset{|}{C}}-\overset{}{C}-)^{\cdot-} \quad &\longrightarrow \quad Ar-\overset{\overset{O}{||}}{\underset{|}{C}}-\overset{}{C}- \; + \; ArX^{\cdot-}
\end{aligned} \tag{23}
$$

The photostimulated process has been used to synthesize α-phenyl cyclic and acyclic ketones [149, 292, 293]. The preparation of phenylacetone in 73% isolated yield by irradiation of a solution of bromobenzene in liquid ammonia in the presence of the potassium enolate of acetone provides an example of this reaction [Eq. (24)] [149].

$$
\underset{\substack{| \\ CH_3-C=CH_2}}{\overset{OK}{}} \ + \ C_6H_5Br \quad \xrightarrow[NH_3]{h\nu} \quad \underset{\substack{CH_3-C-CH_2C_6H_5}}{\overset{O}{\underset{}{\|}}} \tag{24}
$$

A similar reaction involving the use of a 4:1 ratio of the potassium enolate of acetone and iodobenzene in DMSO gave an 81% yield of phenylacetone [294]. However, liquid ammonia is still considered to be the solvent of choice for preparative work [294].

An intramolecular variant of this type of process has been employed by Semmelhack et al. [296] as a key step in their recently reported synthesis of the alkaloid cephalotaxine [Eq. (25)].

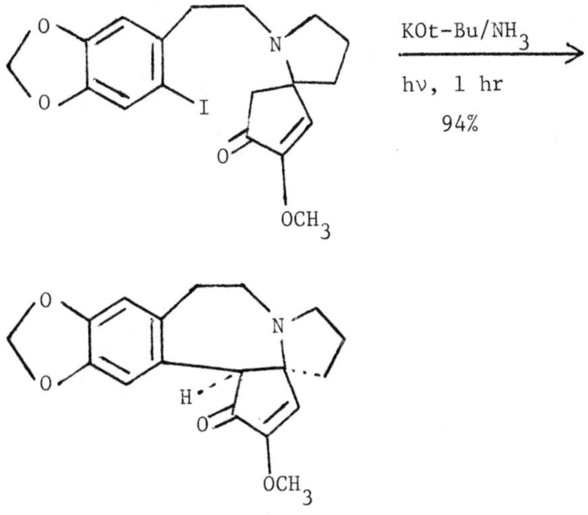

KOt-Bu/NH₃

h\nu, 1 hr

94%

(25)

These $S_{RN}1$ processes do not provide a completely adequate solution to the problem of enolate phenylation. Fully substituted enolates, such as the enolate of diisopropyl ketone, give only low yields of α-phenylation products [293], and it is not clear that this process

can be used for the regiospecific arylation of kinetically formed
enolates. Indeed, subjection of the metal enolates of 2-butanone
and 3-methyl-2-butanone to the α-phenylation procedure apparently
leads to enolate equilibration [292].

The potassium enolate of acetone also undergoes photostimulated
reaction with vinyl halides to give α-vinylated products [295].
Isomerization of the initially formed β,γ-enone into the α,β isomer
occurs to some extent under the reaction conditions [Eq. (26)]. It
has been suggested that these reactions also occur by an $S_{RN}1$
mechanism.

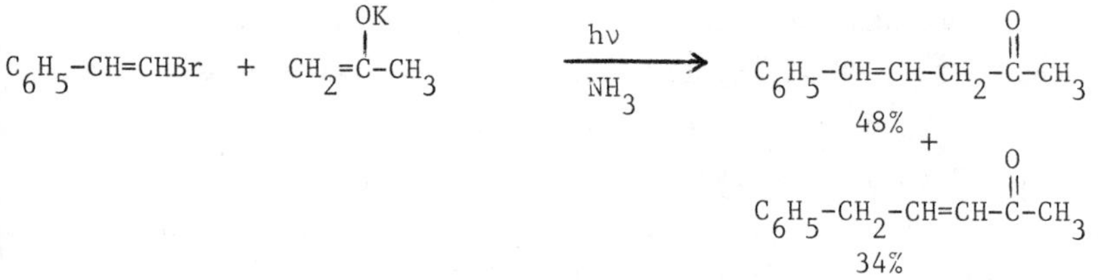

$$
C_6H_5-CH=CHBr \;+\; CH_2=\overset{\overset{\textstyle OK}{|}}{C}-CH_3 \;\xrightarrow[\text{NH}_3]{h\nu}\; C_6H_5-CH=CH-CH_2-\overset{\overset{\textstyle O}{\|}}{C}-CH_3
$$

48%
+

$$
C_6H_5-CH_2-CH=CH-\overset{\overset{\textstyle O}{\|}}{C}-CH_3
$$

34%

By reaction of halobenzenes with tetrakis(triphenylphosphine)-
nickel(0), activated σ-aryl nickel complexes are formed. These intermediates react with metal enolates such as the lithium enolate of
acetophenone to give α-phenylated products on warming [Eq. (27)]
[296, 297].

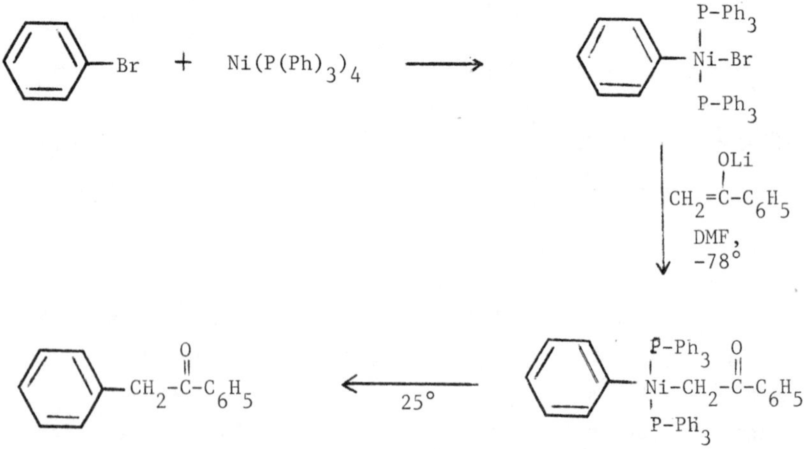

$$
\text{Ph--Br} \;+\; \text{Ni(P(Ph)}_3)_4 \;\longrightarrow\; \text{Ph--}\overset{\overset{\textstyle P\text{-}Ph_3}{|}}{\underset{\underset{\textstyle P\text{-}Ph_3}{|}}{Ni}}\text{--Br} \qquad (27)
$$

$$
\Big\downarrow \begin{array}{c} \overset{\overset{\textstyle OLi}{|}}{CH_2=C-C_6H_5} \\ \text{DMF,} \\ -78° \end{array}
$$

$$
\text{Ph--}CH_2-\overset{\overset{\textstyle O}{\|}}{C}-C_6H_5 \;\xleftarrow[\;25°\;]{}\; \text{Ph--}\overset{\overset{\textstyle P\text{-}Ph_3}{|}}{\underset{\underset{\textstyle P\text{-}Ph_3}{|}}{Ni}}-CH_2-\overset{\overset{\textstyle O}{\|}}{C}-C_6H_5
$$

The use of chromium tricarbonyl complexes for the activation of aryl halides toward nucleophiles has also been reported [298]. However, these reagents have been found to be useful for the arylation of weakly basic anions but not of strongly basic species such as metal enolates.

An interesting new indirect procedure has been developed for the synthesis of α-arylated ketones [299]. This involves the conjugate addition of lithium diphenylcuprate or phenylcopper to p-toluenesulfonylazocyclohex-1-enes (prepared by reaction of tosylhydrazones of α-haloketones with sodium carbonate) to produce α-phenyltosylhydrazones, which are converted to α-phenylketones by carbonyl exchange with acetone in the presence of an acid catalyst [Eq. (28)].

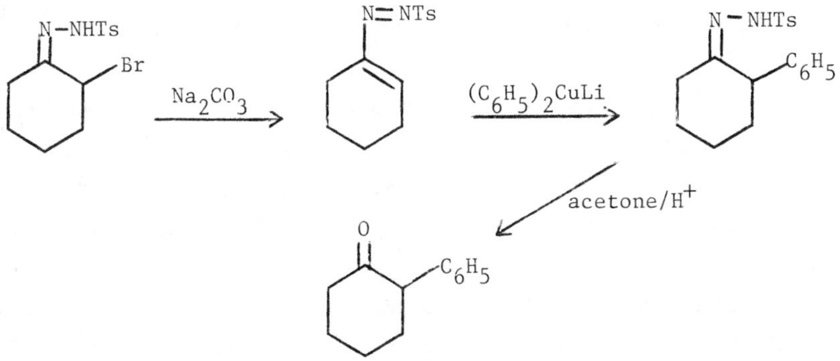

$$(28)$$

5. Enolate Equilibration and Polyalkylation

Enolate equilibration and di- and polyalkylation are the major side reactions which may lead to reduced yields of desired products in enolate alkylations. These processes occur as a result of equilibration of the starting enolate or enolate mixture with the neutral monoalkylation product or products via proton transfer reactions. Polyalkylation is a particularly troublesome problem when methylation or ethylation reactions are being conducted because the mono-, di-, and trialkylated materials have very similar physical properties.

This makes separation of the desired product difficult, and, often, various chemical procedures [300] must be employed to achieve this objective.

As indicated in Eq. (6), with the methylation of the enolates of 2-methylcyclohexanone as an example, a number of factors determine the composition of enolate alkylation mixtures:

1. The rate of alkylation of the starting enolate or, if a mixture is employed, the rates of alkylation of the structurally isomeric enolates

2. The rates of equilibration and thermodynamic stabilities of the isomeric enolates

3. The absolute and relative rates of enolization of the mono-alkylation products

4. The rates of equilibration and thermodynamic stabilities of the enolates of the alkylation products

5. The rates of alkylation of the enolates of the monoalkylation products

Polyalkylation also occurs when, in addition to the starting enolate, other bases which are capable of promoting enolization of monoalkyla-tion products are present in the medium. Whenever possible, it is clearly desirable to generate metal enolates in the absence of other strong bases.

In order to successfully regiospecifically monoalkylate a parti-cular enolate of an unsymmetrical ketone, the rate of the alkylation reaction must be rapid in comparison with competing proton transfer processes. Alkylation rates can be enhanced by the use of highly reactive alkylating agents and conditions which favor the existence of the enolate in the form of solvent-separated ion pairs, e.g., large electropositive metal cations and highly polar solvents. How-ever, in general, the same factors which favor increased reactivity of metal enolates toward alkylating agents also enchange enolate basicity and hence rates of proton transfer reactions. When metal cations which form tight ion pairs or relatively covalent bonds with the oxygen atom of the enolate anion are used, alkylation rates are reduced; but, fortunately, rates of proton transfer are reduced even more [39]. As the previous discussion indicates, lithium enolates in

solvents such as liquid ammonia, ether, THF, or DME or combinations
of these are the most generally useful conditions for performing
regiospecific monoalkylations of ketones. Halomagnesium [207-211,
254], tributyltin, and complex lithium triethylaluminum enolates
and tributyltin enol ethers [273] have also been used occasionally in
regiospecific alkylations.

Although cyclohexanone derivatives are not highly prone to
enolization, enolate equilibration and polyalkylation still occur to
a very significant extent unless the proper reaction conditions are
chosen. The results of reaction of kinetic mixtures of potassium
[41, 59] and lithium enolates [50, 59] of 2-methylcyclohexanone with
methyl iodide in DME at room temperature show the influence which
the metal cation may exert upon the composition of the monoalkylation
products and upon the extent of polyalkylation [Eq. (29)].

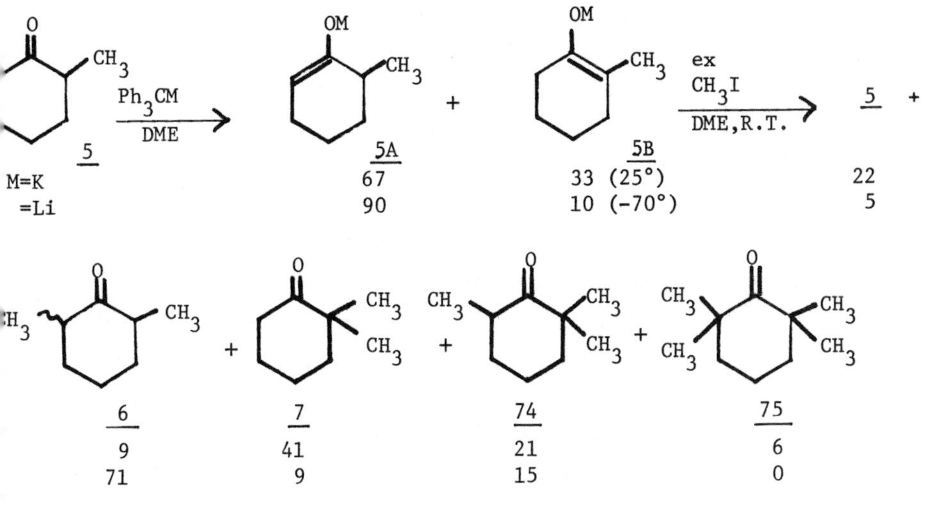

$$M=K$$
$$=Li$$

	5A	5B	
	67	33 (25°)	22
	90	10 (-70°)	5

	6	7	74	75
	9	41	21	6
	71	9	15	0

(29)

For the potassium enolate mixture, proton transfer reactions are
much faster than alkylation rates. Although the kinetic mixture
contained 67% of the less-substituted enolate 5A (M = K) and 33% of
the more-substituted enolate 5B (M = K), the major monoalkylation
product was 2,2-dimethylcyclohexanone (7) and significant amounts of

tri-(74) and tetramethyl cyclohexanone (75) were observed. (Similarly, a significant amount of polymethylation was observed on methylation of the sodium enolates of 5 [59, 65].) When the lithium cation
was used the ratio of 6 to 7 reflected closely the 5A/5B ratio and
polyalkylation products were formed in more limited quantities.

In addition to 5, use of lithium rather than sodium or potassium
cations has also been shown to provide an effective means of controlling enolate equilibration and reducing the extent of polyalkylation
in a number of other systems, including 2-decalone [38], trans-3-
methallyl-5-methylcyclohexanone [228], and 2-heptanone [119].

The data given in Eq. (29) suggest that when potassium enolates
are involved, 2,6-dimethylcyclohexanone (6) undergoes more extensive
subsequent alkylation than does the 2,2 isomer [41]. It has also
been shown that in Michael reactions of the potassium enolates of
5 with α,β-unsaturated and acetylenic esters the 2,6 isomer is converted to disubstituted material more rapidly than the 2,2 isomer
[53]. It has been suggested that a more rapid rate of conversion of
6 to its enolate, 6A (M = K), compared with the rate of conversion
of 7 to 7A (M = K) could account for this [41, 53]. However, the
potassium enolate 6A (M = K) undergoes alkylation with n-butyl iodide
in DME 2.8 times faster than the corresponding enolate 7A (M = K),
and equilibration studies indicate that the two enolates have comparable thermodynamic stabilities [48]. Since proton transfer reactions
are rapid with potassium enolates, it would be expected that equilibration of 6A (M = K) and 7A (M = K) would occur under the reaction
conditions. This, coupled with a more rapid methylation of 6A (M =
K), seems to afford a more likely explanation for the preferential
consumption of 6.

These data do not permit the determination of the source of 2,2,6
trimethylcyclohexanone (74) in the alkylation of the lithium enolates
derived from 5. (Indeed, it is possible that some of this material
results from the presence of excess base in the reaction medium.)
However, in other studies the ketones (6 or 7) were produced in the
presence of lithium t-butoxide by methylation of the lithium enolate

5A (M = Li) or 5B (M = Li) formed from lithium-ammonia reduction of
the corresponding enones [191]. In these reactions it appeared that
the 2,2 isomer 7 was converted into the trimethylketone 74 more
rapidly than the 2,6 isomer. Since 7 has no substituents at the 6-
position it would be expected to undergo more rapid enolization than
6. It has also been shown that, in contrast to the potassium case,
the lithium enolate 7A (M = Li) undergoes somewhat more rapid alkylation than the corresponding enolate 6A (M = Li) [48]. This greater
reactivity may result from the higher ground-state energy and possibly
a smaller degree of association of 7A (M = Li) than 6A (M = Li) [48].

Magnesium forms tighter metal-oxygen bonds than lithium, and
halomagnesium enolates have been used to a limited extent for regiospecific alkylations. The utility of these enolates is limited
because it is difficult to find conditions under which alkylation
occurs at an acceptable rate and, at the same time, proton transfer
reactions are minimized. Acceptable alkylation rates have been
achieved by the addition of the aprotic solvent HMPA to the enolate
in an ethereal solvent (usually THF) and successful regiospecific
alkylations of halomagnesium 2-enolates of 3-alkyl- and 3,3-
dialkylcyclohexanones with allyl bromide [207, 210] and 3,5-dimethyl-
4-chloromethyl isoxazone [209] have been reported [cf. Eq. (15)].
The methylation of the halomagnesium enolate of cyclohexanone with
methyl iodide in a THF-HMPA mixture gave a relatively small amount
of the dimethylation products [Eq. (30)] [54]. However, considerable

$$
\underset{5}{\text{(enolate structure)}} \xrightarrow[\text{THF:HMPA}]{\text{CH}_3\text{I}} \underset{\underline{5}}{\text{(ketone structure)}} \quad + \quad \underline{6} + \underline{7} \tag{30}
$$

72%   7%

quantities of dimethylation or diallylation products were obtained
when the halomagnesium enolate 76 was treated with methyl iodide or
allyl bromide in THF-HMPA [208]. The monoalkylation product 77 could

be obtained in excellent yield by alkylation of the lithioenamine
derivative of the ketone corresponding to 76 [208].

Procedures such as the kinetic enolization of ketones with
strong bases and the cleavage of silyl enol ethers allowing the
generation of lithium enolates in the absence of additional strong
bases should be employed whenever possible because the additional
base can cause enolization of the monoalkylation product or dehydro-
halogenation of the alkylating agent, when this is structurally
possible. However, even if additional bases are absent, high concentrations of reactive alkylating agents and short reaction times must
be used. Otherwise, the unreacted enolate itself may promote a significant amount of polyalkylation by enolizing the monoalkylation
product.

Monoalkylation of lithium enolates generated in the presence of
additional bases requires special precautions. When lithium-ammonia
reduction of enones is used to generate a specific enolate, it has
been found that if 1 eq of water rather than 1 eq of t-butyl alcohol
is employed as a proton donor, the extent of polyalkylation is normally
reduced [Eq. (31)]. Clearly, the weaker base lithium hydroxide is
much less effective than lithium t-butoxide in promoting enolization
of 2,2-methylcyclohexanone (7).

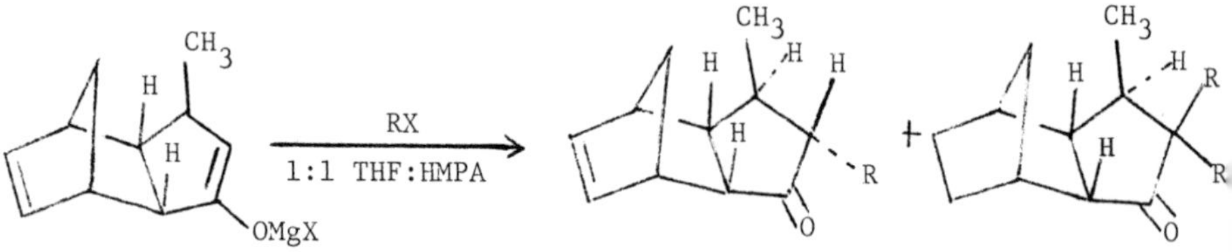

$$R = CH_3$$

$$= CH_2CH=CH_2$$

	77	78
$R = CH_3$	77%	13%
$= CH_2CH=CH_2$	50%	20%

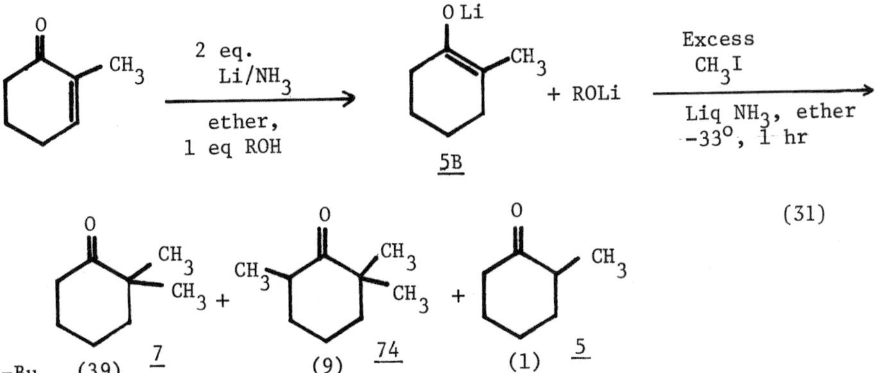

$$(31)$$

R=t-Bu	(39) $\underline{7}$	(9) $\underline{74}$	(1) $\underline{5}$
= H	(60)	(1)	(1)

Lithium t-butoxide, which is produced along with the lithium enolate when enol acetate cleavages are used, is significantly less reactive with common alkylation agents such as methyl iodide than the enolate itself [119]. Therefore, it remains in solution in relatively high concentration as the monoalkylation product is formed and may engender a significant amount of polyalkylation [4, 119]. The effect of this base can normally be minimized by carefully controlling the reaction time, carrying out the reaction at relatively low enolate anion and alkylating agent concentrations, and employing highly reactive alkylating agents. The effect of the use of a highly reactive alkylating agent is illustrated in Eq. (32) for the ethylation of the lithium enolate of 4-t-butylcyclohexanone. Triethyloxonium tetrafluoroborate not only reacts with the enolate much faster than ethyl iodide but probably also consumes lithium t-butoxide so rapidly that it does not have a chance to enolize the monoethylcyclohexanone being formed [301]. Trimethyloxonium 2,4,6-trinitrobenzenesulfonate is another highly reactive alkylating agent which has been used with similar results [119]. Unfortunately, reagents such as these do not provide complete solutions to the problem because significant amounts of 0-alkylated materials are obtained when they are used [119, 301].

The alkylation of halomagnesium enolates using only HMPA as the solvent leads to complete enolate equilibration and to the formation of substantial amounts of polyalkylation products [110, 111].

$$
\underset{\text{t-Bu}}{\text{OAc}} \quad \xrightarrow[\text{DME}]{\text{2 eq. CH}_3\text{I}} \quad \underset{\text{t-Bu}}{\text{OLi}} \quad + \quad (\text{CH}_3)_3\text{COLi} \tag{32}
$$

$$
\xrightarrow[\text{solvent}]{\text{alkylating agent}} \quad \underset{\text{t-Bu}}{\overset{\text{O}}{\bigcirc}}\text{C}_2\text{H}_5 \quad + \quad \underset{\text{t-Bu}}{\overset{\text{O}}{\bigcirc}}\begin{matrix}\text{C}_2\text{H}_5\\\text{C}_2\text{H}_5\end{matrix} \quad + \quad \underset{\text{t-Bu}}{\overset{\text{OC}_2\text{H}_5}{\bigcirc}}
$$

$\text{C}_2\text{H}_5\text{I/DME}$	88	12	<1
$(\text{C}_2\text{H}_5)_3\text{O}^+\text{BF}_4^-/\text{CH}_2\text{Cl}_2$	77	<1	23

Apparently, the highly polar medium leads to the extensive formation of solvent-separated ion pairs. For example, treatment of the chloromagnesium enolates, formed by the reaction of 5 with n-butylmagnesium chloride, with methyl tosylate in HMPA gave 49% of 7 and 11% of 2,2,6-trimethylcyclohexanone (74) along with 24% of recovered 5. The extent of polyalkylation was equal to or greater than monoalkylation when halomagnesium enolates of various methyl ketones were alkylated with methyl tosylate and other alkylating agents in HMPA [110].

The use of tributyltin enol ethers or, preferably, complex lithium triethylaluminum enolates has proved to be an effective means of reducing the extent of polyalkylation [273]. These species may be prepared by treatment of lithium enolates with the appropriate reagents [273]. A comparison of the results obtained for the methylation of the lithium and lithium triethylaluminum enolates of cyclohexanone [Eq. (33)] and cyclopentanone [Eq. (34)] is shown.

Small-ring cyclic ketones such as cyclobutanones and cyclopentanones [3] and acyclic ketones, and particularly methyl ketones, have relatively large kinetic acidities. Consequently, enolate equilibration [44, 181, 226] and polyalkylation [3, 41, 119, 192, 193, 254] are generally a much more serious problem in the alkylation of metal enolates of these compounds than for enolates of cyclohexanone derivatives. For example, attempted monobenzylation of the

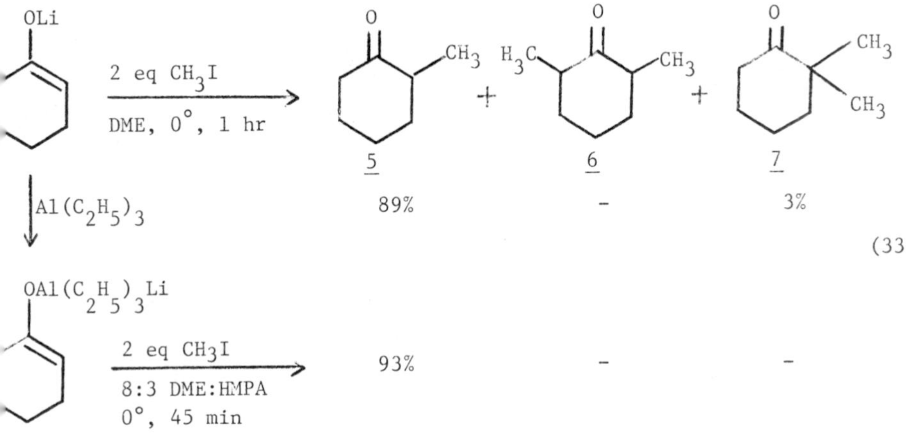

$$\text{(33)}$$

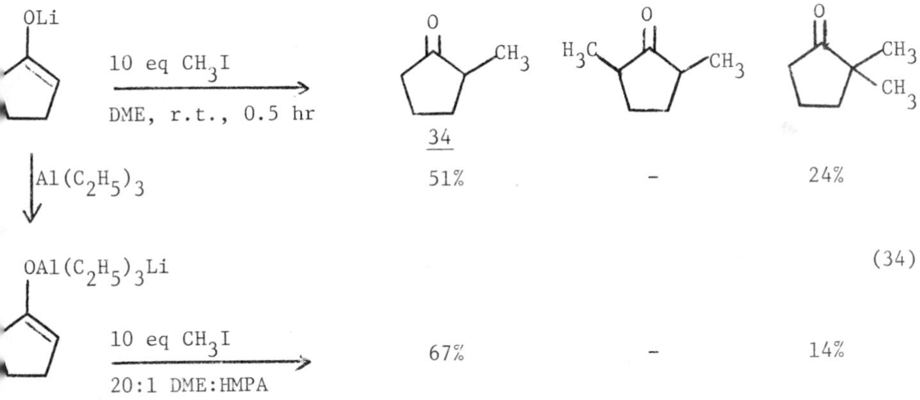

$$\text{(34)}$$

sodium enolate of cyclobutanone in benzene gave only the tetrabenzyl-
ated derivative, and attempted monomethylation of the sodium enolate
of cyclopentanone with dimethylsulfate in benzene gave a mixture
containing only di- and tetramethylcyclopentanone. As shown in Eq.
(34), methylation of the lithium enolate of cyclopentanone gave
extensive polyalkylation. Likewise, methylation of the more-
substituted lithium enolate *34A* of 2-methylcyclopentanone (produced
from the enol acetate) gave a 70:30 mixture of mono- and dimethylation
products under mild conditions [Eq. (35)] [120].

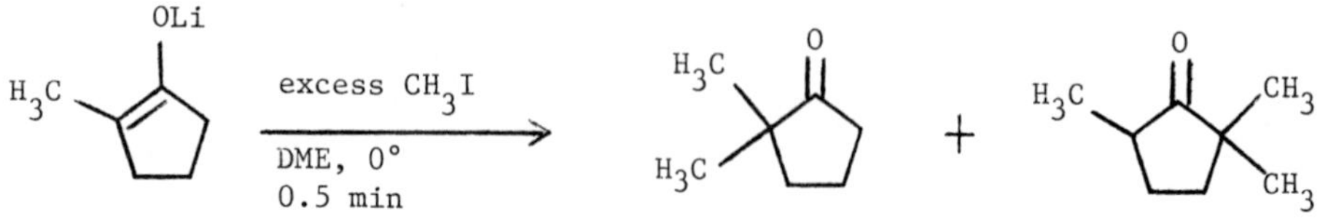

(35)

Mixtures of potassium or lithium enolates of methyl ketones
such as 2-heptanone (27) react with methyl iodide to give substantial
amounts of dimethylation products [Eq. (36)] [41]. In this case

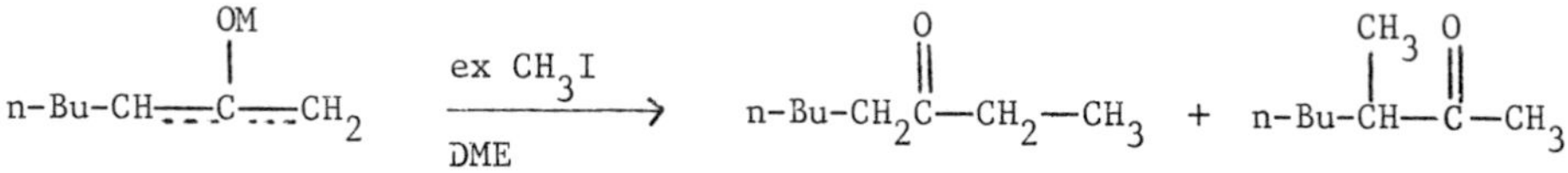

27A-C		
M=K	20%	35%
M=Li	42%	10%

M=K 20%	22%	
M=Li 25%	23%	

approximately the same amount of dialkylated ketone was produced
whether lithium or potassium was employed as the metal. Polyalkylation of enolates of methyl ketones is particularly troublesome when
excess bases are present in the medium. In the reduction-methylation
of trans-4-phenyl-3-butene-2-one (79) the extent of polyalkylation
of the lithium enolate intermediate 79A was shown to be directly
dependent on the strength and quantity of the base present in the
medium [192]. In this case and in related systems polyalkylation
could be minimized by the use of triphenylmethanol rather than t-butyl
alcohol as the proton donor in the reduction [Eq. (37)] [192]. Also,
addition of acetone along with the methyl iodide provided an interesting method of reducing the quantity of polyalkylation.

$$\underset{79}{\overset{\displaystyle H_5}{\underset{H}{>}}C=C\overset{H}{\underset{C\diagdown CH_3}{<}}\overset{O}{}} \xrightarrow[\text{1 eq. ROH, 30 min}]{\text{Li/NH}_3} \underset{79A}{C_6H_5CH_2CH=\overset{\overset{\displaystyle OLi}{|}}{C}-CH_3} \xrightarrow{\text{6 eq. CH}_3\text{I}}$$

R = t-Bu
R = $(C_6H_5)_3C$

$$\underset{\substack{37\% \\ 50\%}}{C_6H_5CH_2\overset{\overset{\displaystyle CH_3}{|}}{CH}-\overset{\overset{\displaystyle O}{||}}{C}-CH_3} \quad + \quad \underset{\substack{11\% \\ 5\%}}{C_6H_5CH_2\overset{\overset{\displaystyle CH_3}{|}}{CH}-\overset{\overset{\displaystyle O}{||}}{C}-CH_2-CH_3} \quad +$$

$$\underset{\substack{7\% \\ 0\%}}{C_6H_5CH_2\overset{\overset{\displaystyle CH_3}{|}}{CH}-\overset{\overset{\displaystyle O}{||}}{C}CH(CH_3)_2} \quad + \quad \underset{\substack{15\% \\ 14\%}}{C_6H_5CH_2CH_2-\overset{\overset{\displaystyle O}{||}}{C}-CH_3} \qquad (37)$$

When easily enolizable ketones are involved, the best yields of monoalkylation products are generally obtained when the alkylation time is minimized [119, 193] and when strong bases other than the starting enolate are not present in the medium. The use of lithium enolates will often allow minimization of polyalkylation, but with reactive systems such as cyclopentanone, complex lithium triethyl-aluminum enolates or tributyltin enol ethers seem to be preferred [Eq. (34)]. In highly reactive systems, the use of halomagnesium or lithium salts of imines [17] or lithium salts of N,N-dimethylhydrazones [24-26] in place of metal enolates seems to have definite advantages.

6. Regiospecific Alkylations of Metal Enolates

a. *General.* Metal enolates of cyclohexanone derivatives have been utilized extensively for regiospecific alkylations. Selected examples of these reactions are recorded in Table III. In most cases lithium enolates have been employed, but halomagnesium enolates have been used also. As noted earlier, recent evidence indicates that the metal enolates produced by conjugate addition of lithium dialkylcuprates

Table III. Regiospecific Alkylation of Metal Enolates of Cyclohexanone Derivatives

Enolate Structure	Method of Enolate Formation[a]	Alkylating Agent and Conditions	Product(s), %[b]			Ref
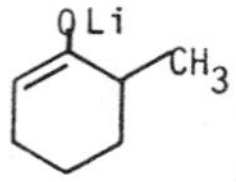 5A (M=Li)	1	Excess CH₃I, Liq NH₃-ether, 1 hr, -33°	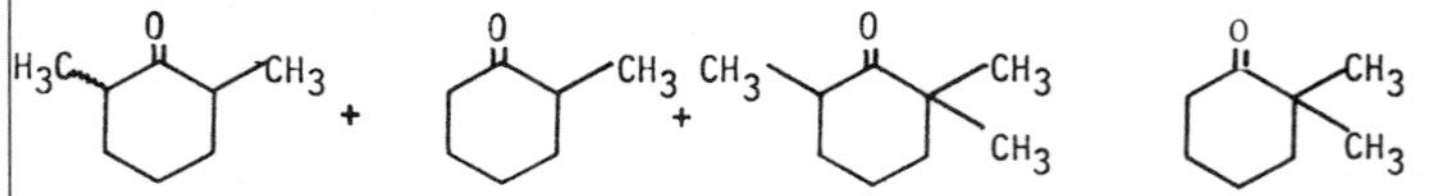 (37)	(5)	(~1)	191
"	2	Excess CH₃I, DME, 1 hr, reflux	72	16	12	59
"	3	Excess CH₃I, DME, 1min, 0°	76	17	3 + tetra-methyl-cyclohexanone (4)	150
"	4	C₆H₅CH₂Br, DME, 30°, 6 min	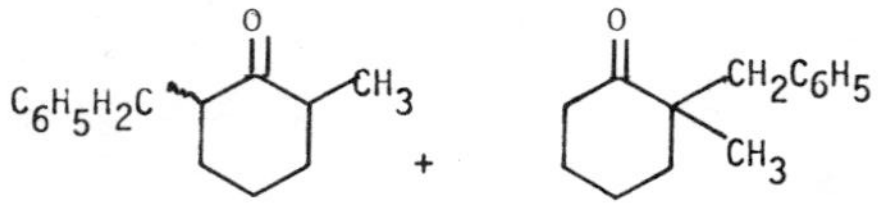 (~53)	(~8)	--	44
"	5	Excess n-BuI, THF-NH₃(4:6), 2 hr, -33°	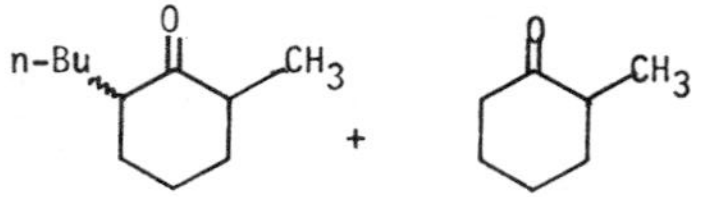 (~75)	(~15)	--	228

			Products (yield %)				Ref.
5A	6	THF, 1 eq., 0° (t-BuO$_2$C, I)	(t-BuO$_2$C structure, CH$_3$) (90)				302
"	7	THF, 1 eq., 0° (t-BuO$_2$C, I)	(t-BuO$_2$C structure, CH$_3$) (85)				302
5B (M=Li) O Li CH$_3$	8	Excess CH$_3$I, Liq. NH$_3$-ether, 1 hr, -33°	(60)	(~1)	(1)	--	191
"	9	Excess CH$_3$I, Liq. NH$_3$-ether, 30 min, -33°	(83)	--	2	--	193
"	10	Excess CH$_3$I, DME	(67)	--	(8)	--	254
"	11	Excess CH$_3$I, Liq. NH$_3$-ether, 30 min, -33°	(69)	--	--	--	176

Enolate Structure	Method of Enolate Formation[a]	Alkylating Agent and Conditions	Product(s)[b]		Ref
5B	12	Excess CH_3I, Liq. NH_3-ether, 60 min, -33°	(83)	--	50
"	13	Excess CH_3I,DME, 1 min, 0°3	76	18 + 3% 2,6-dimethylcyclohexanone	150
				3	
"	14	Excess CH_3I, DME, 1 min, 0°	60	13 + 5%, 2,6-dimethylcyclohexanone	150
				22	
"	15	Excess CH_3I,DME, 0°, 2 min	80	7 + 2% 2,6-dimethylcyclohexanone	150
				11	
"	16	Excess CH_3I,DME, 25°, 2 min	91.5	2	150
				6.5	
5B(90%)+5A(10%)	17	$BrCH_2CO_2CH_3$,DME, 30 sec., 0°	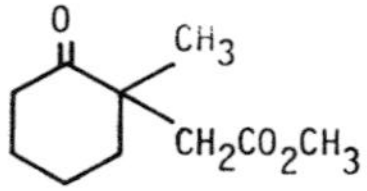 (54) [structure: O, CH_3, $CH_2CO_2CH_3$ cyclohexanone]	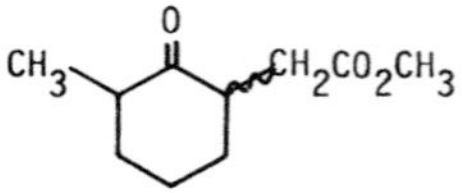 6 [structure: O, CH_3, $CH_2CO_2CH_3$ cyclohexanone]	119
5B(>95%)+5A(<5%)	18	$BrCH_2C_6H_5$,DME, 10°, 2.5 min	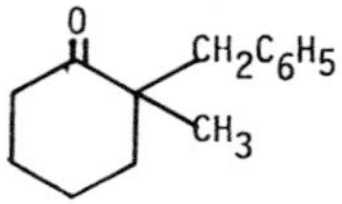 (54-58%) [structure: O, $CH_2C_6H_5$, CH_3 cyclohexanone]		44

$\underline{5B}$(>95%)+$\underline{5A}$(<5%)	18	$BrCH_2C_6H_5$,DME,5 min., 25°	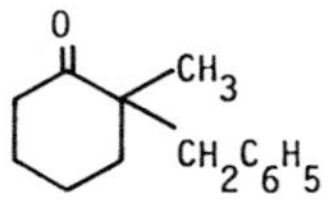 (48)	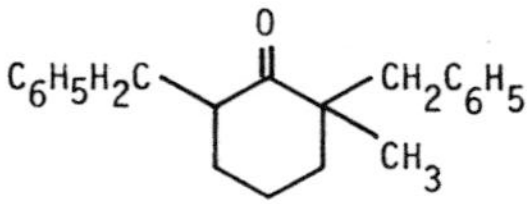 (41)	44
"	19	$BrCH_2C_6H_5$,DME, 25°, 5 min	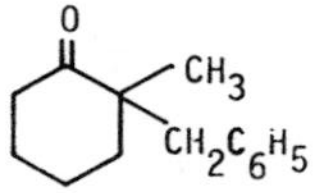 (84)	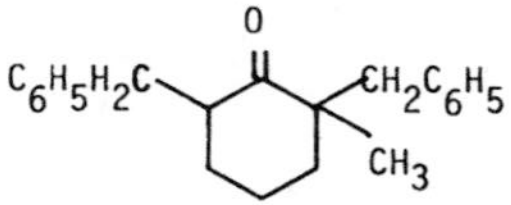 (7)	44
"	9	Excess $BrCH_2C_6H_5$ liq NH_3-ether, 20 min, -33°	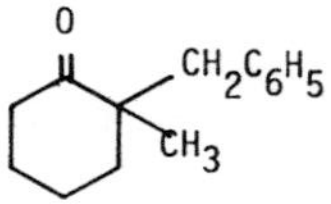 (82)		193
"	9	Excess C_2H_5I,Liq. NH_3-ether, 60 min -33°	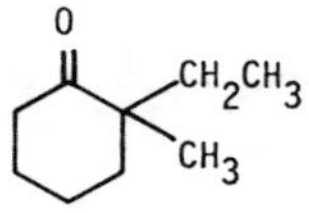 (75)	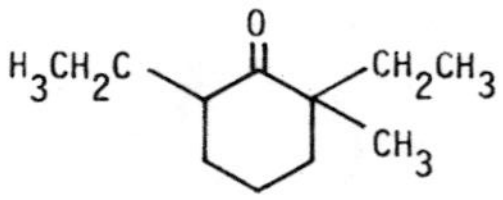 (10)	193

Enolate Structure	Method of Enolate Formation[a]	Alkylating Agent and Conditions	Product(s)[b]	Ref
5B(>95%)+5A(<5%)	9	Excess BrCH$_2$CH=CH$_2$, Liq. NH$_3$–ether, 5 min, -33°	(85) and (5)	193
"	20	Excess n-BuI, Liq. NH$_3$–THF(3:2), 2 hr, -33°	(94)	228
"	21	[I ... CO$_2$-t-Bu ... R] THF, 0°	R = H(90), R = CH$_3$(93)	302
"	21	[I ... (CH$_3$)$_3$Si] THF, 25°	(91)	233

172

Entry	Substrate	Conditions	Products	Ref.
21	"	I–CH₂C(CH₃)=CH₂–Si(CH₃)₃, THF	ketone product	233
22	"	 I–CH₂CH=CH–CO₂CH₃	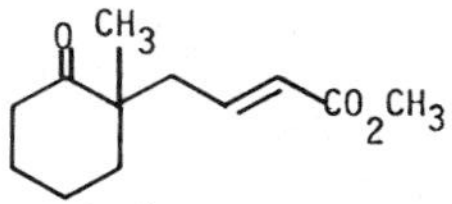 (60)	250
23	OLi-cyclohexene-CH₃	Excess CH₃I, NH₃–ether, –33°, 60 min	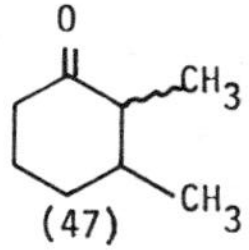 (47) 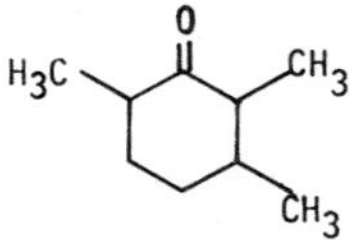 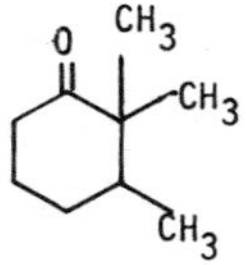 + (7)	191
24	"	BrCH₂CH=CH₂, Liq. NH₃–ether, 6 min, –33°	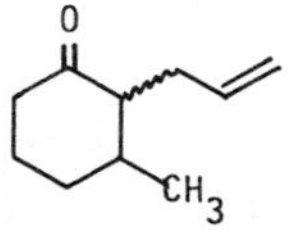 (47) ~20:1 trans:cis mixture	194
25	"	CH₃I, THF–HMPA	product	221

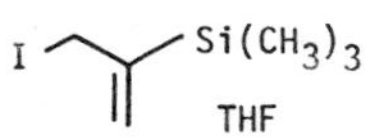

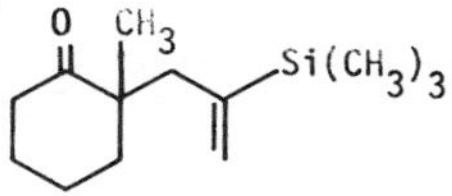

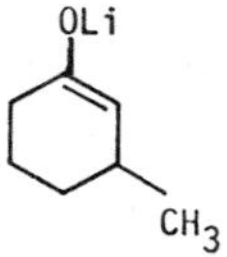

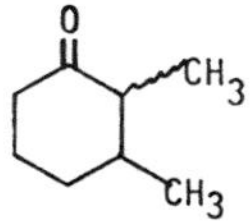

Enolate Structure	Method of Enolate Formation[a]	Alkylating Agent and Conditions	Product(s)[b]	Ref
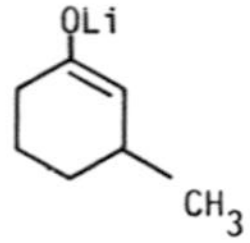 OLi cyclohexene (3-CH3)	25	CH₂=CH-CH₂I THF-HMPA	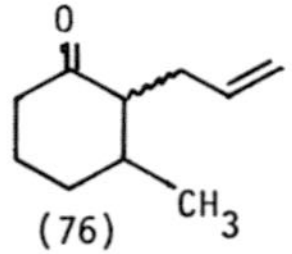 allyl/CH₃ cyclohexanone (76)	221
"	25	Excess CH₃I, DME, 0°, 10 min	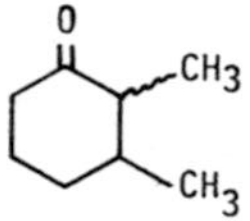 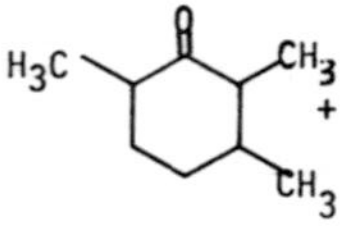 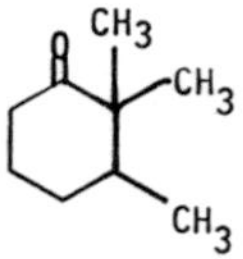 CH₃ cyclohexanone (4); 2,3-diCH₃ cyclohexanone (64) 4:1 trans:cis mixture 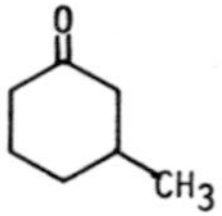 + gem-diCH₃/CH₃ cyclohexanone (10)	225
"	25	Excess CH₂=CH-CH₂I DME, 0°, 10 min	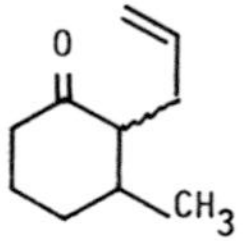 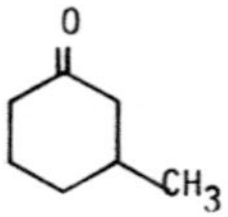 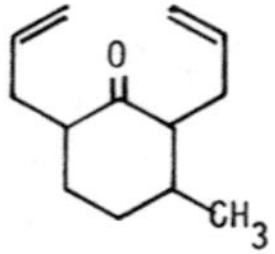 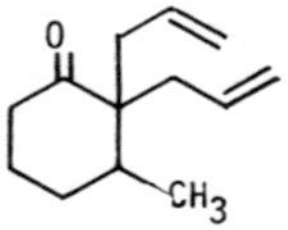 allyl/CH₃ cyclohexanone (74); CH₃ cyclohexanone (10); bis-allyl/CH₃ cyclohexanone (6)	225
"	25	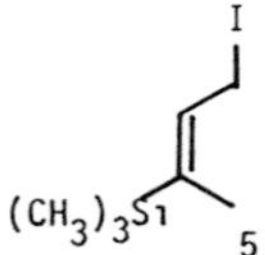 (CH₃)₃Si-C(CH₃)=CH-CH₂I THF:ether (2:1), 5 hr, 25°	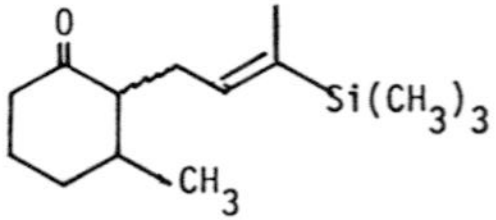 Si(CH₃)₃-containing product (75)	233

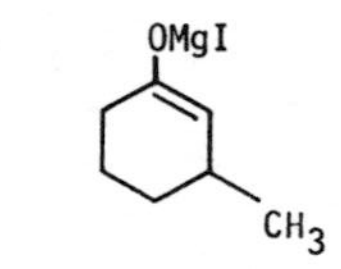 26	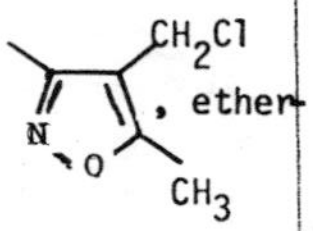 CH_2Cl, ether, HMPA, 0° 1 hr, 25° 13 hr	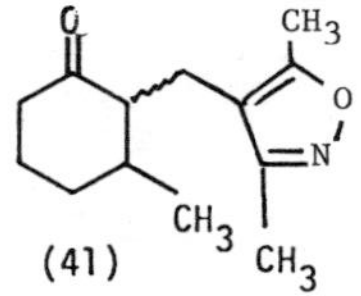 (41)	209
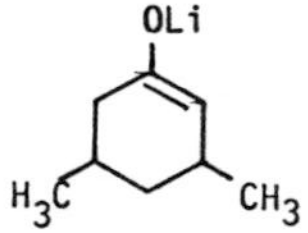 27	Excess n-BuI THF-NH$_3$(4:6) 2 hr, -33°	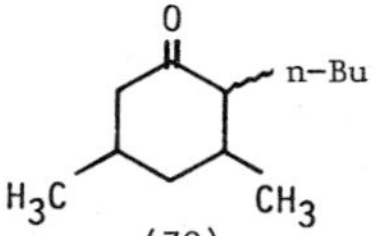 (70)	228
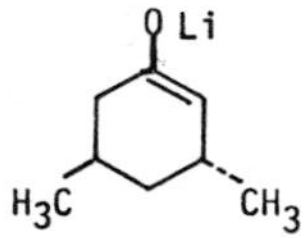 28	"	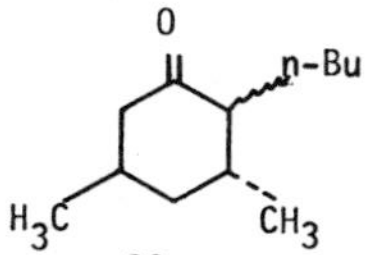 83	228
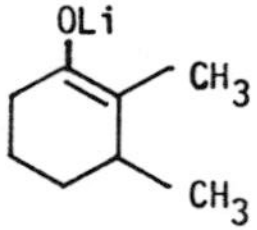 29	Excess CH$_3$I, HMPA-THF	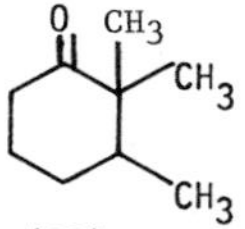 (92)	221

Enolate Structure	Method of Enolate Formation[a]	Alkylating Agent and Conditions	Product(s)[b]	Ref
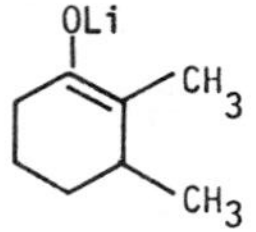 OLi, CH₃, CH₃ (cyclohexene)	29	Excess allyl iodide, HMPA–THF	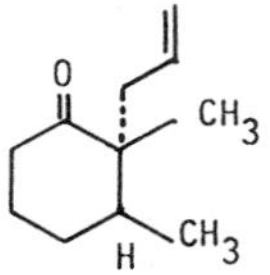 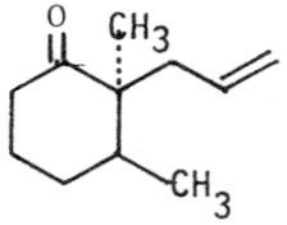 (allyl/CH₃/CH₃ cyclohexanone isomers) 9:1 (75)	221
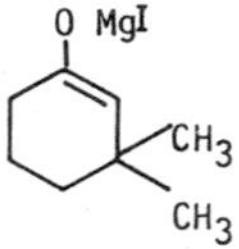 O MgI, CH₃, CH₃ (cyclohexene)	30	 Br⟶ allyl, ether–HMPA 1:1	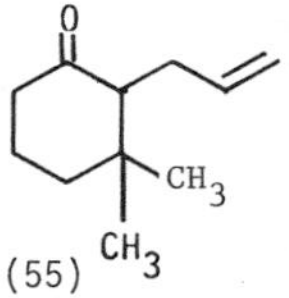 (allyl, CH₃, CH₃ cyclohexanone) (55)	210
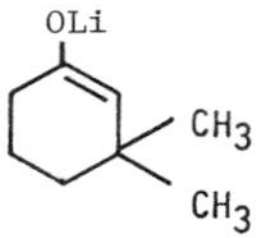 OLi, CH₃, CH₃ (cyclohexene)	31	 Br⟶ allyl, Liq. NH₃, DME, −33°	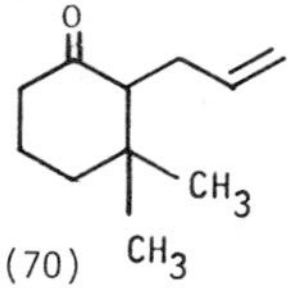 (allyl, CH₃, CH₃ cyclohexanone) (70)	205
"	31	CH₃I, Liq. NH₃, DME, −33°	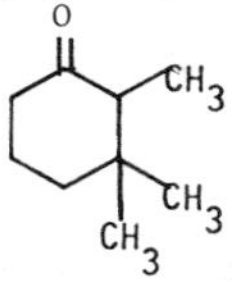 (CH₃, CH₃, CH₃ cyclohexanone) (73)	205

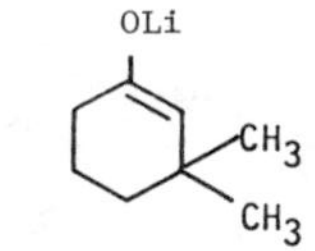

	Enolate	Conditions	Products (yield %)	Ref.
31	OLi, CH$_3$, CH$_3$	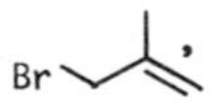 Br(methallyl), Liq. NH$_3$–DME, –33°	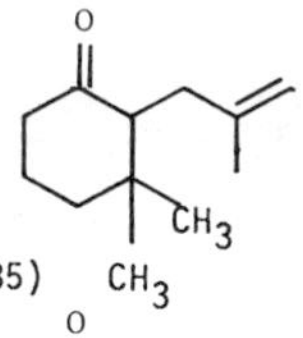 (35)	205
32	OLi, CH$_3$, CH$_3$, H$_3$C	Excess CH$_3$I, Liq. NH$_3$–ether, 1 hr, –33°	(57)	191
33	OLi, CH$_3$, CH$_3$, H$_3$C	Excess CH$_3$I, Liq. NH$_3$–ether, 1hr, –33°	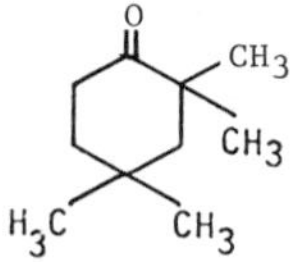 (43), 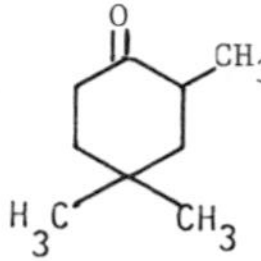 (3)	191
34	OLi, CH$_3$, CH$_3$, CH$_3$	Excess CH$_3$I, DME, 10 min. 0°	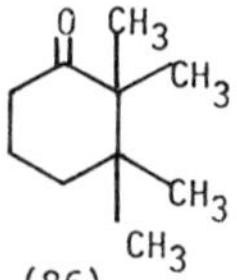 (86), 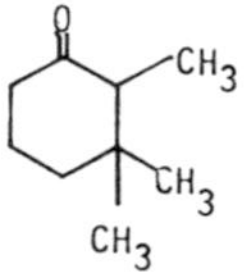 (3)	225
35	OLi, isopropyl	Excess CH$_3$I, DME, 0°, 5 min	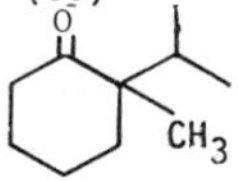 (86)	225

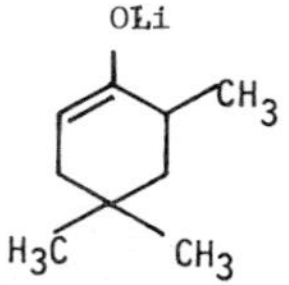

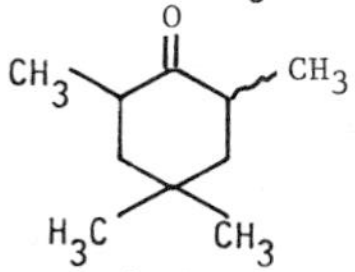

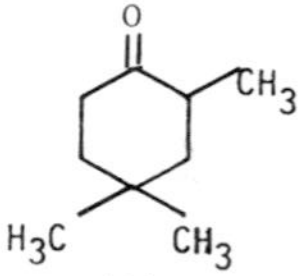

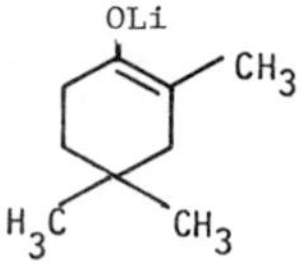

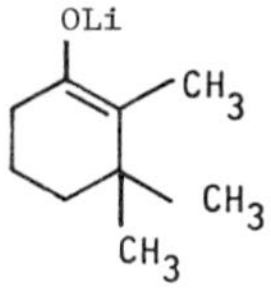

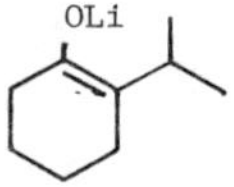

Enolate Structure	Method of Enolate Formation[a]	Alkylating Agent and Conditions	Product(s)[b]	Ref
OLi (2-methyl-5-isopropenyl-cyclohexenyl)	36	CH_3I, THF, $-70°$, 1 hr, $25°$, 1 hr	(95)	249
"	36	I—...—CO_2CH_3	CO_2CH_3 product (88)	250
OLi, n-Bu (cyclohexenyl)	37	CH_3I, THF, 30 min. $25°$	(61) + (6)	252
OLi, n-Bu (cyclohexenyl)	38	Excess CH_3I, THF, $0°$, 1 hr	(37) + (8) + (9)	226
OLi, n-Bu (cyclohexenyl)	39	Excess CH_3I, ether HMPA, $-30°$, 2 hr	(84) (7:1 trans:cis mixture)	226

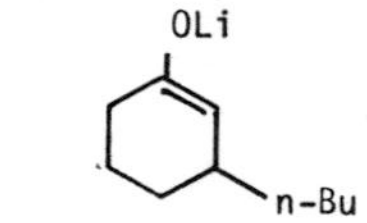	39	n-BuI, ether-HMPA 25°, 30 min	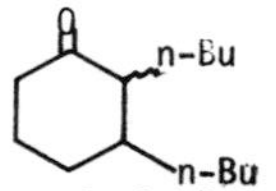 (50) (4.5:1 trans: cis mixture) 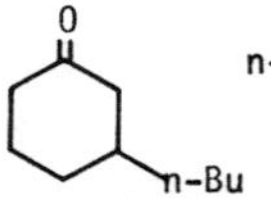 (3) 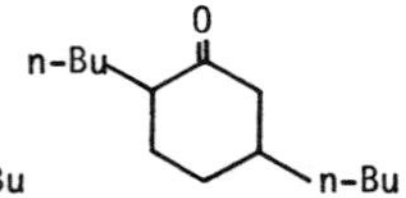	226
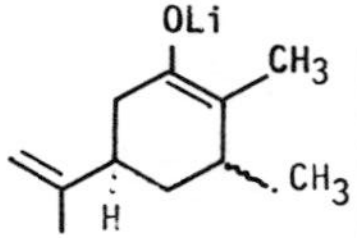	40	Excess CH_3I, DME, 0°, 5 min	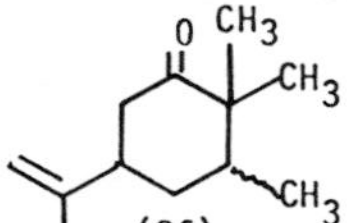 (86)	225
"	40	Excess C_2H_5I, DME, 4 hr, 25°	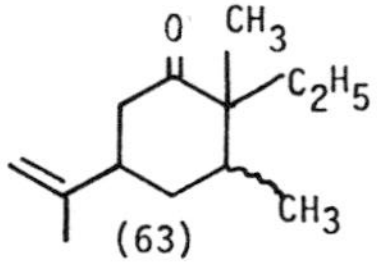 (63) 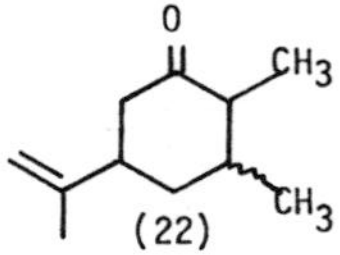 (22) 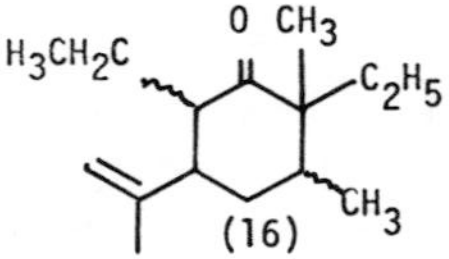 (16)	225
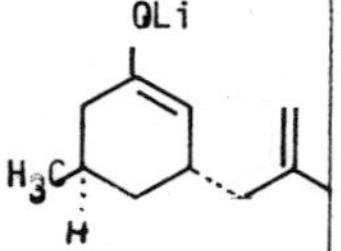	41	n-BuI, Liq. NH_3: THF(3:2) 6 hr, -33°	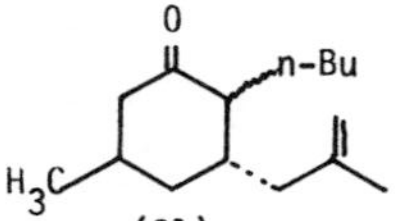 (91)	228
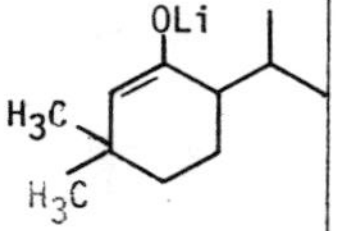	42	CH_3I, THF	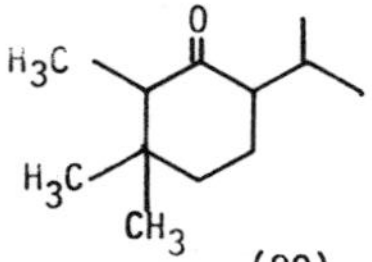 (80)	47

Enolate Structure	Method of Enolate Formation[a]	Alkylating Agent and Conditions	Product(s)[b]				Ref
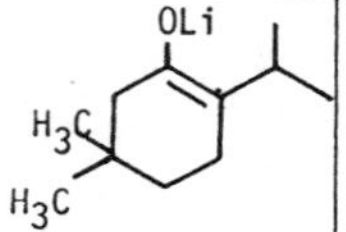 (structure, O Li, isopropyl, H_3C, H_3C)	43	CH_3I, THF	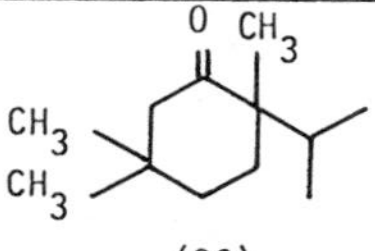 (structure, O, CH_3, isopropyl, CH_3, CH_3) (80)				47
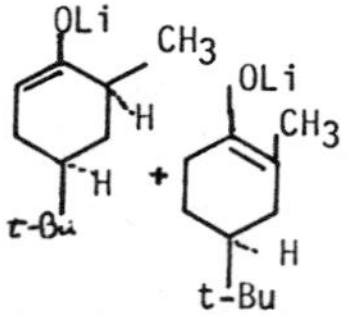 (structures, OLi CH_3 + OLi CH_3, H, t-Bu, H) (86%) (14%)	44	Excess CH_3I, DME, 30 min, 25°	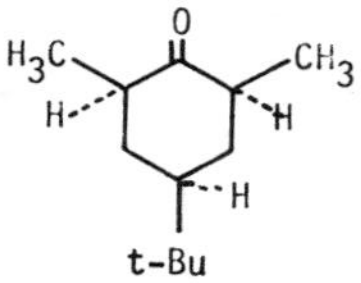 H_3C H·· (structure) CH_3 H t-Bu 28	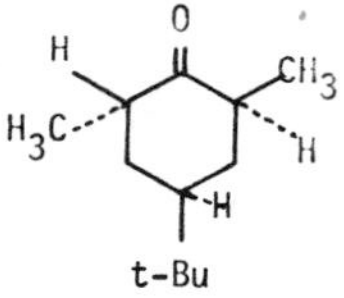 H_3C H (structure) O CH_3 H t-Bu 52	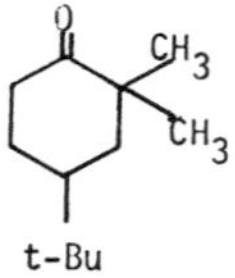 (structure) CH_3 CH_3 O t-Bu 11	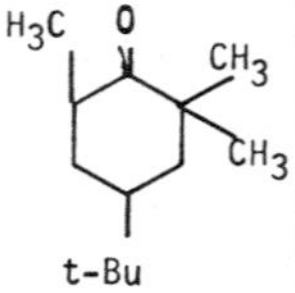 H_3C (structure) O CH_3 CH_3 t-Bu 13	49
(structure OLi CH_3 H, H t-Bu) (11%) + (89%) +25% cis-2-methyl-4-t-butylcyclohexanone	45	Excess CH_3I, DME, 30 min, 25°	5	53 (+28% 2-methyl-4-t-butylcyclohexanone)	--		49
(structure OLi CH_3 H, H t-Bu) (20%) + (5%) 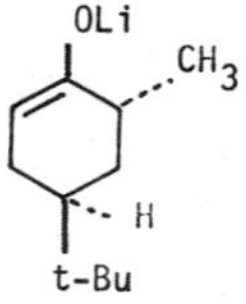 + 77%	46	Excess CH_3I, DME, 30 min, 25°	9	55 (+19%cis-2,6-dimethyl-trans-4-t-butylcyclohexanone)	3	14	49

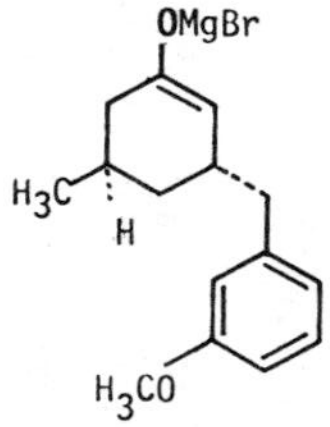	47	Excess BrCH$_2$-CH=CH$_2$, HMPA(40%)-ether-THF 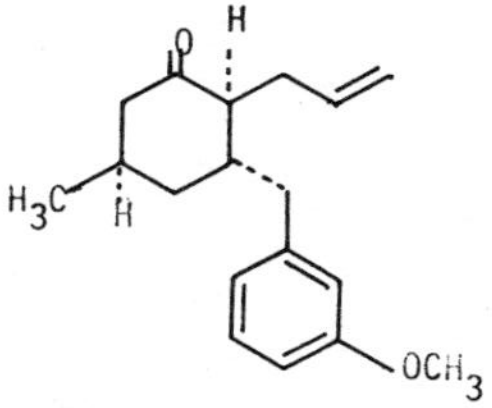

a) Listed on pages following the Table.

b) The numbers in ( ) refer to product yields. The numbers not in ( ) refer to the composition of the isolated product mixtures.

Method of Enolate Formation (notes for Table III)

1. Li/NH$_3$ reduction (1 eq t-BuOH as proton donor) of 6-methylcycloh 2-enone.

2. Li/HN$_3$ reduction of 6-chloromercuri-2-methylcyclohexanone.

3. Cleavage of diethyl 1-(6-methyl)cyclohexenylphosphate with methyllithium in DME for 20 min at 25°C.

4. Kinetic enolization of 2-methylcyclohexanone with LDA at -20° to 0°C (< 1% of the more-substituted lithium enolate 5B present)

5. Cleavage of 1-trimethylsiloxy-6-methylcyclohexene with lithium amide in THF-NH$_3$ for 30 min at -33°C.

6. Cleavage of 1-trimethylsiloxy-6-methylcyclohexene with methyllithium in THF at 50°C.

7. Kinetic enolization of 2-methylcyclohexanone with LDA in THF at 25°C.

8. Li/HN$_3$ reduction (1 eq H$_2$O as proton donor) of 2-methylcyclohex 2-enone.

9. Li/NH$_3$ reduction (2 eq H$_2$O as proton donor) of 2-n-butylthiomethylenecyclohexanone.

10. Cleavage of 1-tri-n-butylstannyloxy-2-methylcyclohexene with phenyllithium in DME.

11. Reductive cleavage of 2-methyl-2-thiophenylcyclohexanone with Li/NH$_3$.

12. Reductive cleavage of 2-chloro-2-methylcyclohexenone with Li/NH

13. Cleavage of dimethyl 1-(2-methyl)cyclohexenyl phosphate with methyllithium in DME for 1 hr at 0°C.

14. Cleavage of diisopropyl 1-(2-methyl)cyclohexenyl phosphate with n-butyllithium in DME at 50°C for 24 min.

15. Cleavage of butyl-1-(2-methyl)cyclohexenyl phenylphosphonate with n-butyllithium at 25°C for 20 min.

16. Cleavage of 1-(2-methyl)cyclohexenyldiphenylphosphinate with n-butyllithium in DME at 0°C for 1 hr.

17. Cleavage of a 9:1 mixture of 1-acetoxy-2-methylcyclohexens and 1-acetoxy-6-methylcyclohexene with 2 eq methyllithium in DME at 25°C.

18. Cleavage of a mixture of >95% 1-acetoxy-2-methylcyclohexene and <5% 1-acetoxy-6-methyllithium in DME at 25°C.

19. Cleavage of 1-trimethylsiloxy-2-methylcyclohexene with methyllithium in DME.

20. Cleavage of 1-trimethylsiloxy-2-methylcyclohexene with lithium amide in liquid ammonia.

21. Cleavage of 1-acetoxy-2-methylcyclohexene with methyllithium (2 eq) in THF at 25°C.

22. Reduction of 2-methylcyclohex-2-enone with lithium tri-sec-butylborohydride (L-Selectride) in THF at -78°C.

23. Li/NH_3 reduction (1 eq t-BuOH as proton donor) of 3-methylcyclohexenone.

24. Li/NH_3 reduction (1 eq H_2O as proton donor) of 3-methylcyclohexenone.

25. Conjugate addition of lithium dimethyl cuprate to cyclohex-2-enone in ether.

26. Cuprous chloride-catalyzed conjugate addition of methylmagnesium iodide to cyclohex-2-enone in ether.

27. Cleavage of 1-trimethylsiloxy-cis-3,5-dimethylcyclohexene with lithium amide in liquid ammonia.

28. Cleavage of 1-trimethylsiloxy-trans-3,5-dimethylcyclohexene with lithium amide in liquid ammonia.

29. Conjugate addition of lithium dimethylcuprate to 2-methylcyclohexenone in ether.

30. Cuprous iodide-catalyzed conjugate addition of methylmagnesium iodide to 3-methylcyclohex-2-enone in ether at 0°C.

31. Li/NH_3 reductive cleavage (1 eq t-BuOH as proton donor) of 2,3-methylene-3-methylcyclohexanone in DME.

32. Li/NH_3 reduction (1 eq t-BuOH as proton donor) of 4,4,6-trimethylcyclohex-2-enone.

33. Li/NH_3 reduction (1 eq t-BuOH as proton donor) of 2,4,4-trimethylcyclohex-2-enone.

34. Conjugate addition of lithium dimethylcuprate to 2,3-dimethylcyclohex-2-enone in ether.

35. Conjugate addition of lithium dimethylcuprate (2 eq) to 2-n-butylthiomethylenecyclohexanone in ether.

36. Reduction of carvone with lithium tri-sec-butylborohydride (L-Selectride) in THF at -70°C.

37. Cleavage of the corresponding vinyloxyborane with n-butyllithium in THF at 25°C.

38. Reaction of t-butoxy(n-butyl)cuprate with 2,6-dibromocyclohexanone in THF at -78°C and warming to 0°C for 15 min.

39. Conjugate addition of lithium di-n-butylcuprate to cyclohex-2-enone at -78°C for 30 min.

40. Conjugate addition of lithium dimethylcuprate to carvone.

41. Cleavage of 1-trimethylsiloxy-trans-3-methallyl-5-methylcyclohexene with lithium amide in liquid ammonia.

42. Cleavage of 1-trimethylsiloxy-3,3-dimethyl-6-isopropylcyclohexene with methyllithium in THF.

43. Cleavage of 1-trimethylsiloxy-2-isopropyl-5,5-dimethylcyclohexene with methyllithium (2 eq) in THF.

44. Kinetic enolization of cis-2-methyl-4-t-butylcyclohexanone with trityllithium at room temperature.

45. Equilibration of the lithium enolate mixture from cis-2-methyl-4-t-butylcyclohexanone by treatment with 25% excess ketone.

46. Kinetic enolization of a 77:23 mixture of trans and cis-2-methyl-4-t-butylcyclohexanone with trityllithium in DME at room temperature.

47. Cupurous-ion-catalyzed conjugate of m-methoxybenzylmagnesium bromide to 5-methylcyclohex-2-enone in ether-THF.

to enones are actually lithium enolates rather than copper(I) species, at least at temperatures above 25°C [238].

Successful regiospecific alkylations have been carried out with methyl, n-alkyl, and benzyl halides and with allylic halides of a variety of structural types. The most common solvents are DME, THF, or mixtures of liquid ammonia and ethereal solvents. Occasionally mixtures of diethyl ether or THF and HMPA are used. The use of the polar additive HMPA is necessary with halomagnesium enolates. The reaction times and temperatures are dependent on the reactivity of the alkylating agent and the solvent; with methyl, allyl, and benzyl halides, reaction times of only a few minutes are needed even at low temperatures, while with less reactive alkylating agents, such as n-butyl iodide, extended reaction times are required.

n-Butyl iodide, which is reported to be 50-100 times less reactive than methyl iodide [3, 44], is the least reactive alkylating agent which has been used successfully in regiospecific alkylations. The reaction conditions and enolate structure seem to be of critical importance when this alkylating agent is used. For example,

2-methyl-6-n-butylcyclohexanone (*80a*) was obtained in good yield
(75%) when the thermodynamically unstable 2-methylcyclohexanone
lithium enolate *5A* (formed by cleavage of the corresponding silyl
enol ether with lithium amide in liquid ammonia) was alkylated with
n-butyl iodide in liquid ammonia-THF [228]. However, reaction of
the same enolate (obtained by cleavage of the corresponding diethyl
enol phosphate with methyllithium) with n-butyl iodide in DME gave
only 2-n-butyl-2-methylcyclohexanone (*80*) [150]. (In this reaction
the enolate was added to an excess of the alkylating agent.) The
absence of *80a* in this reaction indicated that enolate equilibration
was much more rapid than alkylation. The complete equilibration of
5A to *5B* which leads to the product *80* under these conditions is

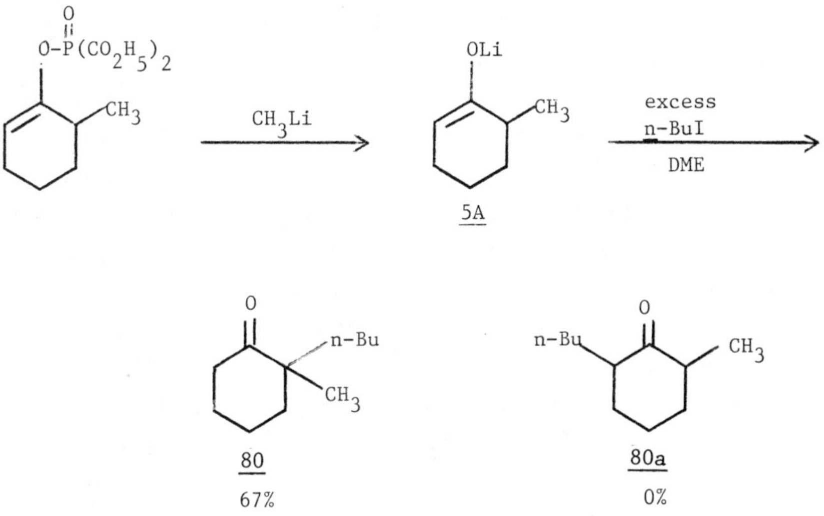

somewhat surprising. Likewise, quenching of the enolate *61*, obtained
by reaction of 2,6-dibromocyclohexanone with lithium t-butoxy(n-butyl)-
cuprate, with n-butyl iodide in THF-HMPA gave a mixture of products,
indicating significant enolate equilibration and polyalkylation
[Eq. (38)] [226].

OM ⟍ n-Bu
          ──n-BuI──→   THF:HMPA

61   28%   17%   13%

(3

Regiospecific alkylations (including n-butylations) of 1-enolates of 3-alkyl and 3,5-dialkylcyclohexanones have been achieved without difficulty. For example, under conditions similar to those described for the alkylations of 61, the 1-enolate of 3-n-butylcyclohexanone gave a mixture of trans- and cis-2,3-di-n-butylcyclohexanone in 50% yield with only 4% of 2,5-di-n-butylcyclohexanone being obtained [226]

The results of treatment of β,β-disubstituted decalone or cyclohexanone enolates with allylic and methyl iodides in THF-HMPA or ether suggested that the rates of alkylation of these species are slower than the rates of enolate equilibration [221]. However, as indicated in Table III, successful 2-alkylation (allyl iodide) or methylation of both lithium [205] and halomagnesium [210] enolates of 3,3-dimethylcyclohexanone have been reported. The lithium enolate alkylations were carried out in liquid ammonia-DME, and the halomagnesium enolate was alkylated in ether-HMPA. Equilibration was found to occur when 1,3-dichloro- and 1-iodo-3-chloro-2-butene were employed as alkylating agents for the bromomagnesium 1-enolate of 3,3-dimethylcyclohexanone [210]. Here, and in the β,β-disubstituted systems already mentioned, it seems possible that the utilization of a different solvent might have allowed successful regiospecific alkylation.

Because cyclopentanones apparently undergo relatively rapid enolizations and aldol condensations, proper conditions for regio-specific alkylation of enolates of unsymmetrical cyclopentanones are difficult to achieve. Nevertheless, several examples of regiospecific alkylations of 2- and 3-alkylcyclopentanones have appeared. The results of methylation of the more-substituted lithium enolate 34A of 2-methylcyclopentanone, which was produced from the reaction of the corresponding enol acetate with methyllithium, wich methyl iodide have been discussed [Eq. (35)] [119]. This enolate has also been prepared by lithium ammonia-reduction (water as the proton donor) of 2-n-butylthiomethylene cyclopentanone and alkylation with methyl iodide, allyl bromide, and isopropyl iodide in liquid ammonia-ether [193] [Eq. (39)]. In these runs 2,2-dialkylcyclopentanones were the only monoalkylation products obtained. The results show that by using liquid ammonia-ether as the solvent and a short reaction time, polyalkylation can be minimized.

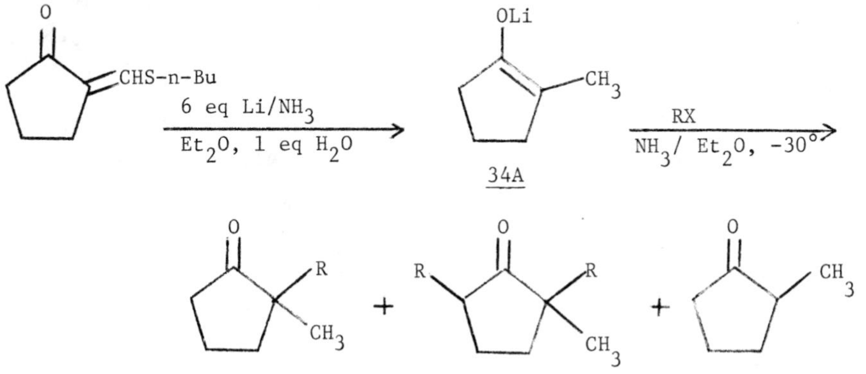

X (time)			
H_3I (1 min)	70%	5%	—
H_3I (15 min)	51%	22%	—
$H_2=CH-CH_2Br$ (1 min)	62%	15%	—
$(CH_3)_2CHI$ (45.5 hr)	40%	—	15%

(39)

It does not appear that the less-substituted enolate *34B* of 2-methylcyclopentanone has been regiospecifically prepared for alkylation studies. However, a 78:22 mixture of *34B* and *34A*, prepared by kinetic enolization of *34* with trityllithium in THF, undergoes alkylation with methyl iodide to provide a 73:27 mixture of 2,5- and 2,2-dimethylcyclopentanone [51].

The attachment of alkyl substituents at the 2- and 3-positions of cyclopentanones is of interest in connection with the synthesis of prostaglandins [223, 226, 244]. Lithium 1-enolates of 3-substituted cyclopentanones have been obtained by conjugate addition of lithium vinylcuprates [223, 226] and lithium salts of protected cyanohydrins derived from α,β-unsaturated aldehydes [244] to cyclopentenone. Regiospecific trappings of these species with various alkylating agents have been reported. For example, the enolate *81*, which was obtained by conjugate addition of lithium methyl(vinyl)-cuprate to cyclopentenone, has been regiospecifically alkylated with allyl bromide and ethyl bromoacetate [Eq. (40)] [226]. Satisfactory conditions for regiospecific alkylation of *81* with trans-1-iodo-2-bute or n-butyl iodide could not be found, and mixtures of products derived from alkylation at the 2-position and the 5-position as well as substantial amounts of polyalkylated materials were obtained with these reagents. Reaction of *81* with methyl 7-iodo-5-heptenoate in THF at -20°C gave the 2-alkylation product in 10-20% yield [226].

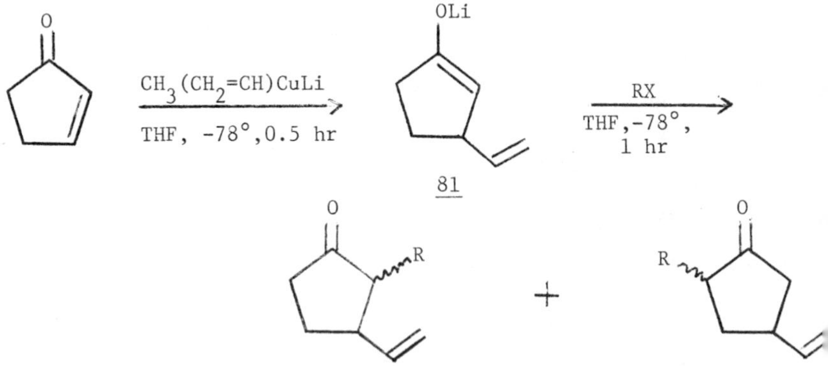

RX		
BrCH$_2$CH=CH$_2$	trans 69% + cis 3%	0.5%
BrCH$_2$CO$_2$C$_2$H$_5$	trans 46% + cis 4%	<1.%

(40)

A mixture of 11-deoxyprostaglandin E_2 methyl ester (*82*) and 11-deoxy-8,12-epi-PGE$_2$ methyl ester was obtained by alkylation of the lithium enolate *83* with excess methyl Z-7-bromo-5-heptenoate (*84*) in liquid ammonia-THF followed by acidic work-up [223]. The enolate *83* was prepared by conjugate addition of the chiral cuprate *85* to cyclopentenone, the trapping of the initial enolate with trimethylsilyl chloride, and cleavage of the enol ether with lithium amide in liquid ammonia. Attempts to trap the initially formed enolate derived from conjugate addition of an achiral cuprate, related

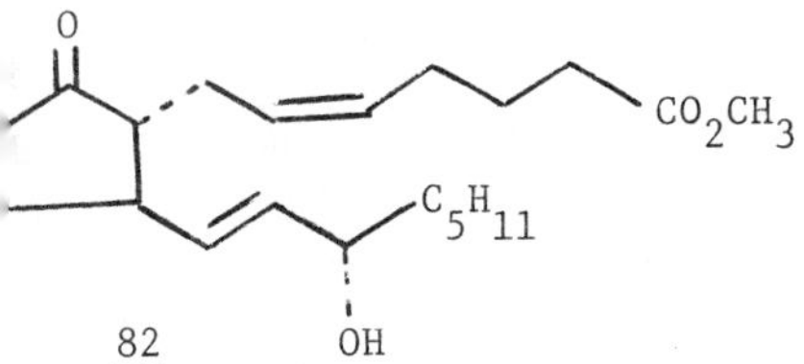

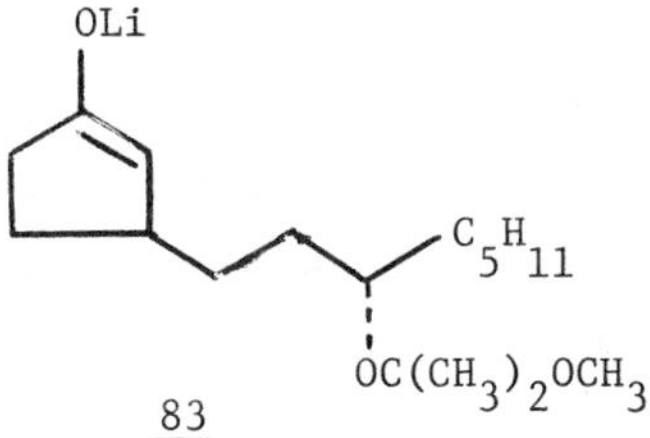

to *85*, with the allylic iodide *84* failed. Racemic 5,6-dehydro-11-deoxyprostaglandin E_2 and 11,15-deoxyprostaglandin E_2 methyl esters were prepared by an approach similar to the synthesis of *82*. Conjugate addition of the lithium salt of the "protected" cyanohydrin *86* to cyclopentenone followed by alkylation with methyl 7-iodo-5-heptynoate gave a mixture of products from which the 2,3-disubstituted cyclopentanone *87* was isolated in 20% yield after appropriate transformations [244]. Partial enolate equilibration occurred in this reaction, since a 3,5-disubstituted cyclopentanone was also isolated [244].

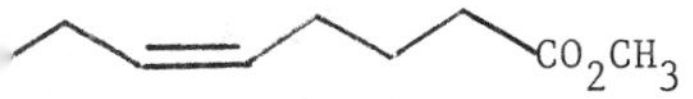

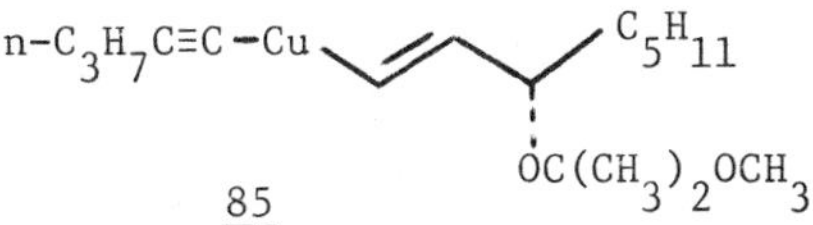

2,3-Disubstituted cyclohexanones or cyclopentanones may also be obtained by addition of lithiated bis(methylthio)(silyl)- and bis(methylthio)(stannyl)methanes to the corresponding enones followed by addition of an alkylating agent [242].

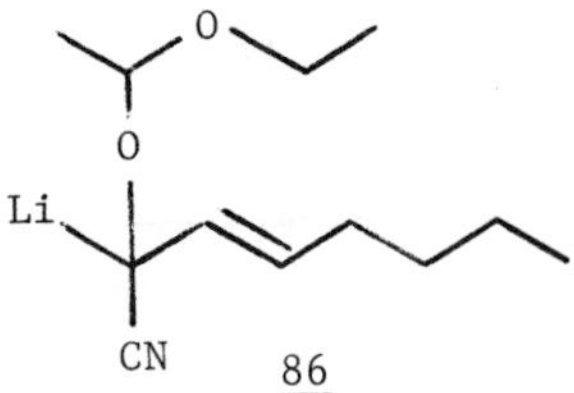

86

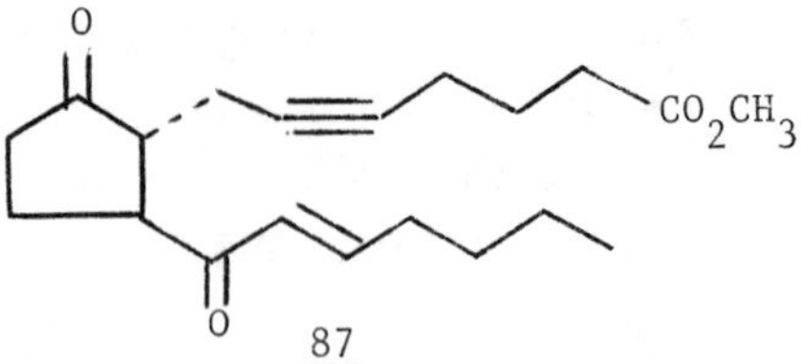

87

Regiospecific alkylations of larger-ring ketones such as cyclo-
heptanones have been attempted only rarely. Conjugate addition of
lithium dimethylcuprate to 2-n-butylthiomethylene cycloheptanone gave
the more substituted enolate of 2-isopropylcycloheptanone which was
alkylated in 93% yield with methyl iodide in DME [Eq. (41)] [225].
However, the structural integrity of the enolate derived from the
conjugate addition of lithium dimethylcuprate to 3,7-
dimethylcyclohepten-2-one was not maintained when alkylation with
methallyl iodide in THF-HMPA was attempted [225].

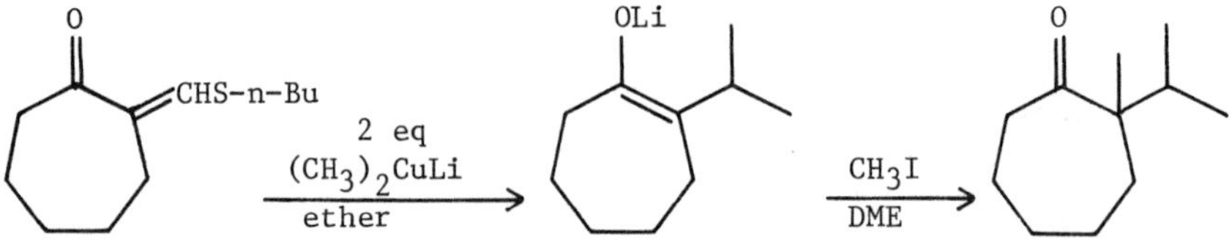

93%

(41)

Regiospecific alkylations have been employed in the synthesis
of a large number of alkylated decalones, hydrindanones, and tricyclic and steroidal ketones [39, 129, 303]. Lithium-ammonia
reductions of α,β-unsaturated ketones have been widely employed to
generate the appropriate enolates. The thermodynamically unstable
1-enolates of trans-2-decalones such as 8A, which are obtained on
reduction of 1(9)-octalin-2-ones, have been trapped in good yield
with methyl and n-butyl iodide and with activated halides such as
allyl, benzyl, and halomethylisoxazole derivatives to produce the
corresponding 1-substituted trans-2-decalones such as 9 [132].

1,1-Dialkyl trans-2-decalones have also been obtained by reduction-
alkylation of the corresponding 1-alkyl-1(9)-octalin-2-ones [39]. In
these systems the best results were obtained if the liquid ammonia
were replaced by THF before addition of the alkylating agent [39].
The fully substituted lithium enolates produced in these reactions
are more soluble in THF than in liquid ammonia and are apparently
more rapidly alkylated in the former solvent. As an example of this
procedure, the trimethyl decalone 89 was obtained in 57% yield by
reduction-methylation of the dimethyl octalone 88 [39]. Lithium
3-enolates of 4-methyl-3-keto steroids can be produced and alkylated
similarly.

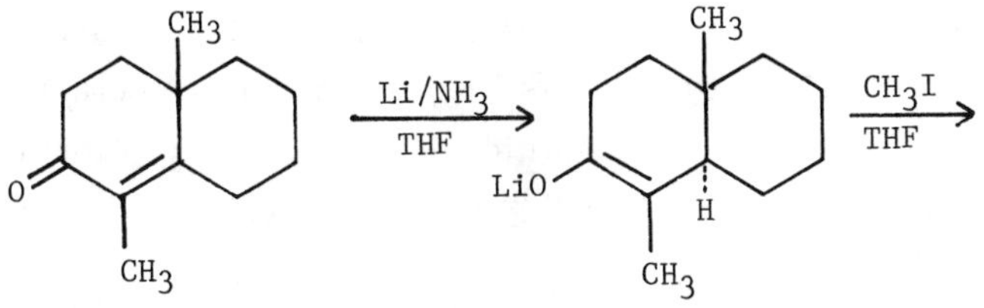

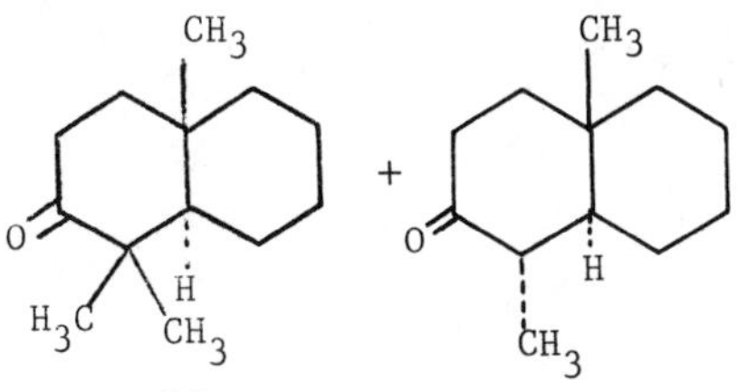

The stereochemistry of alkylation of reductively formed lithium
1- and 2-enolates in the trans-2-decalone series [304-306] and of
lithium 3-enolates similarly prepared from 4-methylcholest-4-en-3-one
[307] has been established. These results will be discussed in
Sec. II.D.9.

Angular alkylations of bicyclic and polycyclic ketones have
provided challenging problems for synthetic organic chemists for
a number of years. The direct base-promoted methylation of ketones

such as 1-decalone (37) occurs largely via the 1(2)-enolate 37B, rather than at the angular position [34, 308]. As will be discussed later, the application of blocking groups [28-35] at the 2-position of 37 provides a possible solution to this problem, but the most direct approach is to generate the desired enolate 37A and to regiospecifically alkylate it. Such enolates have been obtained by lithium-ammonia reduction of 9(10)-octalin-1-one [39] or by cleavage of the 1(9)-enol acetate or trimethylsilyl enol ether of 37 with methyllithium in DME [44, 119]--actually a 91:10 mixture of 37A and 37B was produced from a similar mixture of the silyl enol ethers [44].

In a thorough study involving the methylation of 37A, generated from the enol acetate, best results were obtained by using relatively low enolate concentration and a short reaction time [119]. Under these conditions an 83:17 mixture of cis- (90a) and trans-9-methyl-1-decalone (90b) was obtained in 86% yield along with 7% 2,9-dimethyl-1-decalone when the alkylation was performed with methyl iodide. The

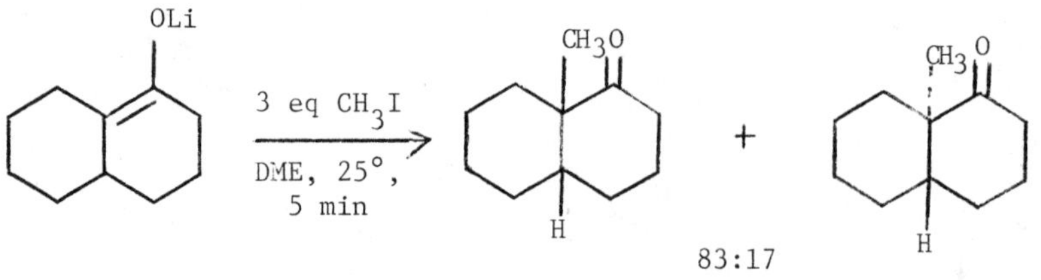

proportion of 90a increased to >95% when the highly reactive alkylating agent trimethyloxonium 2,4,6-trinitrobenzenesulfonate in methylene chloride was employed, but O-methylated material was obtained as well.

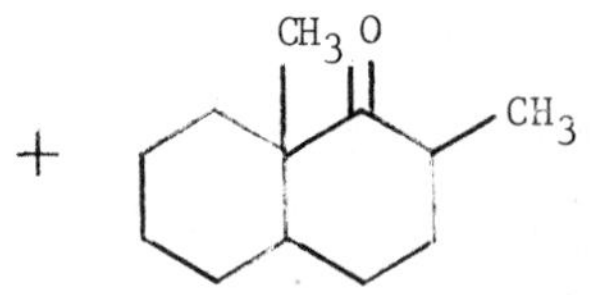

The hexahydroindanone enolate *91a* and the tetrahydroindanone enolate *91b*, derived from cleavage of the corresponding enol acetates with methyl lithium in DME, gave mixtures of monoalkylation products containing 98 and 96%, respectively, of the cis-fused isomers *92a* and *92b* in about 60% yield on alkylation with methyl bromoacetate in DME [309]. In each case ∿10% of the dialkylated ketone was obtained.

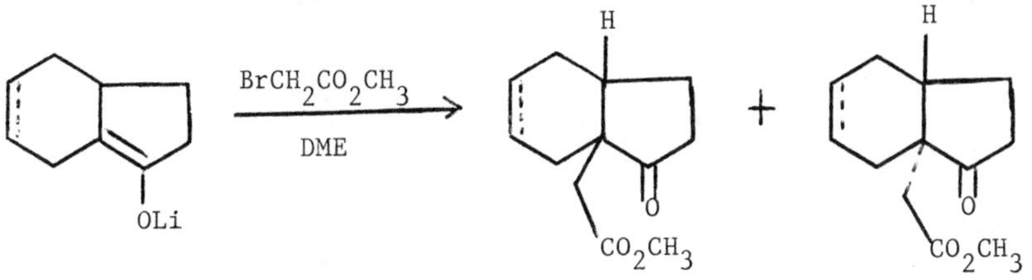

91 a; Sat'd 5,6 bond   92 a; 98:2 (58%)   93 a
   b; 5,6 double bond   b; 96:4 (60%)   93 b

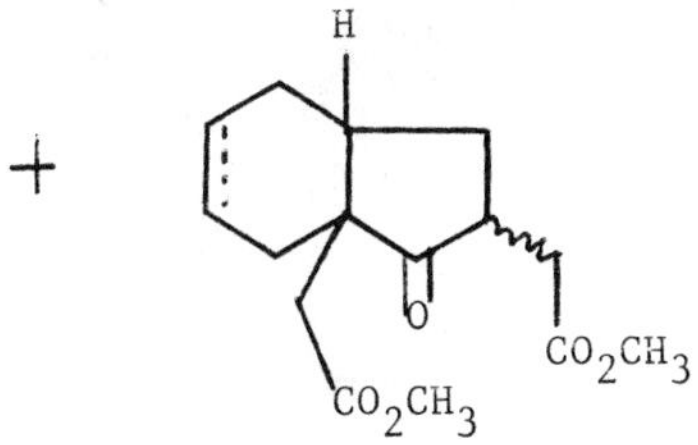

10%

Reactions involving regiospecific formation and alkylation of lithium enolates have been utilized as the key step (or steps) in the synthesis of several complex natural products. Examples of such transformations include *94 → 95* (dl-progesterone) [310], *96 → 97* and *98 → 99* (lupeol) [204], and *100 → 101* (dl-germanicol) [311].

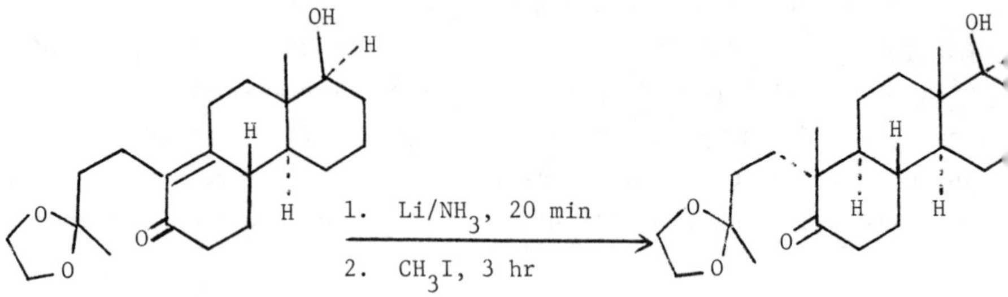

$$\underline{94} \xrightarrow[\text{2. CH}_3\text{I, 3 hr}]{\text{1. Li/NH}_3\text{, 20 min}} \underline{95}$$

nearly quantitative

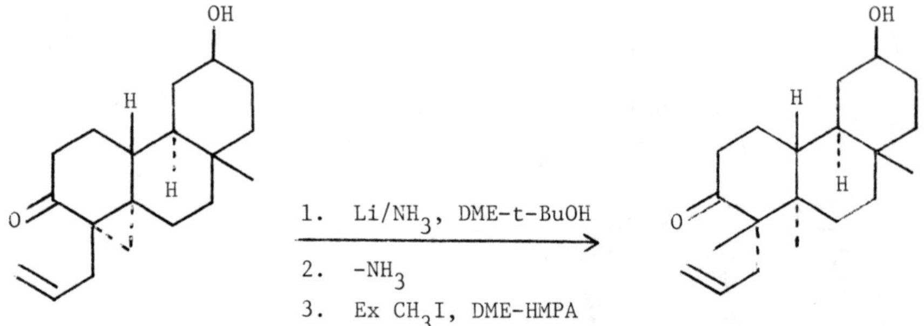

$$\underline{96} \xrightarrow[\substack{\text{2. } -\text{NH}_3 \\ \text{3. Ex CH}_3\text{I, DME-HMPA}}]{\text{1. Li/NH}_3\text{, DME-t-BuOH}} \underline{97}$$

60%

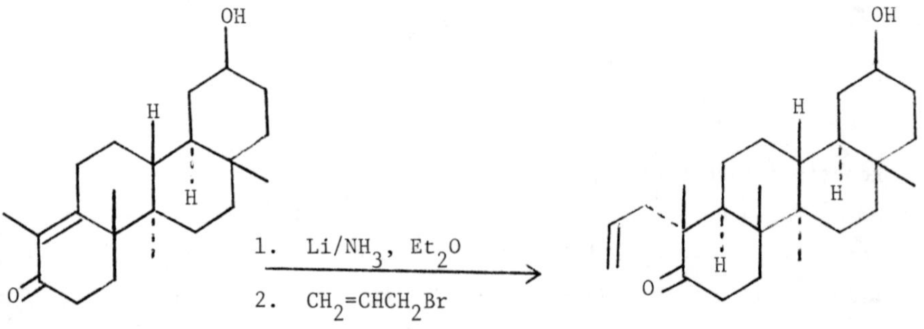

$$\underline{98} \xrightarrow[\text{2. CH}_2=\text{CHCH}_2\text{Br}]{\text{1. Li/NH}_3\text{, Et}_2\text{O}} \underline{99}$$

80%

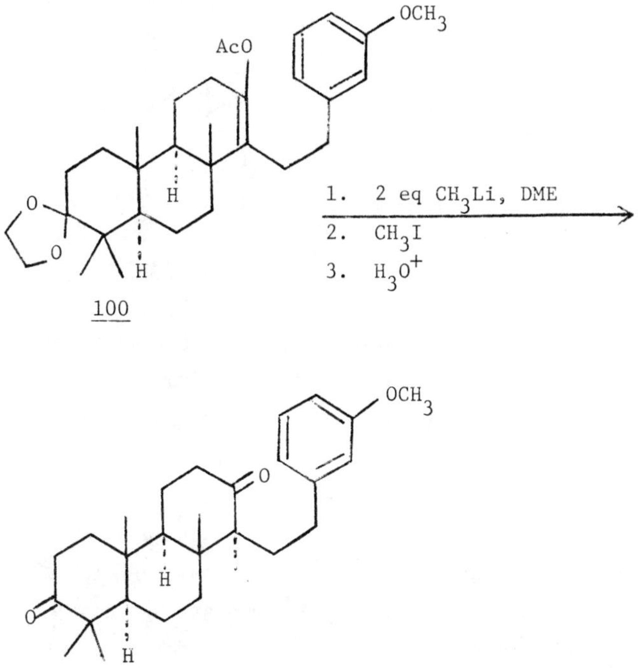

100

101

61%

The synthesis of 17α-alkyl-pregn-20-ones (103) is of interest
because these compounds have relatively high progestational activity
[196]. The generation of pregn-20-one 17-enolate anions, 102, has
been carried out by metal-ammonia reduction of 16-dehydropregn-
20-one derivatives [196, 312] or related compounds having 17α-
acetoxy, α-hydroxy, or α-bromo substituents [195, 196]. Although
other metals have been employed, lithium is again the metal cation
of choice for carrying out regiospecific alkylations of 102 at C17.
Reasonably good yields of 17α-methyl- (103a) (65%) and ethylpregn-
20-ones (103b) (43%) were obtained when the enolate 102 (M = Li),
obtained from lithium-ammonia reduction of the 3-ethylene ketal of
pregna-5,16-diene-3,20-dione, was alkylated with methyl or ethyl
iodide. However, significant quantities of 21-mono- and diethylated
products were obtained with the latter reagent. Higher primary
alkyl iodides, allyl bromide, and benzyl chloride gave yields of

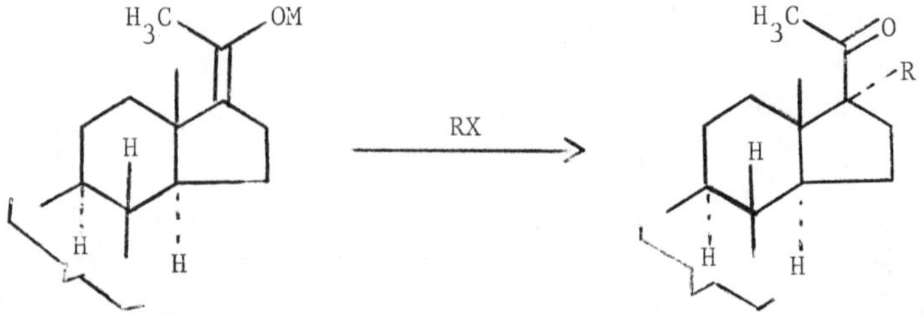

102   103 a; R = Me (65%)

                                                      b; R = Et

products of the type 103 in the 7-24% range. Similar results were
obtained on alkylation of the 17-enolate, produced from lithium-
ammonia reduction of 16-dehydropregnenone acetate, using an excess
of the alkylating agent in liquid ammonia-THF [312].

Unless a proton donor is added, 1,4-addition of amide ion to
pregn-16-en-20-ones apparently occurs as a side reaction during metal
ammonia reductions [195]. The amide adducts are resistant to reduction
and undergo elimination of ammonia to yield the starting enone on
work-up [191, 195]. For this reason the method of choice for production of enolates such as 102 (M = Li) is reductive cleavage of the
corresponding 17α-acetoxy compounds. Thus, treatment of the tetrahydropyranyl ether of 17α-acetoxy-pregn-5-en-3-ol-20-one with lithium
in liquid ammonia containing a THF-toluene mixture, followed by
reaction with excess methyl iodide for 1 hr, gave the 17α-methyl
ketone 103a in 88% yield [195].

Several examples of alkylations of enolates of unsymmetrical
acyclic ketones have been reported. The thermodynamically unstable,
more-substituted lithium enolates 28A and 29A derived from methyl
isobutyl (28) and ethyl isopropyl ketone (29), respectively, have
been regiospecifically alkylated. The enolate 28A has been prepared
by two routes: lithium-ammonia reduction (1 eq of triphenylcarbinol
as proton donor) of mesityl oxide [192] and conjugate addition of
lithium dimethylcuprate to 3-pentene-2-one [225]. The results of
methylation of 28A in liquid ammonia-ether and in DME are shown in
Eq. (42).

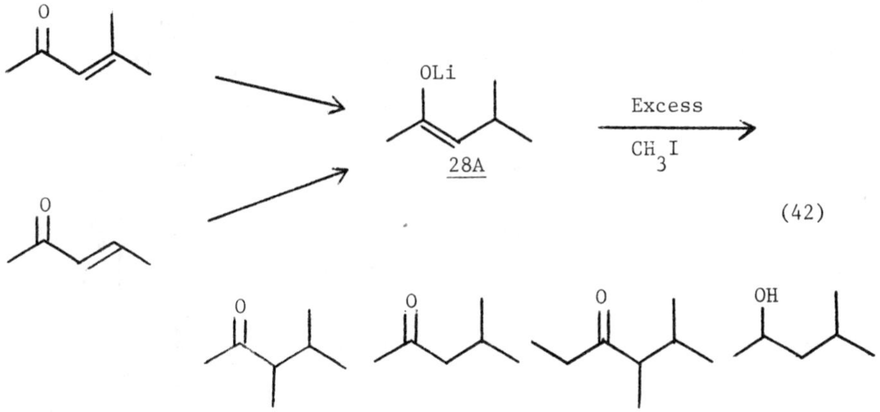

(42)

/Et$_2$O, -33°, 20 min	75%	9%	8%	6%
, 0°, 5 min	46%	2%	3%	–

The enolate *29A*, which was formed by lithium-ammonia reduction
(water as proton donor) of the n-butylthiomethylene derivative of
diethyl ketone, was alkylated regiospecifically in good yield with
methyl iodide or allyl bromide [Eq. (43)] [193]. Methylcyclohexyl
ketone (*33*) undergoes base-promoted alkylation primarily at the methyl

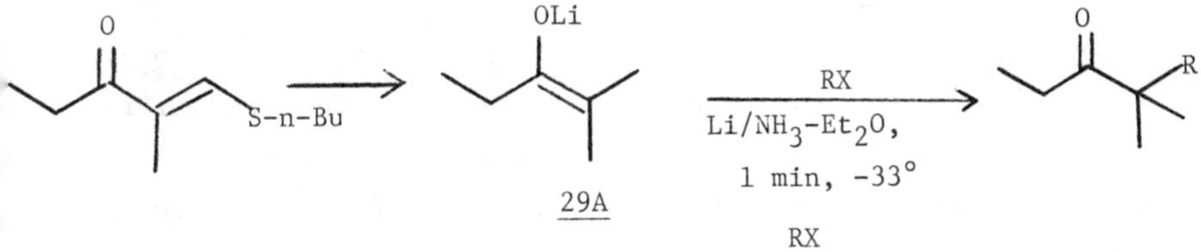

RX	
CH$_2$=CHCH$_2$Br	82%
CH$_3$I	69%

(43)

group [3]. However, by lithium-ammonia reduction of the 1-
cyclohexenylmethyl ketone the internal enolate *33A* was prepared and
methylated in low yield [Eq. (44)] [313]. The stereochemistry of
the methylation of the internal lithium enolate of

methyl-4-t-butylcyclohexyl ketone, obtained by treatment of the
corresponding enol acetate with methyllithium, has been reported
[314]. These results will be discussed in Sec. II.D.9.

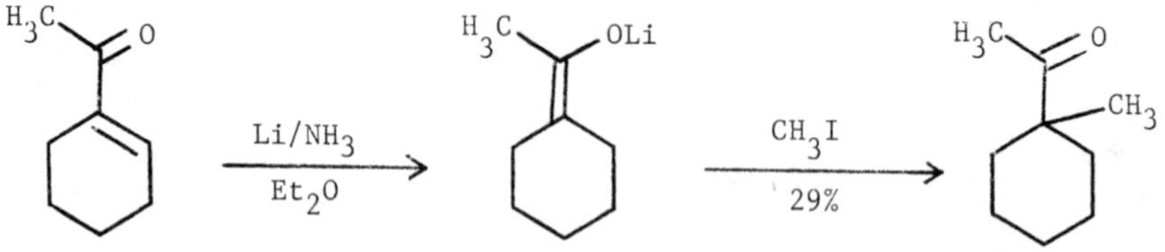

(44)

33A

Because the rates of alkylation of lithium enolates of acyclic
ketones may be quite similar to the rates of enolate equilibration,
the possibility of regiospecific alkylation has been shown to be
markedly dependent on enolate structure. For example, the Z-enolate
27B (M = Li) (p. 109), which was prepared by cleavage of the correspond
ing enol acetate with methyllithium, was regiospecifically alkylated
with benzyl bromide to give the 3-benzyl ketone in good yield using a
short reaction time [44]. However, a mixture of the lithium enolates
of 2-heptanone (27), which was prepared by kinetic enolization of the
ketone with LDA in DME and contained 84% of the terminal isomer 27A
and 14% of a mixture of the internal isomers 27B and 27C, also gave
largely the 3-benzylated product. This indicated that a significant
amount of equilibration of 27A to 27B or 27C occurred during alkylation. The results suggested that the terminal enolate is 5-10 times
less reactive than the internal enolates [44]. The lower reactivity
of less-substituted enolates of acyclic ketones such as 27 and cyclic
ketones such as 2-methylcyclohexanone (5) has been attributed to a
greater degree of aggregation of these species [44, 48]. Partial
equilibration was also observed in the methylation of the internal
lithium enolate of α,α-bis-n-amyl acetone [181]. Steric hindrance
to alkylation at the disubstituted α position could account for a
relatively slow rate of alkylation.

b. *Experimental procedures*. *Caution:* All the following procedures should be performed under an anhydrous, oxygen-free nitrogen
atmosphere. All reagents and solvents should be anhydrous, and all
transfers should be carried out with a hypodermic syringe or with a
flask-to-flask cannular arrangement. While the enolate solutions
prepared as described below may be stable under an inert atmosphere
at low temperatures for extended time periods, it is recommended
that they be prepared immediately before use whenever possible.

i. Reactive alkylation agents (methyl, allylic, or benzylic
halides) in liquid ammonia-ether [193, 194]. A reactive alkylating
agent (0.6 to 2.4 mol) in 100 ml of anhydrous ether is added over
1 to 5 min with rapid stirring to approximately 0.2 mol of the lithium
enolate in 1500 ml of approximately 2:1 liquid ammonia-ether at -33°C.
The reaction mixture is stirred at -33°C for 5 to 30 min. Solid
ammonium chloride (30 g) is added as rapidly as possible, and the
liquid ammonia is allowed to evaporate. The residue is partitioned
between ether and water; the ether layer is separated, washed with
5% hydrochloric acid, dried, and concentrated to yield the crude
alkylated ketone.

ii. Reactive alkylating agents in DME [44, 45, 225]. A
reactive alkylating agent (0.4-2.0 mol) is added rapidly with vigorous
stirring to a solution of approximately 0.2 mol of lithium enolate in
400 ml of anhydrous DME at 0° to 30°C. The mixture is stirred for
2 to 10 min, poured into 500 ml of cold saturated aqueous sodium
bicarbonate, and extracted with three 150-ml portions of pentane.
The combined pentane extracts are dried and concentrated to yield the
crude alkylation product.

iii. Unreactive alkylating agents (ethyl or n-butyl halides)
in liquid ammonia-THF [228]. A solution of the alkylating agent
(10 mmol) in 5 ml of THF is added rapidly with stirring to approximately 2.5 mmol of the lithium enolate in approximately 25 ml of 3:2
liquid ammonia–THF at -33°C. The mixture is stirred for 2 to 6 hr at
-33°C and quenched with excess ammonium chloride. The ammonia is

allowed to evaporate, and the residue is partitioned between ether
and water. The ether layer is dried and concentrated to give the
crude alkylated ketone.

   7. Regiospecific Annulations of Unsymmetrical
       Cyclic Ketones

   Annulations of cyclic ketones, often termed *Robinson annulations*,
have generally been carried out by Michael additions of cyclic ketone
enolates to α,β-unsaturated ketones, either used directly or produced
in situ from some precursor, followed by aldol cyclizations of 1,5-
diketone intermediates [11, 12, 315, 316]. As illustrated by the
reaction of methyl vinyl ketone (MVK) with 2-methylcyclohexanone (5)
to produce the octalone 8 via an intermediate ketol, the conventional
procedure, when the enone is used directly, involves the use of a
catalytic amount of a weak base in a protic solvent [317]. Such

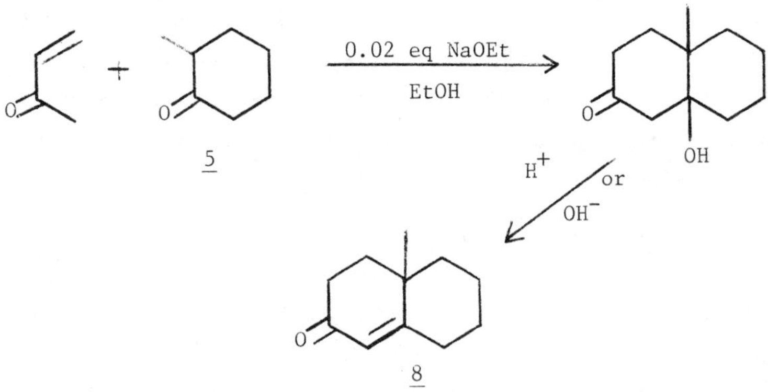

procedures have the following disadvantages: yields are only fair
because of competing polymerization of the Michael acceptor and
there is no possibility of effecting regiospecific annulations of
kinetically unstable enolates of unsymmetrical ketones [315].

   Recently, the reactions of the more-substituted preformed
lithium and chlorozinc enolates of 5 with 1.5 eq of MVK at about -60°C
in diethyl ether were investigated [318]. Using the enolate 5B (M =
Li), a 14% yield of the diketone 104a was obtained, but by using the

104 a; R = H

b; R = t-butyl

enolate *5B* (M = ZnCl), the yield was improved to 35%. Also, the 5-
t-butyl derivatives of *5B* (M = Li) and *5B* (M = MgBr) were found to
yield the diketone *104b* in the 42-54% and 41-55% ranges, respectively,
when reacted with 1.1 eq of MVK in ether at $-30°$ to $-40°$C. As in the
conventional Michael addition procedure, extensive polymerization of
the Michael acceptor was observed with these preformed enolates.
This study and others [68, 319] have revealed that preformed metal
enolates may react with enones in aprotic solvents to give kinetically
favored aldol adducts or thermodynamically favored Michael adducts.
α,α-Disubstituted enolates of the type *5B* apparently form relatively
unstable aldol adducts which dissociate rapidly and yield Michael
adducts even at low temperatures [318]. The literature apparently
contains no examples of Michael additions of kinetically unstable
preformed metal enolates to simple α,β-unsaturated ketones.

The vinyl ketone *105* has been developed to avoid the difficulties
associated with the use of simple enones as Michael acceptors in
annlation reactions [187, 188, 222]. When *105* is employed in Michael
additions, an enolate intermediate which is stabilized by the α-
trimethylsilyl group is obtained upon conjugate addition of the ketone
enolate [187]. Thus the Michael addition step is faster than enolate
equilibration and extensive polymerization does not take place.
Treatment of the initially formed 1,5-diketone with base leads to
removal of the trimethylsilyl group and completion of the cyclization
reaction. This annulation sequence is illustrated by the conversion
of the lithium enolate *8A* to the tricyclic ketone *106* [188].

The enolate obtained directly from lithium-ammonia reduction of
the octalone *8* was used in the sequence. However, much better results
were obtained if a "clean" enolate was prepared by trapping the
initial product of the reduction with trimethylsilyl chloride,
isolating and purifying the trimethylsilyl enol ether derivative, and
cleaving the latter with methyllithium in DME [188].

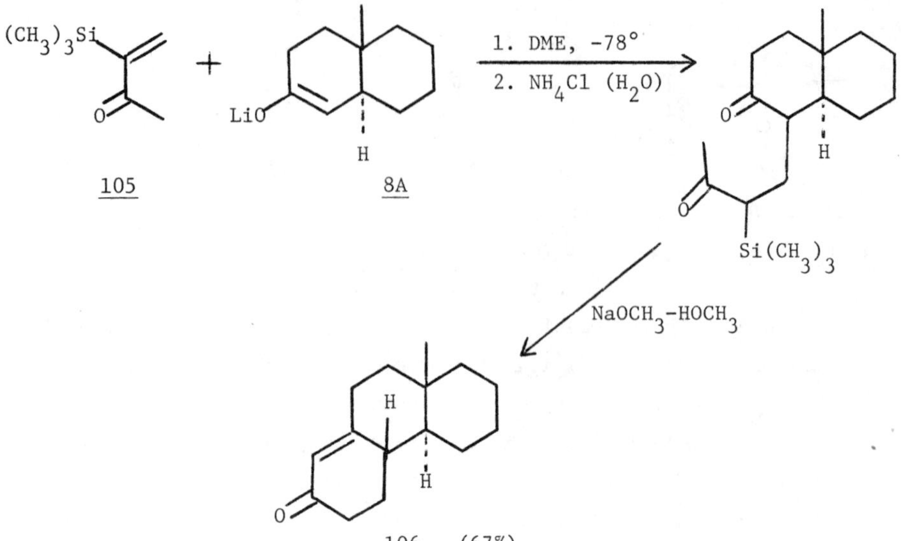

The enolate *5B* (M = Li) [188], the enolate *38B* (M = Li), and
the lithium 1-enolates of 2,3-dimethylcyclohexanone and 2,3-
dimethylcyclopentanone as well as 2-methyl-3-vinylcyclohexanone and
cyclopentanone derivatives have been regiospecifically annulated in
a similar manner [220]. All of these but *5B* (M = Li) were prepared
by 1,4-addition of lithium dimethyl- or divinylcuprate to the
appropriate enones [220].

The α-trimethylsilyl Michael acceptor *107* has been utilized as
a bis-annulating agent [188, 222]. For example, reaction of the
lithium enolate *108* with this reagent followed by base-catalyzed
cyclization gave the tricyclic enone *109* in 74% yield. Similar
routes have been used to prepare other enones closely related to
109 [222]. Compounds of this type have been converted into steroidal
enones by routes involving reduction-methylation (*94* → *95*) [310].

Lithium enolates of cyclohexanone derivatives also have been
found to undergo regiospecific Michael additions to MVK complexed
with $C_5H_5Fe(CO)_2^+$ in acetonitrile at -78°C to yield adducts convertible
to annulated products upon base treatment [320].

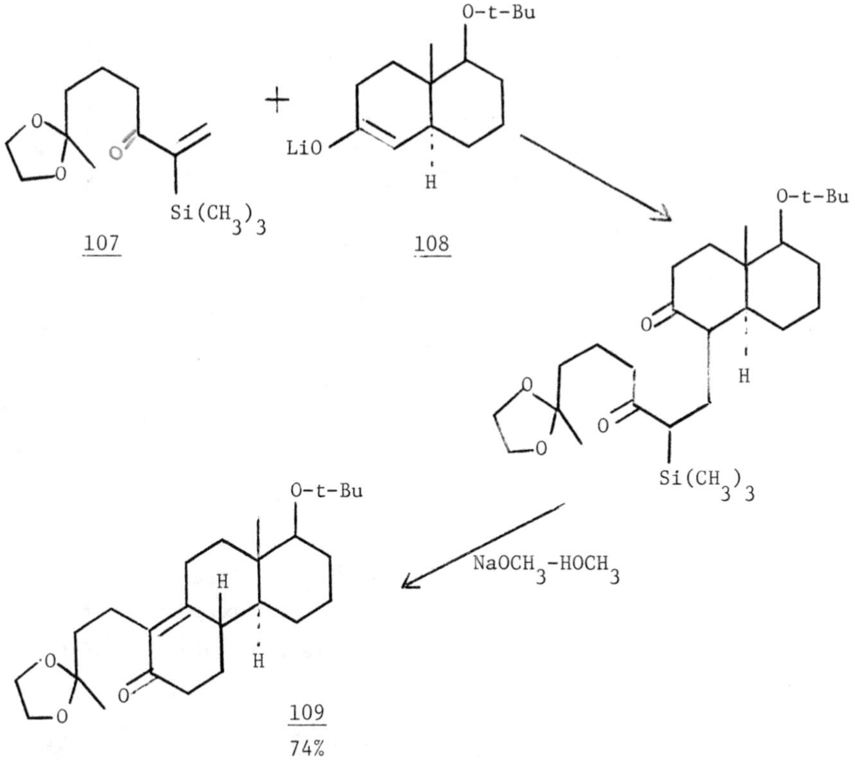

Another solution to the problem of regiospecific annulation is to introduce a substituent containing a latent 3-ketoalkyl side chain α to a carbonyl function via alkylation. The use of 1-halo-3-alkanones for direct alkylations is not possible. Under basic conditions these reagents undergo elimination to vinyl ketones, which then participate in Michael reactions [316-318]. Several simple ketal or enol derivatives of 1-halo-3-butanones have been employed for the alkylation of conjugate enolates of α,β-unsaturated ketones with a moderate degree of success [315]. However, these compounds are normally too unreactive or too unstable for use in regiospecific enolate trapping reactions [315].

The reagents 110-113 are sufficiently reactive for regiospecific alkylation of kinetically formed unstable lithium enolates, and the products of alkylation contain latent 3-ketoalkyl side chains, which may be unmasked under appropriate conditions.

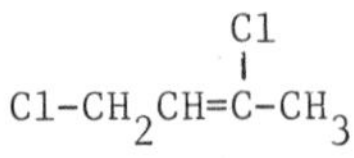

Cl-CH$_2$CH=C-CH$_3$
|
Cl

110

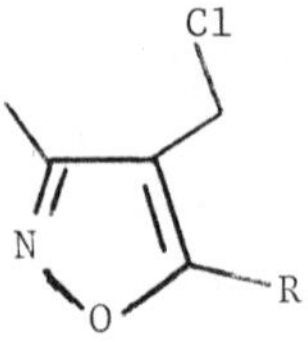

111 a; R = CH$_3$
   b; R = H
   c; R =

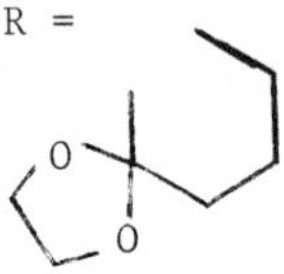

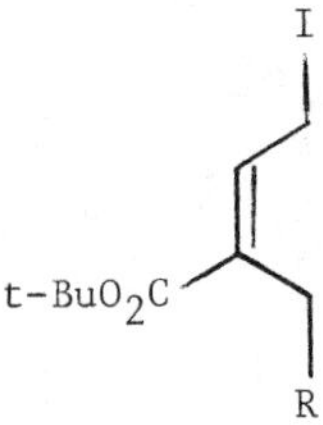

112 a; R = H
   b: R = CH$_3$

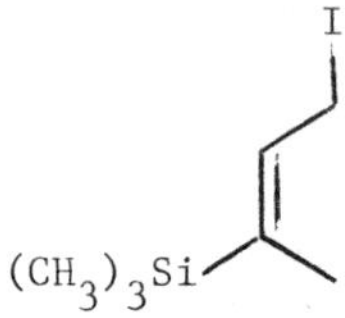

113

1,3-Dichloro-2-butene (*110*), the Wichterle reagent [321], is the oldest and most readily available of the four reagents listed above. However, there are some serious drawbacks to the use of this compound:

1. Alkylations of simple ketones with this reagent occur only in moderate yields [60, 321]

2. Strongly acidic [61] or strongly basic [322] conditions, which may be too severe for other functional groups in the molecule, are required for conversion of the vinyl halide function into a carbonyl group

3. Under acidic conditions undesired cyclization products may be obtained [61].

The simplest method of converting a 3-chloro-2-butenyl side chain into a 3-ketobutyl group has been found to be hydrolysis with concentrated sulfuric acid [61]. However, under these conditions acid-catalyzed aldol cyclization of 1,5-diketone intermediates occur and may lead to desired cyclohexenone derivatives and/or to bridged ketone products. For example, treatment of the ketone *114a* with concentrated sulfuric acid at 0°C gave the octalone *115a* and the bicyclic enone *116a* in a 5:1 ratio in about 60% yield. Similarly the octalone *115b* and the bridged compound *116b* were obtained in a 5:1 ratio from *114b*.

However, the dimethylcyclohexanone derivative *114c* gave the bridged compound *116c* as the only product in high yield [61].

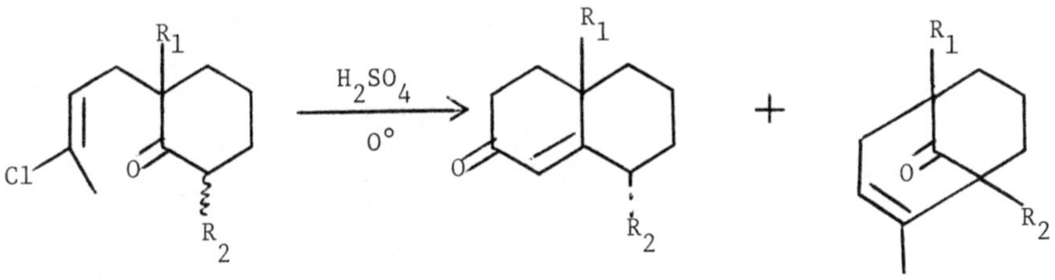

	114	115	116
a;	$R_1 = CH_3$, $R_2 = H$	47%	11%
b;	$R_1 = H$, $R_2 = CH_3$	47%	10%
c;	$R_1 = R_2 = CH_3$	0%	93%

To circumvent this problem, the diketone *117* which undergoes facile cyclization to the octalone *115c* under basic conditions, has been prepared from *114c* by two indirect routes. The first approach involved the introduction of a 5,6-double bond into the cyclohexanone ring by bromination-dehydrobromination to give the enone *118*, the sulfuric acid-catalyzed hydrolysis of the vinyl chloride function to give the enedione *119*, and then catalytic hydrogenation to give the diketone *117* [61]. The second approach [322] involved the dehydro-halogenation of *114c* with sodium amide in liquid ammonia to give the internal acetylene *120*, the isomerization of the triple bond to the terminal position to give *121*, and then mercuric ion-catalyzed hydration to give *117*. Both of these methods suffer from the disadvantage that drastic acidic or basic conditions are required.

Halomethylisoxazole derivatives such as *111a*, which were introduced and developed by Stork and co-workers, have proved to be useful annulating agents [310, 323-325]. These reagents are somewhat less reactive than benzyl bromide as alkylating agents [323]. Therefore, best results are obtained when they are used to alkylate conjugate enolates of α,β-unsaturated ketones. (The alkylated products

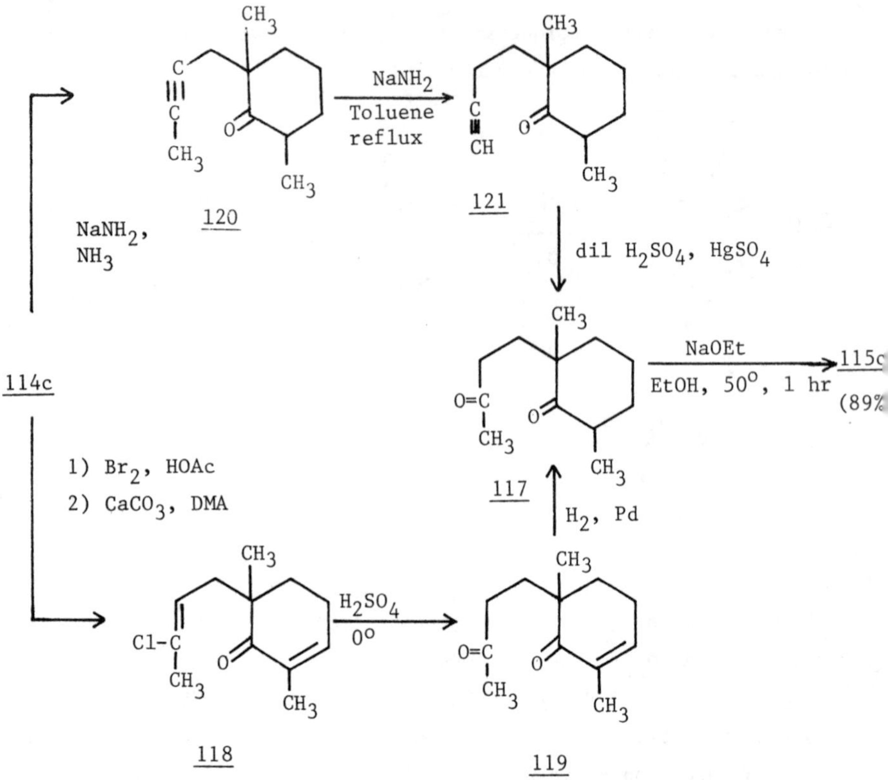

CH$_3$
C
|||
C
CH$_3$
O
CH$_3$
120

NaNH$_2$
Toluene
reflux

CH3
C
||
CH
O
CH$_3$
121

NaNH$_2$,
NH$_3$

114c

dil H$_2$SO$_4$, HgSO$_4$

CH$_3$
O=C
CH$_3$
O
CH$_3$
117

NaOEt
EtOH, 50°, 1 hr

115c
(89%)

H$_2$, Pd

1) Br$_2$, HOAc
2) CaCO$_3$, DMA

CH3
Cl-C
CH$_3$
O
CH$_3$
118

H$_2$SO$_4$
0°

CH3
O=C
CH$_3$
O
CH$_3$
119

can then be selectively reduced by hydrogenation over palladium to
α-substituted saturated ketones [323].) However, in a few cases
trapping of kinetically formed unstable metal enolates with reagents
such as *111* have been reported to occur in fair yield [209, 326, 327]
The reaction of enolate *36B* (M = MgI) with *111a* in ether-HMPA to produce the 2-substituted 3-methylcyclohexanone *122* is an example [209].

Raney nickel reduction of the isoxazole ring followed by reaction with sodium methoxide or ethoxide in the corresponding alcohol
and then aqueous base was the most successful among the early methods
investigated by the Stork group [323-325] for unmasking the carbonyl
group and completing the annulation sequence. This sequence is
illustrated by the conversion of compound *122* to the enone *123* [209].
It was also established that intermediates such as the vinylogous

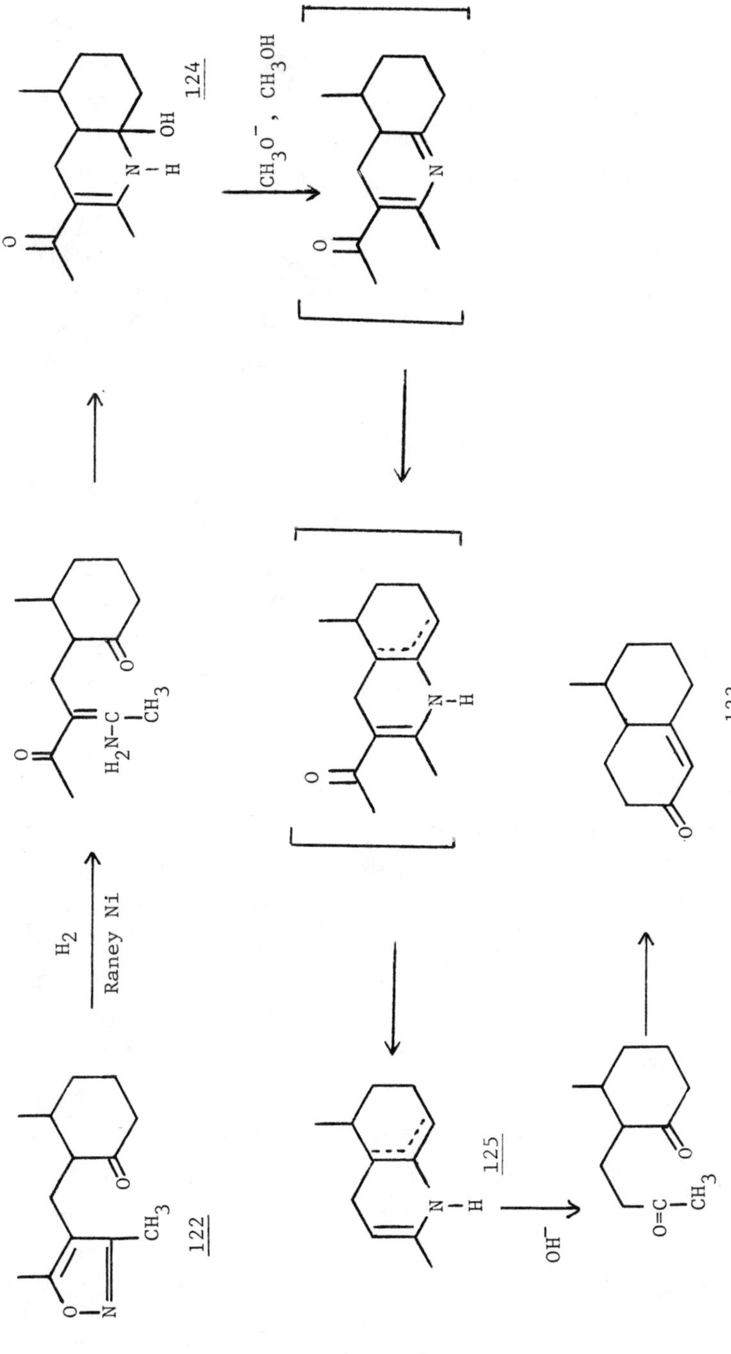

122
H2
Raney Ni
124
CH3O−, CH3OH
125
OH−
123

carbinolamide *124* and the enimine *125* are intermediates in the
sequence [323-325]. In several different systems the overall yields
for the conversion of the alkylated ketone to the enone product (for
example, *122* → *123*) have been found to be about 50%. A somewhat more
efficient method of accomplishing such conversions has been reported
recently [328].

The most generally useful reagents for the regiospecific annulation of saturated ketones are the t-butyl-γ-iodotiglates *112*, developed by Stotter and Hill [302], and the halomethyl vinylsilanes *113*,
developed by Stork and Jung [233]. These reagents are reasonably
easy to prepare, they trap kinetically unstable lithium enolates in
high yield, and the 3-ketoalkyl side chains in alkylated products
can be unmasked in high yield under relatively mild conditions.

Both *5B* (M = Li) and *5A* (M = Li) have been trapped with the
iodotiglates (*112*) in excellent yields (see Table III) [302]. The
conversion of the 2,6-disubstituted cyclohexanone *126* to the enone
115b illustrates the procedure for completion of the annulation
sequence. The first step involves heating the carbonyl compound in
benzene containing PTSA to convert the t-butyl ester to the free
acid *127*. The Weinstock modification of the Curtius reaction
is used for degradation of the tiglic acid side chain. This
involves synthesis of the mixed anhydride *128* by reaction with triethylamine followed by ethyl chloroformate and conversion of this
intermediate to the acyl azide *129* by reaction with sodium azide.
When the azide is heated in methanol, the vinyl urethane *130* is
obtained. It is readily hydrolyzed and cyclized to *115b* with aqueous
potassium carbonate in methanol. This sequence has the disadvantage
that it is lengthy, but overall yields for the conversion are
generally within the 70-80% range.

E-3-Trimethylsilyl-2-butenyl iodide (*113*) is an extremely useful reagent for regiospecific annulations because of the ease with
which the vinylsilane side chain of α-alkylated ketones is converted
to a 3-ketoalkyl group. Alkylations of kinetically generated enolates
of cyclohexanone derivatives (Table III) and other ketones have been

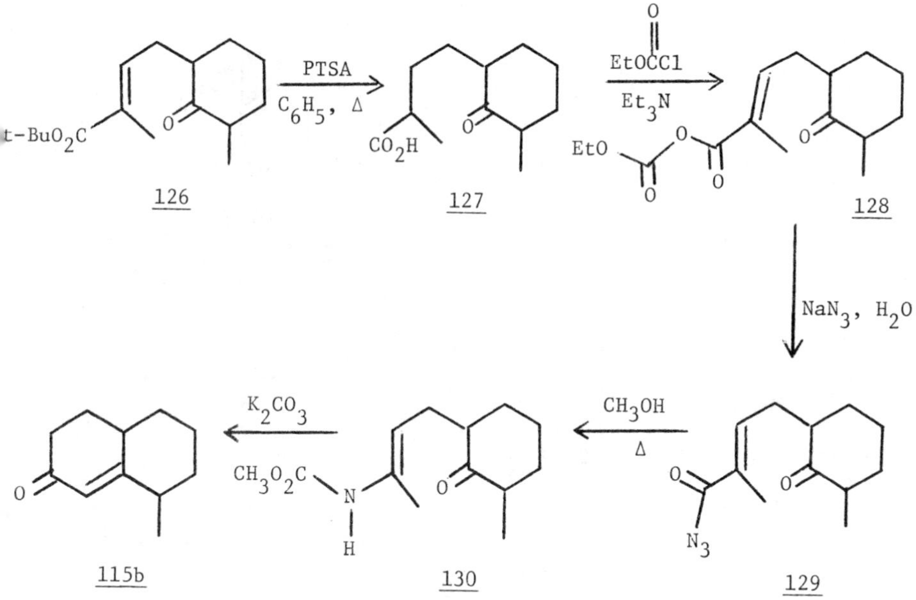

t-BuO$_2$C **126** $\xrightarrow[\text{C}_6\text{H}_5, \ \Delta]{\text{PTSA}}$ CO$_2$H **127** $\xrightarrow[\text{Et}_3\text{N}]{\text{EtOCCl}}$ EtO **128**

NaN$_3$, H$_2$O ↓

115b $\xleftarrow[\text{CH}_3\text{O}_2\text{C} \diagdown \text{N}]{\text{K}_2\text{CO}_3}$ **130** $\xleftarrow[\Delta]{\text{CH}_3\text{OH}}$ N$_3$ **129**

performed in high yield with it [233]. The synthesis of the octalone
8 from the lithium enolate *5B* (M = Li) illustrates this annulation
sequence. The alkylation product *131* was converted to the diketone
132 in 89% yield by treatment with m-chloroperbenzoic acid in methylene
chloride. This transformation is believed to involve an epoxysilane
intermediate which undergoes intramolecular nucleophilic ring opening
by the carbonyl group [233]. Treatment of the diketone *132* with base
provided the enone *8*, in 90% yield.

8. Use of Activating and Blocking Groups
    in Ketone Alkylations

Prior to the discoveries that lithium and other enolates having
relatively covalent oxygen-metal bonds could be alkylated regiospecifically, the best methods of synthesis of specific α-alkylated
ketones involved the use of *activating* or *blocking* groups. In these
approaches a saturated ketone is activated or blocked in an appropriate
manner; enolization and alkylation at the desired site is then performed; afterward, the activating or blocking group is removed.
Because these approaches require a number of steps, overall yields

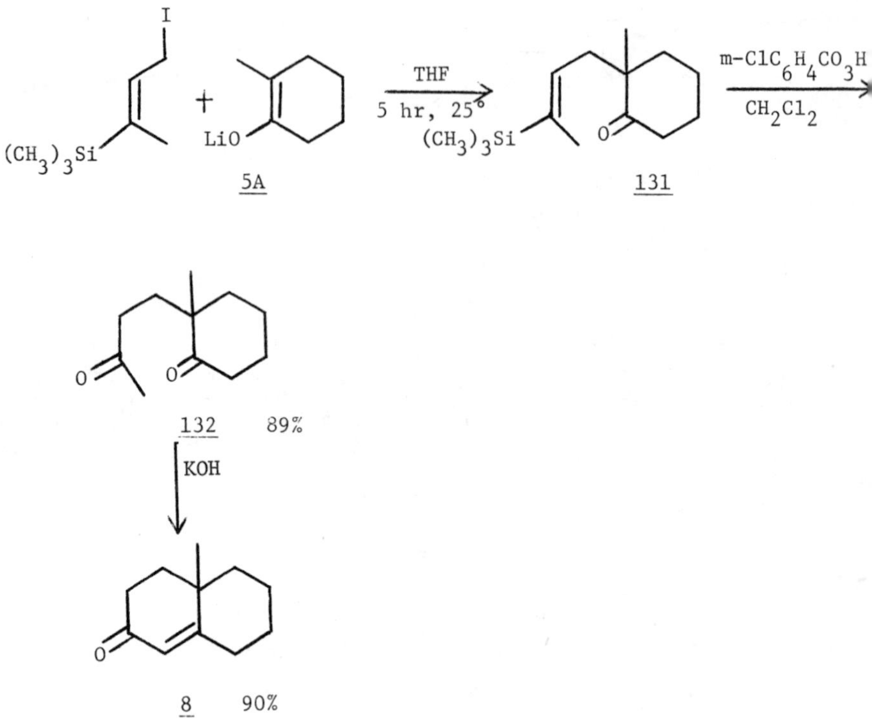

are usually lower than those obtained in direct alkylation processes. However, when the desired enolate anion cannot be formed from the ketone in high yield under kinetically or thermodynamically controlled conditions, or when various precursors to specific metal enolates are not readily available, the use of blocking or activating groups may be required. An important advantage of these methods is that the structures of enolate anions derived from blocked or activated ketones generally permit the introduction of only one alkyl group. Thus polyalkylation is usually not a problem, even when methylations or ethylations are involved. Of course, the problem of enolate equilibration is also completely avoided when the activating- or blocking-group techniques are used.

Formyl [36, 303], carboalkoxyl [329], and ethoxyoxalyl [36, 303] groups are often used as activating groups for ketone alkylations.

These groups are normally introduced by Claisen condensations with appropriate esters. These reactions are thermodynamically controlled, and products which can yield a stable anion of a 1,3-dicarbonyl compound are obtained [6-8]. Thus, unsymmetric cyclic ketones such as 2-methylcyclohexanone (5) undergo formylation [330] and ethoxyoxalation [331] at the less-substituted position to give the corresponding 6-substituted derivatives 133 or 134, respectively [94]. Methyl n-alkyl ketones may yield two possible condensation products on reaction with esters, but the enolate anion intermediate derived from condensation at the methyl group, i.e., the less-hindered species,

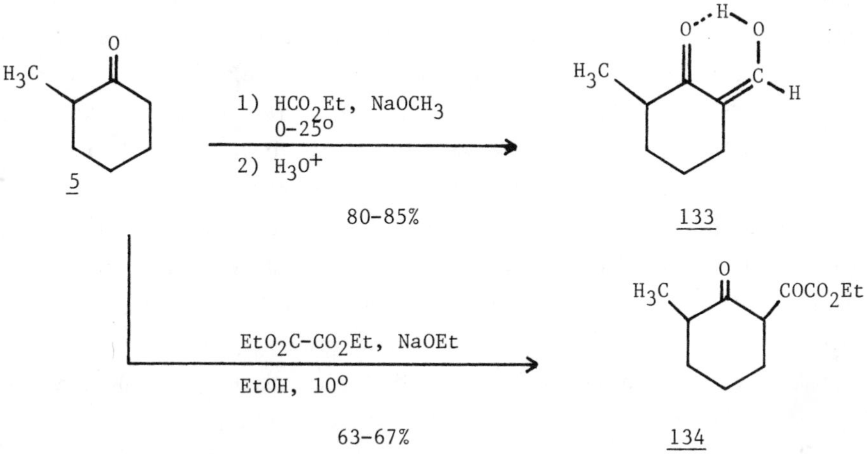

$$\underline{5} \xrightarrow[\text{2) } H_3O^+]{\text{1) } HCO_2Et, \ NaOCH_3 \ 0-25^\circ}$$

80-85%   133

$$\xrightarrow[\text{EtOH, } 10^\circ]{EtO_2C-CO_2Et, \ NaOEt}$$

63-67%   134

is generally significantly favored [332]. Carboalkoxy groups may be introduced α to ketone functions by Claisen condensations with dimethyl or diethyl carbonate, using sodium hydride as the base in aprotic solvents [333]. Excellent results were obtained in the carbomethoxylation of cyclohexanone when a small amount of potassium hydride was used to initiate the reaction [334]. Other methods of synthesis of α-carboalkoxy ketones include:

1. Decarbonylation of α-ethoxyoxalyl derivatives using iron and powdered soft glass catalysts [329, 331, 335]

2. Condensation of ketones with ethyl diethoxyphosphinyl formate using sodium hydride in di-n-butyl ether followed by reaction with sulfuric acid in anhydrous ethyl alcohol [336]

3. Reaction of ketones with methylmagnesium carbonate (Stiles reagent) in DME and esterification [337-339]

4. Carbonation of metal enolates with carbon dioxide followed by acidification and esterification with diazomethane [28, 60, 340].

After the activating group has been introduced, C-alkylation at the position bearing the group is effected under the conditions usuall employed for alkylation of enolate anions of relatively acidic active methylene compounds [1]. Such reactions will be covered in detail in another volume in this series. Briefly, C-alkylation, as opposed to O-alkylation, is favored when 1,3-dicarbonyl compounds which have low enol content, e.g., β-keto esters, are employed. Also, reaction conditions which favor association of enolate anions and metal cations (e.g., lithium and magnesium, and solvents of lower polarity) and the use of soft alkylating agents such as alkyl iodides favor C-alkylation Under conditions which favor high C-alkylation/O-alkylation ratios, rates of enolate alkylations will normally be slow.

Once the desired alkyl group has been introduced, activating groups such as formyl or ethoxyoxalyl are generally removed via a reverse Claisen process, which may be catalyzed by acids or bases. Saponification and decarboxylation is the usual procedure for removal of carboalkoxy groups. However, 2-alkyl-2-carboalkoxycyclopentanones normally undergo cleavage of the five-membered ring, rather than saponification, under basic conditions [1, 341]. These compounds are normally converted into 2-alkylcyclopentanones by acidic hydrolysis followed by decarboxylation or by one of several possible nonhydro-lytic methods [1].

The synthesis of pure 2α-methyl-5α-cholestan-3-one (135) from 5α-cholestan-3-one (136) via the ethoxyoxalyl derivative 137 and the methylated intermediate 138 provides an example of the activating-group method [276]. Several examples of the synthesis of 2α-alkyl-3-keto steroids via synthesis of 2-hydroxymethylene (2-formyl) deriva-tives, alkylations of enolate anions of these derivatives, and deformylations have been reported [303].

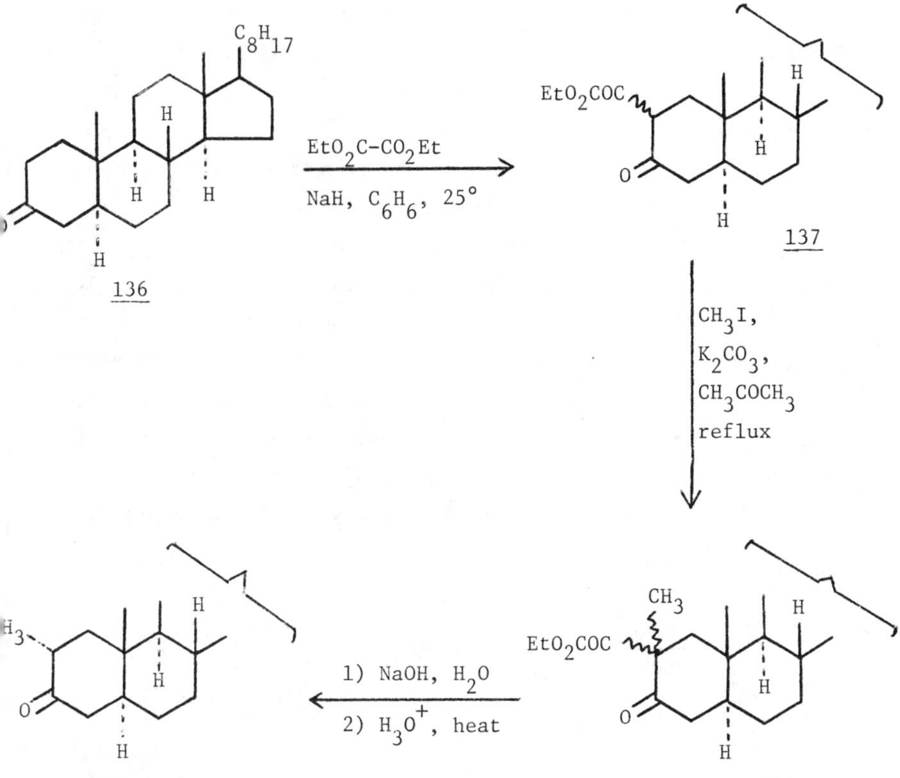

α-Thiophenyl [176] and α-thioalkyl [181] substituents may also be used as activating groups for enolate alkylations. After the alkylation step, these groups are readily removed by reduction with lithium in liquid ammonia.

Methyl or methylene groups of unsymmetrical ketones may be blocked by introduction of a group (which can later be removed) to prevent enolization toward the position it occupies. The ideal blocking group should (a) be easily introduced, (b) be stable under the basic conditions required for alkylation at the free α position, (c) not deactivate the ketone toward alkylation, and (d) be easily removed after the alkylation step. Benzylidine or furfurylidine groups, which may be introduced by aldol condensations [9] of ketones

with benzaldehyde or furfural, have been used to prepare blocked 1-
decalones and tetracyclic ketones for use in angular alkylation studies [29, 33]. Their main drawback is that they are difficult to
remove after alkylation.

2-Hydroxymethylene ketones may be transformed into one of severa
derivatives which are stable to alkylation conditions. For example,
the hydroxymethylene derivative *139* of 1-decalone (*37*) may be converted to the enamine *140* by reaction with N-methylaniline [30], the
enol ether *141* by O-alkylation with isopropyl iodide [31], or the n-
butylthiomethylene derivative *142* by reaction with n-butyl mercaptan
[34]. The enamine function in compounds related to *140* appears to
deactivate the ketone toward alkylation, and enol ethers such as
141 are rather sensitive to moisture [34]. These problems are
avoided by the use of the n-butylthiomethylene blocking group. In

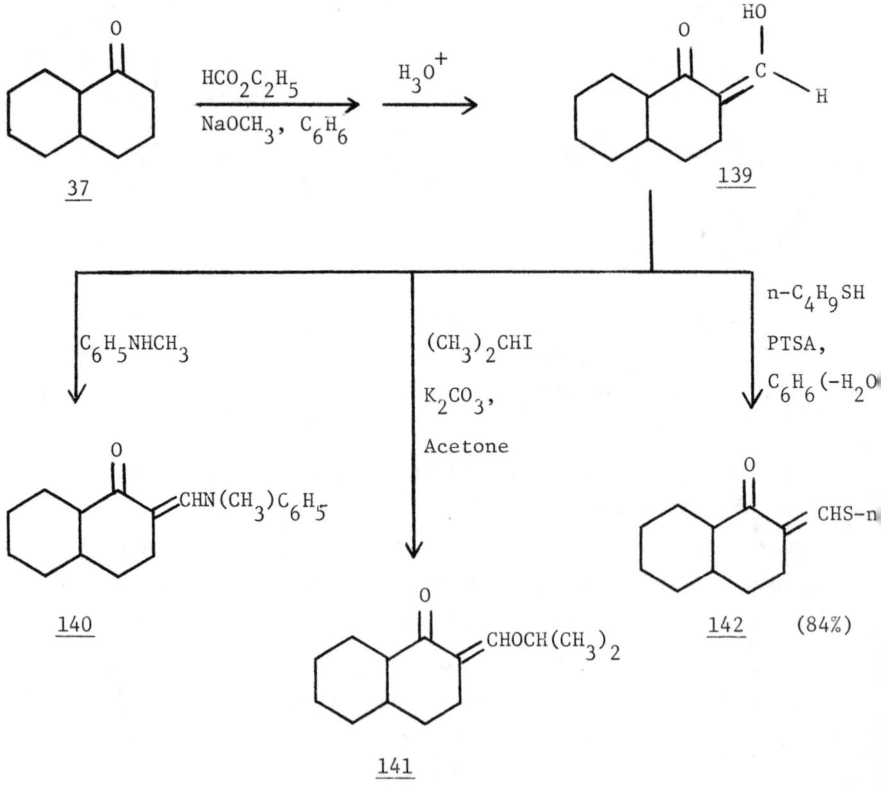

addition to being easily removed by reaction with aqueous bases, this
group offers the advantage of being readily reduced to a methyl group
by Raney nickel in ethanol or with alkali metals in liquid ammonia
[34]. The regiospecific formation and alkylation of lithium enolates
by reduction of 2-n-butylthiomethylene ketones with lithium in liquid
ammonia have been discussed previously [193].

The versatility of n-butylthiomethylene derivatives in synthesis
is illustrated by the preparation of pure 2,6- (6) or 2,2-
dimethylcyclohexanone (7) as well as 2,2,6-trimethylcyclohexanone
(74) from the n-butylthiomethylene derivative 143 of 2-
methylcyclohexanone.

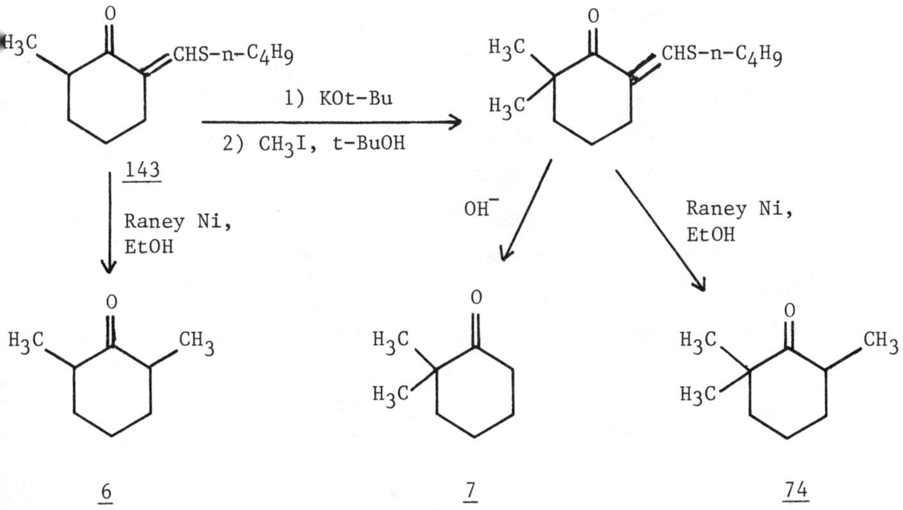

The 1-decalone derivative 142 was angularly alkylated in 85%
yield by treatment with potassium t-butoxide in t-butyl alcohol
followed by addition of excess methyl iodide [34]. Removal of the
blocking group by heating with aqueous potassium hydroxide in
diethylene glocol gave a 1:1 mixture of the decalones 90a and 90b
in 78% yield [34].

142

90a 90b

1:1

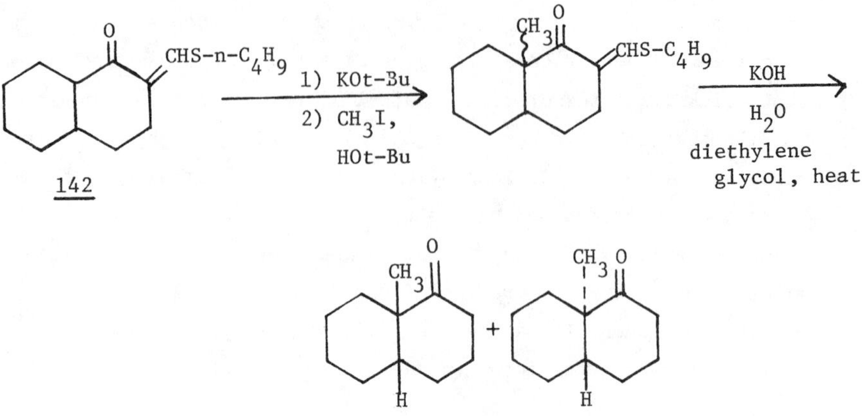

Alkylations of other cyclic ketones containg n-butylthiomethylene blocking groups has been used as an important step in the synthesis of several natural products. For example, methallylation of the blocked ketone *144* followed by removal of the blocking group gave a 4:1 mixture of cis-2,3-dimethyl-2-methallylcyclohexanone (*145*) and the corresponding trans isomer *146* [342]. Compound *145* has been converted to the sesquiterpenes (±)-aristolone [342] and (±)-bakkenolide-A [343].

144 145 4:1 146 (84%)

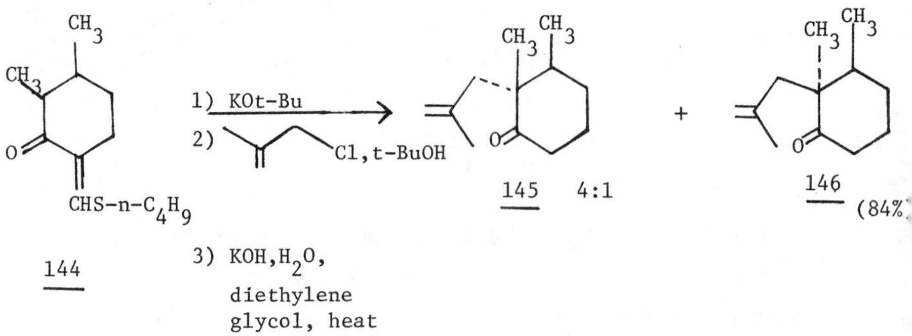

A key step in a synthesis of (+)-valeranone required the introduction of a second angular methyl group into the 1-decalone

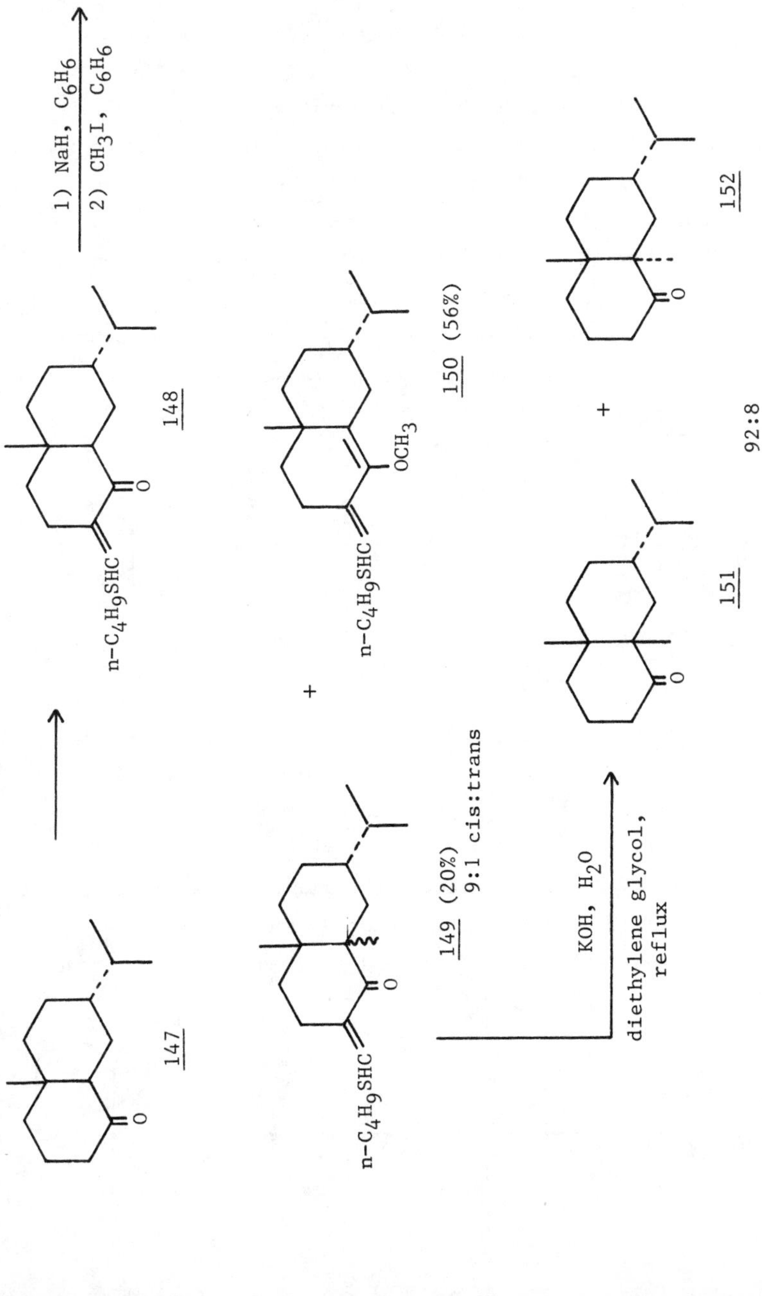
147
n-C_4H_9SHC
148
1) NaH, C_6H_6
2) CH_3I, C_6H_6
n-C_4H_9SHC
O
149 (20%)
9:1 cis:trans
+
n-C_4H_9SHC
OCH_3
150 (56%)
KOH, H_2O
diethylene glycol,
reflux
O
151
+
O
152
92:8

derivative *147* [344]. This was accomplished via the n-butylthiomethylene derivative *148* which yielded a 9:1 mixture of the cis and trans isomers *149* in 20% yield on alkylation of the sodium enolate with methyl iodide in benzene. The major product of the reaction was the O-alkylated compound *150*. (Compound *150* was the sole alkylation product when DMSO was employed as the solvent.) Steric factors appear to be responsible for the unusually large amount of O-alkylation observed in this case. After separation from *150* by chromatography, the C-alkylation products *149* were converted to (+)-valeranone (*151*) and its stereoisomer (*152*) in a 92:8 ratio.

Angular allylation of the blocked tetracyclic ketone *153* to yield the cis-fused product *154* also provided a key step in a synthesis of the alkaloid garryine [345]. When the unprotected ketone corresponding to *153* was alkylated, the allyl group was introduced exclusively at the 13-position [345].

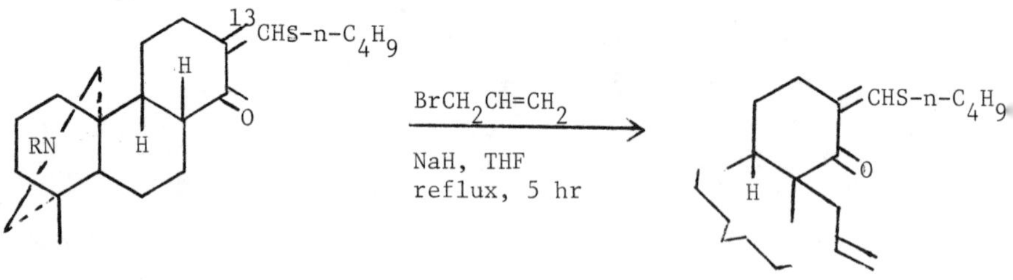

$$BrCH_2CH=CH_2$$
$$NaH, THF$$
$$reflux, 5 hr$$

153   154

2-Hydroxymethylene ketones are converted into α-dithioketals upon reaction with 1,3-propanedithiol di-p-toluenesulfonate in alcoholic potassium acetate [Eq. (45)] [32, 346]. These groups may be removed by desulfurization with Raney nickel. However, the presence of the α-dithioketal function reduces the reactivity of the ketone toward alkylation [34], and little use has been made of this particular blocking group.

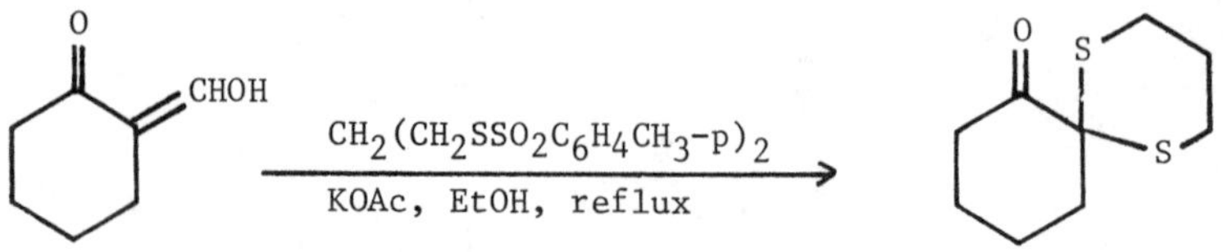

$$CH_2(CH_2SSO_2C_6H_4CH_3-p)_2$$
$$KOAc, EtOH, reflux$$

(45)

β-Dicarbonyl compounds (155) may be converted into dimetal dianions (156) by treatment with 2 eq of a strong base such as potassium amide in liquid ammonia or by reaction of their monometal salts (usually sodium) (157) with 1 eq of a similar base [35].

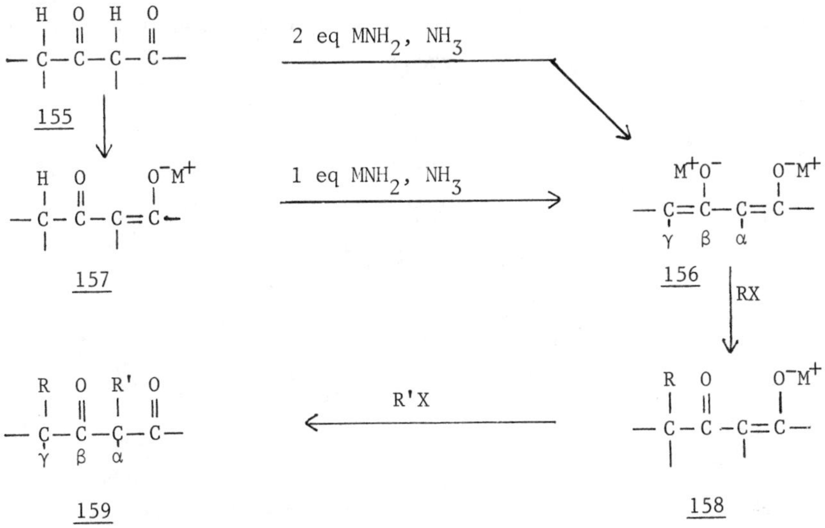

On treatment with 1 eq of an alkylating agent, these species undergo reaction at the more reactive γ position to yield monoalkylated products (158). On addition of a second equivalent of the same or a different alkylating agent, reaction can occur at the α position to yield an α,γ-dialkylated compound, 159 [35].

As previously mentioned, 2-hydroxymethylene ketones may be readily prepared from base-catalyzed reaction of methyl or methylene ketones with ethyl formate. Application of the dianion alkylation procedure to these compounds provides a novel method of preparing monoalkylated ketones [35]. In these cases the α position bearing the hydroxymethylene group is effectively "blocked," and alkylation is directed to the alternative α position. After monoalkylation the monometal enolate may be easily deformylated by treatment with aqueous base [35].

As an example of this procedure, the potassium sodium dianion of 2-methyl-6-hydroxymethylenecyclohexanone (*160*) has been prepared by treatment of the monosodium salt with 1 eq of potassium amide in liquid ammonia. Treatment of this species with 1 eq of methyl iodide, n-butyl bromide, or benzyl chloride and deformylation with aqueous base gave the corresponding 2-alkyl-2-methylcyclohexanone derivatives *161* in 60, 72, and 55% yields, respectively [330, 347].

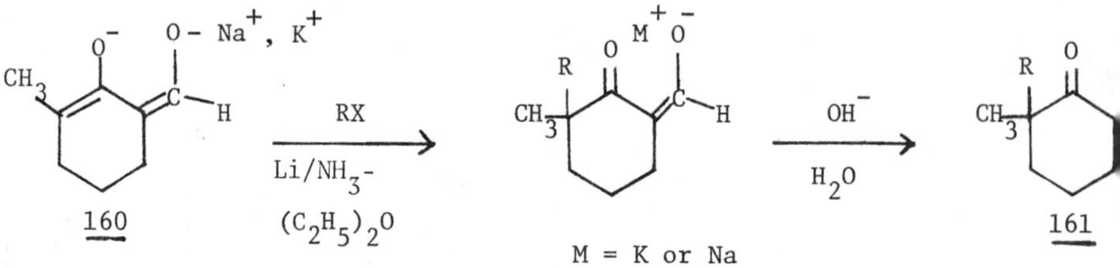

$$M = K \text{ or } Na$$

R = CH$_3$	60%
= n-Bu	72%
= CH$_2$C$_6$H$_5$	55%

A similar sequence has been used for the synthesis of 9-alkyl-1-decalones with the exception that the enolates of the angularly alkylated products were isolated as the corresponding copper chelates. Acidification of these compounds gave the corresponding hydroxymethylene ketones, which were deformylated with aqueous base. When methyl iodide was employed as the alkylating agent, a 44:56 mixture of cis (*90a*) and trans-9-methyl-1-decalone (*90b*) was produced in above 50% yield overall [347].

9. Stereochemistry of Alkylation of Metal Enolates of Cyclic Saturated Ketones

The synthetic organic chemist is concerned not only with regiospecificity in enolate alkylations but also with the stereochemistry of the alkylation process. This latter factor is often of crucial importance in the successful synthesis of complex molecules such as terpenoids, steroids, etc.

Metal enolates of simple cyclohexanone derivatives such as 4-t-butylhexanone (*162a*) may undergo energetically favorable perpendicular

attack by an alkylating agent from either of two directions, path A
or path B [Eq. (46)] [226]. Path A attack leads to the "chair"
conformation 163a of the product having the new alkyl group axial
to the ring and trans to the t-butyl group. Path B attack leads to
the "twist-boat" conformation 164a which may undergo conformational
interconversion to the chair conformation 165a having the new group
equatorial to the ring and cis to the t-butyl group. Paths A and B
are often referred to as *axial* and *equatorial* alkylation, respectively.
When a hydrogen atom is present at the carbon atom occupied by the
new alkyl group, as in 163a, isomerization of the initial product to
the thermodynamically more stable epimer (163a → 165a) is possible
and often occurs under the basic reaction or work-up conditions.
However, lithium enolates in DME or liquid ammonia-ether undergo
proton transfer reactions slow enough that kinetically formed α-
alkylated products also containing an α-hydrogen atom may be isolated
under carefully controlled conditions [49, 194, 301, 306].

The chair conformation 163 of a cyclohexanone derivative is
expected to be significantly more stable than the twist-boat conformation 164. Therefore, the general assumption has been made that, in
the absence of opposing steric and/or conformational factors, stereoelectronic factors would favor the formation of axial alkylation products [138-143, 265-269]. This argument requires that the formation
of the new C-C bond be relatively far advanced in the transition
state of the alkylation reaction [49, 301]. However, mechanistic
studies [48, 348] as well as the stereochemical results of alkylations
of metal enolates of a variety of structural types [29, 33, 49, 301,
314] indicate that the transition state for enolate alkylation resembles the reactants much more closely than the products. Thus structural features associated with the metal enolate exert an overwhelm-
ing influence on the stereochemistry of the reaction.

Ethylation of the enolate 162a with ethyl iodide in DME has been
shown to yield a 1:1 mixture of the products 163a (R = C_2H_5) and
165a (R = C_2H_5) [301]. In 162a, steric interactions associated with
either path A or path B attack by the alkylating agent appear to be

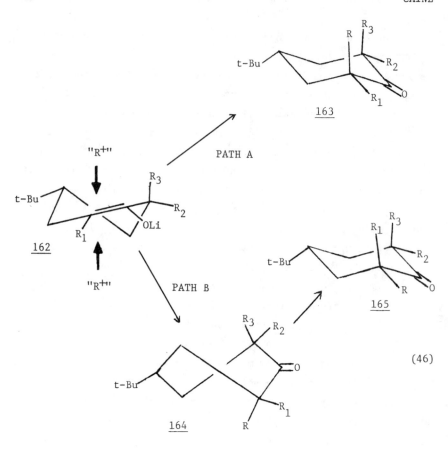

a; $R_1 = R_2 = R_3 = H$

b; $R_1 = R_2 = H$, $R_3 = CH_3$

c; $R_1 = R_3 = H$, $R_2 = CH_3$

d; $R_1 = CH_3$, $R_2 = R_3 = H$

about equal. The enolate *162b*, derived from trans-2-methyl-4-t-butylcyclohexanone on reaction with methyl iodide in DME, yielded the products *165b* (R = CH_3) and *163b* (R = CH_3) in an approximate 2:1 ratio [49]. A 2:1 mixture of *163c* (R = CH_3) and *165c* (R = CH_3) was obtained when the enolate *162c*, derived from cis-2-methyl-4-t-butylcyclohexanone, was treated similarly [49].

It is clear from models that path A should be less sterically favorable than path B for the alkylation of *162b* because of the 1,3 interaction which would exist between the approaching alkylating agent and the quasi-axial 6-methyl group in the former. In the enolate *162c*, a significant OLi-quasi-equatorial 6-methyl interaction would be expected [54, 143]. This could cause the enolate to undergo a change in conformation toward a "half-boat" form [49] or lead to a distortion in the C_1-OLi and C_2-H dihedral angle [349]. Either of these possibilities would cause an increased preference for path A attack in comparison with the 6-unsubstituted system *162a*.

Alkylations of the lithium enolate *162d* [49], the corresponding sodium enolate and related 2-alkyl species [329, 350], and the lithium enolate (*166*) derived from 2-methyl-5-t-butylcyclohexanone [349] have been shown to yield mixtures containing about 60-80% of the axial alkylation products and about 20-40% of the equatorial alkylation products. For enolates of the type *162d*, changes in the nature of the cation, the solvent, or the size of the alkylating agent did not exert a significant effect on the stereochemical results. These studies and many others [304-306, 351] have demonstrated that the presence of a substituent at the α position of cyclic metal enolates increases the stereoselectivity of the reaction in favor of axial alkylation products in comparison with related enolates having no α substituents. As depicted in *167*, it has been suggested that 2-substituted enolates undergo distortion in order to relieve interaction between the R group and the OM substituent and to avoid the eclipsing ($A^{(1,2)}$-strain) of the OM group with the 6-quasi-equatorial group (hydrogen) and of the R group with the 3-quasi-equatorial group (hydrogen). This type of distortion would be expected to cause partial rehybridization of the p orbital at C2 toward an sp^3 orbital. A group entering from the top side of such a distorted enolate would thus experience favorable orbital overlap and decreased steric interference in comparison with a group entering from the bottomside. Clearly, equatorial alkylation of such a deformed enolate would be a higher energy process, since the R substituent would have to move

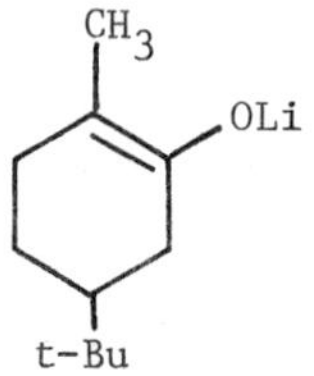

166 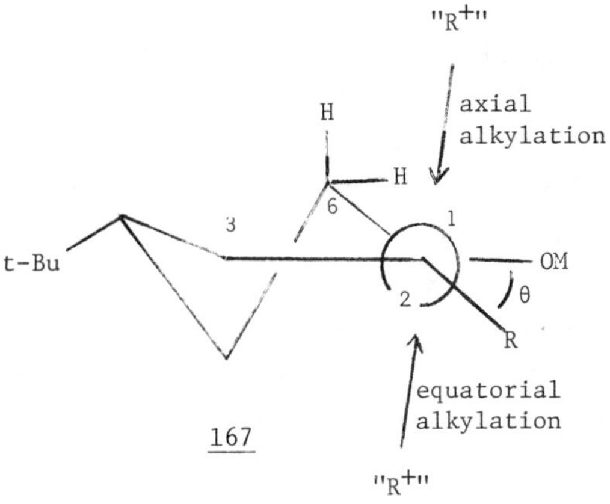   167

past the OM group as the new C-C bond is being formed. Enolates of
cyclohexanone derivatives which have quasi-equatorial substituents
at C6 [351] or C3 [304, 306, 352], as well as 2-substituents, show
an enhanced tendency toward axial alkylation. The enolate distortion
idea nicely accounts for these results, since the C_1-OM and C_2-R
dihedral angle (in structure 167) should increase even more with the
placement of bulky quasi-equatorial groups at C6 and/or C3 [349].

A relatively high degree of stereoselectivity has been observed
in alkylations of conformationally mobile lithium 1-enolates of 3-
methyl [199] and 2,3-dimethylcyclohexanone [221]. Allylation of the
enolate 36B (M = Li) in liquid ammonia-ether gave a 20:1 mixture of
the trans- (168) and cis-2-allyl-3-methylcyclohexanone (169), whereas
allylation of the enolate 170 in DME gave a 9:1 mixture of the products 171 and 172, having the new group trans and cis, respectively,
to the 3-methyl group [221]. A somewhat lower stereoselectivity
(4:1) was observed when the potassium enolate of ketone 144 was
methallylated [342].

Enolates such as 36B and 170 can exist as an equilibrium mixture
of conformations 173 and 174. Conformation 174, having the 3-methyl
group quasi-axial, is likely to be quite important, particularly when
a methyl group is present at C2, as in 170, because $A^{(1,2)}$-strain

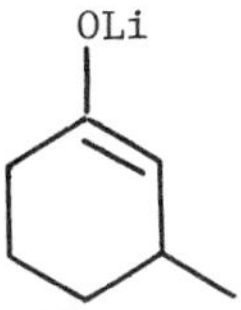

36B 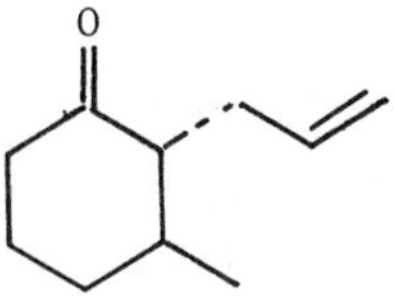   168 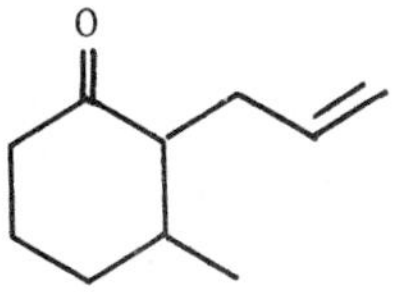   169

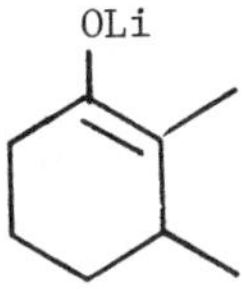

170 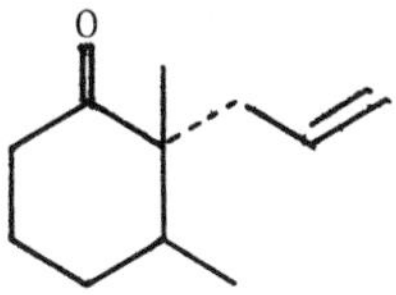   171 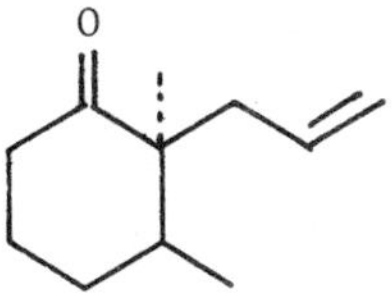   172

would destabilize conformation *173*, having the 3-methyl group quasi-equatorial [143]. A consideration of models reveals that steric interactions between the 3-methyl group and an approaching alkylating agent would be minimized in either of the two possible transition states which lead to the introduction of the new group trans to the 3-methyl group. It is surprising that the stereoselectivity for the alkylation of *36B* (M = Li) is apparently greater than that for *170*. Conformation *174*, which should strongly favor trans alkylation, might be expected to be more important in the 2-substituted enolate. However, it should be noted that in conformation *173* the 2-substituted enolate may undergo distortion in a manner which favors axial approach, i.e., introduction of the new group cis to the 3-methyl group [349].

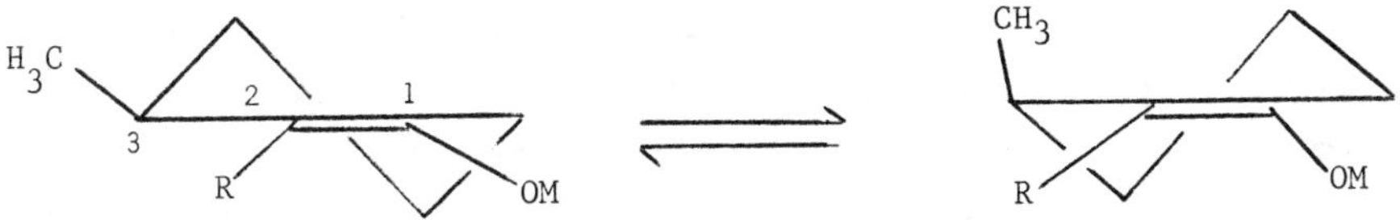

173   174

The stereochemistry of alkylation of enolates of 1- and 2-decalones and related compounds has been a matter of considerable

interest in connection with the synthesis of steroids and terpenoids.
Reaction of the lithium 1(9)-enolate (37A) of 1-decalone with methyl
iodide in liquid ammonia-ether [39] or DME [119] has been shown to
yield mixtures of products in which the cis isomer 90a is favored by
4:1 or 5:1 over the trans isomer 90b.

A high degree of stereoselectivity (>95%) in favor of the cis
product 90a was observed when the enolate 37A was treated with the
highly reactive alkylating agent trimethyloxonium 2,4,6-
trinitrobenzenesulfonate [119]. Apparently, in this case the transition state for the reaction resembles the reactants even more closely
than when methyl iodide is employed for the alkylation [119]. Lower
cis/trans ratios have been observed in angular methylations of
potassium enolates of 37 bearing benzylidine [31], furfurylidine
[31], or, especially, n-butylthiomethylene [33, 34] blocking groups
with methyl iodide, but the cis isomers were still favored. These
results have been interpreted by considering that transition states
such as 175 [33, 353] and 176 [353], which lead to the cis-decalone
90a and involve introduction of the new group equatorial to the nonoxygenated ring, are sterically more favorable than a transition
state such as 177, which would produce a trans-decalone, e.g., 90b.
An alternative transition state leading to the cis product but
involving the introduction of the new group axial to the nonoxygenated
ring has been ruled out [353]. It was found that methylation of the
lithium 1(9)-enolate of syn-6-t-butyl-trans-1-decalone, in which
such a transition state would be extremely unfavorable, gave the
same product mixture as did 37A [353]. In the transition state 177
the approaching alkylating agent would experience a 1,3 interaction
with the axial hydrogen atoms at C4, C5, and C7. Such interactions
would not be involved in the transition states 175 or 176. Top-side
attack on a conformation analogous to that shown in 176 which would
lead to trans products would also be sterically unfavorable.

Because angular alkylations of 1-decalones provide important
models for angular alkylations of 18-nor-D-homo steroids, a thorough
investigation of the manner in which structural modifications

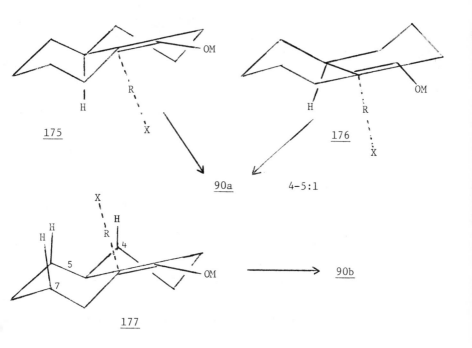

influenced cis/trans ratios in 1-decalones having blocking groups at C2 has been conducted [33]. In such systems, it was noted, the effect of a substituent at any available position would be either negligible or increase, but not decrease, the cis/trans ratio [33]. However, it was found that the introduction of a 6,7-double bond greatly reduced the cis/trans ratio. For example on methylation of the potassium enolate *178* (X_2 = CHC_4H_3O) with methyl iodide in t-butyl alcohol, the trans product *179* (X_2 = CHC_4H_3O) was isolated in 56% yield [33]. The corresponding potassium enolate *178* (X_2 = CHS-n-C_4H_9) gave the pure trans product *179* in 80% yield [34]. The structural change of introducing the 6,7-double bond not only makes the transition state related to *177* more favorable by relieving a 1,3 interaction between the entering group and the axial hydrogen atom at C7 but also requires movement of the axial hydrogen at C5 into a

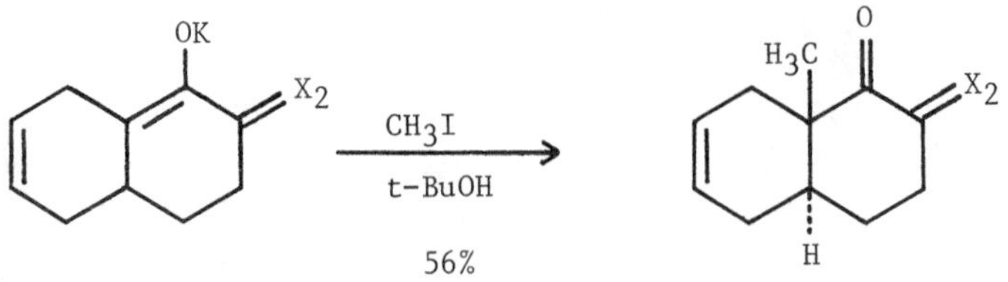

178   179

quasi-axial position, which places it out of the pathway of the
incoming group to some extent. Surprisingly, the introduction of a
3,4- or 5,6-double bond into the blocked decalone system actually
leads to a significant increase in the stereoselectivity of the
alkylation in favor of cis products [33]. It was pointed out that
while such structural modifications would remove 1,3 interactions
involving the hydrogen atoms at C4 and C5 in transition states related
to 177, they would also lead to distortions of the systems that would
bring the quasi-axial hydrogen atoms at C5 and C7 nearer the pathway
of the entering group [33].

The information gained from the study of the alkylations of
enolates such as 178 was used for a stereoselective synthesis of
estrone (180). In this synthesis the key step involved angular
alkylation of the potassium enolate of the 9,11-dehydro-18-nor-D-homo
steroid 181 to produce the trans-D-homosteroid 182 in 56% yield [354].
The cis isomer corresponding to 182 was also obtained in 33% yield
in this reaction.

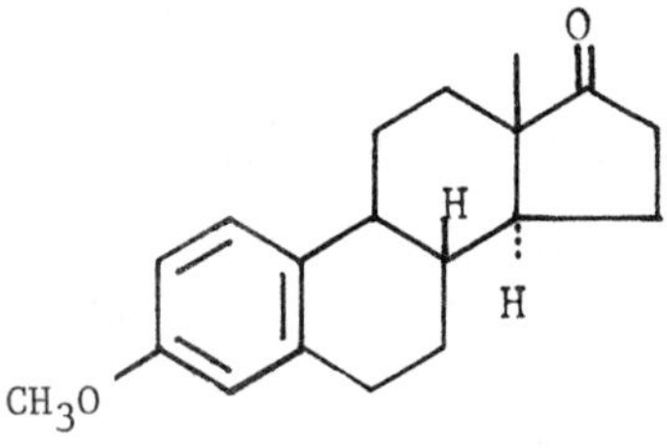

180 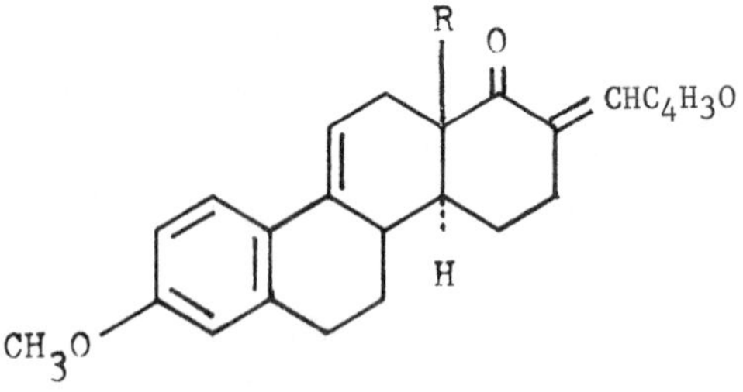   181   R = H

                                 182   R = CH$_3$

Alkylations of tetrahydroindanone enolates such as *91* have been shown to yield even higher proportions of cis-fused products than do the corresponding 1-decalone species [309]. Again, the preferred transition state involves introduction of the new group equatorial to the nonoxygenated ring. The tendency for equatorial alkylation of metal enolates which are exocyclic to six-membered rings is also manifested in the alkylation of the lithium enolate *183* derived from 4-t-butylcyclohexylmethyl ketone [314, 348]. Here reaction with methyl iodide in DME gave a mixture of the equatorial and axial alkylation products *184* and *185* in a 5.7:1 ratio.

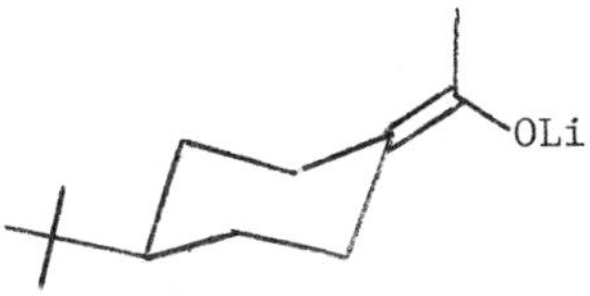

183

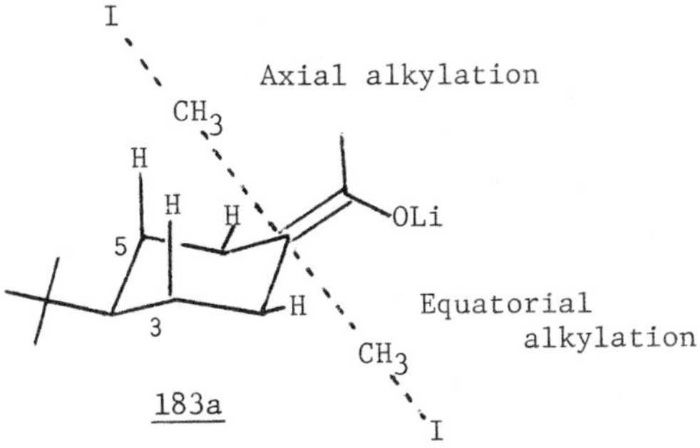

183a

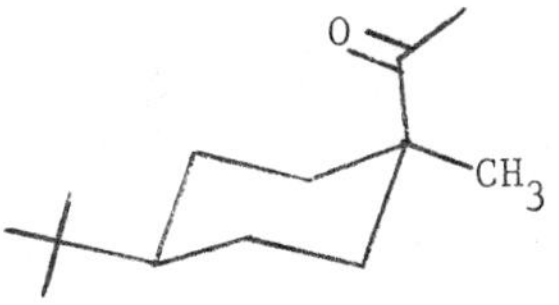

184

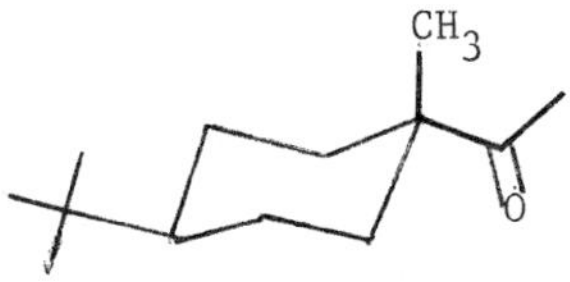

185

Possible transition states for alkylation of *183* are shown in structure *183a*. Here again, in a reactant-like transition-state axial attack on the exocyclic enolate would be hindered by the axial hydrogens at C3 and C5 but equatorial attack is relatively unencumbered [314].

Another possible explanation for the predominance of equatorial alkylation of enolates such as *183* has been published recently [355]. While the details cannot be presented here, the essence of the

argument is that interaction of the filled p orbital at Cl with the
vacant symmetrical antibonding σ orbitals of the β C-C bonds, i.e.,
the 2,3 and 5,6 bonds in *183a*, causes an increase in the electron
density on the equatorial side of the α-carbon atom (Cl) of the
enolate. Consequently, an electrophilic reagent should prefer to
attack from the equatorial direction. Such an orbital distortion
could also be applied to explain the preference of the formation of
cis-fused products in alkylation of enolates such as *37A* or *91*. The
importance of this electronic effect, as opposed to steric and/or
torsional strain factors, in controlling the stereochemistry of these
reactions has not been thoroughly assessed.

The stereochemistry of the alkylation of several lithium enolates
of trans-2-decalones prepared by lithium-ammonia reduction of the
corresponding 1(9)-octalin-2-ones has been determined [304-306].
The results of alkylations of the enolates *186* [305] and *187*
[306] shown in Eqs. (47) and (48) indicate that, as in the case
of t-butylcyclohexanone derivatives such as *162*, the presence of a
substituent at the α position of the enolate significantly increases
the stereoselectivity of the reaction in favor of the axial alkylation
products. This may be attributed to distortions of the enolate in
order to relieve eclipsing interactions of the substituent (methyl
group) with the OLi group and the quasi-equatorial 8-methylene group
[349]. The methylation of the enolate derived from lithium-ammonia
reduction of the tricyclic enone *94* provides a case in which essentially exclusive axial alkylation of a 2-substituted enolate was
observed [310].

A comparison of the data for the alkylation of *186* with that
obtained for the alkylation of the corresponding 10-methyl derivative
188 [304] clearly shows that if axial alkylation involves the development of a 1,3 interaction with an axial methyl group, equatorial
alkylation will be the major pathway [Eq. (49)]. Alkylation of the
3-enolate of 4-methylcholestan-3-one gave stereochemical results
similar to those obtained for *188* [306]. The 2α product (equatorial
alkylation) was obtained exclusively from the lithium 2-enolate of

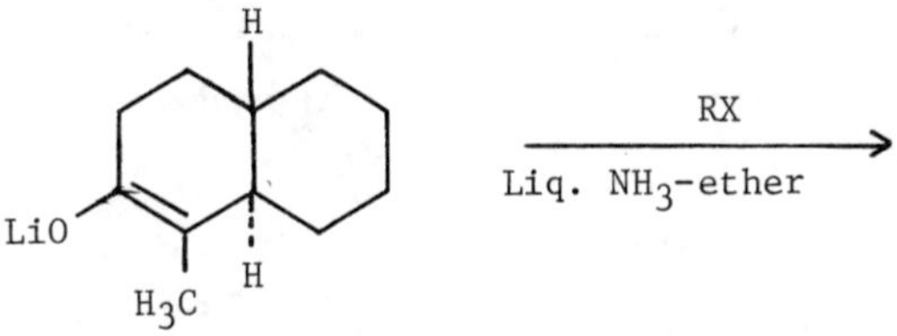

$$RX = CD_3I \qquad 83 \quad : \quad 17$$
$$EtI \qquad >95 \quad : \quad <5$$

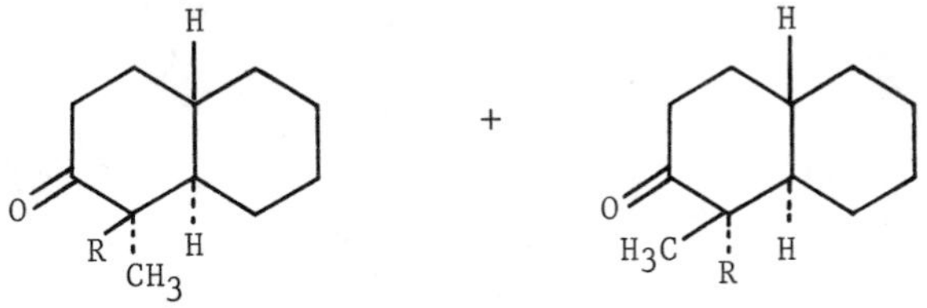

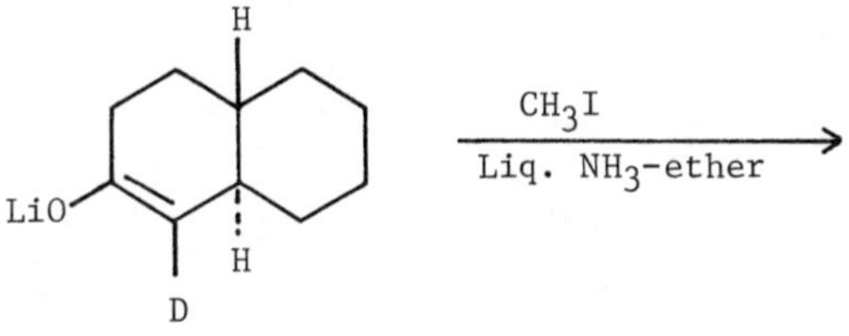

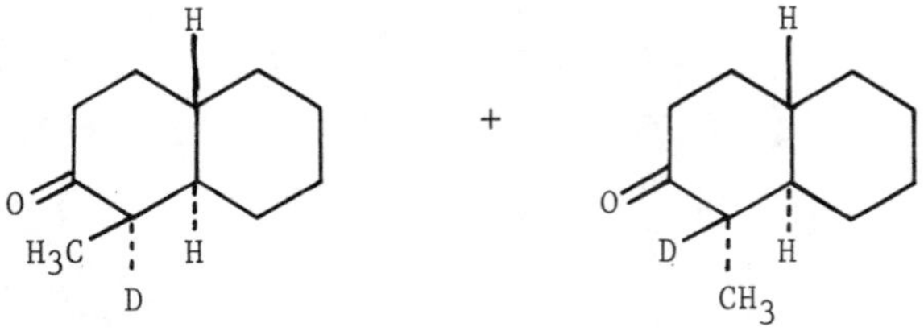

$$40 \quad : \quad 60$$

3,10-dimethyl-trans-2-decalone [304]. A 1,3 interaction of an
entering alkylating agent with a hydroxyl group can also force the
reaction to occur in the equatorial manner. This is indicated by
the conversion of the tricyclic ketone *189* to the α-alkylated product
190 [356].

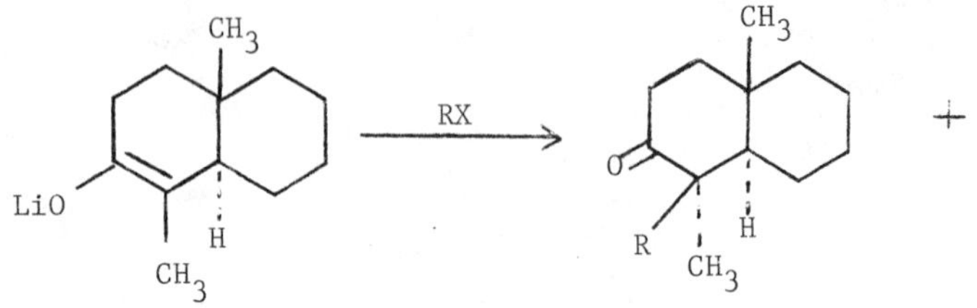

            188   RX = CD_3I
                                                          = EtI

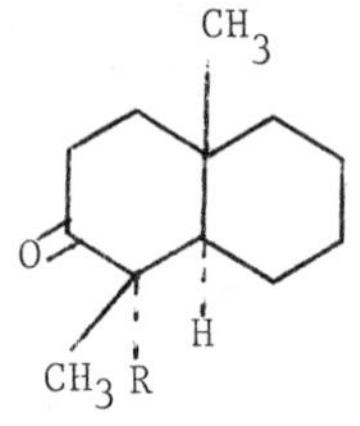

                                                                      (49)

            7:93
            5:95

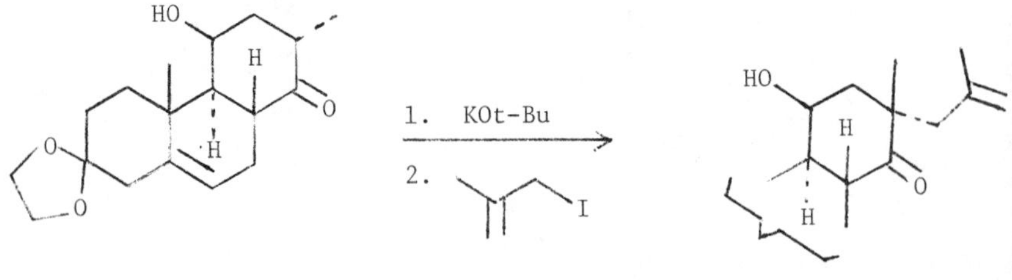

            189   190

The reactions of 96, 98, and 100 provide previously noted examples of alkylation reactions which occur stereoselectively from the less hindered side of the enolate anion. Additional examples which illustrate this point may be found in a review which covers reduction-alkylation reactions of α,β-unsaturated ketones [132].

10. Cycloalkylations Involving Metal Enolates
     of Saturated Ketones

   a. Introduction. Cycloalkylation reactions may be accomplished by the generation of a metal enolate of a ketone containing a suitably disposed leaving group (halide, tosylate, etc.). Such reactions have been utilized to synthesize a variety of cyclic compounds including fused-ring, bridged, and spirocyclic systems. Possible modes of cycloalkylation are represented in Eq. (50). These reactions may be of the endo-type 191 → 192 in which the carbonyl group is present in the new carbocyclic ring or of the exo-type 191 → 193 in which the carbonyl group is attached to the new carbocyclic ring. The corresponding O-cycloalkylation processes 191 → 194 and 191 → 195 may compete with the C modes when structural features permit. Of course, if the ketone is capable of enolization in only one direction, a single C-cycloalkylation mode and perhaps the accompanying O-process will be possible.

   The two steps required for base-catalyzed cycloalkylations, the formation of the enolate anion and its attack on the carbon atom bearing the leaving group, must occur faster than the reactions in which the base causes intermolecular nucleophilic displacement or E2 elimination of the leaving group. Although cycloalkylations have been generally carried out via sodium or potassium enolates, some recent studies involving the use of lithium enolates have appeared. A wide variety of bases and solvents, ranging from potassium hydroxide in protic solvents to dimsyl sodium in DMSO, have been employed.

   It is difficult to make specific recommendations concerning optimum conditions for these reactions. Obviously, these will be quite dependent on the exact nature of the substrate involved. However, as the following discussion will indicate, best results are

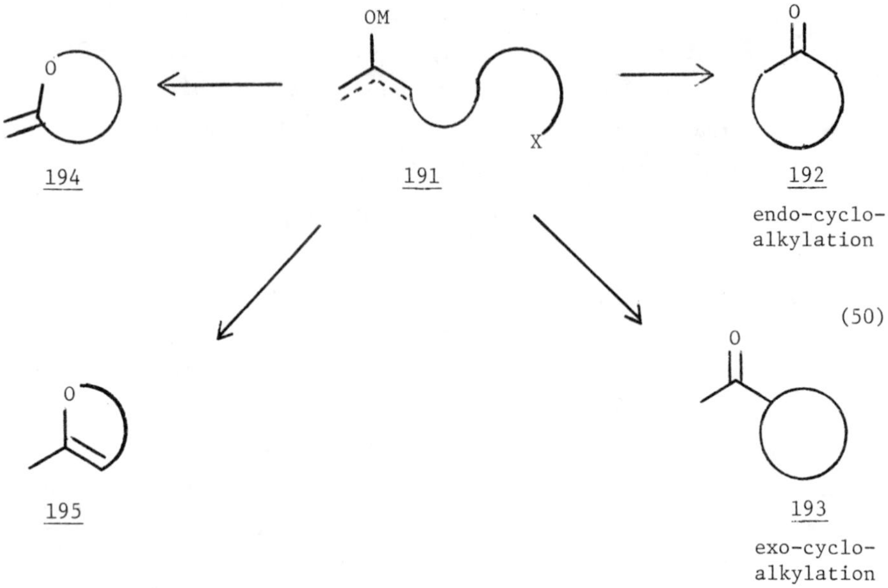

194   191   192
                                                          endo-cyclo-
                                                          alkylation

                                                                    (50)

195   193
                                                          exo-cyclo-
                                                          alkylation

normally obtained when strong bases having low nucleophilicity are
employed. Lithium, sodium, or potassium bases in solvents of low
to medium polarity generally give good results when the substrate
is a keto halide or tosylate having the leaving group at a primary
carbon atom. Good results have been obtained in cycloalkylations
of keto epoxides and tosylates (or mesylates), particularly those
having the leaving group at a secondary position, using dimsyl
sodium in DMSO or other strong bases in dipolar aprotic solvents.

     b. *Formation of three membered rings*. The well-known Favorskii
rearrangement of a α-halo ketones [357-359] provides the simplest
possible example of an endo-cycloalkylation process [Eq. (51)]. Strai
ed cyclopropanones *196* commonly considered to be involved as intermediates undergo nucleophilic cleavage to carboxylic acid derivatives
under the reaction conditions. This reaction has been reviewed in
detail elsewhere [357-359], and its discussion is outside the scope
of this chapter.

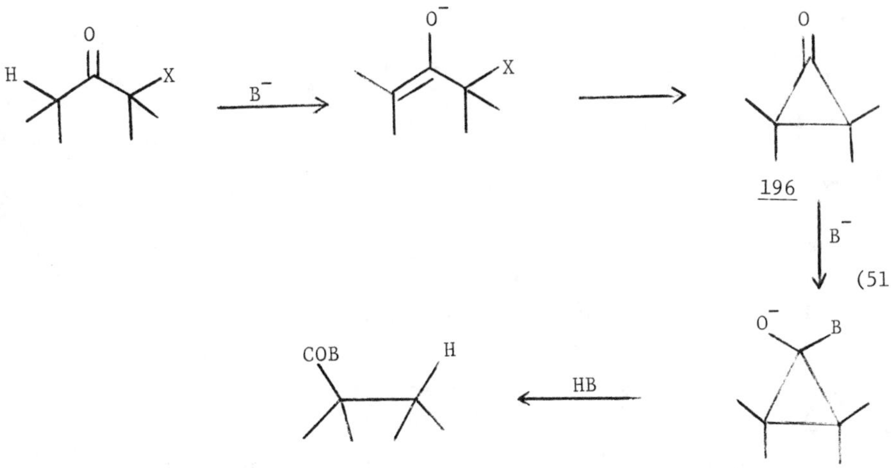

(51)

Several examples of exo-cycloalkylations which lead to three-membered rings are known. 5-Chloropentan-2-one (*197*) has been found to yield exclusively methylcyclopropyl ketone (*198*) under a variety of conditions [360]. The alternative mode of cyclization which would lead to cyclopentanone has not been observed. In intramolecular reactions involving carbanions it has been shown that three-membered rings are formed much faster than larger rings, including five-membered rings [361]. Entropy factors favor the formation of three-membered rings over five-membered rings, and in the case of the terminal enolate derived from *197*, the appropriate geometric alignment for perpendicular attack of C5 (which bears the leaving group) on C1 is difficult to achieve. Also, the reaction conditions which have been employed for the cyclization of *197*, i.e., relatively weak bases such as sodium or potassium hydroxides or alkoxides and/or protic or polar aprotic solvents [360], would be expected to favor rapid enolate equilibration. If equilibration is more rapid than cyclization, the more stable internal enolate of *197*, which leads to *198*, would be expected to be present in high concentration.

The conversions of 2(2-bromoethyl)cyclopentanone (*199*) to the spiro ketone *200* [362, 363], of 4-tosyloxycyclohexanone (*201*) to bicyclo[3.1.0]hexan-2-one [364], of 3-tosyloxymethylcyclohexanone (*202*) to bicyclo[4.1.0]heptan-2-one [365], of 3-tosyloxycyclooctane

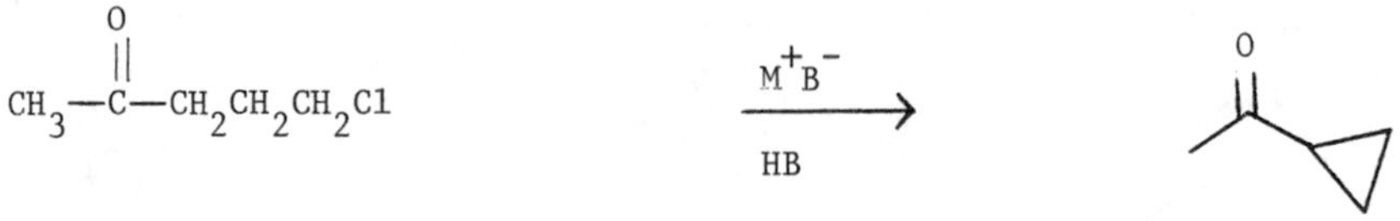

$$CH_3-\overset{\overset{\displaystyle O}{\|}}{C}-CH_2CH_2CH_2Cl \qquad \xrightarrow[HB]{M^+B^-} \qquad$$

197   198

                                                                        85-100%

(203) to bicyclo[5.1.0]octan-2-one [98], and of exo-5-chlorobicyclo-
[2.1.1]hexan-2-one (204) to the highly strained ketone tricyclo-
[2.2.1.1^{2,6}]-hexan-3-one [366] provide examples of exo-cycloalkylation
leading to the formation of three-membered ring.

Compounds 199, 202, and 203 are structurally capable of yielding
bridged systems containing new five-membered rings by an endo-
cycloalkylation mode. However, such products were not observed [98],
nor was enol ether formation, which is possible for compound 199
[362].

The conversion of the keto tosylate 205 into the tetracyclic
ketone 206, which contains the tetracyclic ring skeleton of the
sesquiterpene ishwarane, provides an example of the use of cycloalkylation for the formation of three-membered rings in the natural
products field [367].

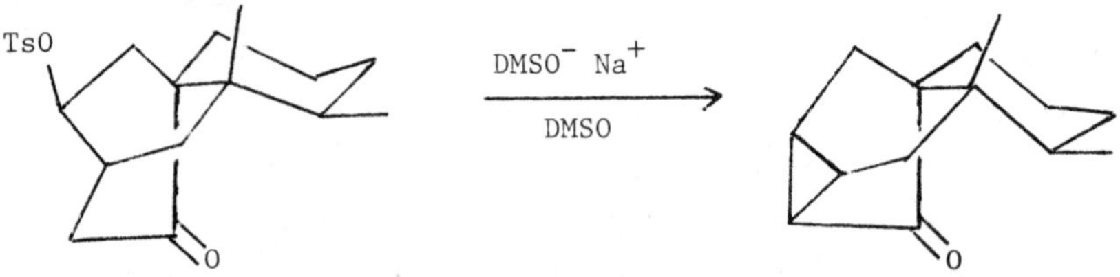

$$\xrightarrow[DMSO]{DMSO^-\ Na^+}$$

                                                                        (60-70%)

205   206

    c. Formation of four-membered rings.   Several examples of
cycloalkylations in which cyclobutanones are formed by the endo-
cyclization mode are known. The keto tosylate 207a yields mixtures
of the cyclobutanone 208a and the α-methyleneoxetane 209a on treatment with metal hydrides in aprotic solvents [368]. The use of

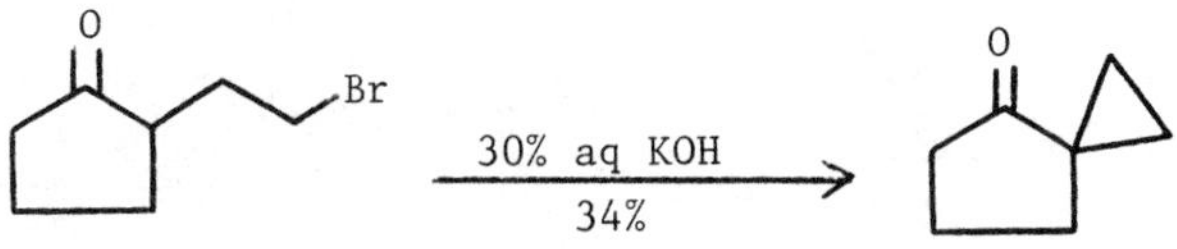

199     $\xrightarrow[34\%]{\text{30\% aq KOH}}$ 200

201     $\xrightarrow[64\%]{\begin{array}{c}\text{NaH}\\\text{dioxane}\end{array}}$

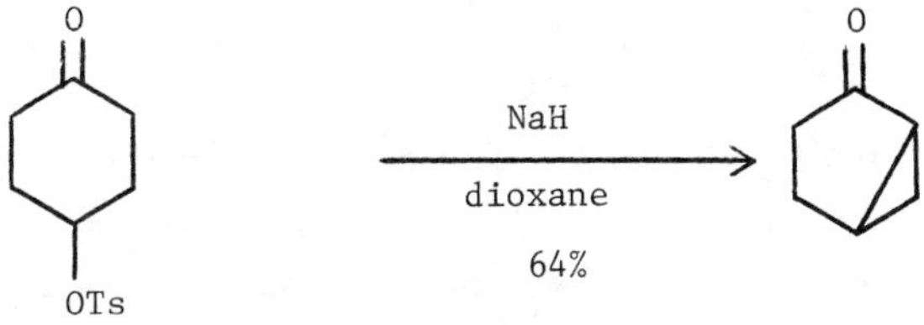

202     $\xrightarrow[\text{low yield}]{\text{NaH} \quad \text{THF}}$

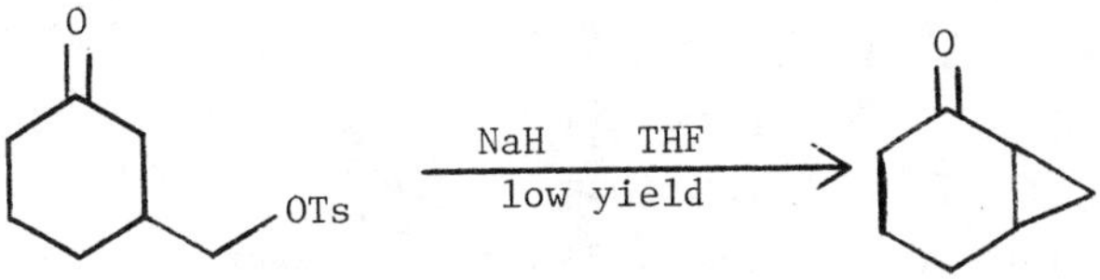

203     $\xrightarrow[90\%]{\text{KOt-Bu, ether}}$

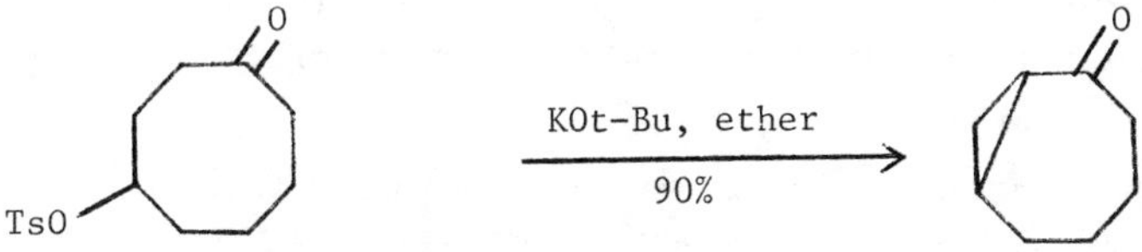

204     $\xrightarrow[50\%]{\text{Kot-Bu, ether}}$

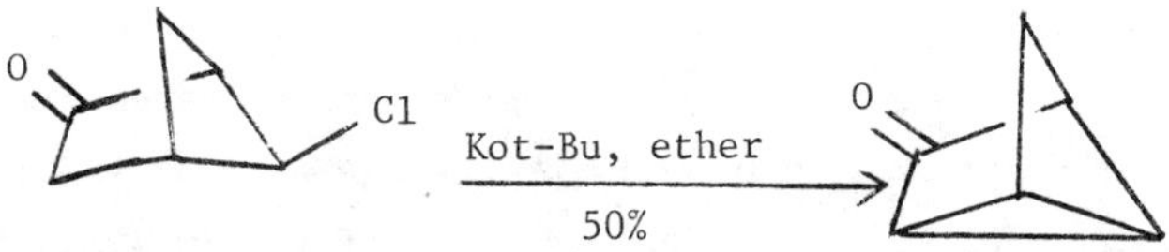

207 a. R=C$_6$H$_5$,X=OTs   208 a. R=C$_6$H$_5$ 209 a. R=C$_6$H
     b. R=H,X=Br   b. R=H   b. R=H

                      MH/Solvent:KH/THF   25%   36%

                      NaH/THF   50%   24%

potassium hydride in THF favored the formation of the enol ether
209a, but with sodium hydride, the cyclobutanone 208a predominated.
The use of more polar solvents to favor the formation of solvent-
separated ion pairs of the metal enolate did not change the C-
cycloalkylation/O-cycloalkylation ratio significantly. Likewise,
the bromo ketone 207b has been shown to yield a 50:50 mixture of
the cyclobutanone 208b and the α-methyleneoxetane 209b on addition
to a solution of LDA in THF [69]. The extent of O-alkylation in
compounds such as 207 is greater than is normally found in intermolecular reactions conducted in THF [368]. However, it would be
expected that the transition state leading to compounds such as 208
would be highly strained because bond formation perpendicular to the
plane of the enolate would be required, while the transition state
leading to α-methyleneoxetanes (209) might not have a rigid stereoelectronic requirement. In view of this the amount of C-
cycloalkylation observed is surprisingly large.

        Cycloalkylations of compounds such as 2-methyl-2-tosyloxy-
methylcyclopentanone (210a) [369], 2-methyl-2-tosyloxymethylcyclo-
hexanone (210b) [370] and 2-methyl-2-chloromethylcyclohexanone
(210c) [371] with aqueous or alcoholic bases have been found to
yield the corresponding bridged cyclobutanones 211. In the case of
210b, the bridged ketone 211 (n = 3) is accompanied by about an
equal amount of the isomeric bicyclic ketone 212 (n = 3) [370].
Several possible mechanisms for the formation of 212 (n = 3) have
been proposed [372]. The pathway shown in Eq. (52) seems most
reasonable [371].

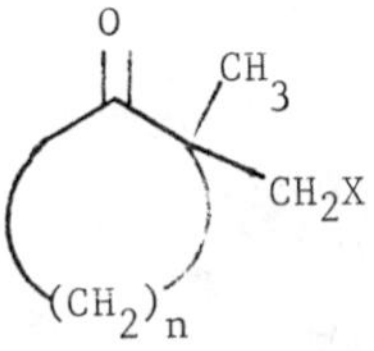

210 a; X = OTs, n = 2   211   212

  b; X = OTs, n = 3

  c; X = Cl, n = 3

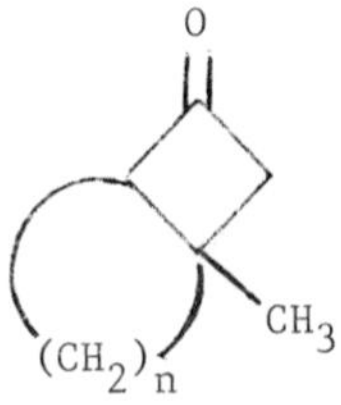

210b

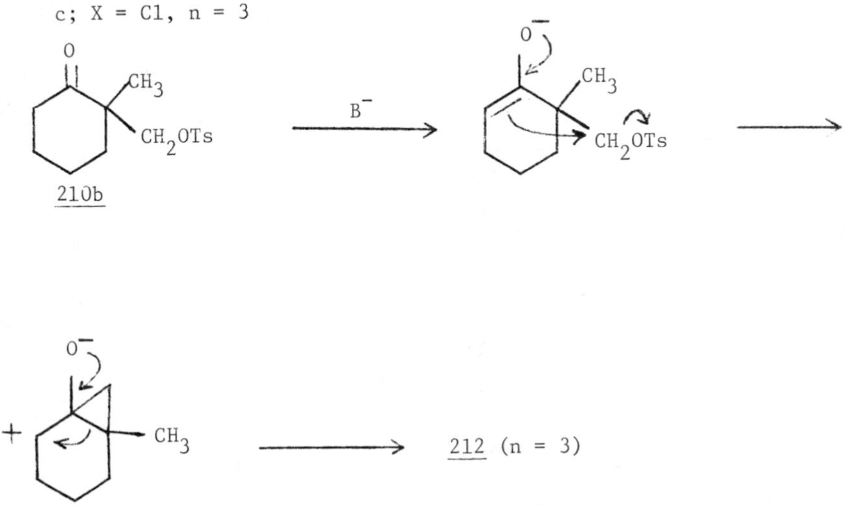

(52)

Cycloalkylation reactions of 2(3-bromopropyl)cyclopentanone (213a) and 2(3-bromopropyl)cyclohexanone (213b) have been studied using potassium t-butoxide in benzene or potassium hydroxide in benzene or water [362, 373, 374]. From 213a, three products (the bridged ketone 214a, the spiro ketone 215a, and the enol ether 216a) were obtained. The reaction conditions employed would be expected to lead to enolate equilibration. Thus, it is not surprising that the major products 215a and 216a are derived from the more stable 1-enolate of 213a via the exo-cycloalkylation mode. However, it is somewhat surprising that the C pathway which leads to a strained four-membered ring is of comparable importance to the O pathway which might involve a relatively unstrained transition state. The bridged ketone

214a was a minor product except when potassium t-butoxide in benzene
was employed. These reaction conditions may favor production of
the kinetic 5-enolate of *213a* in a rather high concentration, and
this species can lead to *214a*.

In the case of *213b*, the enol ether *216b* was by far the major
reaction product, and the bridged system *214b* was not formed at all.
Nonbonded interactions seem to be quite important in the transition
state for the formation of *214b*, although from the point of view of
angle strain it appears that the formation of *214b* would perhaps be
more favorable than *214a*, which has the smaller bridge.

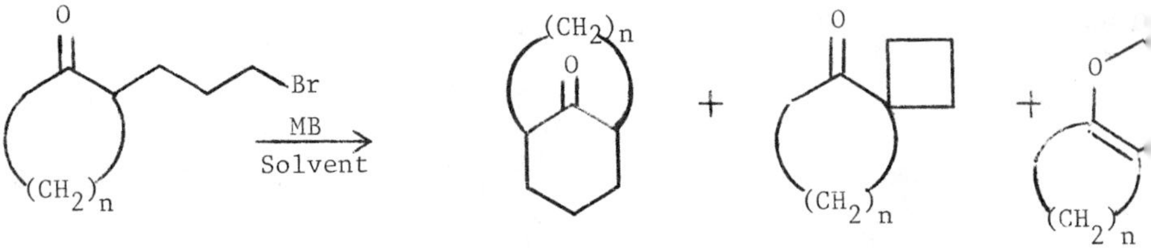

		214	215	216
a. n = 2				
	KOt-Bu/C$_6$H$_6$	19%	30 %	22 %
	KOH/H$_2$O	6	13	15
	KOH/C$_6$H$_6$	2	37	61
b. n = 3	KOtBu/C$_6$H$_6$	–	8	67

An important example of a cycloalkylation leading to a four-
membered ring is the conversion of the cis-2-decalone derivative
217 to the tricyclic ketone *218*, using dimsyl sodium in DMSO [96].
Compound *218* contains the carbon skeleton of sesquiterpenes such
as copaene and ylangene, and the ring closure is a key step in their

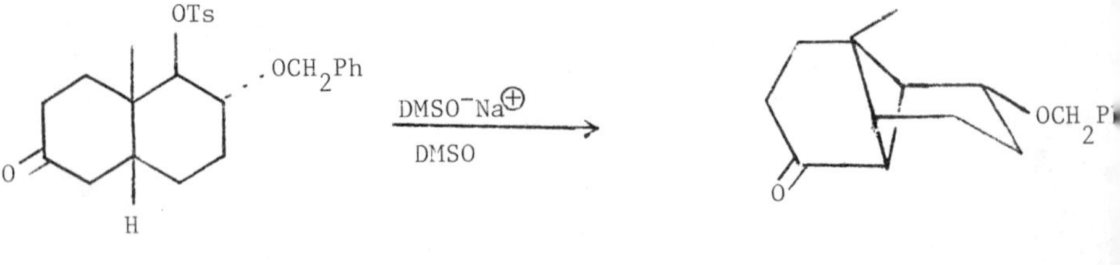

217   218

synthesis. A somewhat different approach to the copaene skeleton
involved intramolecular cyclization of the keto tosylate *219* [375].
The keto tosylate *220a*, which is related to *217* and *219*, undergoes
a similar ring closure [375]. A report [376] that the related compound *220b* having a gem-dimethyl grouping at C7 underwent fragmentation to a cyclodecadienone on base treatment has been shown to be
incorrect [377].

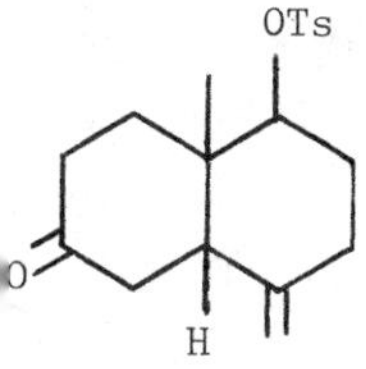

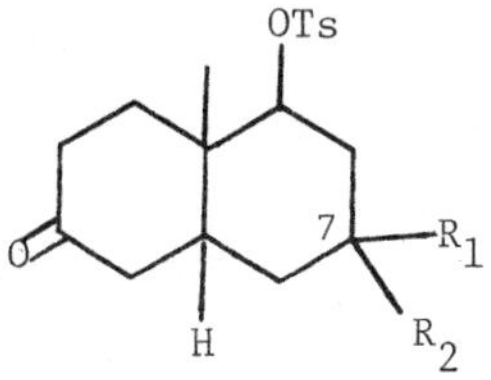

$$\underline{220} \text{ a. } R_1 = R_2 = H$$
$$\text{b. } R_1 = R_2 = CH_3$$

The tosylate derivatives *221* and *222* of trans- and cis-10-
hydroxymethyl-2-decalone yield the bridged products *223* and *224*,
respectively, upon base treatment [378]. These modes of cyclization
are consistent with the direction of enolization expected for these
two types of ring systems. Four-membered ring formation to give

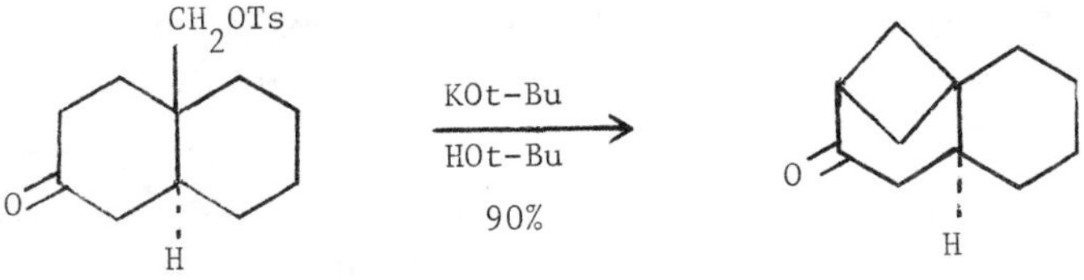

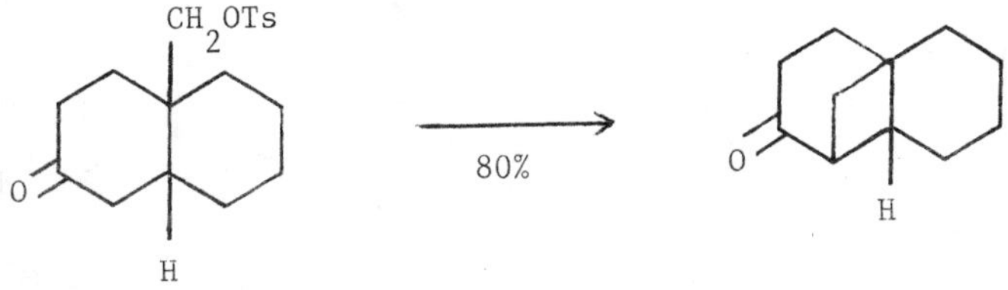

bicyclo[4.2.0]octan-2-one also occurred in high yield when 5-tosyloxycyclooctanone was treated with sodium hydride in DMSO or potassium t-butoxide in ether [98].

    d. Formation of five-membered rings. As noted above, five-membered ring ketones are not readily formed by the endo-cyclization mode. This is clearly indicated by the fact that treatment of the bromoketone *225* with LDA in THF leads exclusively to the enol ether *226* [69]. Examination of models indicates that a transition state for intramolecular C-alkylation of the 1-enolate of *225* would be highly strained. On the other hand, a much less strained transition state for intramolecular O-alkylation leading to *226* is possible if attack of the five-carbon atom on oxygen is allowed in the plane of the enolate system rather than perpendicular to it. The solvolysis

$$CH_3-\overset{\overset{\textstyle O}{\|}}{C}-C(CH_3)_2CH_2-CH_2-Br \xrightarrow[\text{THF}]{\text{LDA}}$$

           225   226

of 10-acetyl-1α-halo-trans-decalins in aqueous or alcoholic base also involves participation by the oxygen atom and leads to a tri-cyclic system containing a five-membered ring enol ether rather than a ketone [379].

    Exo-cycloalkylations leading to five-membered rings are well known. The spiro ketone *228* has been prepared in approximately 70% yield by treatment of 2(4-bromobutyl)cyclohexanone (*227*) with potassium t-butoxide in benzene [380]. Spiro products were also obtained when cyclopentanone and cycloheptanone derivatives related to *227* were treated in a similar manner [362]. Direct reaction of cyclohexanone with 1,4-dibromobutane in the presence of potassium or sodium alkoxides in benzene or toluene has also been reported to produce *228* in high yield [362].

    The formation of five-membered carbocyclic rings by intramolecul alkylation has been widely used in the synthesis of complex natural

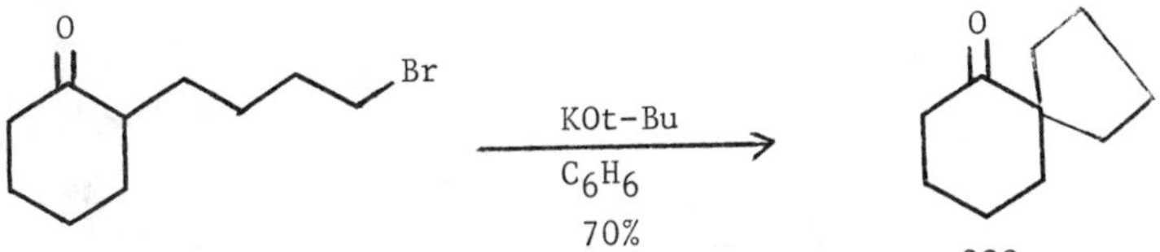

227   228

products. Cycloalkylation of the keto tosylate *229* has been used to
close the D ring in a steroid total synthesis [381]. The keto

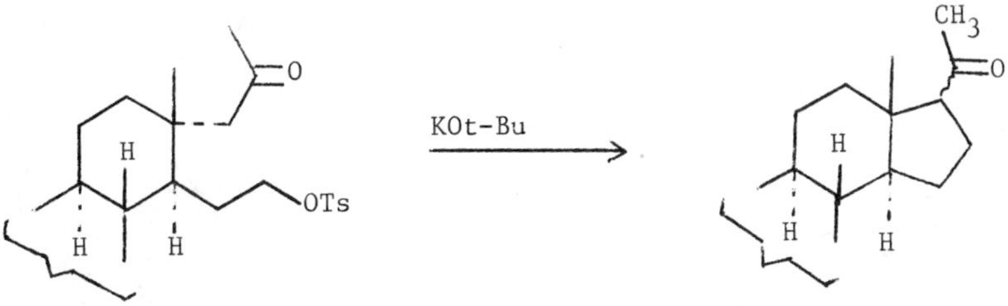

229

tosylate *230* undergoes base-catalyzed ring closure to give a tricyclic ketone, which has the ring skeleton of sativene, on base
treatment [95].

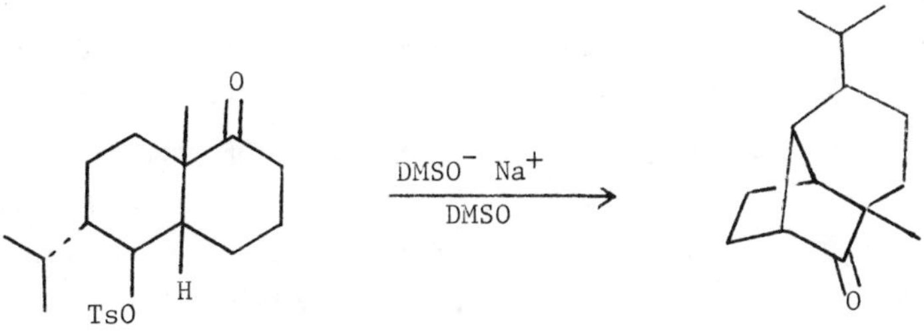

230

Epoxy ketones which can yield five-membered ring products may
undergo base-catalyzed cycloalkylations in high yields. Intramolecular
cyclizations of the epoxy ketones *231* and *232* using dimsyl sodium in
DMSO have been used to prepare the hydroxy ketones *233* and *234* which
have the carbon skeletons of the sesquiterpenes longifoline [93] and
copacamphene [94], respectively.

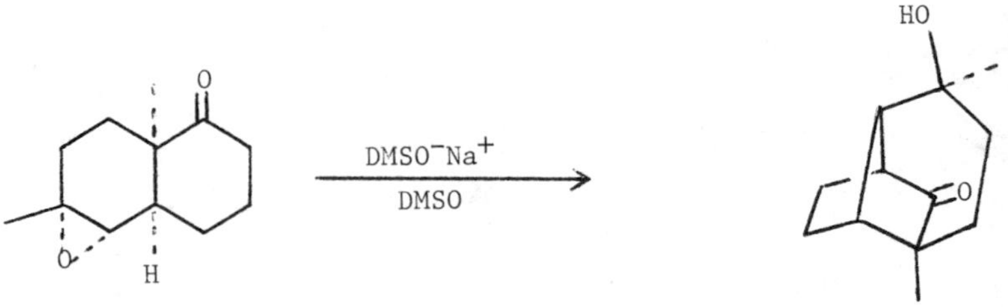

231   →   233

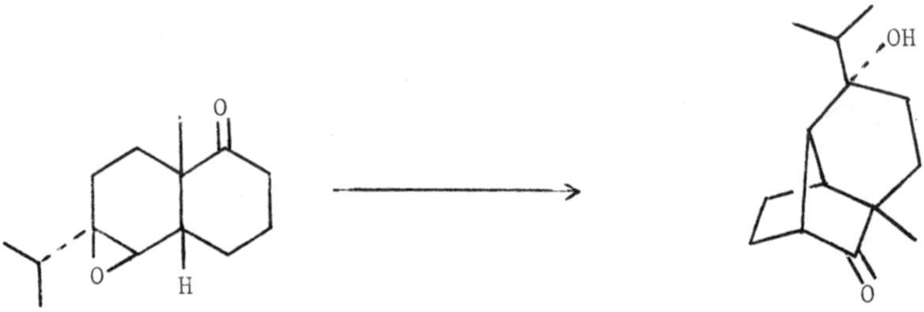

232   →   234

e. *Formation of six-membered rings.*   Six-membered rings may be readily prepared by either the endo- or exo-cycloalkylation modes. The preparation of 2,2-dimethylcyclohexanone (7) in 55-60% yield by reaction of the 6-bromohexan-2-one *235* with LDA in THF provides a simple example of an endo-cycloalkylation leading to a six-membered ring ketone [69].   Additional examples of this type of process are illustrated by the conversions of 3(2-tosyloxyethyl)cyclohexanone (*236*) to bicyclo[2.2.2]octanone (*237*) [373] and of 3(2-tosyloxyethyl) [382] and 3(2-bromoethyl)cyclopentanones [383] such as *238* to the corresponding bicyclo[2.2.1]heptanones *239* with strong bases. Compounds *236* and *238* are structurally capable of yielding four-membered ring products by an exo-cycloalkylation process, but such compounds were not isolated.

The spiro ketone *240* has been obtained in 87% yield by reaction of 2(5-bromopentyl)cyclooctanone (*241*) with sodium hydride in DME

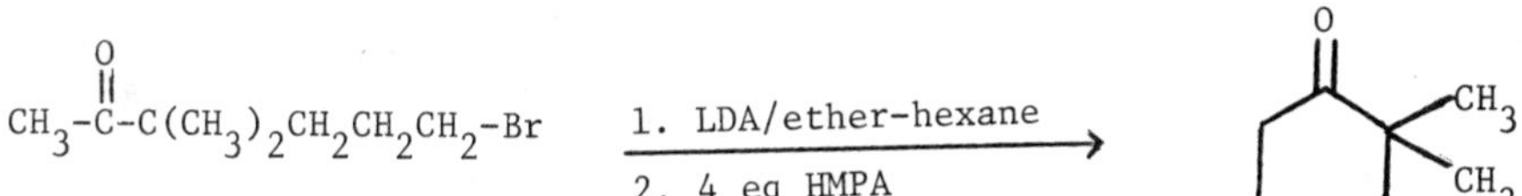

$$CH_3-\overset{\overset{\textstyle O}{\|}}{C}-C(CH_3)_2CH_2CH_2CH_2-Br$$

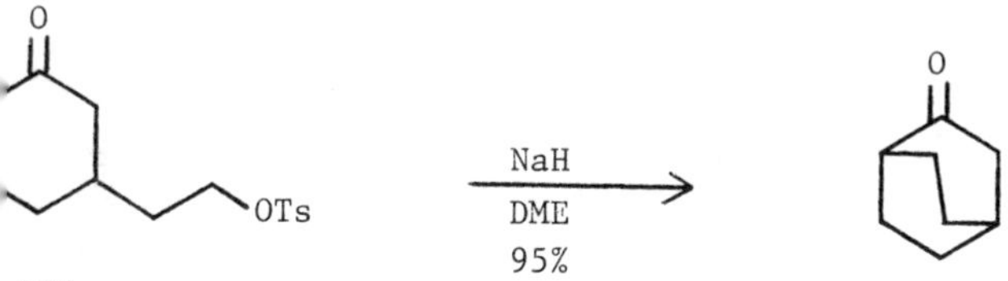

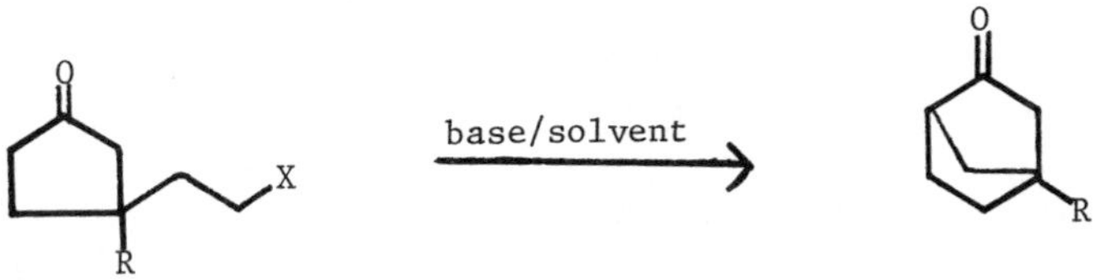

235

1. LDA/ether-hexane
2. 4 eq HMPA

7

236

NaH
DME
95%

237

238

base/solvent

239

base/solvent

a.   R = H, X = OTs   NaH/THF   –

b.   R = CH$_3$, X = Br   KOt-Bu/HOt-Bu   65%

[384], and bicyclo[3.3.1]nonan-2-one (*242*) has been prepared in 42%
yield from 4(3-tosyloxypropyl)cyclohexanone (*243*) using potassium
t-butoxide in THF [385].

Examples of the synthesis of cis-1-decalones by cycloalkylations
involving metal enolates of cyclohexanone derivatives have been
reported. Treatment of 2-methyl-3(4-tosyloxybutyl)cyclohexanone
(*244*) with sodium t-amylate in benzene gave cis-9-methyl-1-decalone
(*90a*) exclusively in 60% yield [386]. The 1-enolate intermediate
derived from *244* would be expected to prefer a conformation having
the 3-bromobutyl side chain quasi-axial because A$^{(1,2)}$-strain [143]

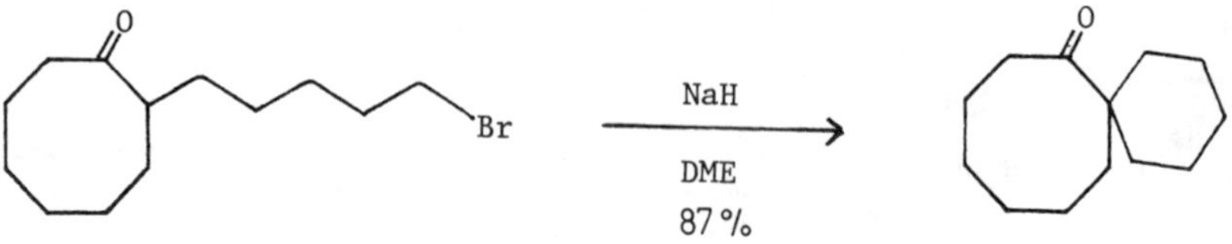

241   240

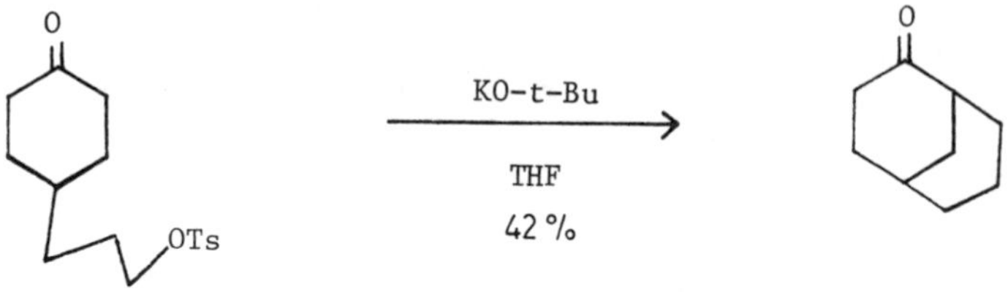

243   242

would be large in the conformation having the bromobutyl side quasi-
equatorial. In the former conformation, equatorial cycloalkylation
would yield 90a. The cis-1-decalone 245 has been prepared in 27%
yield by conjugate addition of lithium dimethylcuprate to 3(4-
bromobutyl)-5-isopropyl-2-methylcyclohex-2-enone (246) in benzene
followed by addition of HMPA to allow cyclization of the enolate
intermediate 247 [226]. Removal of the carbonyl group in 245 gave
the bicyclic sesquiterpene (±)-valerane. A key step in the synthesis
of the tricyclic sesquiterpene (±)-seychellene involved the cycloalkylation of the keto tosylate 248 to the ketone 249 using dimsyl

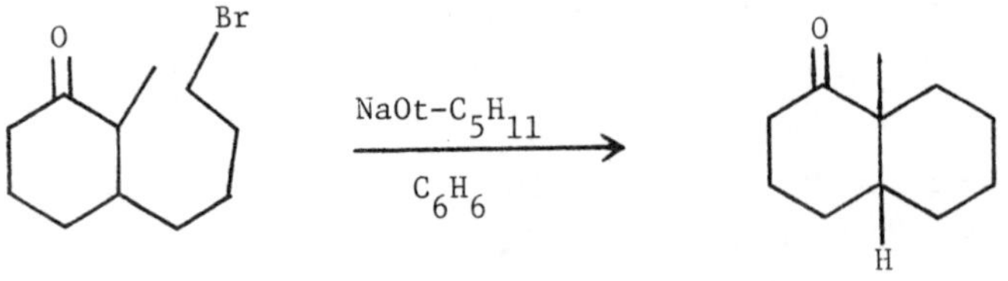

244   90a

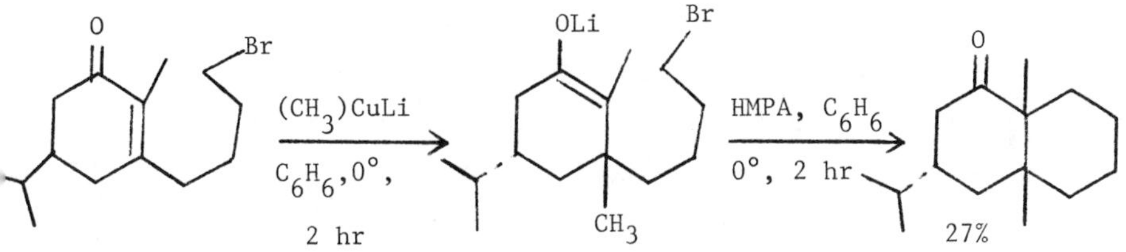

246   247   245

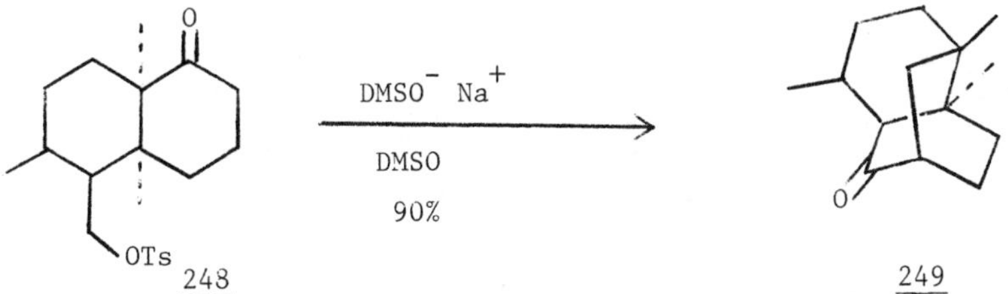

248   249

sodium in DMSO [97]. Two syntheses of twistone (250) by intramolecular ring closures using the keto mesylates 251 [64] and 252 [387] provide interesting examples of the use of cycloalkylation reactions for the synthesis of other complex ring systems.

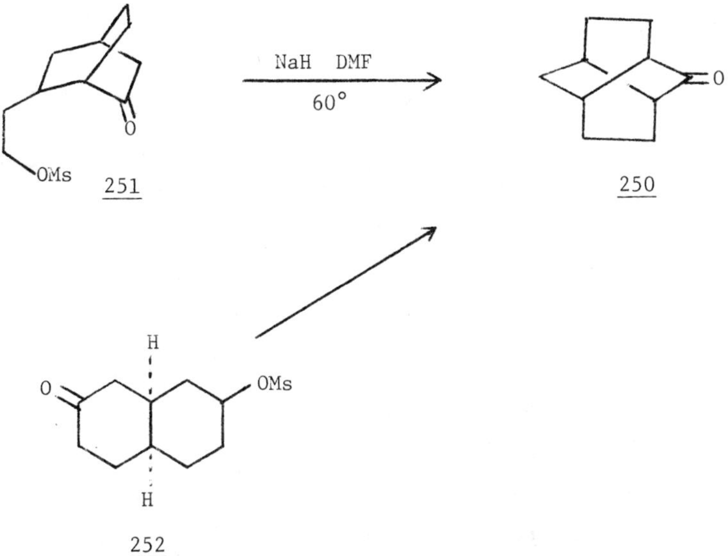

251   250

252

f. Formation of seven-membered rings. In intramolecular
alkylations which can lead to seven-membered rings, the O- and C-
cycloalkylation modes have been shown to be competitive. For example,
2-carbethoxy-2(4-bromobutyl)cycloalkanones such as *253* may yield
bridged ketones such as *254* or enol ethers such as *255* upon base
treatment [388-390]. The C-cycloalkylation/O-cycloalkylation ratios
are dependent on the size of the cycloalkanone ring; for example,
the cyclopentanone derivative *253a* yields 21% of *254a* and 31% of
255a on reaction with sodium hydride in DMF-benzene [388], the cycloheptanone derivative *253b* gives the corresponding products *254b* and
255b in a 1:9 ratio on reaction with sodium hydride in DMF-benzene
[389], and the cyclododecanone *253c* gives the C-alkylation product
254c exclusively, under similar conditions [390]. It is not clear
what factors are responsible for these changes in product ratios
with ring size. It may be that greater nonbonded interactions are
present in the transition state for C-cycloalkylation of the seven-
membered ring system *253b* than in the five- or 12-membered ring
systems *253a* and *253c*, respectively.

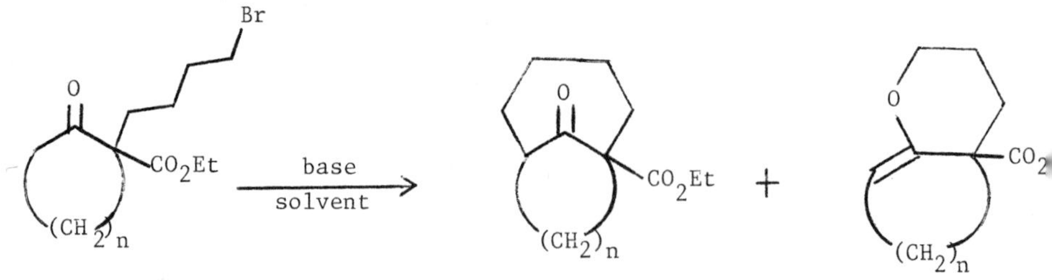

	253		254	255
a.	n = 2	NaH/DMF-C_6H_6	21%	31%
b.	n = 4	NaH/DMF-$C_6H_5CH_3$	10	90
c.	n = 9		100	--

Compared with the cycloheptanone *253b*, the 7-methyl-substituted
cycloheptanone *256* undergoes significantly more C-cycloalkylation
upon base treatment [389]. Thus reaction of *256* with sodium hydride

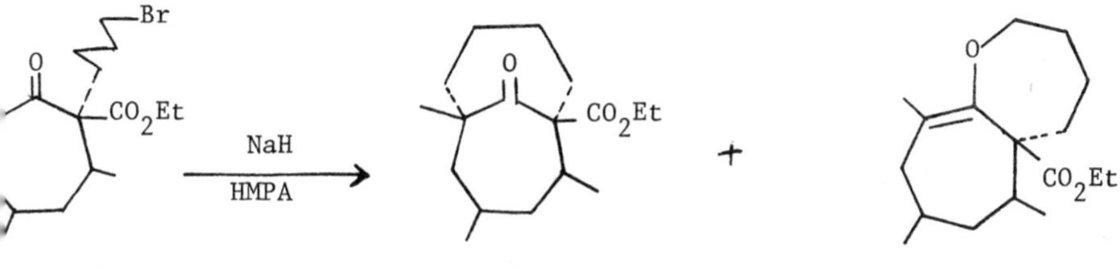

256   257   258

in HMPA gave a 4:6 mixture of the bridged ketone *257* and the enol
ether *258*. Greater nucleophilicity, inhibition of solvation, or
aggregation of the methyl-substituted enolate of *256* could account
for this [389].

Several examples of the synthesis of bicyclo[4.3.1]decan-10-ones
by cycloalkylations involving metal enolates of cyclohexanone derivatives have been reported [391]. Reaction of the more-substituted
lithium enolate *5B* (M = Li) of 2-methylcyclohexanone (prepared by
cleavage of the corresponding enol acetate with methyllithium in DME)

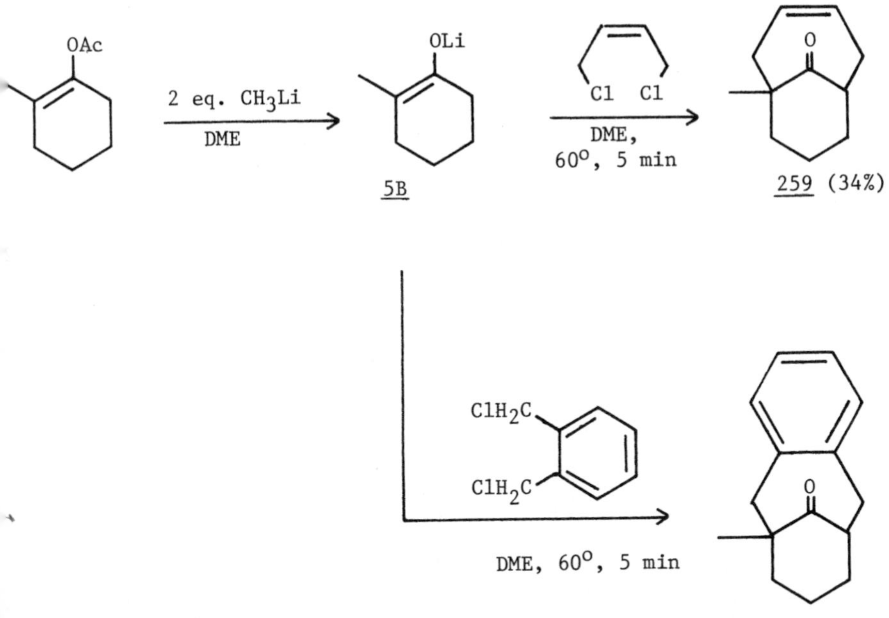

5B   *259* (34%)

260 (95%)

with cis-1,4-dichloro-2-butene or bis-1,2-chloromethylbenzene gave compounds *259* and *260* in 34 and 95% yields, respectively. The lithium t-butoxide produced along with *5B* (M = Li) in the enol acetate cleavage reaction served as the base for the second step of the cycloalkylation reaction.

A related reaction involving cycloalkylation of the 2-tetralone derivative *261* with 2,3-dichloromethyl-1,4-dimethoxybenzene (*262*) using excess sodium hydride in DME to give the bridged ketone *263* has been used as a key step in the total synthesis of the mold metabolite (+)-byssochlamic acid [392].

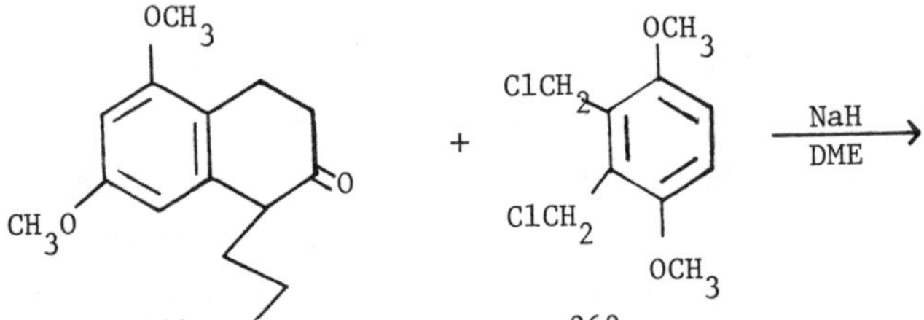

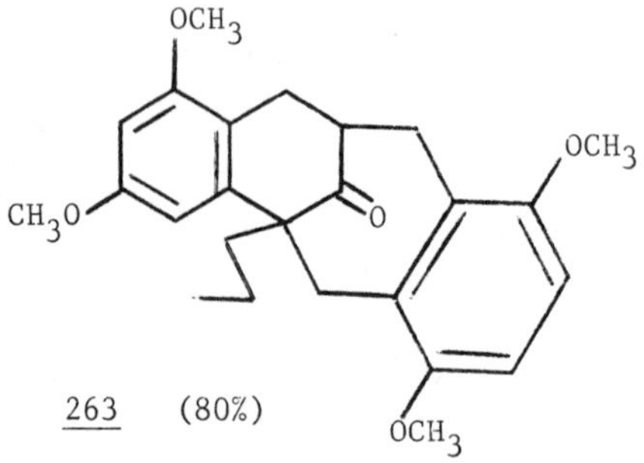

E. Acylation of Saturated Ketone Enolates

1. General

A number of carboxylic acid derivatives act as electrophilic reagents for the acylation of ketone enolates. The well-known Claisen reaction involves the base-promoted C-acylation of ketones with esters to product β-dicarbonyl compounds [6-8]. By the appropriate choice of the ester, β-ketoaldehydes, β-diketones, or β-keto esters may be prepared by this reaction. Acylations with the more reactive esters, i.e., formates and oxalates, are generally carried out using weak bases such as sodium methoxide in aprotic solvents; but for less reactive esters, i.e., alkanoates and carbonates, strong bases such as sodium amide, sodium hydride, tritylsodium, or dimsyl sodium must be employed to obtain high yields [6-8]. Since 1 eq of metal alkoxide is produced along with the β-dicarbonyl compound in the C-acylation step, these reactions are reversible. Thus the acylation product that can yield the more stable anion will be obtained from an unsymmetrical ketone. This usually means that the acyl group will be introduced at the less-substituted α position of an unsymmetrical ketone. If both α positions are substituted to the same degree, the less sterically hindered of the two possible enolizable β-dicarbonyl products will result. O-Acylation by the ester component may compete kinetically with the C-acylation process, but owing to the reversibility of the reaction, only trace amounts of O-acylated products are observed even under carefully controlled conditions [117]. (Full details concerning conditions for the Claisen acylation of ketones will be covered in another volume in this series.)

Irreversible acylations may be carried out by the reaction of ketone enolates with a variety of electrophilic carboxylic acid derivatives such as acid anhydrides [6-8, 89, 117], acid halides [6, 7, 117], ketones [117, 393], and carbon dioxide [28, 39, 60, 132, 193]. Various inorganic acid derivatives such as alkyl nitrites [394] and nitrates [395-397], N,N,N',N'-tetramethyldiamidophosphorochloridate [398], diethyl phosphorochloridate [218], and cyanogen chloride [399] may also be used to prepare α-substituted or enol derivatives of ketones.

Metal enolates for use in acylation reactions may be generated by base-promoted enolization of ketones, as described in Sec. II.A, or by the various indirect methods of enolate formation, e.g., metalammonia reduction of α,β-unsaturated ketones [132], or conjugate addition of Grignard reagents [158, 210] or lithium dialkylcuprates to α,β-unsaturated ketones [94, 215, 216, 230, 234] (see Sec. II.C.).

In many instances the metal enolate may be acylated in the solvent in which it is prepared. An exception to this is liquid
ammonia; it should be exchanged for an unreactive solvent (ether,
THF, DME, etc.) before performing the acylation reaction [132]. A
procedure for the preparation of enol acetate by the trapping of
metal enolates with excess acetic anhydride has been described in
Sec. II.B.3.

Actually, when acid derivatives are employed as acylating
agents, O-acylation and/or C-acylation products may be obtained
depending on the reaction conditions. When kinetic conditions are
employed, i.e., when the metal enolate solution is added to a large
excess of the acylating agent, it has been shown that the O-acylation,
C-acylation ratio will depend upon (a) the nature of the metal cation
(b) the solvent, (c) the structure of the enolate anion, and (d) to
some extent, whether an acid anhydride or an acid halide is employed.
In general, it appears that O-acylation is strongly preferred under
conditions which favor the existence of metal enolates as solvent-
separated ion pairs (15) or which favor α-metalated ketones (16;
Eq. (7)] [117]. The presence of solvent-separated ion pairs will be
favored when alkali metal cations (Li^+, Na^+, K^+) and certain divalent
metals (Zn^{2+}) are employed and when DME or more polar solvents are
used. However, when cations such as $Mg^{2+}/2$ or $MgBr^+$ are used in
less polar solvents, such as diethyl ether, conditions are favorable
for the existence of the metal enolates in the form of tight ion pairs
(14) or covalently bonded species [13; Eq. (7)]. When the latter
conditions obtain C-acylation products are often preferred kinetically. The manner in which a change of metal cation and solvent
influences the composition of the products of acylation of the
cyclohexanone enolates 54 with acetic anhydride is shown in Eq. (53)
[117]. When 54 (M = Li) was prepared by cleavage of the trimethylsilyl enol ether of cyclohexanone and reacted with acetic anhydride
in DME, 49% of the enol acetate 264 and 16% of 2-acetylcyclohexanone
(265) were obtained. However, when the $Mg^{2+}/2$ enolate (54, M = $Mg^{2+}/2$
was prepared by reaction of 264 with dimethylmagnesium and reacted

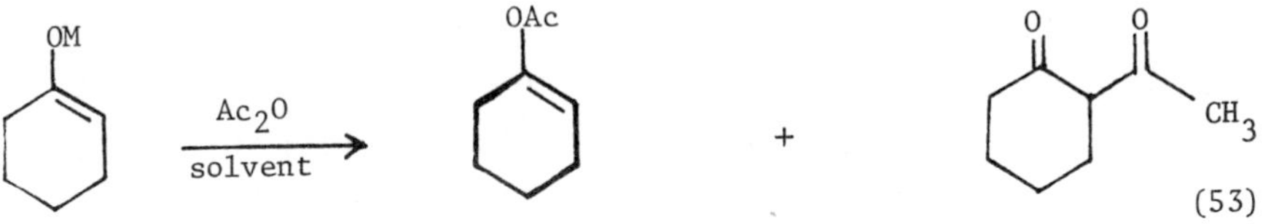

(53)

M^+/solvent:Li^+/DME	49%	16%
:Mg^{+2}/2/Et_2O	25%	43%

with acetic anhydride in ether, the percentages of O- (*264*) and C-alkylated (*265*) materials obtained were 25% and 43%, respectively. A very dramatic solvent effect on the O-acylation/C-acylation ratio has been noted in the reaction of the bromomagnesium 1-enolate of 3,3-dimethylcyclohexanone (*52A*, M = MgBr) with acetyl chloride. In DME, only the O-acylation product was obtained in 63% yield, but in ether a 62:38 distribution of C- and O-acylated materials was obtained in 39-53% yield [210].

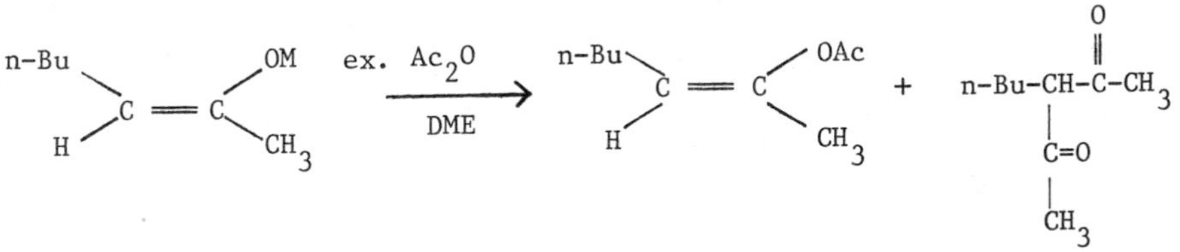

27B (M = Li)

72-75%

7-4%

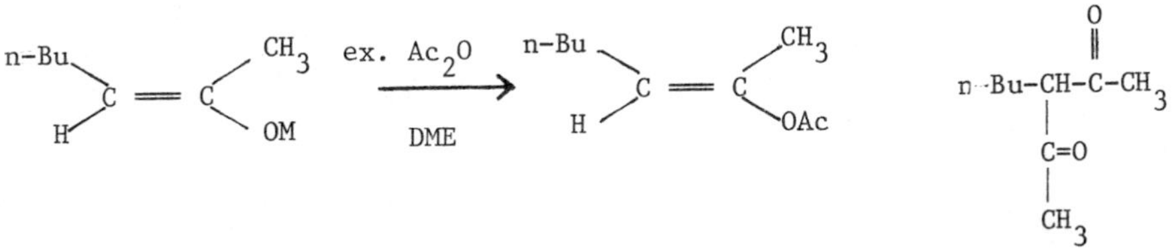

27C (M = Li)

24-38%

43-28%

As illustrated by the results of acetylation of the internal
Z- (27B, M = Li) and E-enolates (27C, M = Li) of 2-heptanone (27)
with acetic anhydride in DME, the O-acylation/C-acylation ratio is
normally larger for Z than for E geometric isomers [117]. These
results as well as spectroscopic studies indicate that with the same
cation and solvent, Z-enolates, such as 27B, exist more in the form
of solvent-separated ion pairs than the corresponding E species.
This would be expected if the interaction of the solvated oxygen
anion is less than that of the solvated tight ion pair with the
adjacent alkyl group on the double bond in the Z isomer [117, 118].

It has been suggested that the preference for C-acylation of
metal enolates as tight ion pairs may result from shielding of the
oxygen atom by the metal or from the involvement of a transition
state, such as 266, in which the metal cation acts as an electrophilic
catalyst [117]. The extent of C-acylation normally increases as the

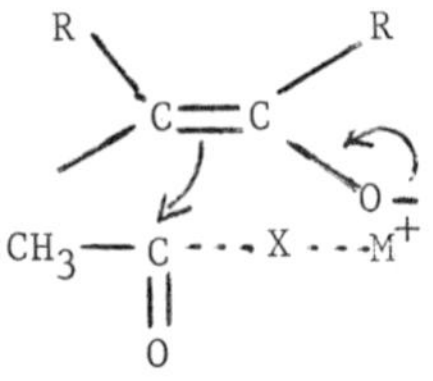

266

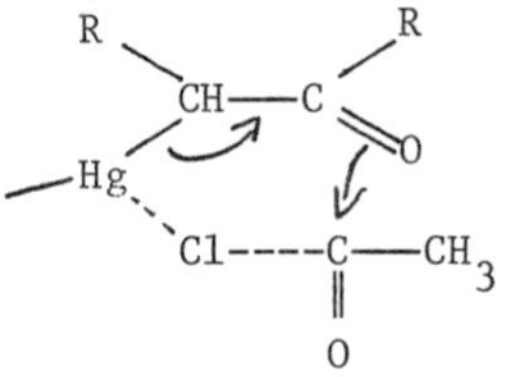

267

acetylating agent is changed from acetic anhydride to acetyl chloride
(or bromide). A transition state of the type 266 would possibly
account for this result, since the formation of a metal halide would
be expected to be energetically more favorable than the formation of
a metal acetate [117].

α-Metalated ketones such as 2-mercuricyclohexanone normally
yield O-acylated products exclusively on reaction with acylating
agents such as acetyl chloride [117]. Either shielding of the α-
carbon atom by the metal atom or the involvement of transition states
such as 267, in which the mercury atom acts as an intramolecular
catalyst has been suggested as a possible explanation for these resul
[117].

A comparison of the regioselectivity of the less reactive
acylating agent methyl chloroformate with that of acetic anhydride
(or acid halides) is of interest [225]. For example, the lithium
enolate *36B* (M = Li), prepared by conjugate addition of lithium dimethyl cuprate to cyclohexenone, has been shown to yield exclusively
the O-acylation product *268* on reaction with excess acetic anhydride
[234] or acetyl chloride [232] in ether. However, under the same
conditions treatment of *36B* (M = Li) with methyl chloroformate gave
the enol carbonate *269* in 58% yield [225]; compound *269* resulted
from C-acylation, enolization (either by methyl copper or by the
unreacted enolate), and O-acylation of the 1,3-dicarbonyl anion. The

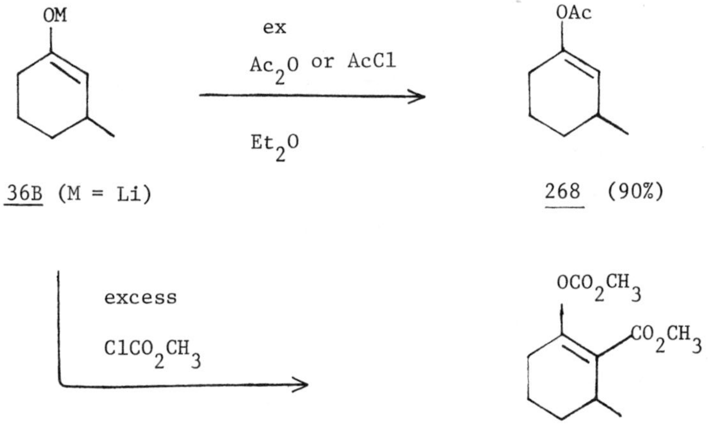

OM

ex

Ac$_2$O or AcCl
─────────────────→
Et$_2$O

36B (M = Li)

OAc

268   (90%)

excess

ClCO$_2$CH$_3$
───────────→

OCO$_2$CH$_3$

CO$_2$CH$_3$

269   (58%)

tendency of less reactive acylating agents to give greater amounts
of C-acylation parallels the situation in alkylation reactions in
which the less reactive (soft) alkylating agents react primarily on
carbon.

The extent of C-acylation with methyl chloroformate has been
shown to be subject to steric hindrance. Thus, while *269* was
obtained from *36B* (M = Li), the lithium 1-enolate of 3,5,5-
trimethylcyclohexanone gave a 7:2 ratio of O-acylation/C-acylation

products, and the lithium 1-enolate of 3,3,5,5-tetramethylcyclohexanone
gave only the O-acylation product under similar conditions. Also,
the enol carbonate 271 was the only product obtained on treatment of
the 1-decalone lithium enolate 270 with excess methyl chloroformate
in ether [400].

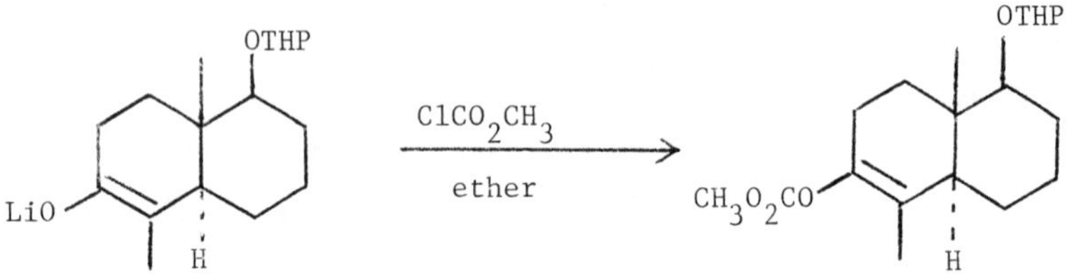

Treatment of the potassium enolate 54 (M = K) with ethyl chloroformate in DME gave the enol carbonate 272 in 39% yield as the only
product [117]. This result indicates that when conditions are favorable for solvent-separated ion pair formation, chloroformate esters
should be expected to give O-acylation products kinetically.

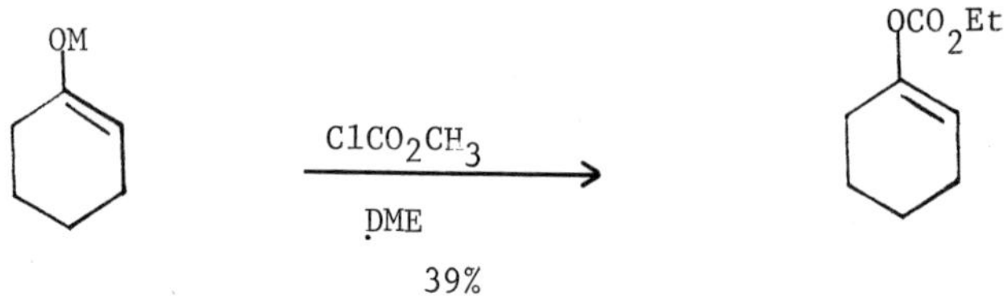

The use of metal enolates obtained by conjugate addition of
lithium di-n-butylcuprate, prepared from reaction of a tri-n-
butylphosphine-copper(I)-iodide complex and n-butyllithium, or other
complex lithium dialkylcuprates to cyclohexenones or cyclopentenones
apparently promotes kinetically controlled C-acylation with acid
chlorides [232]. For example, when the 1-enolate of 3-
butylcyclohexanone was prepared in this way and added to an excess
of acetyl chloride in ether-HMPA, 2-acetyl-3-butylcyclohexanone was

obtained from a 92% yield [232]. As already noted, the related
enolate *36B* (M = Li), prepared from lithium dimethylcuprate and
cyclohexenone, was O-acylated exclusively with acetyl chloride [232].
It has been suggested that the involvement of n-butylcopper(I)-lithium
enolate complexes may account for these C-acylation results [232].
However, the actual nature of the reactive intermediate and the
possible role of the tri-n-butylphosphine ligand in these reactions
have not been established.

The effect of changing the metal enolate/acylating agent ratio
on the course of the benzoylation of the sodium enolate of acetone
is shown in Eq. (54) [401]. These results show that it is possible
to obtain good yields of C-acylated products by reacting acid halides

$$CH_3-CO-CH_3 \xrightarrow[\text{Et}_2O]{NaNH_2} \overset{\overset{ONa}{|}}{CH_3-C=CH_2} \xrightarrow[\text{Et}_2O]{n \ \ eq \ C_6H_5COCl} \xrightarrow{H_2O}$$

(54)

$$\overset{\overset{\overset{O}{\|}}{O-C-C_6H_5}}{\underset{CH_3-C=CH_2}{|}} \quad + \quad \overset{\overset{O}{\|} \quad \overset{O}{\|}}{CH_3-C-CH_2-C-C_6H_5}$$

0.5 equiv. added to enolate	9%	33%
enolate added to 2 equiv.	41%	6%

or anhydrides with an excess of a metal enolate. This is because
kinetically formed enol derivatives (O-acylation products) are
capable of C-acylating ketone enolates [41, 43, 393, 402]. When
preparing C-acylation products by this procedure, 3 eq of metal
enolate are normally employed per equivalent of acylating agent. One
equivalent of the metal enolate is consumed in the reaction with the
product of the C-acylation reaction, i.e., a 1,3-dicarbonyl compound,
and the high enolate/acylating agent ratio prevents further acylation
of the enolate of the 1,3-dicarbonyl compound [402].

2-Benzoylcyclopentanone (273) has been prepared in 53% yield by
reaction of the sodium enolate of cyclopentanone with 1/3 eq of
benzoyl chloride in ether at 0°C [402]. Attempted C-acylations* of
cyclopentanone under the usual condition of the Claisen reaction
gave poor results because of the strong propensity of this ketone
toward self-condensation [58].

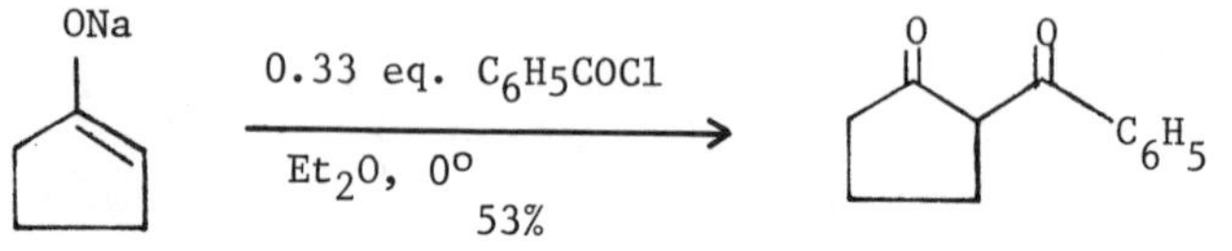

273

Ketenes have been employed infrequently as acylating agents
[117, 245, 393]. These reagents appear to offer no synthetic advantages over acid anhydrides and acid halides for the acylation of
alkali metal enolates, since they also yield predominantly O-acylation
products in relatively polar solvents [117, 303]. As expected,
because of the covalent nature of the O-Al bond, the Z isomer of
dimethylaluminum-4,4-dimethylpent-2-en-2-olate undergoes C-acylation
on reaction with diphenylketene in toluene at 0°C [245].

Metal enolates undergo reaction with carbon dioxide (carboxylation) to yield carboxylate salts of β-keto acids [28, 39, 60, 132,
193, 400, 403, 404]. Rates of carboxylation reactions are fast
compared with the rates of proton transfer between the unreacted enolate and β-keto acid salt product. Thus under proper experimental
conditions carboxylation reactions may be used to determine compositions of kinetic or thermodynamic mixtures of metal enolates of
unsymmetrical ketones [28]. However, enolate mixture compositions
are not usually determined in this way because the acetic anhydride
or trimethylsilyl chloride quenching methods give higher product
yields and are easier to carry out experimentally. There is a
possibility that lithium enolates may be partially carbonated on
oxygen, leading to enol carbonate salts which undergo decomposition
to the starting ketone and carbon dioxide on work-up [69].

Generally carboxylations are performed by treating the ketone enolate in an inert solvent (usually diethyl ether) with a large excess of crushed solid carbon dioxide [39, 403, 404]. Some workers have used the practice of passing carbon dioxide gas into enolate solutions [403]. However, this procedure is not recommended if regio-specific trapping of unstable enolates is desired because if the carboxylation process is slow there is a distinct possibility that enolate equilibration may compete with it. β-Keto acids and their salts undergo decarboxylation very easily. Thus carboxylation products are normally isolated as the corresponding methyl esters, which may be obtained by careful neutralization of carboxylate salts followed by esterification with ethereal diazomethane [39, 400, 403, 404].

It is of critical importance that dry solid carbon dioxide be employed for carbonation reactions. Thus dry ice should be freshly cut and then crushed in an inert atmosphere. The solid should be placed in a cloth bag inside a dry plastic bag while it is being crushed [403, 404], or, ideally, the crushing process and other manipulations could be conducted in an inert atmosphere glove box.

A number of carbomethoxylations (carboxylation plus esterifica-tion) of reductively formed lithium 1(2)-enolates of trans-2-decalone derivatives and related lithium 3-enolates of 5α-3-keto steroids have been reported [132, 404]. These enolates are, of course, kinetically unstable with respect to their structural isomers. As an example of this procedure, the octalone 274 was reduced with lithium in liquid ammonia to the lithium enolate 275. After replacement of the ammonia by diethyl ether, treatment of the enolate with a large excess of solid carbon dioxide at low temperature, acidification, and reac-tion with ethereal diazomethane, the β-keto ester 276 was isolated in 68% yield. This compound proved to be a useful intermediate for the synthesis of diterpenoid products related to abietic acid [403].

Lithium enolates bearing methyl substituents at the α-carbon atom have been successfully carboxylated [193]. For example, a 56% yield of 2-carbomethoxy-2-methylcyclohexanone (277) was obtained upon carbonation of the enolate 5B (obtained by lithium ammonia reduction of 2-n-butylthiomethylenecyclohexanone) in ether followed by acidification and esterification [193]. However, attempted carbomethoxylation of the methyl-substituted lithium enolate 270 gave only a poor yield of the corresponding α-carbomethylation

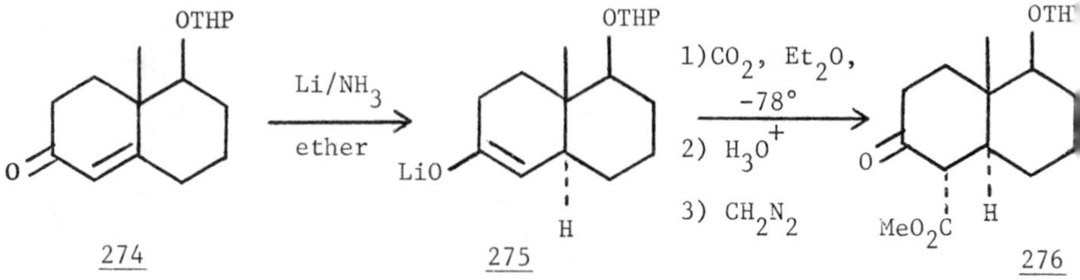

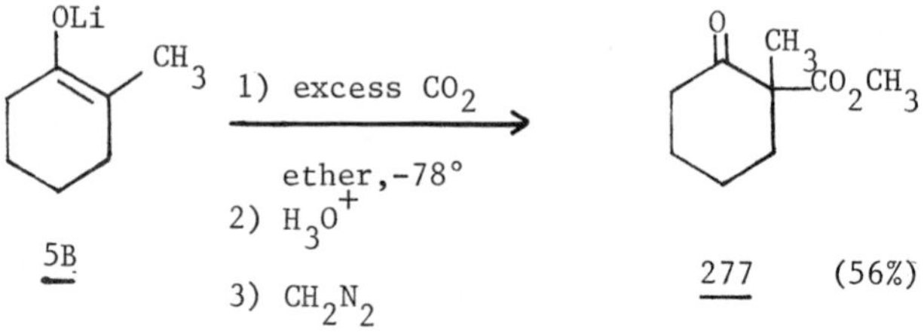

product [400]. It has been suggested that steric crowding of the axial product may have caused decarboxylation to be more rapid than trapping with diazomethane [400].

The hindered base lithium 4-methyl-2,6-di-t-butylphenoxide can be generated by reaction of the corresponding phenol with n-butyllithium in ether at -78°C under an argon atmosphere [340]. This base reacts slowly with carbon dioxide and is especially effective for use in base-promoted carboxylations of symmetrical ketones or ketones which may undergo enolization in one direction only. For example, reaction of 4-t-butylcyclohexanone with this base in ether under an atmosphere of carbon dioxide for 16 hr gave, after careful acidification, 2-carboxy-4-t-butylcyclohexanone in 89% yield [340].

The well-known condensation of ketones with methylmagnesium carbonate (Stiles Reagent) in DMF followed by acidification also provides an excellent method for the synthesis of β-keto acids

[337-339]. When unsymmetrical ketones are used, the reaction normally leads to the β-keto acid, which can form the more thermodynamically stable magnesium enolate.

C-Cyanogenations and O-phosphorylations of metal enolates are reactions related to the acylations discussed above. Reactions of regiospecifically generated lithium enolates with cyanogen chloride in benzene have been used to produce the corresponding α-cyano ketones in moderate yields [399]. For example, the tricyclic lithium enolate 279, which was prepared by reduction of the enone 278 with lithium in liquid ammonia, gave a 42% yield of the α-cyano ketone 280 on treatment with cyanogen chloride in benzene [399]. (O-Cyanogenation of 279 apparently occurred when THF was used as the solvent [399].) Somewhat lower yields of related products were obtained when lithium 1(2)-enolates of 2-decalone derivatives were treated in a similar manner [399].

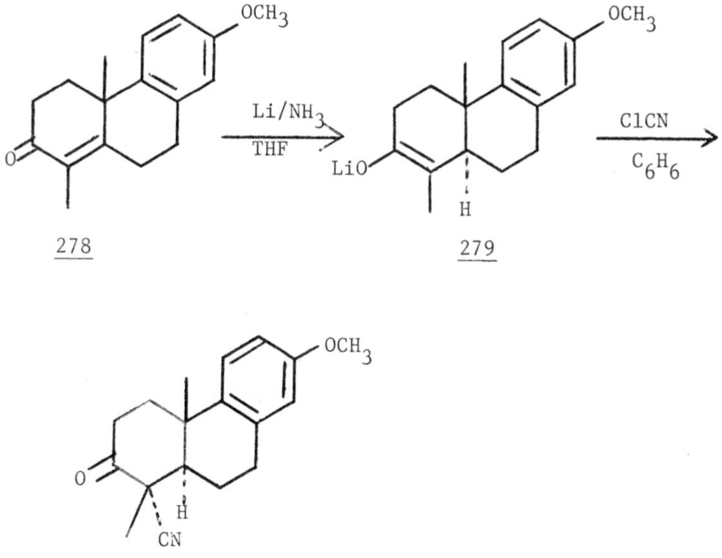

Diethylphosphorochloridate in ether [218, 219] or N,N,N',N'-tetramethyldiamidophosphorochloridate in 4:1 DME- (or THF-) N,N,N',N'-tetramethylethylenediamine [398] regiospecifically traps

lithium enolates to produce the corresponding enol phosphates in
high yield. These enol phosphates may be reduced to the corresponding
olefins by reduction with excess lithium in ethylamine-THF mixtures
containing 2-4 eq of t-butyl alcohol [218, 219, 398]. Lithium-ammonia
reductions of α,β-unsaturated ketones [132] or conjugate additions
of lithium dialkylcuprates have been used to generate the desired
specific lithium enolates. An application is shown in the conversion
of cholest-4-en-3-one (281) to 5-methylcoprost-3-ene (282) [219].

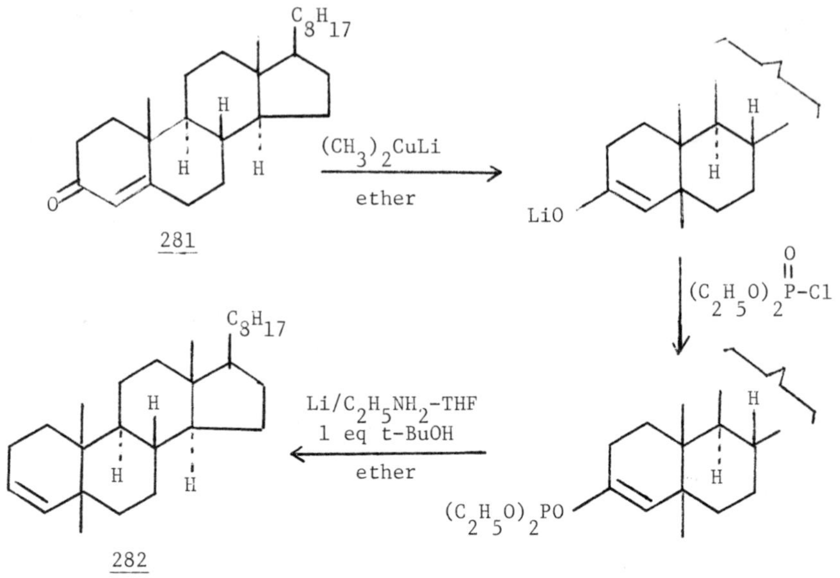

2. Experimental Procedures

Caution: All of the following procedures should be performed
under an anhydrous, oxygen-free nitrogen atmosphere. All reagents
and solvents should be anhydrous, and all transfers should be carried
out with a hypodermic syringe or with a flask-to-flask cannular
arrangement. While the enolate solutions prepared as described
as follows may be stable under an inert atmosphere at low temperatures for extended time periods, it is recommended that they be
prepared immediately before use, whenever possible.

a. *Carbomethoxylation of a lithium enolate of a saturated ketone [400].* The lithium enolate (0.1 mol) in 500 ml of anhydrous ether is cooled to -78°C in a dry ice-acetone bath. Finally powdered, dry solid carbon dioxide (200 g) (prepared in a plastic bag) is then added to the reaction mixture through a powder funnel encased in a larger plastic bag. The reaction mixture is stirred vigorously during the addition of the carbon dioxide. The dry ice-acetone bath is removed, the mixture is stirred for 30 min, brought to room temperature with a water bath, and stirred for an additional 30 min. It is then cooled again to -78°C; dry ice (500 g) is added followed by 500 ml of cold water. The ether layer is separated; the aqueous layer is mixed with 500 ml of cold ether and carefully acidified with a mixture of 50 ml of concentrated hydrochloric acid and 50 g of ice. The ether layer is again separated, and the aqueous layer is extracted with two 250-ml portions of ether. The combined ethereal extracts are cooled, washed with two 250-ml portions of cold saturated aqueous sodium chloride, and filtered into an excess of ethereal diazomethane at room temperature. After 30 min enough acetic acid is added to dispel *partially* the yellow color of the excess diazomethane, and the solvent is removed under reduced pressure to yield the crude α-carbomethoxy ketone.

b. *C-acylation of a sodium enolate of a saturated ketone with an acid chloride [402].* A suspension of 0.3 mol of sodium enolate (prepared by treatment of a ketone with 1 eq of sodium amide in anhydrous ether for about 30 min) in 400 ml of ether is cooled in an ice bath, and 0.10 mol of an acid chloride in 50 ml of ether is added as rapidly as possible with vigorous stirring. Stirring is continued for 5 min in the ice bath and for 15 min with the ice bath removed. The reaction mixture is added carefully to a mixture of 27 ml of concentrated hydrochloric acid and 100 g of crushed ice. The ether layer is separated, the aqueous layer is extracted with ether, and the combined ethereal extracts are washed with a saturated aqueous solution of sodium bicarbonate until the evolution of carbon dioxide

ceases. The ether solution is washed with a saturated solution of
sodium chloride, dried, and concentrated to yield the crude C-
acylated ketone.

F.   *Directed Aldol Condensations via Preformed*
     Metal Enolates of Saturated Ketones

   1.   General

   Traditionally base-catalyzed aldol condensations of saturated
ketones with aldehydes or other ketones have been conducted in protic
solvents under equilibrating conditions.   Such conditions often produce low yields because aldol products are frequently less stable
than the reactants and mixtures of products may be obtained if more
than one mode of condensation is possible [9, 10].   These complica-
tions and possible methods of avoiding them are discussed in Chap. 1.
The discussion here will be confined to examples of directed aldol
condensations in which preformed metal enolates are reacted with
aldehydes or ketones in aprotic solvents.

   2.   Synthetic Applications of Directed
        Aldol Condensations

   Formaldehyde, higher aldehydes, and many ketones react rapidly
with metal enolates of saturated ketones in aprotic solvents to
yield metal chelates of β-hydroxy ketones such as *283*.   Although the
reaction is usually reversible, it is possible to choose conditions
(metal with strong chelating ability and nonpolar aprotic solvents)
which make dissociation of the chelate *283* relatively unfavorable
[151].   Thus high yields of the aldol products *284* may be obtained
by treatment of the chelate with mildly acidic solutions such as
aqueous ammonium chloride.   These directed intermolecular aldol
condensations have considerable preparative value because they make
possible the regiospecific hydroxyalkylations of metal enolates.   In
addition, many of the undesired transformations (such as retroaldol
cleavage and the formation of α,β-unsaturated ketones) which diminish
the value of the classical aldol condensation are avoided.

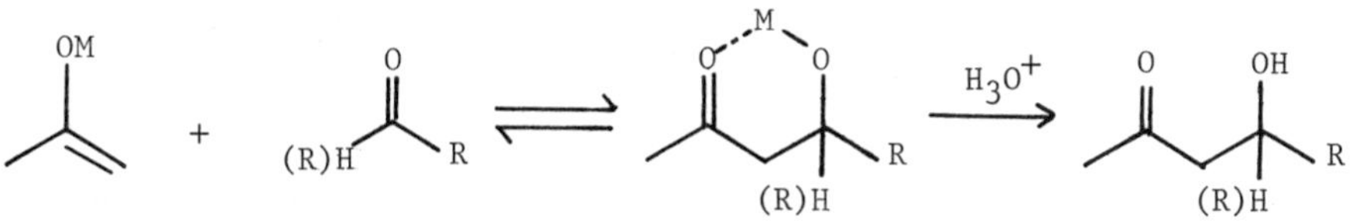

283   284

A number of studies involving directed aldol condensations have
involved the use of preformed lithium enolates, which may be readily
generated by one of the methods described in Secs. II.B and II.C
[81, 151, 154, 212, 221, 229, 253, 352, 405, 406]. Lithium itself
can serve as the chelating metal cation in species such as 283.
Therefore, these enolates may be treated directly with the desired
carbonyl component at low temperatures (usually -20°C or below) in
solvents such as THF, DME, or, preferably, diethyl ether [151]. For
example, the unstable lithium enolate 8A when prepared by cleavage
of the corresponding trimethylsilyl enol ether with methyl lithium,
gave 1-hydroxymethyl-trans-2-decalone (285) in 95% yield on reaction
with dry, gaseous formaldehyde in ether at -78°C [212]. The availability of 1-hydroxymethyl ketones such as 285 is of considerable synthetic significance. They are readily converted to α-methylene ketone
which may undergo Michael additions with electrophilic reagents to
yield a variety of α-substituted ketones [212, 253, 316].

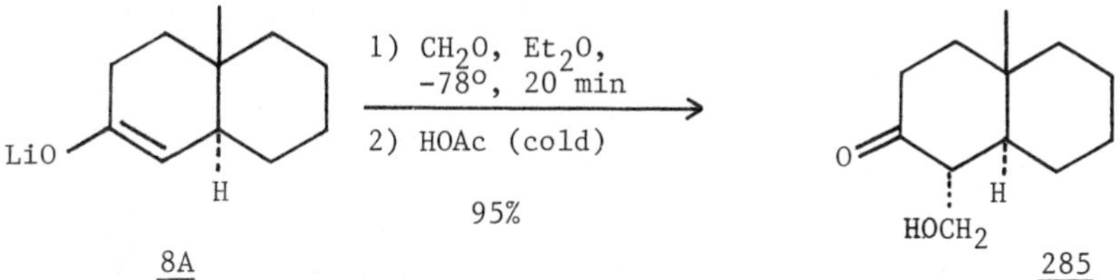

1) CH_2O, Et_2O,
   -78°, 20 min

2) HOAc (cold)

95%

8A   285

The high electrophilicity of aldehydes and ketones makes possible
the trapping of kinetically formed terminal lithium enolates of methyl
ketones [81, 406, 407]. Deprotonation of methyl ketones with bases
such as LDA in THF [406], lithium diethylamide in benzene containing
1 eq of HMPA [81, 407], or other lithium dialkylamides in various
solvents [81] allows the production of these enolates, and the carbonyl

component is then added in THF or ether at low temperature. Successful trapping of the terminal lithium enolate of 2-pentanone (*286*)
with n-butyraldehyde [406], benzaldehyde [406], and acetone [407] to
yield β-hydroxy ketones such as *287* provides examples of these reactions. It will be recalled that the low reactivity and ease of
equilibration of terminal lithium enolates of methyl ketones preclude
the regiospecific alkylation of these species [44].

$$
\underset{286}{CH_3CH_2CH_2\overset{\overset{\displaystyle OLi}{|}}{C}{=}CH_2} \quad \xrightarrow[\text{2) } H^+]{\text{1) } R_1\text{-}\overset{\overset{\displaystyle O}{\|}}{C}\text{-}R_2,\ \text{conditions}} \quad \underset{287}{CH_3CH_2CH_2\text{-}\overset{\overset{\displaystyle O}{\|}}{C}\text{-}CH_2\text{-}\overset{\overset{\displaystyle OH}{|}}{\underset{\underset{\displaystyle R_2}{|}}{C}}\text{-}R_1}
$$

R_1	R_2	conditions	
H	C_3H_7	THF, -78°, 15 min	65%
H	C_6H_5	THF, -78°, 15 min	75-80%
CH_3	CH_3	1 eq HMPA, C_6H_6, Et$_2$O, -70°, 15 min	40%

Divalent metal cations, such as Zn^{2+} and Mg^{2+} and trivalent
cations, such as Al^{3+}, form more stable chelates, such as *283*, than
lithium cations do. Thus it is often experimentally convenient to
add an anhydrous salt containing such cations to a solution of a
preformed lithium enolate prior to the addition of the carbonyl component. Magnesium bromide and aluminium chloride have been employed,
but better results are normally obtained when zinc chloride is used
as the additive [151]. When this salt is used reaction temperatures
in the range of 0° to -20°C may be employed, and good yields of aldol
products have been obtained even in solvents of medium polarity such
as THF or DME [151]. The optimum quantity of zinc chloride which
should be added is that amount required to convert all alkoxide bases
present to zinc alkoxide [$Zn(OR)_2$] derivatives; that is, 0.5 eq of
zinc chloride should be used with a "clean" lithium enolate, but 1 eq

should be employed when an equivalent of lithium alkoxide is present in addition to the enolate [151].

Procedures involving the addition of zinc chloride to lithium enolates have been used to prepare α-hydroxyalkyl acyclic and cyclic ketones of a variety of structural types. The synthesis of threo-4-hydroxy-3-phenyl-2-heptanone (288) by condensation of the lithium Z-enolate 42B of phenylacetone with n-butyraldehyde provides an example of the synthesis of an acyclic system [151, 405].

Regiospecific syntheses of α-hydroxyethyl derivatives of acyclic and cyclic ketones have been carried out by the conjugate addition of lithium dimethylcuprate to the appropriate α,β-unsaturated ketone in ether followed by addition of anhydrous zinc chloride and an excess of acetaldehyde at 0°C [231]. Formaldehyde and other aliphatic aldehydes could be used in similar reactions, but attempted condensations

$$
\underset{\underset{42B}{\text{H}_3\text{C}}}{\overset{\text{LiO}}{\diagdown}}\text{C}=\text{C}\underset{\text{H}}{\overset{\text{C}_6\text{H}_5}{\diagup}}
\quad
\begin{array}{l}
\text{1) ZnCl}_2,\ \text{Et}_2\text{O} \\
\text{2) n--C}_3\text{H}_7\text{CHO, 0--10}^\circ \\
\text{3) NH}_4\text{Cl, H}_2\text{O, 0}^\circ
\end{array}
\longrightarrow
\quad
\underset{\underset{288}{\text{H}_5\text{C}_6\ \ \text{H}}}{\overset{\overset{\text{O}\quad\ \text{OH}}{\|\quad\ \ |}}{\text{CH}_3\text{-C-CH-C-CH}_2\text{CH}_2\text{CH}_3}}
$$

with benzaldehyde and acetone were reported to be unsuccessful [231]. The 3,4-disubstituted cyclopentanone enolate 289 has been condensed with formaldehyde to give the 2-hydroxymethylcyclopentanone derivative 290 in 80-90% yield using zinc chloride as an additive [253]. The enolate 289 was prepared by cleavage of the corresponding diphenyl enol phosphinate with t-butyllithium in 2:1 ether-THF at -78°C. Compound 290 was employed as a key intermediate in a recent total synthesis of prostaglandin $F_{2\alpha}$ [253].

Naturally, it is possible to directly prepare and utilize di- or trivalent metal enolates in aldol condensations. Condensations of bromozinc [199], dimethylaluminum [245], and halomagnesium enolates [10, 112, 113, 202, 210, 212, 214] with various aldehydes and ketones have been reported. For example, the iodomagnesium 1-enolate of 3,3-dimethylcyclohexanone, prepared by copper(I)-catalyzed addition of

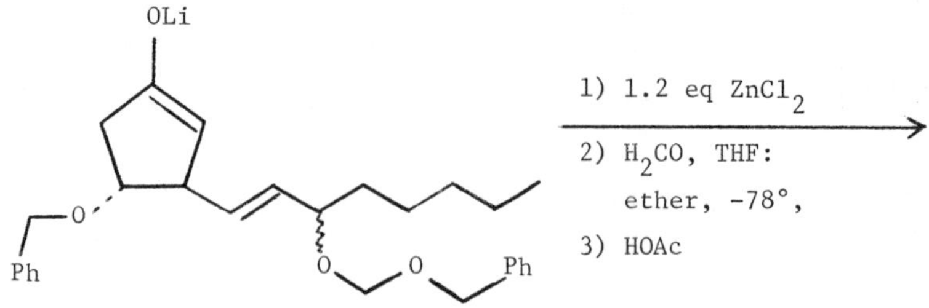

$$289 \xrightarrow{\begin{array}{l} 1) \ 1.2 \ eq \ ZnCl_2 \\ 2) \ H_2CO, \ THF: \\ \quad ether, \ -78°, \\ 3) \ HOAc \end{array}}$$

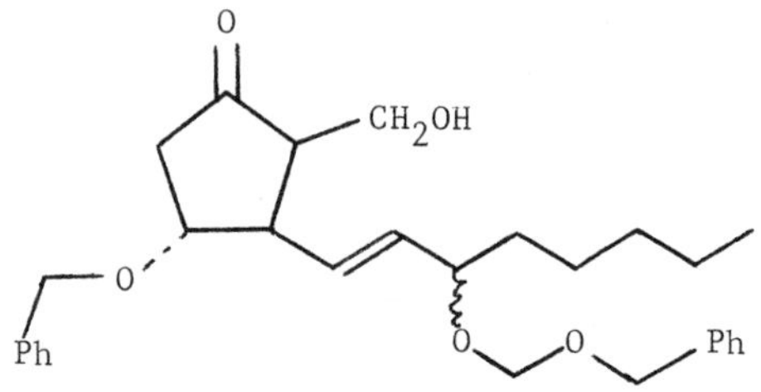

290   80-91%

methylmagnesium iodide to 3-methylcyclohex-2-enone, has been reacted
with acetaldehyde, crotonaldehyde, and citral to give the corresponding 2-hydroxyalkyl-3,3-dimethylcyclohexanones (291) in 75, 90, and 95%
yields, respectively [Eq. (55)] [210].

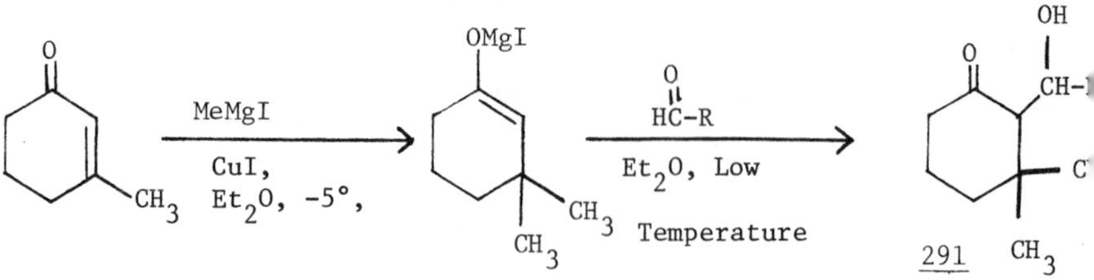

R	
$-CH_3$	75%
$-CH = CH-CH_3$	90%
$-CH = C(CH_3)(CH_2)_2CH = C(CH_3)$	95%

Regiospecific intramolecular aldol condensations have been carried out by conjugate addition of lithium dimethylcuprate to

ζ-oxo-α,β-unsaturated ketones [229]. The synthesis of the spiro
hydroxy ketone *293* from the keto enone *292* provides an example of
this type of reaction [229]. Lithium dimethylcuprate addition to
an α,β-unsaturated ketone followed by intramolecular aldol condensations of the enolate intermediate has also been used in a synthesis
of longifolene [93].

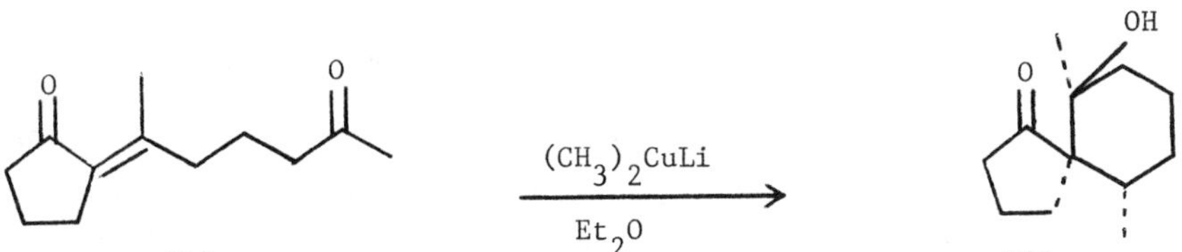

292     $\xrightarrow[\text{Et}_2\text{O}]{(\text{CH}_3)_2\text{CuLi}}$ 293

Recently it has been shown that treatment of 1:1 mixtures of
methyl ketones and aldehydes with 1 eq of the hindered base lithium
1,1-bis(trimethylsilyl)-3-methylbutoxide in THF at -40°C leads to a
regiospecific aldol condensation involving the less substituted
enolate of the ketone and the aldehyde carbonyl function [408].

Preformed metal enolates may react with α,β-unsaturated ketones
in aprotic solvents to yield kinetically favored aldol or thermodynamically favored Michael adducts [68, 318, 319]. The type of product
actually isolated seems to depend upon the degree of substitution at
the α-carbon atom of the metal enolate. As noted earlier, fully
substituted metal enolates such as *5B* yield Michael adducts exclusively upon reaction with MVK at low temperatures. On the other hand,
the reaction of the lithium enolate of methyl t-butyl ketone with
cyclohex-2-enone, or of the lithium, chlorozinc, or bromomagnesium
enolates of methyl-t-butyl ketone with MVK in ether at -50°C, gave
aldol adducts in good yield upon acidification [68, 318]. Moreover,
Michael adducts of these enones could not be isolated on increasing
the temperature. The reaction of the monosubstituted bromomagnesium
enolate of ethyl t-butyl ketone with chalcone provides an example of
an intermediate case [319]. In ether at 20°C a 60:40 mixture of the

aldol adduct and the Michael adduct was isolated upon quenching after 0.5 min, but the Michael adduct was obtained exclusively when the reaction time was extended to 3 hr before quenching. Interestingly, lithium enolates of esters give 1,2-adducts as the major products when reacted with cyclohex-2-enone in THF at -78°C, but conjugate addition products are produced in high yield when the temperature is increased to 25°C [409].

3. The Stereochemistry of Directed
   Aldol Condensations

Metal enolates such as *294* are capable of yielding diastereomeric products in directed aldol condensations with aldehydes and unsymmetrical ketones [151, 155, 156, 245, 405, 410]. Threo- and/or erythro-chelate intermediates such as *295* and *296* may be formed. These species are converted into the corresponding threo- or erythro-β-hydroxy ketones upon acidification. (The alternative chair conformations corresponding to *295* and *296* may also be present in the equilibrium.)

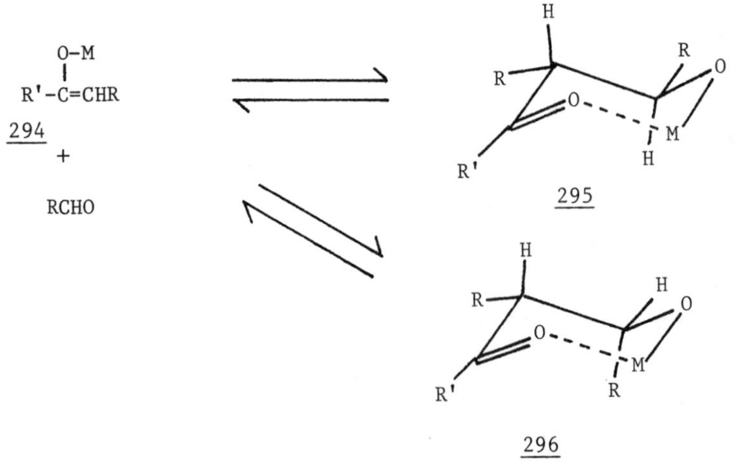

The carbon-carbon bond-formation step in directed aldol condensations is extremely fast. Reactions of lithium enolates of ethyl ketones with benzaldehyde in THF or ether are complete in seconds at -70°C [155]. However, equilibration among possible chelate intermediates is also a rapid process, even when chelates containing divalent metal cations such as Zn^{2+} are present in nonpolar solvents at temperatures of $0^{O}C$ or below [151].

Under equilibrium control, threo isomers are normally preferred because both alkyl substituents may occupy equatorial positions on the six-membered chelate ring as shown in *295*. Under such conditions the stereochemical composition of product mixtures appears to be independent of the geometry of the starting metal enolate [151]. The destruction of enolate geometry should be possible as a result of free rotation about the $OC-C_{\alpha}$ single bond in the β-alkoxy anionic species in equilibrium with chelate intermediates such as *295* or *296*.

The stereochemical outcome of kinetically controlled aldol condensations can be determined by using extremely short reaction times and very low temperatures [155] or by extrapolations of mixture composition-vs-time curves [156]. It has been shown that lithium E-enolates (*24B*, M = Li) of ethyl ketones such as *24* yield larger amounts of threo (*297*) than erythro isomers (*298*) under kinetic conditions [155, 156]. The stereoselectivity in favor of *297* increases significantly when large alkyl groups are present in the enolate *24B*. For example, although the enolate *24B* (M = Li, R = ethyl) yields a 52:48 mixture of *297a* and *298a* kinetically on reaction with pivaldehyde [156], enolate *24B* (M = Li, R = mesityl) gives a 92:8 mixture of the corresponding products *297b* and *298b* on reaction with benzaldehyde [155].

Kinetic aldol condensations are generally pictured in terms of transition states in which the carbonyl component approaches the α-carbon atom perpendicular to the plane of the enolate and the developing β-hydroxy ketone anion serves as a bidentate donor for the metal cation. For E-enolates such as *24B*, transition state *299* which leads to the threo metal chelate should be of lower energy than transition state *300* which leads to the erythro metal chelate. It also seems clear that the preference for *299* over *300* should increase as the size of the R group increases.

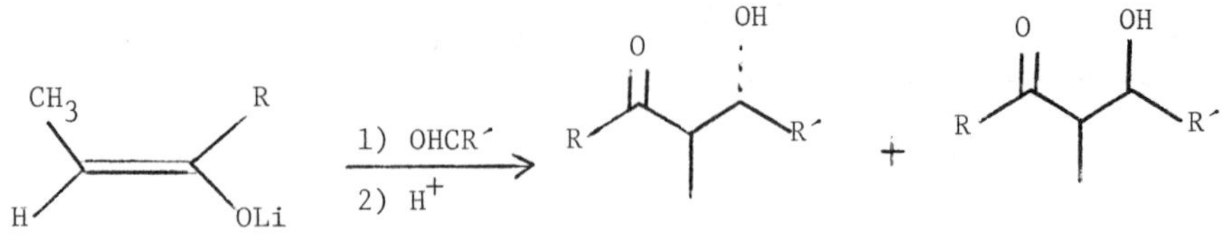

	R	R'		
			297	298
a;	C_2H_5	t–butyl	52%	48%
b;	mesityl	phenyl	92%	8%

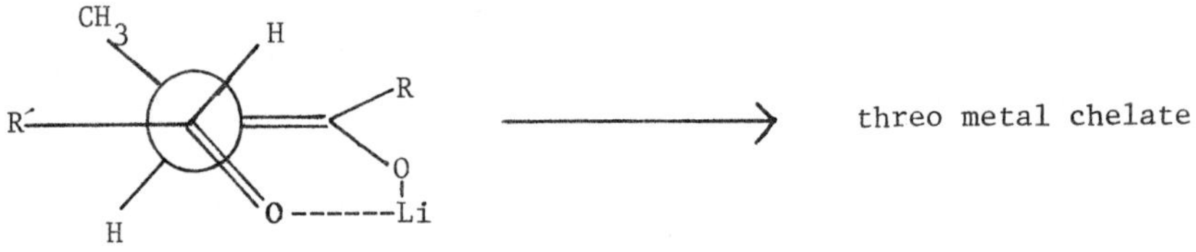

299

300

Bromomagnesium (24A, M = MgBr) [202] and lithium (24A, M = Li) [155] Z-enolates of ethyl t-butyl ketone (24b) give the erythro-aldol product 301 almost exclusively on reaction with benzaldehyde under kinetic control. Little, if any, threo isomer, 302, is obtained. Similar results have been obtained with lithium enolates related to 24A having other bulky alkyl groups [155]. With Z-enolates such as 24A (R = t-butyl), transition states such as 303 which lead to erythro chelates should be less sterically hindered than transition states such as 304 which yield threo isomers.

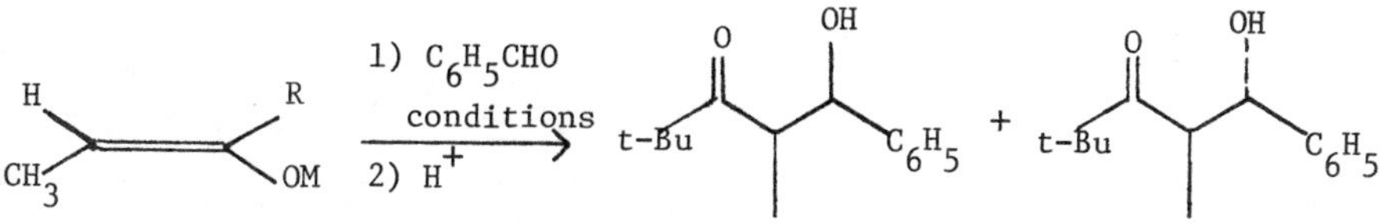

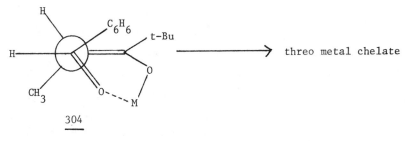

24A (R = t-butyl)		301	302
M	conditions		
Li	ether, -70°	100%	-
MgBr	ether	>95%	<5%

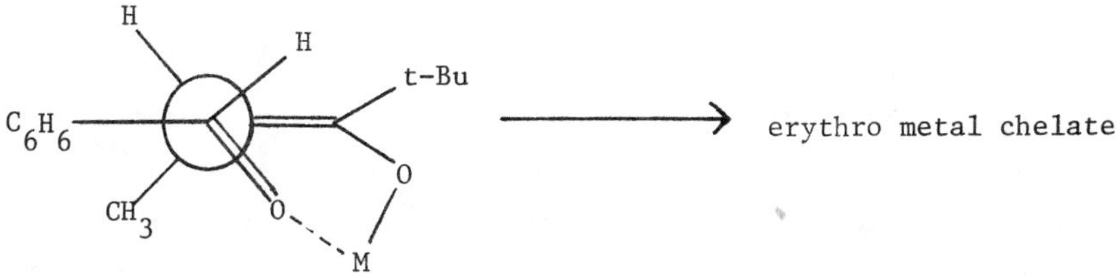

303 → erythro metal chelate

304 → threo metal chelate

The tetraalkylammonium Z-enolate *24A* (M = R_4N^+, R = t-butyl), prepared by reacting the corresponding silyl enol ether with a catalytic quantity of benzyltrimethylammonium fluoride in THF at 25°C, gave the threo-aldol product exclusively on reaction with benzaldehyde [155]. Since the quaternary ammonium cation cannot accept two negatively charged oxygens, it is believed that a transition state of the type *305* in which electrostatic repulsions are minimized should be favored [155].

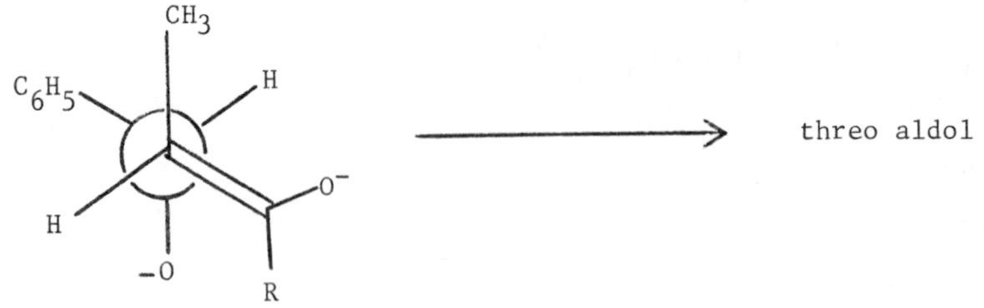

305

The results of condensations of the Z-dimethylaluminum 2-enolate
of 4,4-dimethyl-2-pentanone (*306*) with acetaldehyde or benzaldehyde
are contrary to those obtained for enolates such as *24A* (R = t-butyl)
in that the threo-dimethylaluminum chelate is formed exclusively
[245]. In this system the transition state *307*, which leads to the
threo chelate and has the bulky t-butyl group at the α-carbon atom
and the methyl or phenyl group of the aldehyde as far apart as
possible, is apparently strongly preferred over the alternative transition state *308*, which would yield the erythro species. The erythro
product is obtained using the E-enolate derived from *306* [245]

As the results of the condensations of acyclic enolates suggest,
enolates of cyclic ketones generally yield larger amounts of threo
products under kinetically controlled conditions [410]. Mixtures
containing both possible diastereomers are produced when enolates of
cyclopentanone [410] and cyclohexanone [151] are reacted with aldehydes under equilibrium conditions. When metal enolates such as *166*,
derived from 2-methyl-5-t-butylcyclohexanone, are reacted with

$$CH_3COCH_2\text{-}t\text{-}Bu$$

306

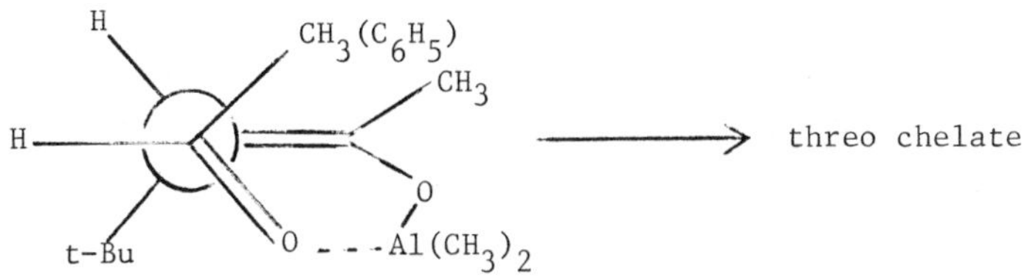

307

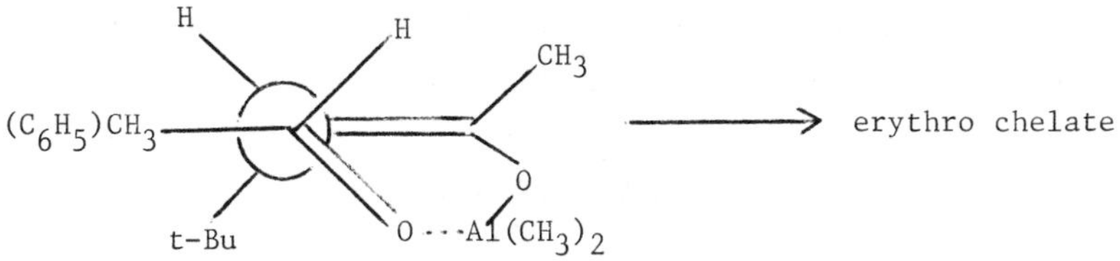

308

carbonyl compounds, four diastereomers, erythro and threo pairs,
derived from both axial and equatorial attack upon the enolate, are
produced under thermodynamic conditions [151].

4. Experimental Procedures

a. *Preparation of a threo-β-hydroxy ketone by condensation of
a zinc enolate with an aldehyde in ether-DME [151, 405].* A solution
of 0.2 mol of a lithium enolate (prepared from cleavage of the
corresponding enol acetate with methyllithium in DME) in 120 ml of
DME is cooled to $-10°C$, and 285 ml of an ether solution containing
0.202 mol of anhydrous zinc chloride is added dropwise with stirring
to the cold ($-10°$ to $10°C$) reaction mixture over 10 min. The mixture
is then stirred at $0°$ for 10 min, and 0.201 mol of a freshly distilled
aldehyde is added over 30 sec to the cold ($-5°$ to $10°C$) reaction mixture. The reaction mixture is stirred for 4 min at $0°-5°C$, and a
cold ($0°-5°C$) mixture of 500 ml of aqueous 4 M ammonium chloride and
200 ml of ether is added. The ether layer is separated, the aqueous
layer is extracted with ether, and the combined etheral extracts are

washed with aqueous 1 M ammonium chloride followed by saturated aqueous sodium chloride. The ether layer is dried and concentrated to yield a mixture containing mainly the crude threo-aldol condensation product.

   b. Aldol condensation of a thermodynamically unstable lithium enolate [406]. A solution of 1 eq of a carbonyl compound in THF is added slowly with stirring to a solution containing a 0.5 M solution of a thermodynamically unstable lithium enolate in THF at -78°C. The solution is stirred for 15-20 min at -78°C and quenched by addition of 1.1 eq of acetic acid at -78°C. The usual work-up procedure leads to the isolation of the crude β-hydroxy ketone.

G. *Reactions of Metal Enolates with Hetero*
   Atom Electrophiles

   1. Trialkylsilylation
   Alkali metal and divalent metal enolates of saturated ketones react with trialkylsilyl chlorides to produce the corresponding trialkylsilyl enol ethers in excellent yield [43, 46, 47, 87]. Trialkylsilyl halides are harder acids than the corresponding alkyl halides, and the reaction occurs exclusively on the oxygen atom of the ambient enolate anion [129]. Trimethylsilyl chloride has been by far the most widely used trialkylsilylating agent, but t-butyldimethylsilyl chloride has been used occasionally to prepare more hydrolytically stable silyl enol ethers [46, 162].
   As discussed previously, the trimethylsilylation of metal enolates provides a basic method for the determination of the compositions of kinetic and thermodynamic enolate mixtures of unsymmetrical ketones. The mixtures of isomeric trimethylsilyl enol ethers produced upon quenching of enolate mixtures are usually conveniently analyzable by chromotographic or spectroscopic methods. Kinetically unstable specific metal enolates produced by methods such as those described in Sec. II.C yield the corresponding trimethylsilyl enol ethers when quenched with trimethylsilyl chloride. Trimethylsilylations are usually conducted in ether, THF, or DME solution. Liquid

ammonia cannot be employed as a solvent because of its rapid reaction
with silyl halides. A procedure for the preparation of trimethylsilyl
enol ethers by the trapping of alkali metal enolates with excess trimethylsilyl chloride has been described in Sec. II.B.3. Trapping of
halomagnesium enolates in ether solutions can be carried out by using
a reaction time of about 1 hr and HMPA as an additive [46].

Procedures for cleaving trialkylsilyl enol ethers to the corresponding metal enolates with methyllithium in THF or DME, lithium
amide-liquid ammonia, or Grignard reagents in ethereal solvents have
been described in Sec. II.C.7. These compounds undergo reaction
with Ag_2O in DMSO to produce silver(I) enolate intermediates which
couple to produce 1,4-diketones [411] and yield alkylated ketones (by
way of quaternary ammonium enolates or pentavalent silicon species)
on reaction with alkylating agents in the presence of benzyltrimethylammonium fluoride (BTAF) [258]. Other important synthetic uses of
trialkylsilyl enol ethers include (a) aldol condensations with carbonyl compound activated by titanium tetrachloride [412], (b) the
preparation of trialkylsilyl cyclopropyl ethers, precursors to
methyl ketones, by reaction with the Simmons-Smith reagent [413],
(c) synthesis of α-halo ketones by reaction with bromine or chlorine
[414], (d) the synthesis of α-benzoyloxy ketones by reaction with
lead tetrabenzoate [415], (e) and the synthesis of Mannich bases by
reactions with dimethyl(methylene) ammonium iodide and further transformations [416].

2. Halogenation

Enolate anions are likely to be intermediates in the well-known
base-promoted halogenations of saturated ketones. However, direct
base-catalyzed halogenations are of little value for the synthesis
of monohaloketones [417]. Also, other methods of synthesis of α-halo
derivatives of unsymmetrical ketones have the drawback of requiring
extensive reaction sequences and/or not being entirely regiospecific
[234, 418]. The quenching of THF solutions of lithium enolates by
rapid addition of 1 eq of bromine in methylene chloride at -78°C is
an excellent method for the regiospecific synthesis of α-bromoketones

[234]. The procedure has been applied to the synthesis of several
2-bromocyclohexanone derivatives. For example, the more-substituted
lithium enolate *5B* of 2-methylcyclohexanone, prepared by cleavage of
the corresponding enol acetate with methyllithium, gave 2-bromo-2-
methylcyclohexanone in 92% when treated with 1 eq bromine in methylene
chloride. Likewise, 2-methyl-6-bromocyclohexanone (*59*) was obtained
in 92% yield when *5A*, prepared by cleavage of the corresponding
trimethylsilyl enol ether, was treated similarly [234]. Lithium
enolate bromination has been used to prepare 2-bromobicyclo[2.2.1]hexan
2-one which could not be obtained by more conventional methods [417].
The synthesis of the α-bromohydrendanone *311* in about 50% yield by
quenching a THF solution of the lithium enolate *310* with bromine in
methylene chloride at -78°C provides another example of this reaction
[419]. The enolate *310* was prepared by lithium-ammonia reduction of
the hydroxy enone *309*.

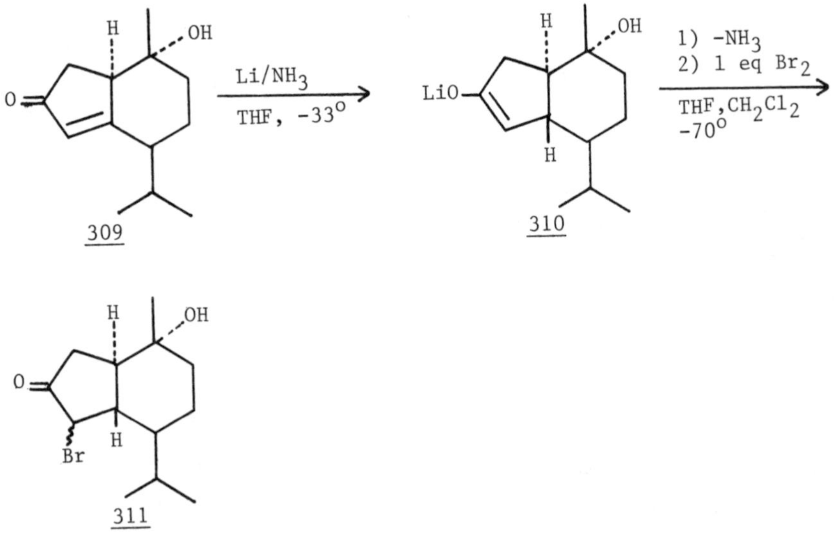

3. Sulfenylation and Selenenylation

In recent years there has been a great amount of interest in the
synthesis of α-phenylthio [147, 148, 182], α-alkylthio [147, 148, 182]
and α-phenylseleno ketones [171]. It has been noted previously

that secondary α-phenyl and α-alkylthio ketones may be alkylated at
the α position via the corresponding metal enolates and that specific
lithium enolates which may be further alkylated may be produced by
lithium-ammonia reduction of these compounds [176, 181]. Enolates
of α-phenylseleno ketones have apparently not been used in alkylation
reactions. However, since α-phenylseleno esters [185] and lactones
[186] readily undergo enolization and alkylation such reactions should
be possible. An important reason for interest in the synthesis of
α-phenylthio and α-alkylthio ketones as well as α-phenylseleno ketones
is that these compounds may be oxidized to the corresponding sulfoxides
[182] or selenoxides [171], which readily undergo thermal syn elimination to form α,β-unsaturated ketones. Various other reactions of
α-alkylthio and α-phenylthio ketones are also possible [182].

An important method of synthesis of α-sulfenylated ketones
involves reaction of preformed lithium enolates with appropriate
sulfenylating agents. Diphenyl disulfide [147, 182], phenyl sulfenyl
chloride [147, 148], and phenyl phenylthiosulfonate [182] have been
reacted with lithium enolates to form α-phenylthio ketones, while
dimethyl disulfide [147, 182] and methylsulfenyl chloride [147] have
been used to prepare α-methylthio ketones. Thiuram disulfide has also
been used as a thiolating agent [148].

In general, sulfenylation reactions have been conducted using
lithium enolates prepared by enolizations of ketones with LDA in THF.
(Obviously, enolates generated by various indirect methods could also
be used.) Diphenyl disulfide reacts readily with lithium enolate
solutions in THF or THF-HMPA mixtures at room temperature [182].
For sulfenylations at tertiary centers a 1:1:1 ketone/base/disulfide
ratio should be employed. For example, treatment of 2,6-
dimethylcyclohexanone (6) with 1 eq of LDA in THF at -78°C for 1 hr
and addition of the cold enolate solution to a slight excess of a
solution of diphenyl disulfide in THF at room temperature, with
stirring for 1.5 hr, gave 2,6-dimethyl-2-phenylthiocyclohexanone in
94% yield [182]. When sulfenylations are to be performed at secondary
centers, 1:2:1 or 1:2:2 ketone/base/disulfide ratios are recommended.

The second equivalent of base apparently deprotonates the kinetically
acidic sulfenylated product and prevents if from participating in
proton transfer reactions with the unreacted metal enolate [182].
Disulfenylated products may be obtained as by-products when 2 eq of
base and 2 eq of disulfide are employed, particularly when THF-HMPA
is the solvent.

Regiospecific sulfenylations of lithium enolates have been carried out [182]. On reaction of the kinetic mixture of lithium
enolates of 2-methylcyclohexanone (prepared by enolization of 5 with
LDA in THF) with diphenyldisulfide, partial enolate equilibration
occurred, and an 80:20 mixture of 2-methyl-6-phenylthiocyclohexanone
(312) and 2-methyl-2-phenylthiocyclohexanone (313) was obtained.
However, when the reactive sulfenylating agent, phenyl
phenylthiosulfonate, was substituted for diphenyl disulfide, the
regioselectivity in favor of 312 was increased to 97% [182]. In the
latter reaction the amount of 312 corresponds closely to the
kinetic selectivity for formation of the less-substituted enolate 5A.

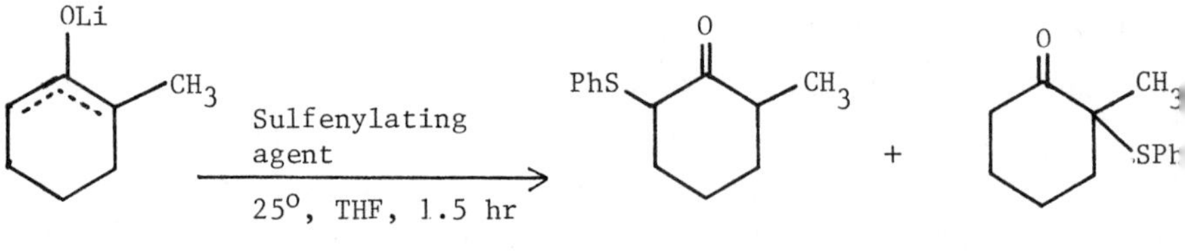

5A (97%)		312	313
+ 5B (3%)	Sulfenylating agent		
	Ph–S–S–Ph	70%	17%
	Ph–S–SO$_2$Ph	>82%	<3%

Phenylsulfenyl chloride is also a highly reactive sulfenylating
agent which has been shown to react smoothly with lithium enolates
in THF at -100°C [147].

Dimethyl disulfide reacts readily with lithium enolates of esters
in THF at room temperature, but it is not sufficiently reactive to

sulfenylate lithium enolates of saturated ketones under these conditions [182]. However, by employing 3:1 THF-HMPA as the solvent,
the enolate reactivity is sufficiently enhanced to permit reaction.
For example, 2-methylthiocyclohexanone was obtained in 75% yield
on reaction of the lithium enolate of cyclohexanone with dimethyl
disulfide under these conditions [182]. Several 2-methylthiocyclo-
alkanones including 2-methylthiocyclohexanone have been prepared,
but in rather low yield, by reaction of the corresponding lithium
enolates with methylsulfenyl chloride at -100°C [147]. As in
alkylation reactions, α-substituted enolates exhibit a preference
for axial sulfenylation [182].

The reaction of ketone enolates with phenylselenenyl bromide or
chloride provides a useful route to α-phenylseleno ketones [171].
Diphenyl diselenide is not useful for reactions with lithium enolates
of ketones (unfavorable equilibrium), but this reagent can be employed
for reactions with ester enolates and more reactive carbanions [171].

Ketone enolates for use in selenenylations have been generated
by deprotonation of ketones with LDA or potassium hydride and by
various indirect methods. Phenylselenenyl bromide is generally preferred over the chloride because it is easily prepared just before
use by reaction of diphenyl diselenide with bromine in THF at room
temperature. A quantity of selenenylating agent sufficient to react
with the total base present should be employed because lithium
alkoxides, amides, and organocopper reagents undergo selenenylation
just as do metal enolates.

A typical procedure for conversion of a ketone to the α-
phenylseleno derivative via the lithium enolate involves treatment
of the ketone with 1.1 eq of LDA at -78°C for 15 min followed by the
rapid addition of 1.2 eq of phenylselenenyl bromide in THF. (Phenylselenenyl bromide is generated in situ in the addition funnel by the
dropwise addition of 0.6 eq of bromine to 0.6 eq of diphenyl diselenide
in THF, followed by brief agitation to dissolve any $PhSBr_3$ which is
formed.) The reaction mixture is worked up by addition to a mixture
of 0.5 N HCl and a 1:1 ether-pentane solution. The product is isolated
in the usual way.

Regiospecific phenylselenenylations are possible. For example, kinetic deprotonation of 2-methylcyclohexanone with LDA in THF followed by reaction with phenylselenenyl bromide gave 2-methyl-6-phenylselenocyclohexanone in 89% yield [171]. The reaction has been employed to prepare α-phenylseleno derivatives of both cyclic and acyclic ketones.

4. Hydroxylation and Acyloxylation

α-Substituted metal enolates can be converted into α-ketohydroperoxide anions by reaction with molecular oxygen at temperatures in the -25° to -50°C range in dipolar solvents [137]. α-Ketohydroperoxide intermediates can be isolated by careful acidification and reduced with zinc dust to the corresponding α-hydroxy ketones [420]. However, it is preferable to oxidize the enolate to the hydroperoxide in the presence of triethylphosphite [421].

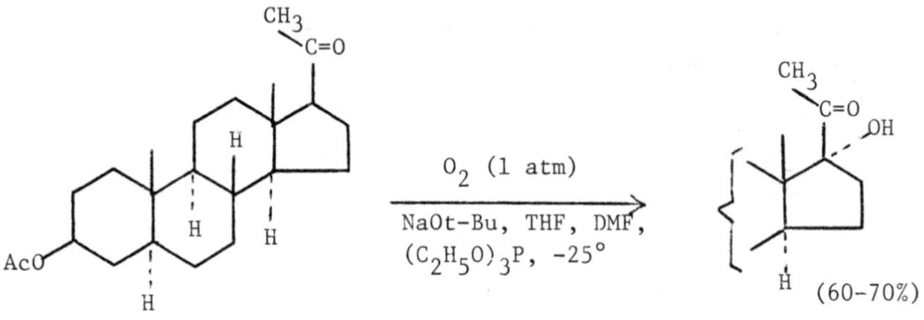

$$O_2 \ (1 \ atm)$$
$$\overrightarrow{NaOt\text{-}Bu, \ THF, \ DMF,}$$
$$(C_2H_5O)_3P, \ -25°$$

314   315

This method has been applied to the conversion of 20-keto steroids (314) to corresponding 17α-hydroxy derivatives (315). The triethylphosphite causes reduction of the α-ketohydroperoxide as it is formed, and possible rearrangement and α-cleavage reactions of this intermediate are minimized. These latter processes occur sometimes exclusively, or at least to a significant extent, when enolates are oxygenated at room temperature or above.

α-Hydroperoxy derivatives of methylene ketones readily undergo fragmentation to α-diketones. (In the absence of excess oxygen these compounds react with excess metal enolate to yield semidione radical anions [422].) Therefore, successful synthesis of α-hydroxy derivatives of methylene ketones by enolate oxygenation reactions have not been reported. Recently, hydroxylations of ketones having enolizable methine or methylene protons have been accomplished [172]. The procedure involves the reaction of preformed lithium enolates with molybdenum peroxide $MoO_5 \cdot Py \cdot HMPA$ (MoOPH) in THF at $-40°$ to $-70°C$. Ketone deprotonations were carried out with LDA in THF according to the usual procedures. Examples of the reaction include the conversion of 2-phenylcyclohexanone to trans-2-hydroxy-6-phenylcyclohexanone (70%), of isobutyrophenone to α-hydroxyisobutyrophenone (65%) of 3-β-tetrahydropyranyloxyandrost-5-en-17-one to 3β, 16α-dihydroxyandrost-5-en-17-one-3-tetrahydropyranyl ether (75%) [172].

Examples of acyloxylation of ketone enolates are rare. 2-Acetoxy-3,3,5,5-tetramethylcyclohexanone was obtained in 60% yield when the bromomagnesium enolate of 3,3,5,5-tetramethylcyclohexanone [prepared by copper(I)-catalyzed conjugate addition of methylmagnesium bromide to 3,5,5-trimethylcyclohex-2-enone] was treated with lead tetraacetate (LTA) in refluxing benzene [423]. It has also been reported that the enolate obtained by lithium ammonia reduction of 10-epi-eudesm-4,11-dien-3-one was trapped by LTA in benzene to give a mixture of 4β-acetoxy-10-epi-eudesm-11-en-3-one and the corresponding 4α-hydroxy compound in low yield [424]. However, the scope of this acetoxylation method has not been thoroughly explored. α-Benzoyloxy lactones have been prepared via reaction of the corresponding lactone lithium enolates with benzoyl peroxide in DME at about $5°C$ [425], but this reaction has apparently not been extended to the synthesis of α-benzoyloxyl ketones. Trans-1,2-diols have been prepared by the hydroboration-oxidation of lithium enolates [426].

5. Enolate Coupling Reactions

1,4-Diketones may be prepared in satisfactory yield by treatment of preformed lithium enolates with 1 eq of copper(II) chloride in THF-DMF at $-78°C$ [427]. Regiospecifically generated lithium enolates couple regiospecifically to produce the corresponding 1,4-diketones.

For example, the terminal enolate of methylisopropyl ketone, prepared
by cleavage of the corresponding silyl enol ether with methyllithium
in THF, gave the 1,4-diketone *316* in 78% yield when treated with
copper(II) chloride in THF-DMF. Mixtures of products are obtained

$$(CH_3)_2CH-\overset{\displaystyle OLi}{\underset{\displaystyle |}{C}}=CH_2 \xrightarrow[\substack{THF:DMF \\ -78°}]{CuCl_2} (CH_3)_2CH-\overset{\displaystyle O}{\overset{\|}{C}}-CH_2-CH_2-\overset{\displaystyle O}{\overset{\|}{C}}-CH(CH_3)$$

316

when metal enolates are produced by deprotonation of unsymmetrical
ketones, but unsymmetrical methyl ketones give primarily products
derived from coupling of the terminal enolates. Unsymmetrical 1,4-
diketones or γ-keto esters also have been prepared by cross-coupling
of enolates of different methyl ketones or of methyl ketone and
ester enolates [427].

III. ALKYLATION AND RELATED REACTIONS OF α,β-UNSATURATED KETONES VIA METAL ENOLATES

A. Base-Promoted Enolization of α,β-Unsaturated Ketones

α,β-Unsaturated ketones of the general structure *317* which contain both α' and γ protons are capable of undergoing enolization to
produce metal dienolates of the cross-conjugated *318*, or extended
319, type. There is a considerable amount of information to indicate
that α' protons of conjugated enones are kinetically more acidic than
γ protons. The first definitive evidence on this point was provided
by Malhotra and Ringold [428], who demonstrated that strong base-
catalyzed H-D exchange of steroidal 4-en-3-ones such as testosterone
takes place much more rapidly at the 2- than at the 6-position. [The
2β(axial) proton was also shown to undergo exchange much more rapidly
than the 2α(equatorial) proton.] Earlier work reported by Birch
[429] which indicated that treatment of cholest-4-en-3-one with
potassium amide in liquid ammonia gave the cross-conjugated 2,4-
dienolate supported these results.

$$\underset{\underline{317}}{\overset{\displaystyle \gamma|\ \ \overset{}{\beta}\ \ \overset{}{\alpha}\ \ \overset{O}{||}\ \ \overset{H}{|}\overset{}{\alpha'}}{-C-C=C-C-C-}}$$

$$\underset{\underline{318}}{\overset{\displaystyle \overset{H}{|}\ \ \ \ \ \ \ \ \overset{OM}{|}}{-C-C=C-C=C\diagdown}}$$

$$\underset{\underline{319}}{\overset{\displaystyle \ \ \ \ \overset{OM}{|}\ \ \overset{H}{|}}{\diagup C=C-C=C-C-}}$$

As in the case for saturated ketones, it appears that the transition state for strong base-promoted enolization of α,β-unsaturated ketones closely resembles the starting materials [428]. Therefore, α' protons which are close to the carbonyl group should have greater kinetic acidity than more remote γ protons [428].

The high kinetic acidity of α' protons allows a variety of strong bases, including LICA [80, 430-432], LHDS [90, 433-435], sodium hexamethyldisilazide [87], LDA [436-445], the LDA-HMPA 1:1 complex [446, 447], and potassium t-butoxide [448] in aprotic solvents, to be used to quantitatively prepare kinetic dienolates of α,β-unsaturated ketones. Dienolate formation is generally carried out by slowly adding the enone to more than 1 eq (about 6 eq in the case of potassium t-butoxide) of the base in THF at low temperature (usually -78°C). As illustrated in Table IV, conjugated enone systems of a variety of structural types have been enolized in this manner.

Although useful for the generation of kinetic enolates of saturated ketones, trityllithium cannot be employed for the kinetic enolization of α,β-unsaturated ketones because of the tendency of this base to participate in electron transfer reactions [78]. On treatment with electrophilic species products derived from the enolates 321 or 319, were obtained when enones (317) were reacted with trityllithium (or sodium). The radical anion 320 of an enone produced

Table IV   Bases for Kinetic Enolization of α,β-Unsaturated Ketones

Base		Ref.
Lithium isopropyl-cyclohexylamide (LICA)	Steroidal 4-en-3-ones, pulegone, alkylcyclohex-2-enones	79, 430, 432
Lithium hexamethyl-disilazide (LHDS)	Steroidal 4-en-3-ones, steroidal 4-en-3,11-diones, α-thioalkylcyclohex-2-enones, β-alkoxycyclohex-2-enones	433 90 434 435
Sodium hexamethyl-dizilazide	Acyclic enones (mesityl oxide)	87
Lithium diisopropyl-amide (LDA)	β-Alkoxyacyclic enones, cyclohex-2-enone, and simple alkyl derivatives β-alkoxycyclohex-2-enones	436 439, 440 437, 444
	β-thioalkylcyclohex-2-enones, β-chlorocyclohex-2-enones β-pyrrolidinocyclohex-2-enones, α-hydroxycyclohex-2-enones cyclohept-2-enones, 1(9)-octalin-2-ones, hydrindenones,	441, 442 443 431 438 171 445 419
LDA:HMPA (1:1 complex)	β-Alkoxycyclohex-2-enones, cyclopent-2-enones	446 447
Potassium t-butoxide	Steroidal 4-en-3-ones	448

by electron transfer from trityllithium or tritylsodium can undergo coupling with a trityl radical to produce a conjugate addition product *321* or a hydrogen atom abstraction reaction to produce an extended enolate *319* [78].

Prior to the introduction of the above bases, a few isolated examples of α' alkylations by treatment of α,β-unsaturated ketones such as 4-methyl-9(10)-octalin-1-one [449] or 2-allylcyclohex-2-enone [450] with sodium hydride in DME or sodium t-amylate in benzene in

$$\underset{317}{\text{317}} \xrightarrow[\text{THF}]{(C_6H_5)_3CM} \underset{320}{\underset{\overset{|}{C}-\overset{OM}{\overset{|}{C}}-C=C-\overset{H}{\overset{|}{C}}-}{\overset{H}{\overset{|}{C}}-\overset{|}{C}-C=C-\overset{|}{C}-}}$$

coupling → $(C_6H_5)_3C\bullet$

$$\underset{321}{-\overset{H}{\overset{|}{C}}-\overset{H}{\overset{|}{C}}-C=\overset{OM}{\overset{|}{C}}-\overset{H}{\overset{|}{C}}-}$$
$$C(C_6H_5)_3$$

hydrogen abstraction →

$$\underset{319}{C=C-C=\overset{OM}{\overset{|}{C}}-\overset{H}{\overset{|}{C}}-}$$

M=Li or Na

the presence of an alkylating agent had been reported. These reactions apparently involved the generation of the respective kinetic dienolates in low concentration.

Metal dienolates such as *319* having extended conjugation are more stable than the corresponding cross-conjugated species *318* and are produced from conjugated enones under thermodynamically controlled enolization conditions [1, 3, 157, 451]. Ringold and Malhotra [451] have shown that the conversion of steroidal 4-en-3-ones to the corresponding heteroannular dienolates is relatively slow and is incomplete when 1 eq of potassium t-butoxide in t-butyl alcohol is employed as the base. For example, 17α-methyltestosterone (*322*) was 40% converted to the deconjugated enone *324* when treated with 1 eq of potassium t-butoxide for 1 hr at room temperature and then quenched with 10% aqueous acetic acid. The amount of *324* produced reflected the extent of formation of the heteroannular enolate *323*. After 16 hr with 1 eq of potassium t-butoxide in t-butyl alcohol, 75% of *324* was obtained. By employing 10 eq of potassium t-butoxide in t-butyl alcohol, 95% of *324* was obtained when quenching was performed after 1.5 hr. Conjugate enolate formation was faster when diethylene glycol dimethyl ether (diglyme) was the solvent and much slower when benzene was used. Similar results were obtained with other steroidal 4-en-3-ones and with 10-methyl-1(9)-octalin-2-one

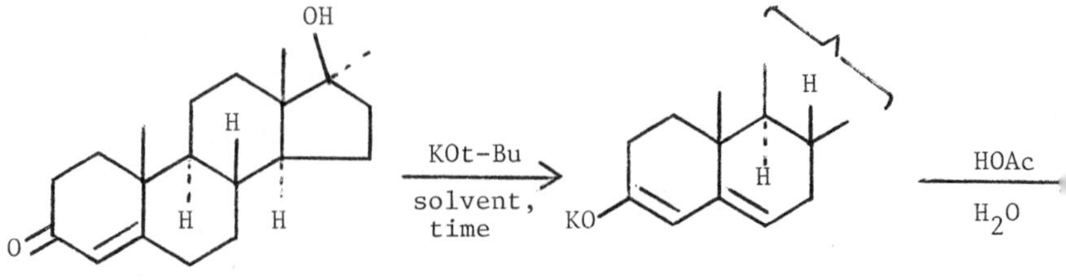

322   323

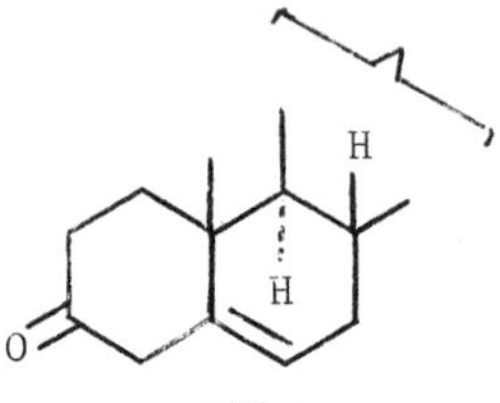

324

Eq. KOt-Bu	Time	Solvent	% of 324 in recovered enone product
1	1.5 hr	HOt-Bu	40
1	16 hr	HOt-Bu	75
10	15 hr	HOt-Bu	95
2	1 hr	diglyme	80
10	1.5 hr	benzene	13

(8). 4-Methyl-4-en-3-ones in the steroid series undergo conjugate enolate formation with potassium t-butoxide in t-butyl alcohol considerably more slowly than the related normal steroids [157].

Extended dienolates of α,β-unsaturated ketones may be produced with a variety of base-solvent combinations in addition to potassium t-butoxide-t-butyl alcohol. These include potassium t-amylate in benzene [3, 452], sodium hydride in dioxane [453, 454], DME [310, 323, 324], or DMSO [455], and potassium hydride in THF [66]. Kinetically generated lithium dienolates of the type 318 (M = Li) may be equilibrated to the corresponding extended systems 319 in the presence of excess ketone. These equilibrations are sometimes slow and require long reaction periods. For example, equilibration of the kinetic dienolate *325* (R = CH_3) to the extended dienolate *326* (R = CH_3) was accomplished using a 9% excess of free 2-methyl-3-pyrrolidinocyclohex-2-enone in a 1:1 THF-HMPA mixture for 21 hr at room temperature and then 3 hr at 40°C [431]. On the other hand, the kinetic dienolate of 3-pyrrolidinocyclohexenone-2-one (*325*, R = H) undergoes equilibration to the thermodynamic isomer *326* (R = H) in the presence of a slight excess of ketone at 0°C in 2 hr [456].

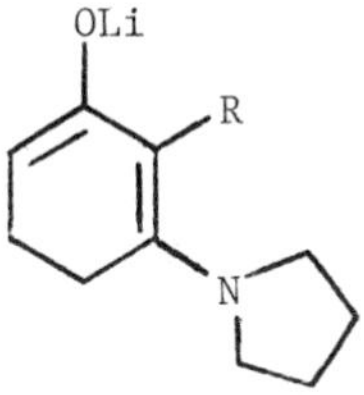

325 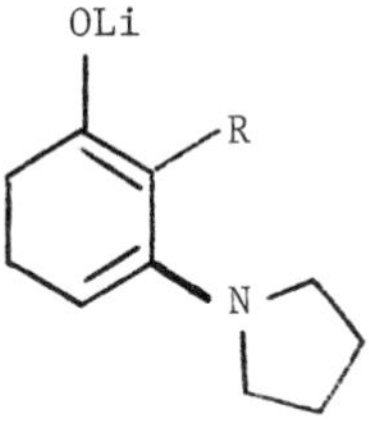   326

Structural features often permit the formation of two different dienolates of the extended type. For example, 3-methylcyclohex-2-enone (*49*) can yield the cisoid enolate *327* or the transoid species *328*. While the rates for formation of the two possible extended dienolates may be competitive under certain conditions [457], it is normally expected that the more stable system, e.g., the transoid system *328*, will be preferred [3]. 1(9)-Octalin-2-one and 19-nor steroidal 4-en-3-ones yield heteroannular dienolates rather than homoannular systems under thermodynamic control [452].

Extended dienolates such as *319* are normally produced directly upon treatment of β,γ-unsaturated ketones with strong bases. The presence of the electron-withdrawing β,γ-double bond enhances the acidity of α protons significantly with respect to α' protons in such systems [157, 303, 449, 458]. On the basis of alkylation results on 4-methyl-4(10)-octalin-1-one (*329*), it has been estimated that the α proton is about four times more acidic than the α' proton [449].

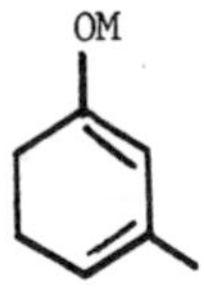

327 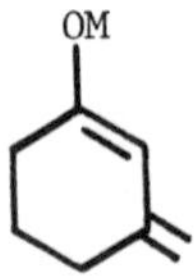   328 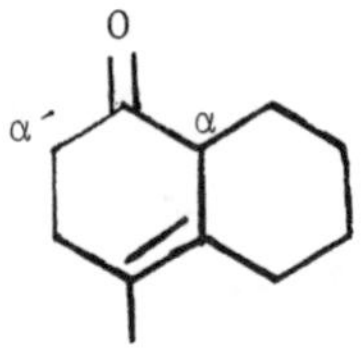   329

Steroidal 1,4- and 4,6-dien-3-ones are not converted in significant concentrations to the corresponding trienolates with potassium
t-butoxide in t-butyl alcohol [451]. Trienolates of 1,4-dien-3-ones
(330) may be obtained using excess potassium t-butoxide if DMSO is
employed as the solvent [459]. Several other bases including sodium
hydride, sodium amide, and sodium acetylide in DMSO are also effective
Kinetic protonation of these trienolates with aqueous acetic acid
leads to the formation of 1,5-dien-3-ones in good to excellent yields
[459]. Linearly conjugated dienones such as the 19-nor steroid 331
have been converted to the corresponding extended trienolates with
sodium methoxide in DMSO [460]. The dienones 332 and 333 have been

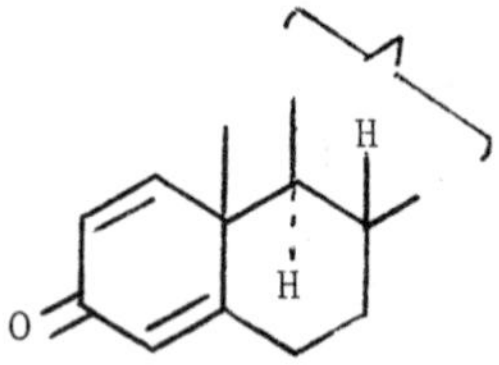

330 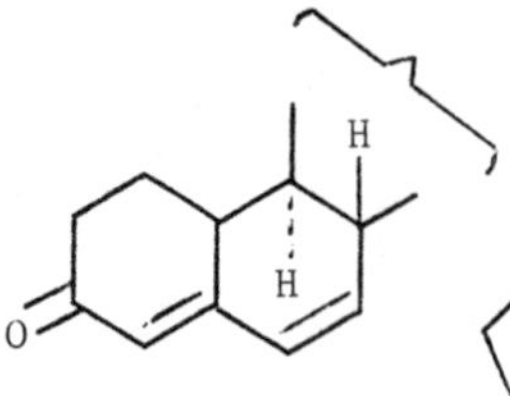   331 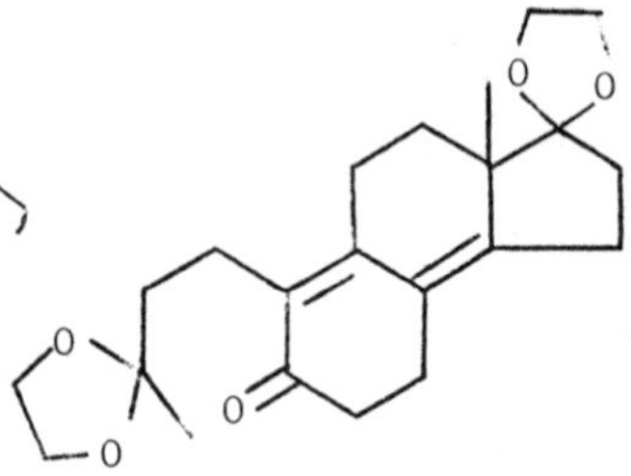   332

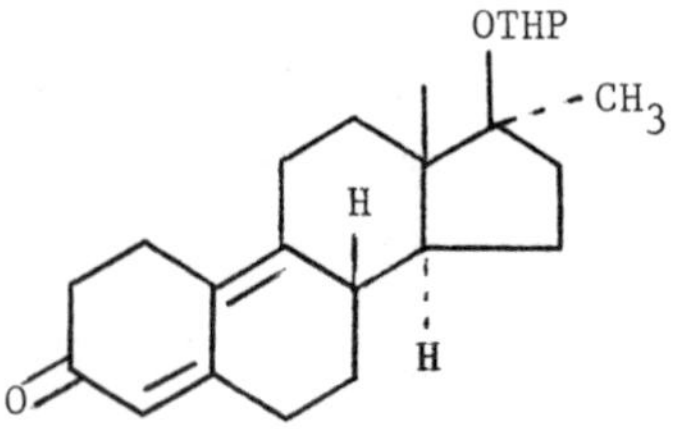

333

alkylated at the α position via the corresponding extended trienolates
using dimsyl sodium in DMSO [461] and potassium t-butoxide in t-butyl
alcohol [448], respectively, for enolization.

Enolization of hexamethyl-2,4-cyclohexadienone (334) with LHDS
in THF at low temperature yields the cross-conjugated trienolate
335 rather than the fully conjugated system 336 [462].

334   335   336

Interesting results have been obtained on treatment of steroidal
1,4-dien-3,11-diones such as 337 with strong bases in aprotic solvents
[88-90]. Under kinetically controlled conditions 337 is converted
to the trienolate 338 using sodium hexamethyldisilazide, LHDS, or
trityllithium in THF at -78°C [89]. In the presence of excess
ketone the sodium enolate 338a undergoes equilibration to the more
thermodynamically stable species 339a. The lithium enolate 338b was
shown to be stable in the presence of excess 337. The structures of
338 and 339 and the compositions of mixtures of these enolates were
determined by quenching reactions with benzoic anhydride [89]. It
is of interest to note that by converting 337 to the kinetic enolate
338, the cross-conjugated dienone system can be protected, while
reactions such as metal hydride reductions are performed at the 11-
carbonyl function [90].

Extended dienolates of steroidal 4-en-3-ones have been generated
by several indirect methods, including reaction of 4,6-dienol acetates with nucleophilic reagents [32, 429], lithium-ammonia reduction
of 6-bromo-4-en-3-ones [463], and lithium-ammonia reduction of 4,6-
dien-3-ones [464, 465]. The latter method has been applied similarly
to generate extended dienolates of 1(9)-octalin-2-ones [465]. Cross-
conjugated dienolates have been prepared by metal-ammonia reduction
of aromatic ketones under appropriate conditions [466].

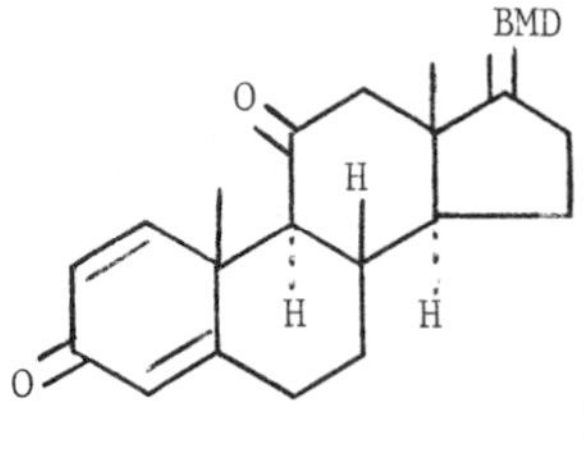

337 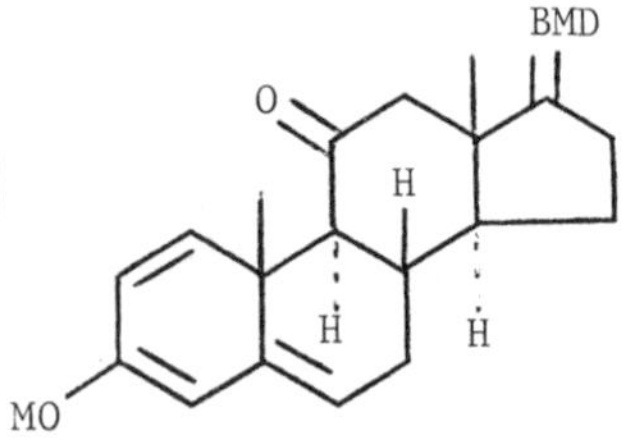   338

a. M = Na

b. M = Li

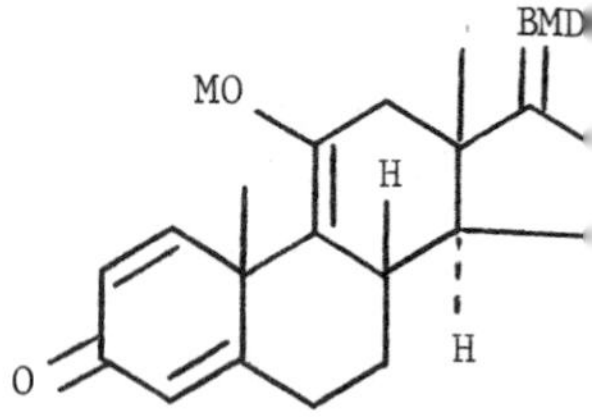

339

(BMD = Bismethylenedioxy
protected side chain)

B. α'-Alkylation and Related Reactions

Kinetically formed cross-conjugated lithium dienolates of cyclohex-2-
enone derivatives [79, 434, 437, 438, 441, 442, 444, 446], certain
cyclopent-2-enones [438, 447], and steroidal 4-en-3-ones [79, 430,
433] may be readily alkylated in good to excellent yields without
equilibration to the corresponding extended dienolates. The procedure developed by Stork and Danheiser [437] for alkylation of lithium
dienolates of 3-alkoxycyclohex-2-enones is generally applicable to
such reactions. This involves the dropwise addition of 1.1 eq (or
an excess) of the alkylating agent to a solution of the lithium
dienolate [prepared by the dropwise addition of the enone to a slight
excess of LDA (most commonly), LICA, or LHDS in THF at -78°C]. When
reactive alkylating agents such as allyl bromide are employed, the
reaction may be completed by allowing the temperature to rise to
about 25°C over a 4-hr period. When less reactive alkylating agents
such as n-propyl bromide are used, 1.1 eq of HMPA is added to increase
the reactivity of the enolate, and the reaction mixture is allowed to
warm to room temperature and stirred for a period of up to 24 hr.
Dialkylation is generally not a serious problem in these reactions,
but it occurs to some extent when less reactive alkylating agents
are used.

6-Substituted enones of the type 342a-d have been prepared by
alkylation of the lithium dienolate 341, derived from 3-
isobutoxycyclohex-2-enone (340), with allyl bromide [437], n-propyl
iodide [437], ethyl bromoacetate [444], and methyl iodide [435].
These derivatives are of considerable synthetic interest because
they are readily convertible to the less easily obtainable 4-
alkylcyclohex-2-enones [437].

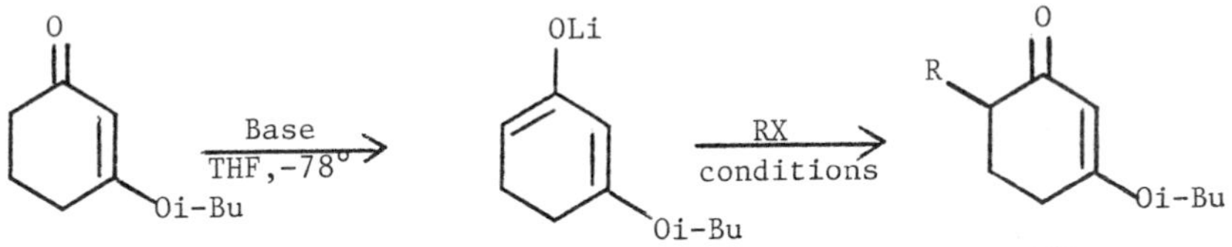

Base	RX/Conditions	
LDA	1.1 eq. allyl bromide, THF -78° to ∿25°, 4 hr	a. R = allyl (98%)
LDA	1.1 eq. n-propyl iodide, THF (1.1 eq. HMPA, ∿25°, 24 hr	b. R = n-propyl (65%)
LDA	1.1 eq. BrCH$_2$CO$_2$Et, THF -78° to ∿25°, 4 hr	c. R = CH$_2$CO$_2$Et (55%)
LHDS	1.2 eq. MeI, THF -78° to ∿25°, overnight	d. R = CH$_3$ (91%)

In addition to 340, other 3-alkoxycyclohex-2-enones [434, 437,
444, 446] as well as 5,5-dimethylcyclohex-2-enone [78], 3-
thiobutylcyclohex-2-enone and 6-methyl-3-thiobutylcyclohex-2-enone
[441, 442], 2-hydroxycyclohex-2-enone (via the dianion) [438], 5,5-
dimethyl-2-thiopropylcyclohex-2-enone [434], and 2-methyl-3(1-
pyrrolidyl)cyclohex-2-enone [432] have been regiospecifically
alkylated with methyl iodide and allylic alkylating agents via the
corresponding lithium enolates. Attempted monoallylation of 3-
chloro-2-methylcyclohex-2-enone with allyl bromide led to the

formation of a significant amount of 6,6-diallylated material [443].
Apparently, β-chloroenones are more acidic than are the alkyl- or
hetero-substituted enones noted above; thus the rates of proton
transfer leading to dialkylation are enhanced [443]. Partial equilibration of the kinetic lithium enolate of pulegone to the thermodynamically more stable isomer occurred during alkylation with methyl
iodide at 25°C, since a 2.4:1 mixture of α' and α methylation products
was obtained [78].

By the use of dihalides as alkylating agents it is possible to
spiroannulate enones such as *340* via kinetic lithium enolates [446].
This method has been used to prepare a key intermediate in a total
synthesis of the sesquiterpene β-vetivone. The kinetic lithium
enolate *343* of 5-methyl-3-ethoxycyclohex-2-enone was reacted with the
dichloride *344* to give initially the intermediate *345*. In the favored
conformation of the kinetic enolate of *345*, the 3-methyl group is
expected to be axial to the ring, and in the second (alkylation) step
the new carbon-carbon bond is formed trans to this group to produce
346 [446].

Methylations of kinetic lithium dienolates of 3-alkylcyclopent-
2-enones (*347*) to produce the 5-methyl derivatives *348* have been
carried out in acceptable yields using methyl iodide [447]. In
these cases deprotonations were performed using the LDA-HMPA complex.
Attempted alkylation of cyclopent-2-enone at C5 gave only polymeric
material. Apparently, the dilithium dienolate of 1,2-cyclopentandione
(*349*) is the only example of a β-unsubstituted cyclopent-2-enone
which has been successfully alkylated [438]. In the monolithio
derivative of *349* or its 5-methylation product, the electron density
at the β position should be much higher than in ordinary cyclopent-
2-enones [438]. This could prevent the occurrence of Michael-type
reactions leading to polymerization [447].

Alkylations of kinetic lithium enolates of steroidal 4-en-3-ones
such as testosterone 17-tetrahydropyranyl ether [433] and cholest-4-
en-3-one [78, 430] with methyl iodide in THF-HMPA or THF yield mixtures containing the 2α- and 2β-methylated products in 80-85% yields.

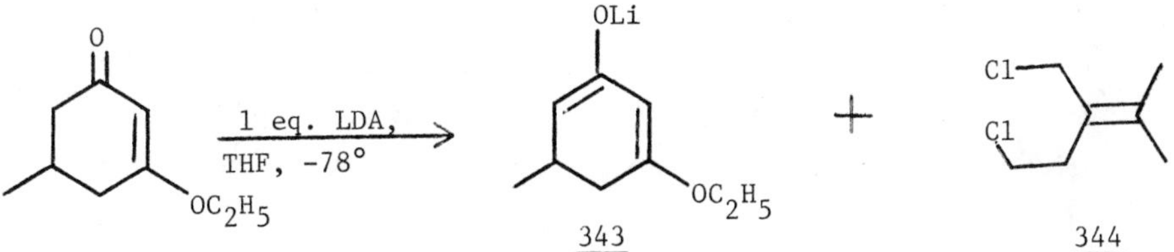

343

344

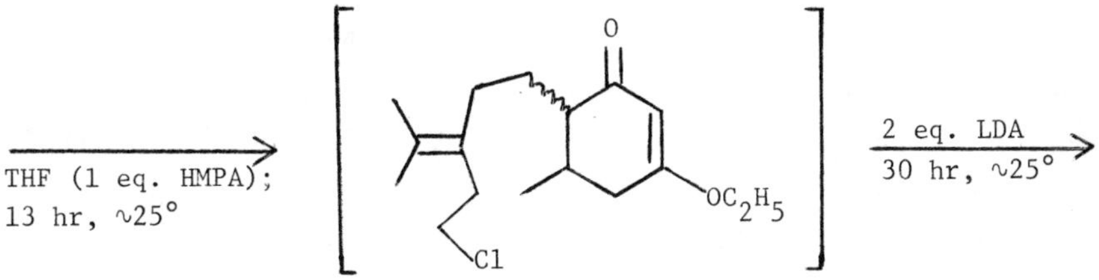

345

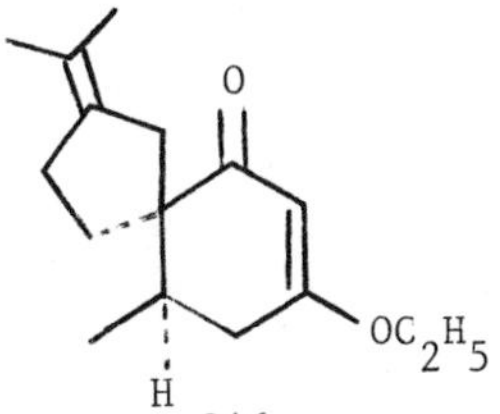

346

(45%)

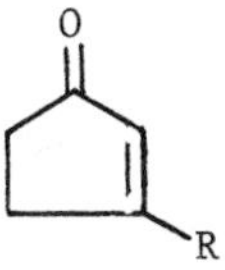

347

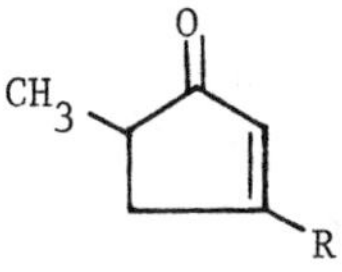

348

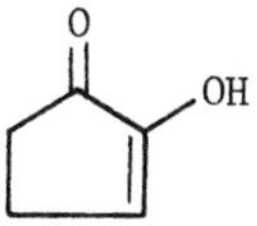

349

Alkylation of the lithium 2-enolate of 2-methylcholest-4-en-3-one with trideuteromethyl iodide produced a 40:60 mixture of products resulting from α and β attack upon the enolate anion [430]. Alkylation of the 2-enolate of trans-3,10-dimethyl-2-decalone gives exclusively the product of α attack [304], and similar results would be expected for the 2-enolate of a steroidal 2-methyl-3-one. Thus,

β attack at C2 of a dienolate system is much more favorable than β
attack at C2 of related saturated ketone enolates. This preference
can be accounted for in terms of steric and torsional angle factors
associated with the 1,3-diplanar conformations of the homoannular
dienolate system [467].

Methylation of kinetic potassium enolates of 17α-methyl-19-
nortestosterone tetrahydropyranyl ether (350) and related compounds
have also been investigated [448]. Treatment of 350 with 5.8 eq of
potassium t-butoxide in THF at -70°C followed by the addition of an
excess methyl iodide gave largely the 2,2-dimethylated product.
However, monoalkylation at C2 occurred when 1.9 eq of a solution of
potassium t-butoxide in THF was added dropwise to a solution of the
enone 350 and methyl iodide in THF or THF-HMPA at -70°C. Again, the
introduction of the 2-methyl group from the β side was preferred
[467].

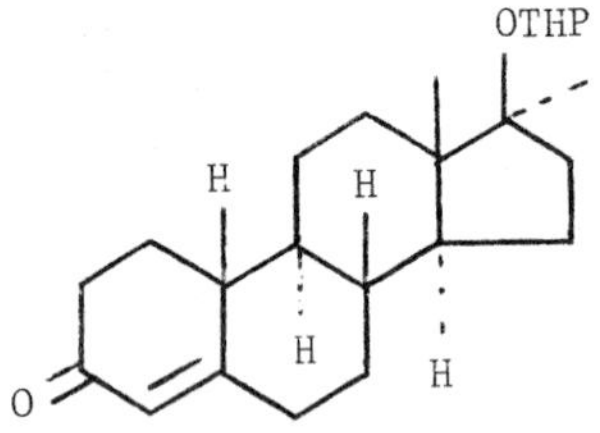

350

        Intramolecular alkylations of the bicyclic α,β-unsaturated
ketones 351-353 having 2-bromoethyl groups at the angular positions
have been studied [468]. Using potassium t-butoxide in t-butyl
alcohol, it was found that the corresponding α'-alkylation products
354-356 were produced with a high degree of selectivity in all cases.
It was suggested that the α'-alkylation products did not arise from
the trapping of the kinetic α' enolates corresponding to 351-353 but
rather from the lowest-energy transition states for intramolecular
alkylation of an equilibrating mixture of cross-conjugated and
extended dienolate anions [468].

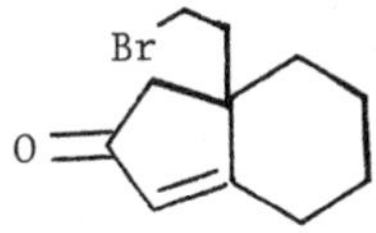

351

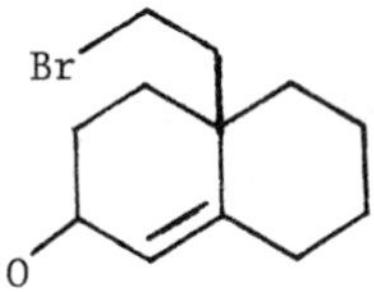

352

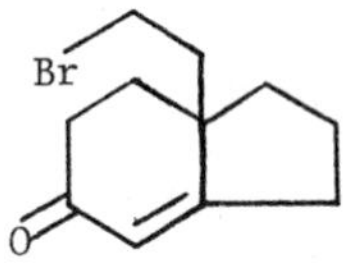

353

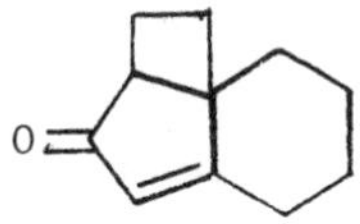

354

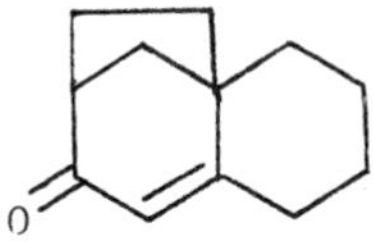

355

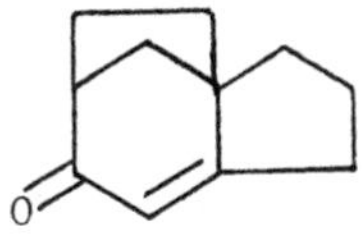

356

Intermolecular additions of the kinetic lithium enolates of cyclohex-2-enone derivatives to methyl acrylate are followed by Michael addition of the enolate intermediate to the enone to produce bicyclo[2.2.2]octan-2-ones in high yields [432]. A recently reported one-step synthesis of tricyclo[3.2.1.0^{2,7}]octan-6-ones involves intramolecular Michael additions of kinetic cyclohex-2-enone enolates to vinyltriphenylphosphonium bromide followed by intramolecular addition of the ylide intermediate to the enone and intramolecular displacement of triphenylphosphine [440].

Thermodynamically controlled acylations normally occur at the α' positions of α,β-unsaturated ketones. Formylations of steroidal 4-en-3-ones [469, 470], 10-methyl-1(9)-octalin-2-one (*8*) [471], and the enone *340* [435] provide examples of such reactions. These acylations are usually carried out with sodium methoxide or sodium hydride as the base.

Attempted carbomethoxylation of the kinetic enolate of the hydrophenanthrone *357* has been reported [472]. The enone was added to an excess of tritylsodium in ether, and carbon dioxide gas was passed into the reaction mixture. After acidification of the mixture and treatment with diazomethane, a 4:6 mixture of the β-keto esters

358 and *359* was isolated in fair yield. The formation of products
derived from both the kinetic (homoannular) and thermodynamic (heteroannular) dienolates of *357* can be explained in two ways. First, the
homoannular enolate could have been produced in the initial deprotonation step, but because the carbon dioxide was introduced slowly,
partial equilibration of the system was brought about by the relatively acidic carboxylate salt of the β-keto acid carbonation product;
or, second, the reaction of tritylsodium with the enone may have led
directly to a mixture of the homoannular and heteroannular enolates
(the former being produced by deprotonation of *357*, while the latter
was formed by an electron transfer mechanism of the type discussed
previously). No examples of carbonations of kinetically formed
lithium enolates of α,β-unsaturated ketones have been reported at
this time.

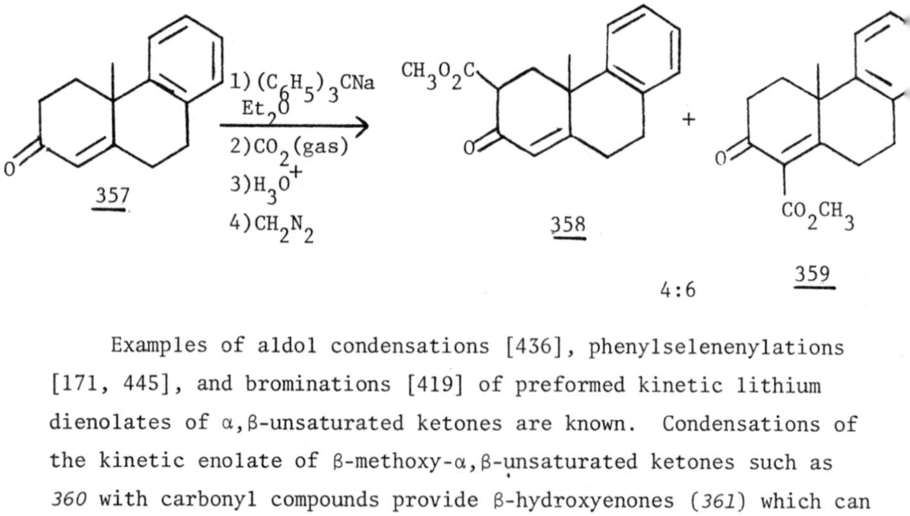

Examples of aldol condensations [436], phenylselenenylations
[171, 445], and brominations [419] of preformed kinetic lithium
dienolates of α,β-unsaturated ketones are known. Condensations of
the kinetic enolate of β-methoxy-α,β-unsaturated ketones such as
360 with carbonyl compounds provide β-hydroxyenones (*361*) which can
be readily converted into aldols and polyenones under appropriate
conditions [436].

Phenylselenenylation of the kinetic lithium dienolate of
cyclohept-2-enone *362* with phenylselenyl bromide in THF at -78°C gave
the selenide *363* in 81% yield [171]. The latter compound was oxidized
and phenylselenic acid was eliminated to yield 2,6-cycloheptadienone

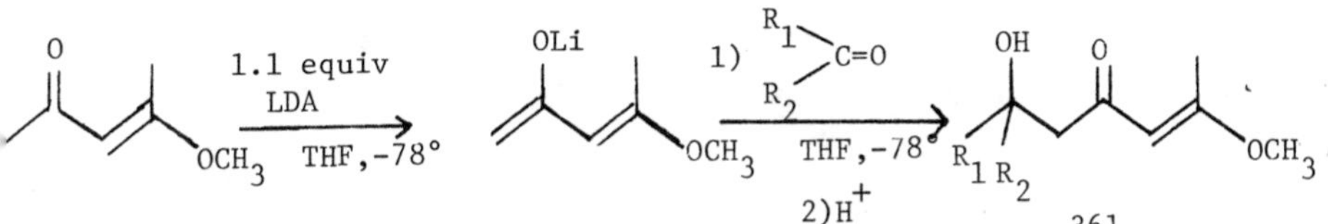

360

361

[171]. The phenylseleno derivative 365 of the cis-dimethyl octalone 364 has been obtained similarly [445].

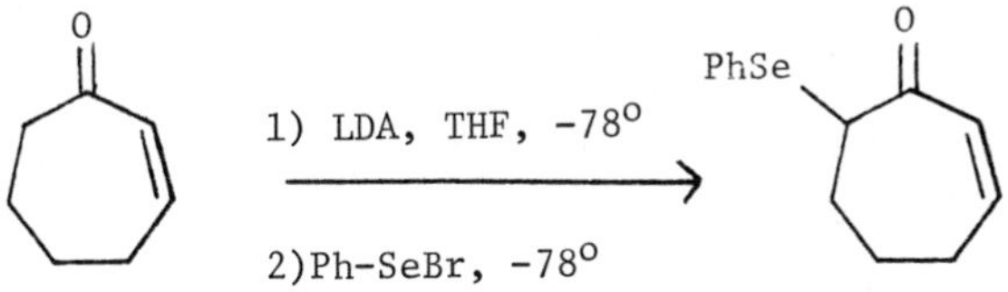

362

363 (81%)

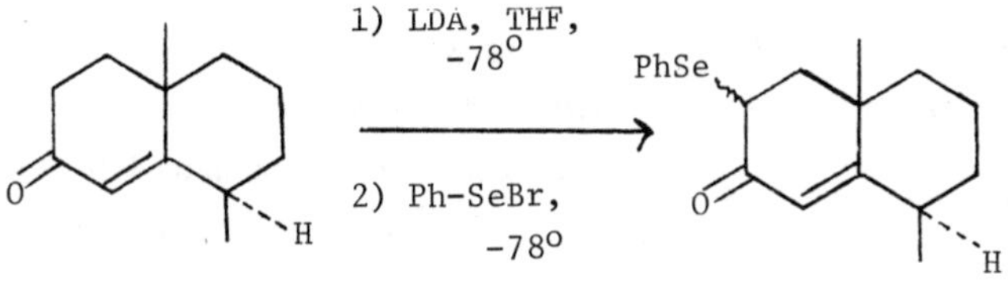

364

365

Bromination of the kinetic lithium enolate of the hydrindenone 366 with molecular bromine in THF at -78°C gave a mixture of the 2-bromo enones 367 in about 80% yield [419].

Aldol condensation of the kinetic bromomagnesium enolate of 1-acetyl-2,6,6-trimethylcyclohexa-1,3-diene (enone plus

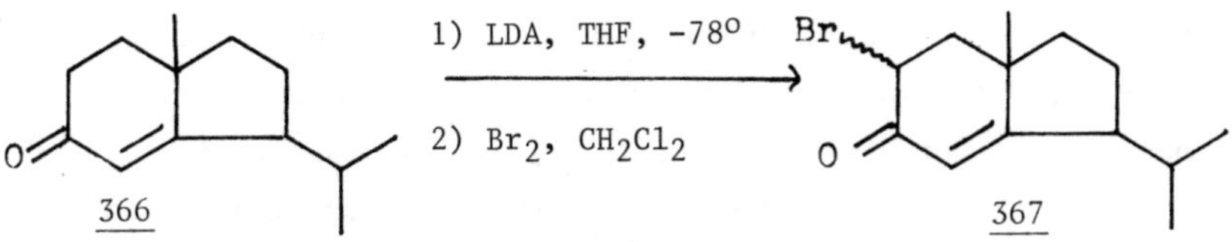

366

367

N-methylanilinomagnesium bromide in ether) with acetaldehyde has recently been used in a synthesis of β-damascenone [473].

Trialkylsilyl dienol ethers are readily obtained by the quenching of kinetic dienolates of α,β-unsaturated ketones with trialkylsilyl halides. Derivatives of this type have been prepared from mesityl oxide [87], cyclohexenone [439], and steroidal 4-en-3-ones [433, 439].

C. α-Alkylation

1. General

Extended metal dienolates of α,β-unsaturated ketones normally undergo C-alkylation at the α position to produce α-alkylated β,γ-unsaturated ketones [1, 3, 474]. If the α carbon of the dienolate bears a substituent, monoalkylation is the only possible reaction. However, if the α carbon is unsubstituted, dialkylation may be an important or, under certain conditions, the major reaction. The ease of dialkylation of extended dienolates is readily accounted for by the mechanism shown in Eq. (56), which was proposed by Ringold and Malhotra [157]. The initial alkylation product 370 is more acidic than the starting α,β-unsaturated ketone 368 or the α-alkylated α,β-unsaturated ketone 372. Thus 370 is readily deprotonated by the base(s) present in the medium to give the alkylated extended dienolate 371. This species may then undergo a second alkylation leading to the α,α-dialkylated produce 373 or protonation at the γ position yielding the α,β-unsaturated ketone 372 at competitive rates. (Once formed, conjugated enones such as 372 are normally converted back to the dienolate 371, slowly. Therefore, monoalkylation products are usually not intermediates in the dialkylation process [157].)

Studies on the methylation of extended potassium dienolates of steroidal 4-en-3-ones have been extensive [32, 157, 475-478]. Reactions of these ketones with excess potassium t-butoxide in t-butyl alcohol (or potassium t-amylate in benzene) followed by rapid addition of excess methyl iodide yield largely 4,4-dimethyl-5-en-3-ones. Although C2 alkylation may also be a problem under vigorous conditions

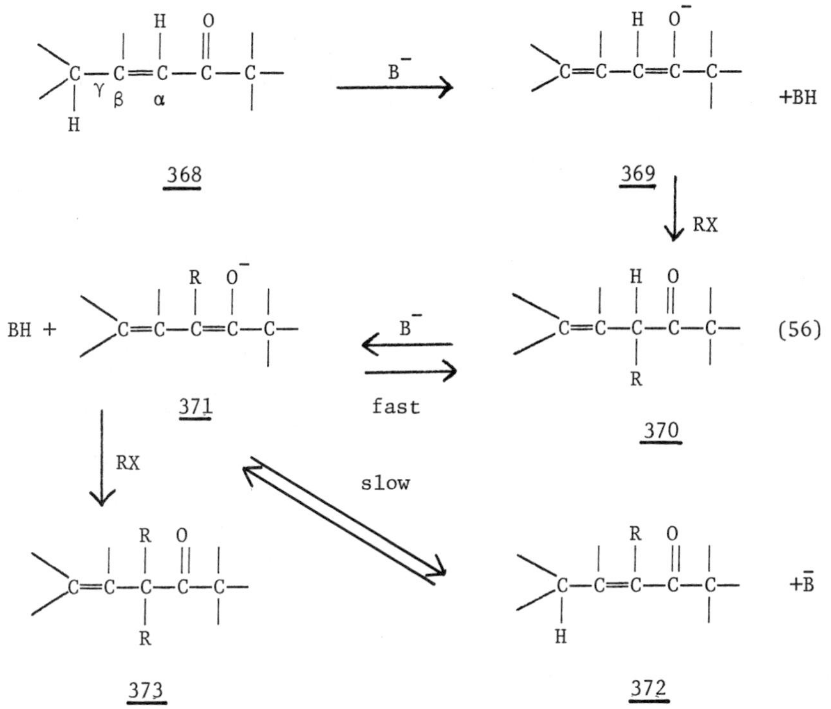

the dimethylation process is normally very efficient and is of
general synthetic value. For example, 4,4-dimethylation of cholestenone was employed as a key step in the synthesis of lanosterol
reported by Woodward, et al. [32].

Even when limited amounts of methyl iodide and base are used
for the methylation of 4-en-3-ones, 4,4-dimethylated products still
predominate over monoalkylated material if the halide is added
rapidly to the reaction medium [157]. Atwater [475] found that slow
addition of 1.25 eq of methyl iodide to a refluxing solution of 17α-
methyltestosterone (322) containing 1.5 eq of potassium t-butoxide
in t-butyl alcohol led to the isolation of 49% of the monomethylated
enone 375 and 15% of the dimethylated product 374. Similar results
were obtained with other 4-en-3-ones and alkylating agents [475].
Thus, when a deficiency of the halide is used, the initial monomethyl
5-en-3-one is converted into the conjugated ketone more rapidly than

it is dimethylated. Later, Ringold and Malhotra [157] showed that
under conditions where use of excess methyl iodide produced a 13:1
mixture of *374* and *375*, methyl chloride, a considerably less reactive alkylating agent, gave a mixture containing 30% of *375* and 11%
of *374*.

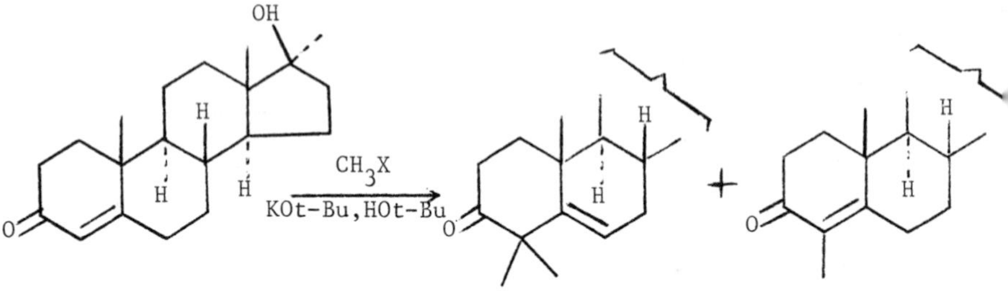

<u>322</u>   <u>374</u>   <u>375</u>

1.5 eq. KOt-Bu, reflux
1.25 eq. CH$_3$I added over 2.5 hr   15%   49%
3 eq. KOt-Bu, 5 hr, 25°,   11%   30%
   excess CH$_3$Cl

Like steroidal 4-en-3-ones, simpler α,β-unsaturated ketones,
such as 10-methyl-1(9)-octalin-2-one (*8*) [479, 480] and cyclohex-2-
enone derivatives [3], are also readily dimethylated via the corresponding potassium dienolates. The preparative value of these reactions is illustrated by the conversion of the enone *8* to the enone
376 in 77% yield [480].

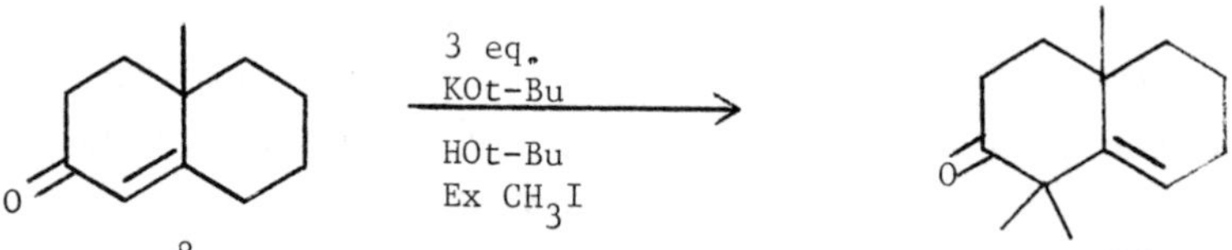

<u>8</u>   <u>376</u>

Studies involving the methylation of extended lithium dienolates
of α-unsubstituted α,β-unsaturated ketones have been limited. However, the conversion of cholestenone to its conjugate lithium dienolate with trityllithium in THF at 0°C followed by addition of excess

methyl iodide has been reported to give the 4,4-dimethylated product
exclusively [78, 79]. Likewise, only 4,4-dimethylated products were
reported when lithium dienolates produced by lithium-ammonia reduction
of steroidal and bicyclic $\alpha,\beta,\gamma,\delta$-dienones were alkylated with methyl
iodide [464, 465].

Because limited success has been achieved so far in the monomethylation of extended metal dienolates, the method of choice for
the conversion of systems such as *368* to *372* (R = CH_3) involves the
preparation and methylation of lithioenamines of the α,β-unsaturated
ketone [481]. This procedure has been employed for the monomethylation of steroidal and simpler enones and is successful because imines
do not undergo rapid proton transfer reactions and dialkylation of
metalloenamines is a slow process [481].

As suggested by the mechanism depicted in Eq. (56), dialkylation
of conjugated dienolates is a less severe problem when the reactivity
of the alkylating agent is diminished. Also, when higher-molecular-
weight alkylating agents are used, it is generally possible to
efficiently separate monoalkylated products from accompanying dialkylated material. Therefore, alkylating agents which are less reactive
than methyl iodide may be reacted with α-unsubstituted extended
dienolates in synthetically useful yields.

In connection with steroid total synthesis there has been
interest in the α-alkylation of bicyclic keto enones such as *377* or
378 (and various derivatives of the saturated ketone functions of
these compounds). α-Alkylations of the corresponding conjugated
dienolates of these enones with alkylating agents such as 1,3-
dichloro-2-butene (*110*) [482], 3,5-dimethyl-4-chloromethylisoxazole
(*111a*) [323], the diethylene ketal of 1-bromo-3-pentanone [455],
and a variety of related methyl vinyl ketone equivalents [315] allow
the introduction of masked 3-ketoalkyl side chains α to the unsaturated
carbonyl function. From such products tricyclic enones of the type
379 or *380* may be generated. Alkylations of systems such as *377* and
378 with bis-annulating agents such as the isoxazole derivative *111c*
[310], the vinylsilane derivative *381*, or with β-phenylethyl bromide

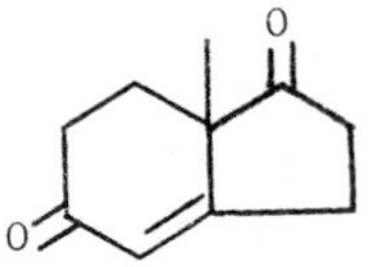

377 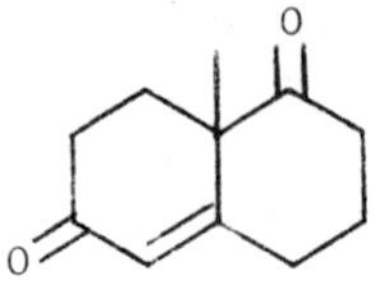   378 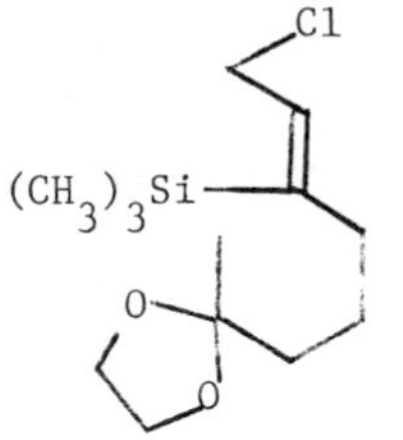   381

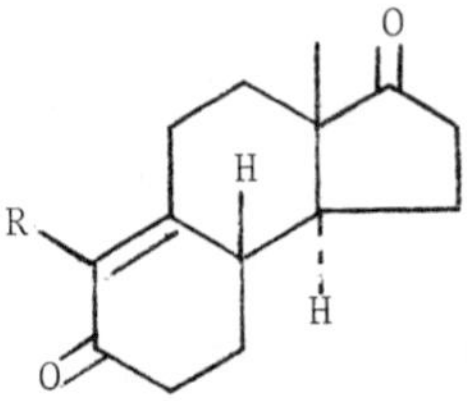

379 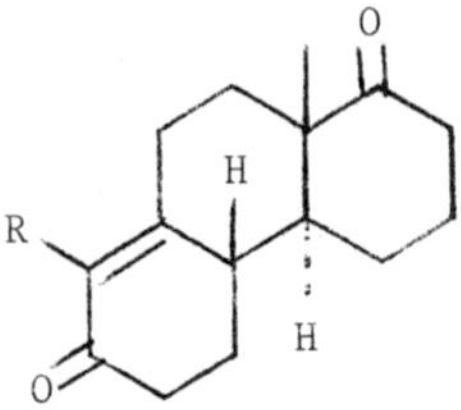   380   332

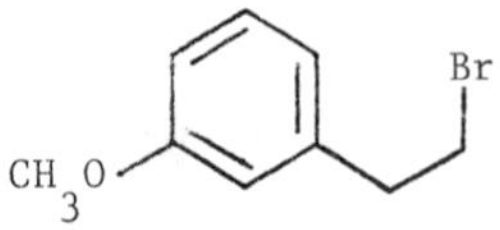

derivatives such as *382* [483] provide products which by appropriate
manipulations allow elaboration of both the A and B rings of the
steroid skeleton.

Normally O-alkylation products are not encountered when extended
potassium or sodium dienolates are treated with the usual alkylating
agents in solvents of low to medium polarity. However, highly reactive benzolyoxymethyl halides yield significant amounts of O-
alkylated products on reaction with extended metal dienolates (*383*)
of 1,10-dimethyl-1(9)-octalin-2-one (*88*) Eq. (57) [285, 489]. The
O-alkylation/C-alkylation ratio is influenced by (a) the polarity of
the solvent, (b) the nature of the halogen, and (c) the metal cation
[285, 484].

When polar aprotic solvents are used, O-alkylation may occur
even with unreactive alkylating agents [455]. For example, treatment
of the extended sodium dienolate of the hydrindenone *384* with the
diethylene ketal of 1-bromo-3-pentanone in DMSO gave a mixture of C-
and O-alkylated products in 56 and 23% yields, respectively [455].
It appears that, in general, dienolates are somewhat more prone to
O-alkylation than are monoenolates.

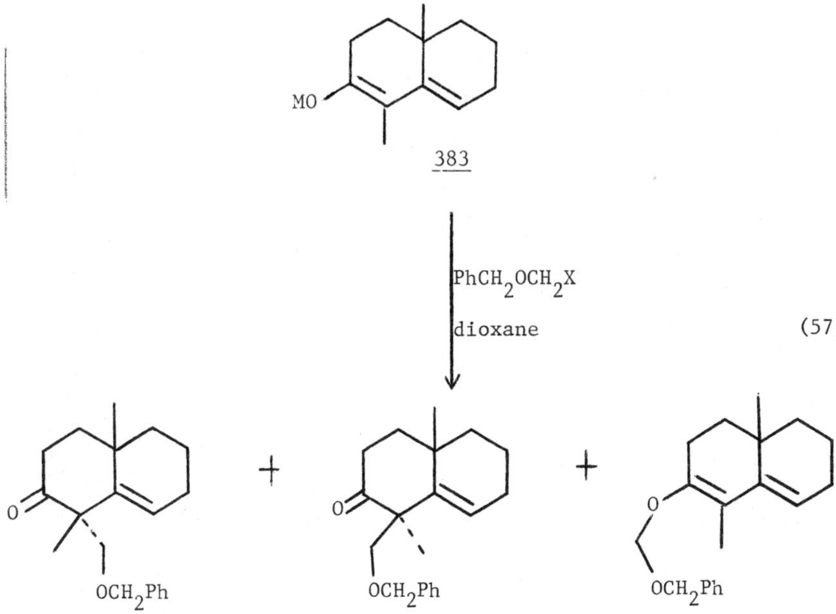

383

$$PhCH_2OCH_2X$$
dioxane

(57)

OCH$_2$Ph + OCH$_2$Ph + OCH$_2$Ph

M	X			
Na	F	52	7	41
Na	Cl	59	11	30
Na	Br	64	9	27
Na	I	70	15	15
Li	Cl	53	7	40
K	Cl	67	10	23
MgBr	Cl	43	12	45

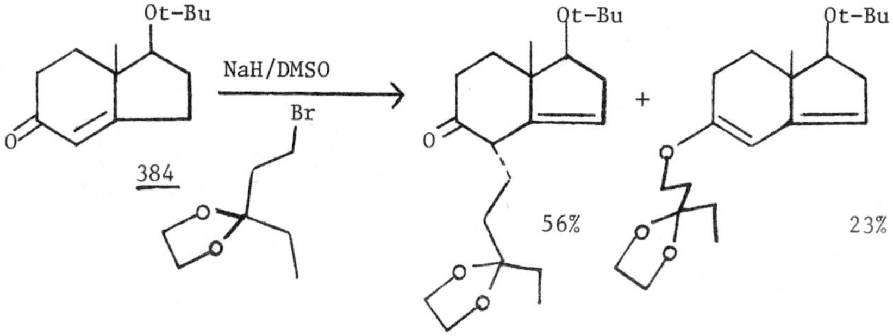

384

NaH/DMSO
Br

56%   23%

The alkylation of α-substituted α,β-unsaturated ketones which
leads to α,α-disubstituted β,γ-unsaturated ketones is readily
accomplished via the corresponding potassium or sodium dienolates.
In cyclic systems these alkylations may yield mixtures of diastereomeric products. As is the case for saturated ketone enolate alkylations, the stereochemistry of dienolate alkylations is influenced
strongly by steric factors within the enolate anion [266, 285, 452,
453, 484-486]. Alkylations of the dienolates of the octalone 385
[452, 484], the hydrindenone 386 [485], and the tricyclic enone 387
[486] provide examples of cases in which steric factors largely or
exclusively favor bottom-side attack of the alkylating agent upon
the dienolate system. Bottom-side attack was also observed when
angular methylations at the 8-, 9-, and 14-positions of steroidal
ketones were carried out via the dienolates 388 [487], 389 [488],
and 390 [32].

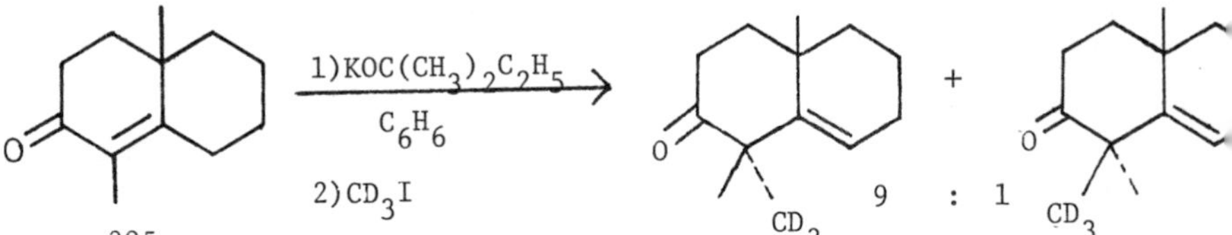

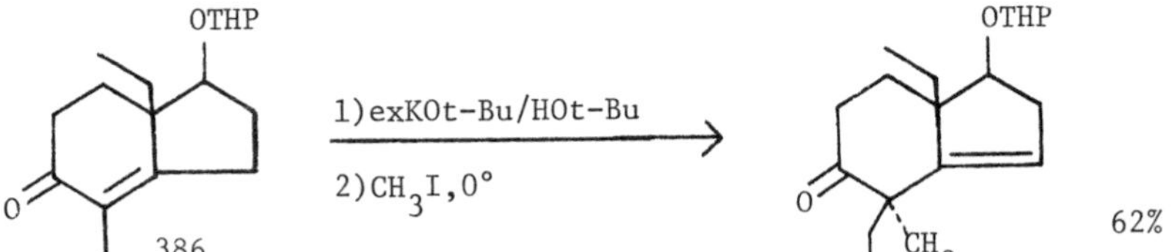

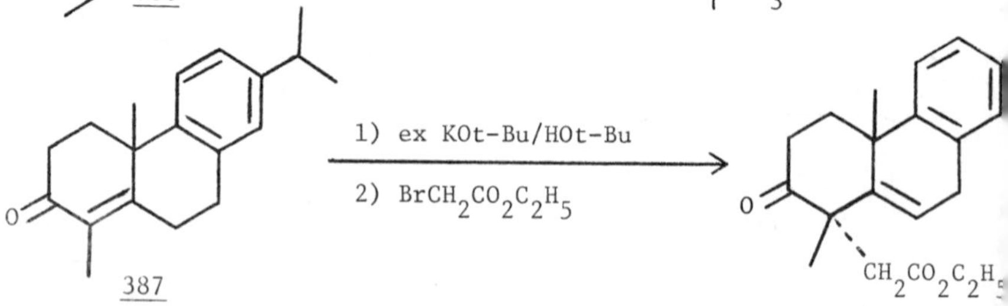

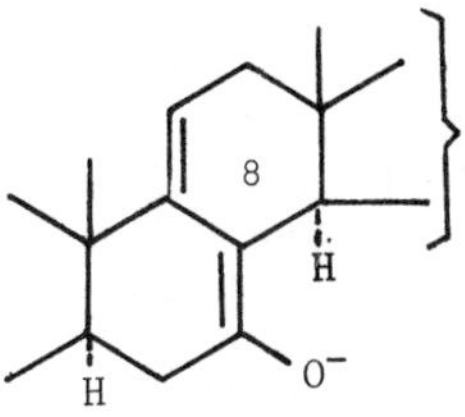

388 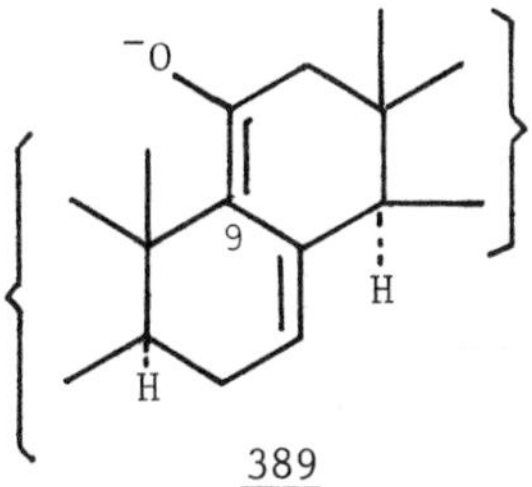   389 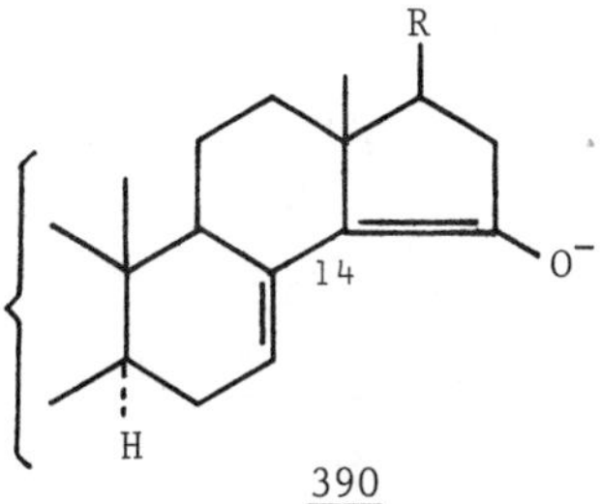   390

Steric factors do not favor highly stereoselective methylations of dienolates of compounds, such as *94*, which contain a latent precursor of the A ring of steroids [310]. The use of the reduction-methylation procedure to produce saturated ketones such as *95* from *94* provides a practical solution to this problem [310].

The stereochemical course of the methylation of 17β-acetoxy-4-methyl-19-nortestosterone (*391*) has been shown to be highly solvent dependent [452]. Trideuteriomethylation of the potassium dienolate in benzene gave a 3:7 mixture of the 4α- and 4β-trideuteriomethylation products *392* and *393*. However, in t-butyl alcohol a 10:1 ratio of products *392* and *393* was produced. Two explanations have been offered for these results:

1. When the reaction is conducted in a nonpolar solvent such as benzene, the transition state has more product character, favoring top-side attack and leading to a chair conformation of the product.

2. The less hindered top-side of the dienolate is specifically solvated by the more polar solvent t-butyl alcohol, favoring bottom-side attack of the alkylating agent.

Sodium dienolates of octalones, such as *394*, which contain polar angular carboethoxy or cyano substituents are alkylated at the α position with benzoyloxymethyl chloride in dioxane predominantly from the β side [453, 489]. [Of course, enone *385* (*394*, R = CH_3) gives predominantly the product of α-side attack under the same conditions.] This effect has been ascribed to the influence of the polar angular substituent in stabilizing the transition state leading to β-alkylation [453, 489].

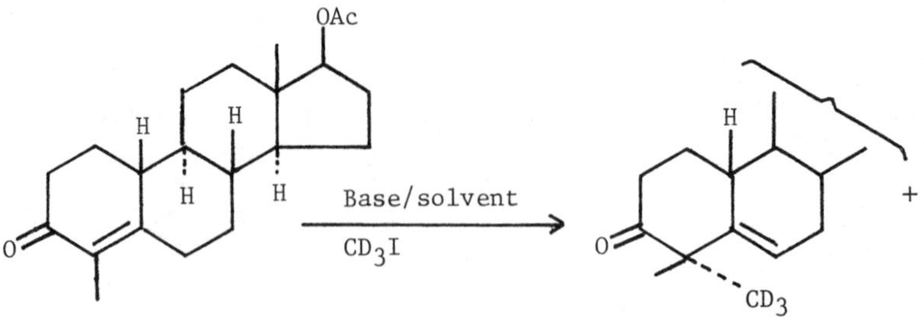
OAc
H
H
H
H
Base/solvent
CD_3I
O
H
O
CD_3
+

391 392

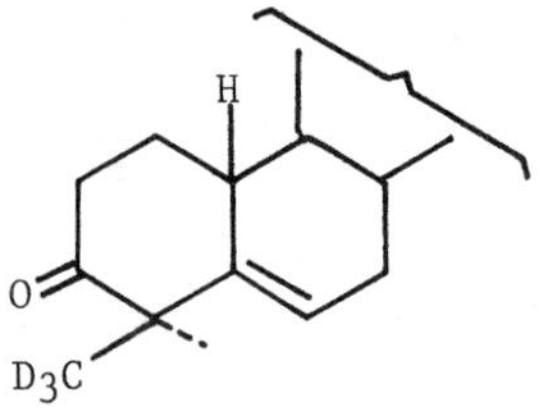
H
O
D_3C

393

$KOC(CH_3)_2C_2H_5$/ C_6H_6	3	:	7
KO*t*-Bu/ *t*-BuOH	10	:	1

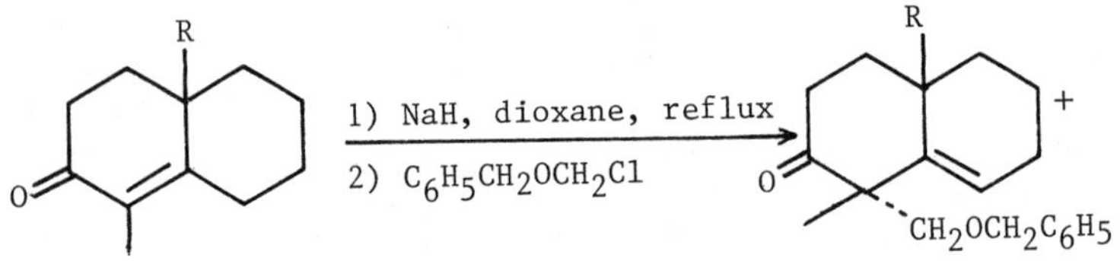
R
1) NaH, dioxane, reflux
2) $C_6H_5CH_2OCH_2Cl$
O
R
O
$CH_2OCH_2C_6H_5$
+

394

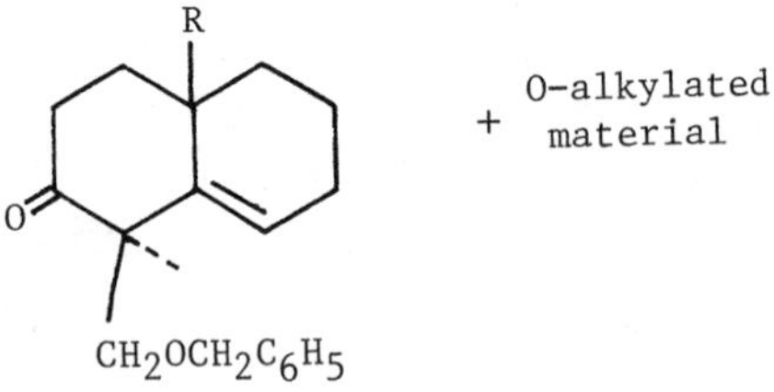
R
O
$CH_2OCH_2C_6H_5$
+ O-alkylated material

R = CO_2Et	11%	89%
= CN	-	major
= CH_3	major	minor

2. Experimental Procedures

a. α,α-Dimethylation of an α,β-unsaturated ketone [32]. A solution of potassium t-butoxide (0.20 mol) in 450 ml of t-butyl alcohol is added to the enone (0.06 mol) in 1 liter of t-butyl alcohol at 40°C under nitrogen. Methyl iodide (0.36 mol) in 100 ml of t-butyl alcohol is added, and the mixture is stirred and refluxed for 1 hr, cooled, and acidified with concentrated hydrochloric acid. After addition of water to dispel turbidity, t-butyl alcohol is removed under reduced pressure. The organic material is partitioned between benzene and water and worked up in the usual way to yield the crude dimethylated β,γ-unsaturated ketone.

b. Monoalkylation at C4 of a steroidal 4-en-3-one [475]. To a boiling solution of potassium t-butoxide (15 mmol) in a 33 ml of t-butyl alcohol is added 10 mmol of the steroidal enone in hot t-butyl alcohol under nitrogen. The alkyl halide (12.5 mmol) in 165 ml of t-butyl alcohol is added dropwise with stirring over a 2.5-hr period to the refluxing solution. The mixture is stirred and refluxed for 0.5 hr, cooled, and acidified with concentrated hydrochloric acid. After addition of water to dispel turbidity, t-butyl alcohol is removed under reduced pressure. The organic material is partitioned between benzene and water. The benzene extracts are washed with water and dried. Concentration of the solvent yields the crude 4-alkyl-4-en-3-one contaminated with some of the corresponding 4,4-dialkyl-5-en-3-one.

D. γ-Alkylation

Intermolecular γ-alkylations of extended dienolates of simple unsaturated ketones have rarely been observed [3]. However, lithium dienolates of cyclic β-amino-α,β-unsaturated ketones such as 395 undergo γ-alkylation exclusively on reaction with methyl iodide and allylic alkylating agents [431, 490]. Similar results have been observed in acyclic systems [491]. The required lithium dienolate is obtained by the addition of a slight excess of the β-amino enone to LDA in

1:1 THF-HMPA at 0°C followed by stirring of the mixture at higher temperatures for a time period sufficient to effect enolate equilibration. After the extended dienolate solution has been cooled to 0°C, 1.1 eq of a reactive alkylating agent is added, and the mixture is stirred at room temperature for 1.5 hr. As discussed previously, in cyclic systems the time required for equilibration is quite dependent on the structure of the substrate [431, 456]. n-Butyllithium has been reported to be a useful base for the enolization of β-amino enones [491], but it gave unsatisfactory results in the hands of other workers [431, 490].

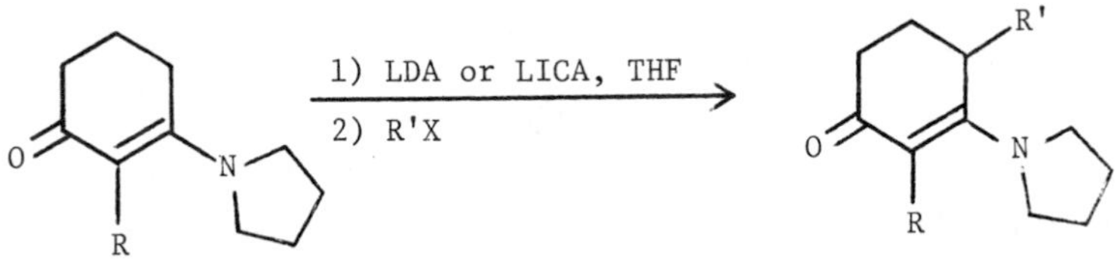

$\underline{395}$ a; R = H

b; R = CH_3

R' = CH_3 or allyl

The extended potassium dienolate of $395a$, prepared by reaction of the enone with 2 eq potassium hydride in THF at 0°C, also yielded the γ-methylation product upon treatment with 1 eq of methyl iodide [492]. Enones related to 395 but bearing alkoxy or thioalkyl groups at the β position gave dimethylation or polyalkylation products under similar conditions [492]. The reason for exclusive γ-alkylation of dienolates of systems such as 395 is unclear. Conjugation of the unshared electron pair on nitrogen with the β,γ-double bond of the extended dienolate possibly leads to a significant increase in the electron density at the γ position. The thermal rearrangement of α-vinyl α-allylketones to produce products of apparent γ-allylation of extended dienolates is a well-known process [3].

E.   *Intramolecular Alkylations Involving Extended*
     Dienolates of α,β-Unsaturated Ketones

Alkylations of extended dienolates of α,β-unsaturated ketones with
polymethylene dihalides lead to spirocyclic ketones except in the
case where a four-membered ring product would be formed [362, 457,
493]. The results of the alkylation of the sodium dienolate of 3-
methylcyclohex-2-enone (*49*) with several dibromoalkanes in liquid
ammonia are shown in Eq. (58) [457]. The second alkylation step
occurred at the α position when 1,2-dibromoethane, 1,4-dibromobutane,
or 1,5-dibromopentane was employed. The six-membered ring enol ether
resulting from O-alkylation in the second step was produced when 1,3-
dibromopropane was used as the alkylating agent. No bicyclic ketones
resulting from intramolecular α'- or γ-alkylation were observed in
these reactions. Spiro ketones containing five-membered rings have
also been produced by alkylation of extended dienolates of cholestenone
[457] and 1(9)-octalin-2-one and related systems [493].

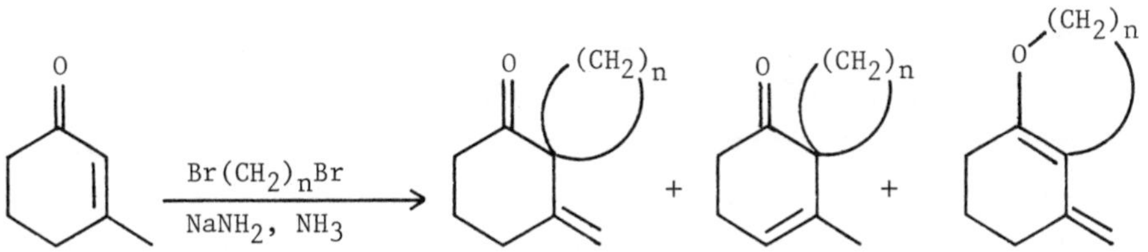

	n = 2	26%	15%	–
	3	–	–	40%
	4	58%	–	–
	4	22%	–	–

$$(58)$$

     The conversion of the keto epoxide *396* to the bridged product
397 [494] and the bicyclic keto mesylate *398* to the tricyclic product
399 [495] provide other examples of intramolecular α-alkylations of
extended dienolate anions.

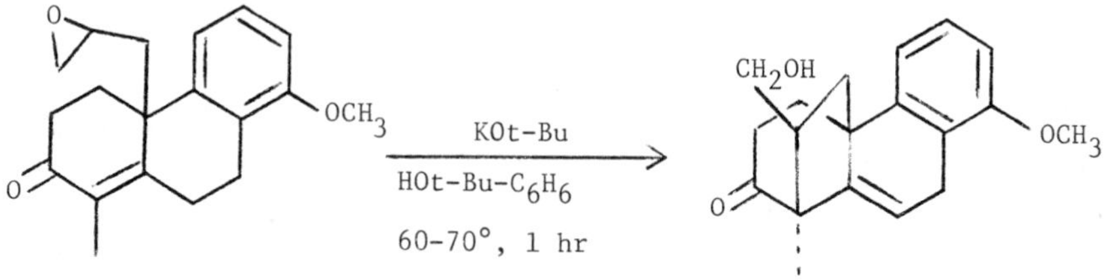

396 → 397

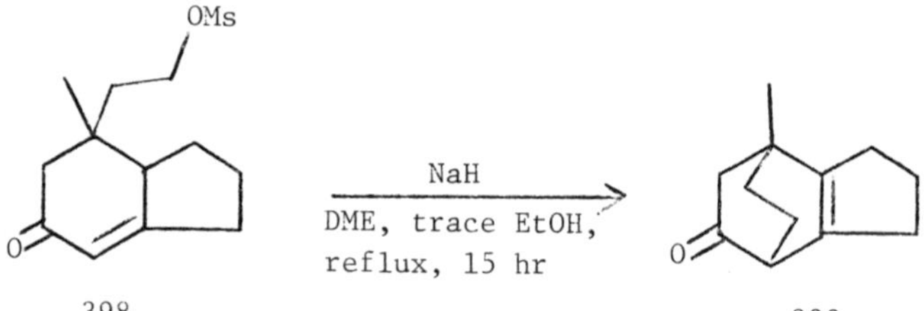

KOt-Bu
HOt-Bu-C$_6$H$_6$
60-70°, 1 hr

398 → 399

NaH
DME, trace EtOH,
reflux, 15 hr

60%

Intramolecular γ-alkylations involving extended dienolates are
well known [362, 496-502]. For example, (-)-maalienone (401) has
been prepared by treatment of the bromo enone 400 with methanolic
potassium hydroxide [496]. Attempts to prepare (-)-epimaalienone
(403) by reaction of the bromo enone 402a or the chloro derivative
402b under similar conditions led only to simple dehydrohalogenation,
apparently because the transition state leading to closure of the
three-membered ring is sterically hindered [497]. However, when
the extended dienolate of 402b was generated using sodium hydride
in DME, the desired ring closure occurred in 74% yield [497].

Spiro enones such as 405 have been prepared in low yields by
reaction of 4-tosyloxyalkylcyclohex-3-enones (404) with sodium hydride
in dioxane containing a trace of t-butyl alcohol [498]. Likewise,
on reaction with sodium hydroxide in aqueous DMSO intramolecular
alkylation of the keto tosylate 406 occurred to yield (+)-β-vetivone
(407) [499]. The correct stereochemistry of the natural product was

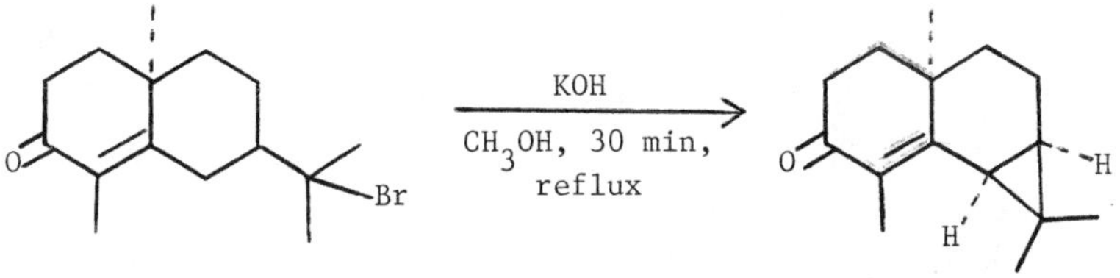
KOH
CH_3OH, 30 min,
reflux
Br
O
H
H

400 401 (60%)

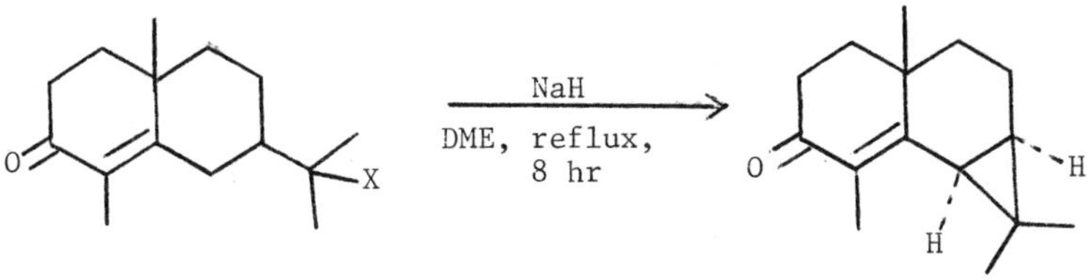
NaH
DME, reflux,
8 hr
X
O
H
H

(74%)

402 a; X = Br 403

b; X = Cl

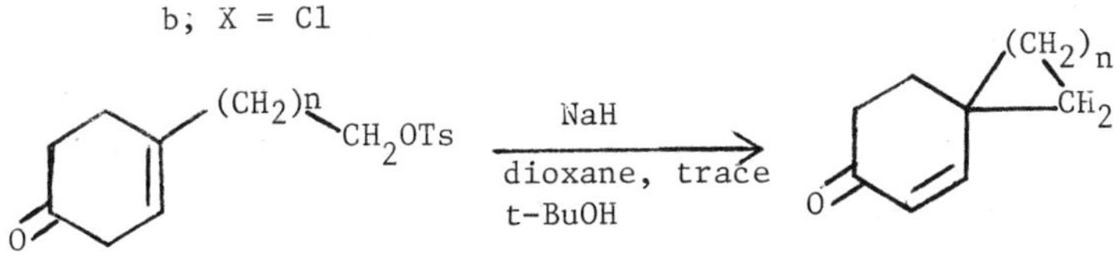
$(CH_2)_n$
CH_2OTs
NaH
dioxane, trace
t-BuOH
CH_2
O

404 405

n = 1 29%

n = 3 28%

obtained because the formation of the new carbon-carbon bond occurred trans to the secondary methyl group on the six-membered ring.

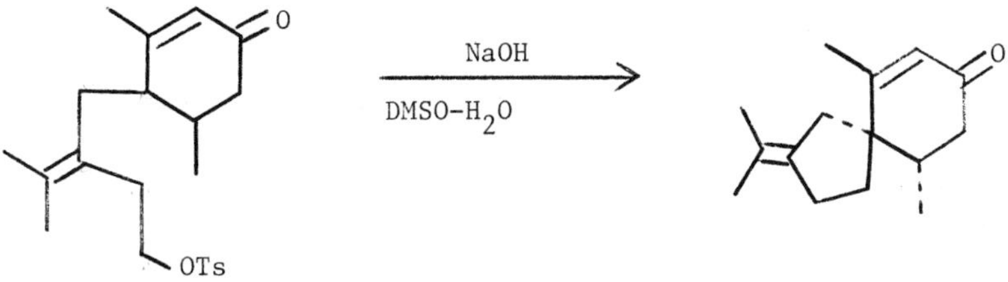
NaOH
$DMSO-H_2O$
OTs
O

406 407

In the preceding systems, cyclization at the γ position is the
only feasible pathway. A priori intramolecular alkylation of the
steroidal enone *408* containing an angular tosyloxymethyl group [500]
or the related octalone derivative [501] can occur at the α or γ
positions (both processes would lead to a new four-membered ring).
However, in these cases only the γ-alkylation products, such as *409*,
were isolated.

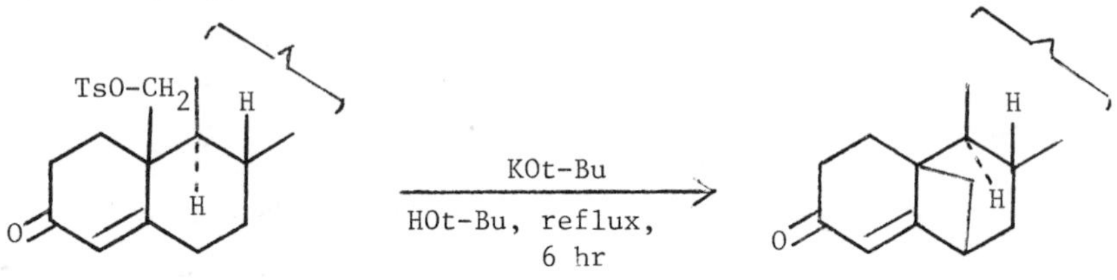

The extended dienolate of the tosyloxy enone *410* is also capable
of undergoing intramolecular cyclization at the α or γ positions.
In fact, treatment of *410* with 1.2 eq of dimsyl sodium in DMSO at
25°C led to the γ-alkylation product *411* in 96% yield [502]. Interestingly, the enone *411* could be converted to the isomeric product
412 by a vinylcyclopropane cyclopentene rearrangement at 450°C [502].

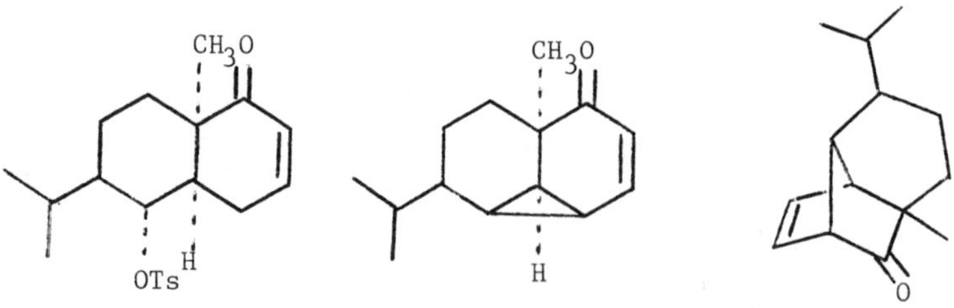

Other examples of intramolecular alkylations of extended dienolates may be found in a recent review [362].

F. *Reactions of Extended Dienolates with*
    Other Electrophilic Reagents

Reactions of extended dienolates of α,β-unsaturated ketones with only a few electrophilic reagents other than alkylating agents have been reported. These species undergo reaction on oxygen with acetic anhydride to yield dienol acetates [157]. α-Carboalkoxyl α,β-unsaturated ketones may be obtained in good to fair yields by carbonation of extended dienolates of α,β-unsaturated ketones [471, 472, 503]. For example, reaction of the octalone *8* with sodium hydride in DME to produce the dienolate, reaction of this with dry, gaseous carbon dioxide followed by brief stirring with aqueous sodium hydroxide to isomerize the β,γ-double bond to conjugation, careful acidification, and reaction with diazomethane gave 4-carbomethoxy-1(9)-octalin-2-one (*413*) in 63% yield [471]. Several related conversions using steroidal 4-en-3-ones have been reported [503]. Carbonation of the hydrindenone *384* with methylmagnesium carbonate in DMF provided the α-carboxy derivative in 64% yield [504]. Under similar conditions the octalone derivative related to *384* gave primarily the α' carboxy derivative [504]. Hydrindenones apparently yield linearly conjugated dienolates more rapidly than the corresponding octalones do.

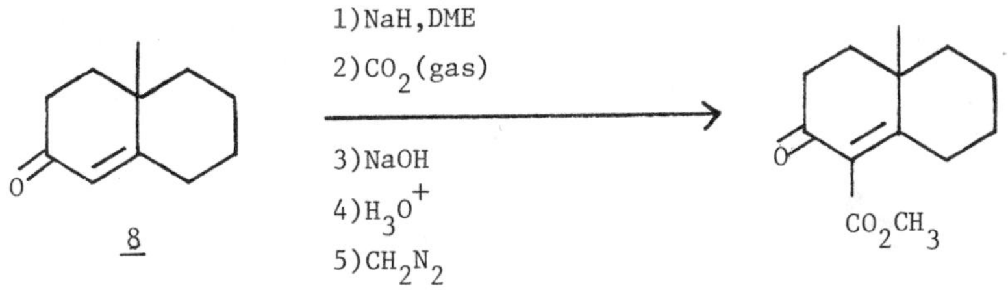

1)NaH,DME

2)CO_2(gas)

3)NaOH

4)H_3O^+

5)CH_2N_2

<u>8</u>

CO_2CH_3

<u>413</u> (63%)

IV. ALKYLATION OF ALDEHYDES

A. *Alkylation of Saturated Aldehydes*

The direct alkylation of simple aldehydes via metal enolates is generally not useful synthetically [13, 17]. In the presence of bases which cause slow or incomplete enolization significant concentrations of metal enolates cannot be produced because of rapid self-(aldol)condensation. Lithium enolates (*415*) of simple aldehydes such as isobutyraldehyde (*414*) may be generated in aprotic solvents such as DME by deprotonation with LDA or by cleavage of enol acetates (*416*) with 2 eq of methyllithium in DME [505]. However, using benzyl

$$(CH_3)_2CHCHO \qquad (CH_3)_2C=C{\overset{OLi}{\underset{H}{\Big\langle}}} \qquad (CH_3)_2C = C{\overset{OAc}{\underset{H}{\Big\langle}}}$$

<u>414</u>   <u>415</u>   <u>416</u>

chloride as the alkylating agent, enolate alkylations are complicated by the formation of Cannizzaro (e.g., *418*) or Tichshenko reaction products which result from reaction of the alkylated aldehyde *417* with various bases, such as the unreacted enolate *415*, LDA, or lithium t-butoxide, present in the medium [505].

$$\underset{DME}{\overset{C_6H_5CH_2Br}{\underset{415}{\longrightarrow}}} \quad C_6H_5CH_2C(CH_3)_2CHO \quad + \quad C_6H_5CH_2C(CH_3)_2CH_2OH$$

                                              <u>417</u>   <u>418</u>

Method of prep of <u>415</u>	T°	Time (min)	% <u>417</u>	% <u>418</u>
Deprotonation of <u>414</u> with LDA	23 -25°	30	40	6
Reaction of <u>416</u> with 2 eq CH$_3$Li	30 -40°	45	19	7
Same	25 -41°	60	12	20

Successful alkylations of tri-n-butyltin enol ethers of aldehydes have been reported [506]. Equation (59) shows some of the results which have been obtained using the tri-n-butyltin enol ether *419*, of isobutyraldehyde and relatively reactive alkylating agents [506]. Ether *419* may be produced by reaction of the enol acetate *416* with tri-n-butyltin methoxide [506]. The synthetic potential of these reactions has not been explored fully at this time.

$$(CH_3)_2C=CHOSn(n-Bu)_3 \xrightarrow{RX} RC(CH_3)_2CHO$$

$$\underline{419} \tag{59}$$

RX	T°	Time (hr)	%
CH_3I	90°	14	86
CH_3CH_2I	120°	14	59
CH_3OCH_2Cl	100°	16	92

Phase transfer alkylations provide a partial solution to some of the problems associated with the preparation and alkylation of simple aldehyde enolates [507]. Aldehydes such as *414* which contain only one α-hydrogen atom may be alkylated in reasonable yields with reactive alkylating agents such as methyl iodide, allyl chloride, or benzyl chloride in an emulsion of benzene and 50% aqueous sodium hydroxide in the presence of a catalytic amount of a tetra-n-butyl-ammonium salt [Eq. (60)] {507]. When less reactive alkylating agents were used, aldol condensation predominated over alkylation.

$$(CH_3)_2CHCHO + RX \xrightarrow[C_6H_6, n-Bu_4N^+Y^-, T°]{50\%NaOH-} R-C(CH_3)_2CHO$$

$$\underline{414}$$

RX	Y^-	T°		(60)
CH_3I	I^-	42°	15%	
$CH_2=CHCH_2Cl$	Cl^-	60°	56%	
$C_6H_5CH_2Cl$	Cl^-	70°	75%	

2-Ethylhexanal, which is more stable toward sodium hydroxide than *414*, gave somewhat higher yields of alkylation products, but both C- and O-alkylations were observed [507]. For example, when the alkylation was performed with benzyl chloride, a mixture of 55% C- and 35% O-alkylated products was obtained. Bulky groups at the α position of aldehyde enolates appear to facilitate O-alkylation significantly [507]. It has also been shown that the sodium enolate of phenylacetaldehyde gave the O-alkylation product, β-methoxystyrene, exclusively on reaction with dimethyl sulfate in DMSO or liquid ammonia [283].

In complex systems, aldol condensations of aldehyde enolates are sufficiently slow for alkylations to be feasible synthetically. There have been several examples of successful aldehyde enolate methylations in the field of natural product synthesis [508-512]. The highly stereoselective conversion of the aldehyde 420 to the methylated compound 421 was an important step in the total synthesis of (±)-rimuene [508]. Potassium t-butoxide in t-butyl alcohol or benzene has usually been employed as the base for alkylations of complex aldehydes [508-510, 512]. An example of a methylation involving a sodium enolate intermediate prepared using tritylsodium in DMF has also been reported [511].

In alkylations of enolates of aldehydes such as 420 and related compounds the new alkyl group is introduced equatorially to the cyclo-hexane ring with a high degree of stereoselectivity [506-510]. This result is similar to that obtained in the alkylation of exocyclic enolates of cyclohexyl ketones such as 183 [314]. Again, in a reactant-like transition state for the alkylation, hydrogen atoms in the 3- and 5-positions with respect to the enolate α carbon hinder axial approach of the alkylating agent [314].

B. *Alkylation of α,β-Unsaturated Aldehydes*

Extended dienolates of α,β-unsaturated aldehydes are more stable than saturated aldehyde enolates [513-515]. The preparation and alkylation of potassium dienolates of α-substituted α,β-unsaturated aldehydes such as cyclohexenecarboxyaldehyde (422) and 2-methyl-2-pentenal have been reported [513]. The aldehyde is added slowly with stirring to a slight excess of potassium amide in liquid ammonia at -60°C, and the mixture is stirred for 2 hr. Then an excess of the alkylating agent is added dropwise over 30 min, and the mixture is stirred at -33°C for 1 hr. The usual work-up procedure gives the alkylated products.

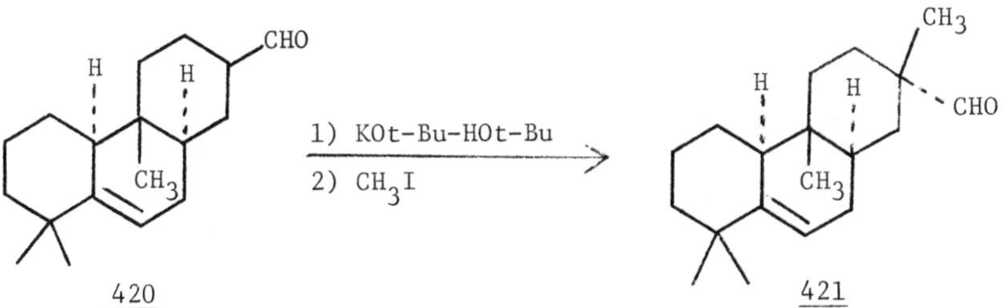

Representative examples of alkylations of the potassium dienolate of *422* are shown in Eq. (61). Varying amounts of methylimine derivatives of the starting aldehyde and the C-alkylated products were obtained when methylating agents were employed in excess [513]. The highly reactive alkylating agent dimethyl sulfate gave exclusively the O-methylation product, and with isopropyl iodide a 1:1 mixture of O- and C-alkylated material was obtained. When crotyl bromide was used for the alkylation, 80% of a mixture of α-alkylated compounds and 8% of the γ-crotyl derivative were obtained. The α-alkylated material contained 8% of a butenyl derivative which probably resulted from a Claisen rearrangement of O-crotylated material. Cope rearrangement of this compound could also have accounted for the formation of the γ-crotylated product.

Examples of alkylations of extended metal dienolates of aldehydes have been reported in connection with the synthesis of (±)-pimaradiene and related compounds [516, 517], (±)-germanicol [518], and 20-methylcholesterol [512].

C. *Alkylations and Directed Aldol Condensations*
   of Aldehyde Derivatives

Since the aldehyde function is highly prone to aldol condensation and Cannizzaro (or Tishchenko) reactions under basic conditions, methods of synthesis of α-substituted aldehydes utilizing aldehyde derivatives or metalated aldehyde derivatives have been developed. Alkylations of aldehyde enamines were among the first reactions of this type to be explored [13-16]. The use of aldehyde enamines derived from simple

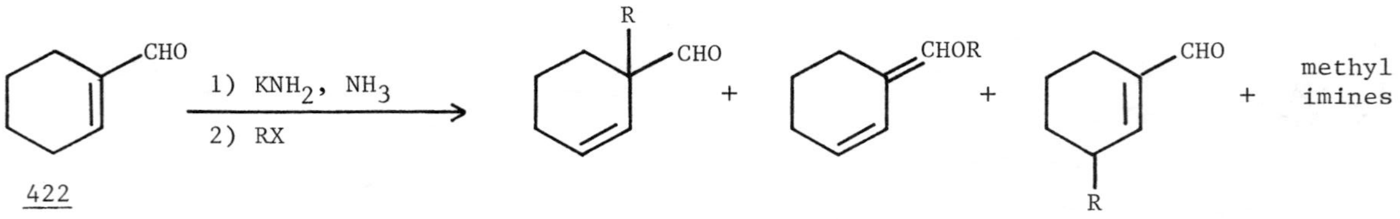

RX				methyl imines
10 equiv CH$_3$I	63%	10%	–	20%
1.1 equiv (CH$_3$)$_2$SO$_4$	–	77%	–	2%
3 equiv CH$_3$CHICH$_3$	33%	33%	–	–
3 equiv CH$_2$=CH–CH$_2$Br	90%	–	–	–

(61)

secondary amines such as pyrrolidine proved to be of very limited value. Saturated halides were found to give almost exclusively N-alkylated products on reaction with these compounds [519]. C-Alkylated products, which could be hydrolyzed to the corresponding aldehydes, were produced in fair yields when allylic [520, 521] or propargylic [522] halides were used. These products were shown to arise via thermal rearrangements of initially formed N-alkylated materials [521]. Later, it was found that the problem of N-alkylation with saturated halides could be overcome by the use of an aldehyde enamine prepared from a bulky secondary amine such as n-butylisobutylamine [523]. For example, the enamine 423 gave the α-ethylated aldehyde 424 in 78% yield upon reaction with ethyl iodide in acetronitrile followed by hydrolysis [523].

$$C_3H_7CH=CH-N\begin{array}{c} \diagup n\text{-}Bu \\ \diagdown i\text{-}Bu \end{array} \quad \xrightarrow[\text{2) } H_3O^+]{\text{1) } CH_3CH_2I, CH_3CN} \quad \begin{array}{c} CH_2CH_3 \\ | \\ C_3H_7CH\text{-}CHO \end{array}$$

423   424 (78%)

Reactions of metalloamines with electrophilic reagents such as alkylating agents or carbonyl compounds provide useful methods for the synthesis of α-alkyl or α,β-unsaturated aldehydes [17-21]. Metallo-enamines can be prepared by treatment of aldimines, derived from t-butylamine or cyclohexylamine, with bases such as ethylmagnesium bromide in THF [17], LDA in etheral solvents [18-21], or lithium diethylamide in HMPA-benzene [524]. The imine derivatives initially obtained on addition of electrophilic reagents are hydrolyzed to the corresponding α-substituted aldehydes. This alkylation sequence is illustrated in Eq. (62). Aldimines derived from aldehydes having one or two α-hydrogen atoms may be employed.

$$\begin{array}{c} H \\ | \\ \text{—C—CH} = NR \end{array} \xrightarrow{MB} \begin{array}{c} M \\ | \\ \text{—C=CH-NR} \end{array} \xrightarrow[\text{2) } H_3O^+]{\text{1) } RX} \begin{array}{c} R \\ | \\ \text{—C—CHO} \end{array} \qquad (62)$$

The original alkylation method developed by Stork and Dowd [17] employed halomagnesium salts of aldimines in THF. However, there are advantages to the use of lithium salts because they are readily prepared in DME and require shorter reaction times in the alkylation step [21]. The lithioaldimine is prepared by the addition of the aldimine to 1 eq of LDA in DME at -10° to 0°C under nitrogen. The lithio imine solution is warmed to 20°C over 1 hr and 1 eq of the alkylating agent is added dropwise with stirring over 15 min while the temperature is maintained within the 20°-40°C range. The mixture is then stirred at about 25°C for 3.5 hr, and the organic product is partitioned between water and ether. The ether solution is dried and concentrated to yield the crude alkylated aldimine. Hydrolysis of the alkylated aldimine is readily carried out by adding 1 M aqueous acetic acid (2.5 eq of HOAc) and hexane (2.5:1) and stirring the mixture under nitrogen for 2 hr. The usual work-up procedure yields the alkylated aldehyde.

Metalloaldimines are alkylated in good yields with allylic and benzylic halides. For example α-benzylisobutyraldehyde (417) was obtained in 65 and 69% yields, respectively, via the bromomagnesium [17] and lithium salts [21] of the t-butyl aldimine of isobutyraldehyde (414). Note that a much lower yield of 417 was obtained when direct alkylation of the lithium enolate 415 was attempted [21]. No rearrangement of allylic halides occurs during the alkylation of metalloaldimines [17]. Yields in the 40-50% range are obtained when saturated primary halides are employed as alkylating agents. Secondary halides give largely elimination products when reacted with lithioaldimines in HMPA-benzene [524].

Lithioaldimines undergo *directed aldol condensations* with aldehydes and ketones in ether solution at low temperatures [Eq. (63)] [18, 19, 525, 526]. (Such reactions are also possible with lithioketimines.) As in the case of the aldol condensations of zinc and lithium enolates of ketones (Sec. II.F), these reactions lead to relatively stable chelate intermediates. Hydrolysis of these chelates produces α,β-unsaturated aldehydes directly. Good results are obtained with acetaldimine and related aldimine salts bearing one α-substituent

$$\diagdown\!\!\!\!\diagup C = O + R' - \overset{Li}{\underset{H}{C}} = CH - NR \xrightarrow[\text{ether } 0^\circ]{}$$

aromatic
or aliphatic
aldehydes
and ketones

$$\diagdown\!\!\!\!\diagup C = C - CHO$$
$$\overset{|}{R'}$$

An efficient method for the synthesis of α,β-unsaturated aldehydes is outlined in Eq. (63) [25]. Lithioaldimines such as 425
(R = H or alkyl) are converted to α-trimethylsilyl derivatives 426
by reaction with 1 eq of trimethylsilyl chloride in THF at $0^\circ C$ [25].
The lithio derivative 427 is then prepared by the reaction of 426
with LDA in THF at $0^\circ C$. On addition of 1 eq of an aldehyde or
ketone at $-78^\circ C$ followed by warming to $-20^\circ C$, the α,β-unsaturated
aldehyde derivative is produced. Hydrolysis of the latter at pH 4.5
then yields the α,β-unsaturated aldehyde. Lithioacetaldimine (425,
R = H) gave the unsaturated aldehydes 429 (R = H, R' = C_5H_{11}) and
429 (R = H, R'R''-$(CH_2)_5$), in 94 and 90% yields, respectively, on
reaction with hexanal and cyclohexanone. Similar results were obtained when lithiopropionaldimine (425, R = CH_3) was reacted with isobutyraldehyde, benzaldehyde, and cyclohexanone [25]. With modifications in the reaction procedures, lithiodimethylhydrazones such as
430 could be used in place of lithioaldimines with similar results.
α-Trimethylsilyl derivatives of carboxylic acids and lactones may be
converted into the corresponding α,β-unsaturated derivatives via a
sequence similar to that employed in Eq. (63) [527].

Lithiated α,β-unsaturated aldimines have been prepared by the
reaction of the corresponding aldimines with LDA in ether. Alkylation
of these species with methyl and allylic halides at low temperature
yields mainly α-alkylated aldimines [528]. Small amounts of dialkylated or γ-alkylated products were also obtained in some cases. γ-
Trimethylsilylation products were produced exclusively when trimethylsilyl chloride was used as the electrophilic reagent [528].

$$R-CH\!=\!\!CH-\overset{\overset{\displaystyle Li}{|}}{N}-t\text{-}Bu \xrightarrow[\text{THF, }0°]{(CH_3)_3SiCl} \overset{\overset{\displaystyle (CH_3)_3Si}{|}}{R-CH-CH}\!=\!N-t\text{-}Bu$$

$$\underline{425} \qquad\qquad\qquad\qquad\qquad\qquad\qquad \underline{426}$$

$$\Big\downarrow \text{ LDA, THF, }0°$$

$$R'CH\!=\!\underset{\underset{\displaystyle R}{|}}{C}-CH\!=\!N\text{-}t\text{-}Bu \xleftarrow[\text{or}]{\overset{\displaystyle \underset{H}{\overset{R'}{\diagdown}}C\!=\!O}{}} \overset{\overset{\displaystyle (CH_3)_3Si}{|}}{R-C}\!=\!\overset{\overset{\displaystyle Li}{|}}{CH}\!-\!N\text{-}t\text{-}Bu$$

$$\underline{427}$$

$$\text{or} \qquad \underset{R''}{\overset{R'}{\diagdown}}C\!=\!O \qquad\qquad\qquad\qquad\qquad (63)$$

$$-78° \text{ to } -20°$$

$$R'R''C\!=\!\underset{\underset{\displaystyle R}{|}}{C}-CH\!=\!Nt\text{-}Bu$$

$$\underline{428}$$

$$\Big\downarrow \text{ Hydrolysis}$$

$$R'CH\!=\!\underset{\underset{\displaystyle R}{|}}{C}-CHO$$

$$\text{or}$$

$$R'R''C\!=\!\underset{\underset{\displaystyle R}{|}}{C}-CHO \qquad\qquad\qquad R-CH\!=\!CH-\overset{\overset{\displaystyle Li}{|}}{N}-N(CH_3)_2$$

$$\underline{429} \qquad\qquad\qquad\qquad\qquad\qquad \underline{430}$$

An elegant method for the synthesis of α-alkyl, α-alkylidene, or α-hydroxyalkyl aldehydes is based on the use of carbanionic species such as *431* or *432*, which may be obtained by treatment of 2-alkyldihydro-1,3-oxazines or 2-alkylthiazolenes with n-butyllithium at -78°C [23]. Electrophilic reagents such as alkyl halides, epoxides, and carbonyl compounds react readily with these carbanions to produce elaborated heterocyclic products, which are converted into masked aldehydes with appropriate reducing agents. The α-substituted aldehyde can then be liberated by hydrolysis. A detailed survey in which these and related methods of synthesis of aldehydes and ketones are covered has appeared recently [23].

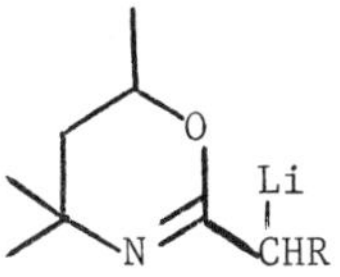

431 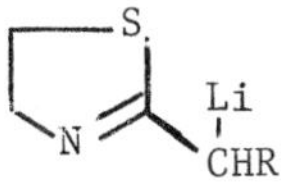   432

D. *Reactions of Aldehyde Enolates with Acylating*
   and Other Electrophilic Reagents

Only a few examples of reactions of aldehyde enolates with acylating and hetero-atom electrophiles have been reported. Diphenylacetaldehyde gave the corresponding enol benzoate in good yield upon treatment with sodium hydride followed by benzoyl chloride [529]. A 6:4 mixture of the E- and Z-trimethylsilyl enol ethers of heptanal was obtained when the aldehyde was treated with sodium anthracene in THF at -30°C and the enolate mixture was quenched with trimethylsilyl chloride [104]. Lithium enolates of aldehydes, prepared either by reaction of the aldehyde with LDA in THF at -30° to 0°C or by cleavage of the corresponding trimethylsilyl enol ethers with methyllithium in THF at room temperature, have been C-sulfenylated in good yield with methyl- or phenylsulfenyl chloride or thiuram disulfide [148].

V. ADDENDUM

Review articles concerning the structure and reactivity of alkali
metal enolates [530] and the factors influencing C- vs. O-alkylation
ratios [531] have appeared recently. The ^{13}C chemical shifts of
sodium, magnesium, and halomagnesium enolates of t-butyl alkyl
ketones have been reported [532]. Detailed instructions for the
preparation of stable 0.5 - 0.6 M solutions of LDA in hexane or
hexane-pentane have been published [533]. It has been shown that
when certain enolizable α-halo- or α-methoxy ketones are treated
with LDA, reduction may compete strongly with deprotonation [534].
The reduction problem was eliminated when LTMP or LHDS was employed
as the base [534]. Deprotonation of 3-pentanone with LTMP in THF
gave slightly more of the E-enolate (84%) [535] than was obtained
when LDA was used as the base [536]. Lithium-arene π-coordination
has been proposed to account for the preferential kinetic deprotonation of 3-arylcyclopentanones toward C2 by strong bases [537].

The lithium enolate of acetaldehyde, generated by the reaction
of n-butyllithium with THF at 25°C, has been trapped by silylation
and O-acylation [538]. A new study on the conversion of α,β-epoxy
ketones to α-substituted-α,β-unsaturated ketones by reduction-
alkylation has been published [539].

New work on the inter- [540] and intramolecular [541] photo-S$_{RN}$1
reactions of metal enolates with aryl halides indicates that hydrogen-
atom transfer to transient phenyl radicals may be the dominant reaction when a β hydrogen atom is present in the metal enolate. Tri-
ethanolamineborate has been shown to be a useful additive for controlling polyalkylation of cyclic and acyclic ketones [542]. Per-
methylations of cyclic and acyclic ketones with methyl iodide have
been accomplished in high yields using potassium hydride as the base
[543]. Magnesium or bromomagnesium enolates of methyl mesityl ketone
have been found to react with chalcone or benzylidene acetone to
give 1,2-addition products after short reaction times [544]. However,
in some cases longer reaction times led to the production of 1,4-
adducts or redistribution products [544]. 2-Enolates of

3-phenylcyclopentanones, formed by conjugate additions of lithium diphenylcuprate to the corresponding enones, have been shown to yield cis-2-alkyl-3-phenylcyclopentanones upon alkylation [545]. Detailed studies on the modes of cycloalkylation of lithium and potassium enolates of ω-bromoketones have been published [533, 546, 547]. Intramolecular cycloalkylations have been employed in connection with the total synthesis of (+)-ishwarone [548] and (+)-damsin and related compounds [549].

Amine-free solutions of lithium enolates in THF have been C-acylated with acid chlorides in good yields at -78°C, or below [550]. Enol carbonates have been obtained by the reaction of lithium enolates, generated with LTMP, with ethyl chloroformate in THF-HMPA [551]. Details for the C-acylation of the metal enolates, produced by conjugate addition of lithium dialkylcuprate complexes to enones, with acid chlorides have been published [552].

The quenching of ketone enolates with dimethyl(methylene)ammonium trifluoroacetate (Eschenmoser's salt) provides a convenient route to α-N,N-dimethylmethylamino ketones (Mannich intermediates) [553, 554]. A study on the influence of enolate structure on the stereochemistry of the aldol condensation has appeared [555]. The products of the aldol condensation of the lithium enolate of 2-methyl-2-trimethylsilyloxy-3-pentanone with aldehydes have been converted to β-hydroxy acids with high diastereomeric purity [556]. A novel synthesis of aldols by reaction of α-bromo ketones with dialkylaluminum chloride and zinc in the presence of carbonyl compounds has been reported [557]. It was suggested that the reaction proceeds via aluminum enolate intermediates [557].

The reaction of acyclic ketones with ethyl trimethylsilyl acetate in the presence of tetrabutylammonium fluoride yields Z-trimethylsilyl enol ethers, which are capable of being cleaved to the corresponding Z-enolates with organometallic reagents [535]. The reaction of β-cyano diethylaluminum enolates, from the addition of diethylaluminum cyanide to enones, with trimethylsilyl chloride gives the corresponding trimethylsilyl enol ethers [558]. The Lewis acid-catalyzed

reaction of silyl enol ethers with tert-butyl chloride has been
shown to provide a good method of synthesis ketones and aldehydes
with tert-alkyl groups at the α position [559]. The synthetic utility
of silyl enol ethers has been reviewed recently [560].

Examples of preparations of α-sulfenylated ketones by reaction
of metal enolates, produced by lithium-ammonia reduction of enones
[561-563], or by conjugate addition of lithium dimethylcuprate or
dimethylaluminum cyanide to enones [562], with sulfenylating agents
have been published. A full paper on the synthesis of α-hydroxy
ketones by the reaction of metal enolates with $MoO_5 \cdot Py \cdot HMPA$ has
appeared [564]. A full paper on the use of metal enolates of conjugated enones as protecting groups for carbonyl functions during metal
hydride reductions has been published [565]. Lithium 6-enolates of
3-substituted cyclohex-3-enones have been generated by lithium-ammonia
reduction of trialkylsilyl aryl ethers followed by cleavage of the
resultant 1,4-dihydroaryl silyl ethers with methyllithium [566].
The kinetic lithium enolate of 2,3-dimethylcyclohex-2-enone has been
found to give a single Michael adduct in low yield with divinyl
ketones [567].

High yields of α-alkylated saturated and unsaturated aldehydes
have been obtained by alkylations of the corresponding potassium
enolates, produced by deprotonation of the parent aldehydes with
potassium hydride, in THF [568]. Details for the use of metallated
dimethyl hydrazones for the regio- and stereoselective alkylation
of carbonyl compounds have been published [569]. Metallated chiral
hydrazones have been used for the asymmetric synthesis of α-alkyl
aldehydes [570] and α-alkyl [571] and α-hydroxyalkyl ketones [572].
Chiral metalloenamines of ketones [573, 574] and aldehydes [575]
have been employed similarly. Lithioenamines have been regiospecifically generated from α,β-unsaturated imines by rearrangements to the
corresponding N-alkenylimines with potassium tert-butoxide followed
by addition of tert-butyllithium [576]. Alkylation and hydrolysis
provided the corresponding α-alkyl ketones. The method has certain
advantages over the reduction-alkylation of enones and can lead to
some relatively inaccessible systems [576].

REFERENCES

1. H. O. House, *Modern Synthetic Reactions*, 2nd Ed., W. A. Benjamin, Menlo Park, Calif., 1972, pp. 492-595.

2. A. C. Cope, H. L. Holmes, and H. O. House, *Org. React.*, *9*, 107 (1957).

3. J. M. Conia, *Rec. Chem. Progr.*, *24*, 43 (1963).

4. H. O. House, *Rec. Chem. Progr.*, *28*, 99 (1967).

5. B. P. Mundy, *J. Chem. Educ.*, *49*, 91 (1972).

6. H. O. House, Ref. 1, pp. 734-765.

7. C. R. Hauser and B. E. Boyd, *Org. React.*, *1*, 266 (1942).

8. C. R. Hauser, F. W. Swamer, and J. T. Adams, *Org. React.*, *8*, 59 (1954).

9. H. O. House, Ref. 1, pp. 629-733.

10. A. T. Nielsen and W. J. Houlihan, *Org. React.*, *16*, 1 (1968).

11. H. O. House, Ref. 1, pp. 595-628.

12. E. D. Bergmann, D. Ginsberg, and R. Pappo, *Org. React.*, *10*, 179 (1959).

13. G. Stork, A. Brizzolara, H. K. Landesman, J. Szmuszkovicz, and R. Terrell, *J. Amer. Chem. Soc.*, *85*, 207 (1963).

14. M. E. Kuehne, *Enamines* (A. G. Cook, ed.), Marcel Dekker, New York, 1969, Chap. 8.

15. H. O. House, Ref. 1, pp. 570-586.

16. P. W. Hickmott, *Chem. Ind.*, 731 (1974).

17. G. Stork and S. R. Dowd, *J. Amer. Chem. Soc.*, *85*, 2178 (1963).

18. G. Stork and S. R. Dowd, *Org. Syn.*, *54*, 46 (1975).

19. G. Wittig, *Rec. Chem. Progr.*, *28*, 45 (1967).

20. G. Wittig and H. Reiff, *Agnew Chem.* (Int. Ed.), *7*, 7 (1968).

21. H. O. House, Ref. 1, p. 682.

22. H. O. House, W. C. Liang, and P. D. Weeks, *J. Org. Chem.*, *39*, 3102 (1974).

23. For a review, see A. I. Meyers, *Heterocycles in Organic Synthesis* (E. C. Taylor and A. Weissberger, eds.), Wiley-Interscience, New York, 1974, Chaps. 9 and 10.

24. E. J. Corey and D. Enders, *Tetrahedron Lett.*, 3 (1976).

25. E. J. Corey, D. Enders, and M. G. Bock, *Tetrahedron Lett.*, 7 (1976).

26. E. J. Corey and D. Enders, *Tetrahedron Lett.*, 11 (1976).

27. W. S. Matthews, J. E. Bares, J. E. Bartmess, F. G. Bordwell,
    F. J. Cornforth, G. E. Drucker, Z. Margolin, R. J. McCallum,
    G. J. McCollum, and N. R. Vanier, *J. Amer. Chem. Soc.*, *97*, 7006
    (1975).

28. H. M. E. Cardwell, *J. Chem. Soc.*, 2442 (1951).

29. W. S. Johnson, *J. Amer. Chem. Soc.*, *65*, 1317 (1943); *66*, 215
    (1944).

30. A. J. Birch and R. Robinson, *J. Chem. Soc.*, 501 (1944).

31. W. S. Johnson and H. Posvic, *J. Amer. Chem. Soc.*, *69*, 1361 (1947

32. R. B. Woodward, A. A. Patchett, D. H. R. Barton, D. A. J. Ives,
    and R. B. Kelley, *J. Amer. Chem. Soc.*, *76*, 2852 (1954); *J. Chem.
    Soc.*, 1131 (1957)1

33. W. S. Johnson, D. S. Allen, Jr., R. R. Hindersinn, G. H. Sausen,
    and R. Pappo, *J. Amer. Chem. Soc.*, *84*, 2181 (1962).

34. R. E. Ireland and J. A. Marshall, *J. Org. Chem.*, *27*, 1615, 1620
    (1962).

35. T. M. Harris and C. M. Harris, *Org. React.*, *17*, 155 (1969).

36. R. B. Woodward, F. Sondheimer, D. Taub, K. Heusler, and W. M.
    McLamore, *J. Amer. Chem. Soc.*, *74*, 4223 (1952).

37. Y. Mazur and F. Sondheimer, *J. Amer. Chem. Soc.*, *80*, 5220 (1958)

38. G. Stork, P. Rosen, and N. L. Goldman, *J. Amer. Chem. Soc.*, *83*,
    2965 (1961).

39. G. Stork, P. Rosen, N. Goldman, R. V. Coombs, and J. Tsuji, *J.
    Amer. Chem. Soc.*, *87*, 275 (1965).

40. J. d'Angelo, *Tetrahedron*, *32*, 2979 (1976).

41. H. O. House and V. Kramar, *J. Org. Chem.*, *28*, 3362 (1963).

42. H. O. House and B. M. Trost, *J. Org. Chem.*, *30*, 1341 (1965).

43. H. O. House, L. J. Czuba, M. Gall, and H. D. Olmstead, *J. Org.
    Chem.*, *34*, 2324 (1969).

44. H. O. House, M. Gall, and H. D. Olmstead, *J. Org. Chem.*, *36*,
    2361 (1971).

45. M. Gall and H. O. House, *Org. Syn.*, *52*, 39 (1972).

46. G. Stork and P. F. Hudrlik, *J. Amer. Chem. Soc.*, *90*, 4462 (1968)

47. G. Stork and P. F. Hudrlik, *J. Amer. Chem. Soc.*, *90*, 4465 (1968)

48. D. Caine and B. J. L. Huff, *Tetrahedron Lett.*, 4695 (1966); 3399
    (1967).

49. B. J. L. Huff, F. N. Tuller, and D. Caine, *J. Org. Chem.*, *34*,
    3070 (1969).

50. B. J. L. Huff, Ph.D. Dissertation, Georgia Institute of Techno-
    logy, 1968.

51. W. J. Powers III, Ph.D. dissertation, Georgia Institute of Technology, 1968.

52. D. E. Pearson and C. A. Buchler, *Chem. Rev.*, *74*, 45 (1974).

53. H. O. House, W. L. Roelofs, and B. M. Trost, *J. Org. Chem.*, *31*, 646 (1966).

54. S. K. Malhotra and F. Johnson, *J. Amer. Chem. Soc.*, *87*, 5513 (1965).

55. C. A. Brown, *Chem. Commun.*, 680 (1974).

56. C. R. Hauser and W. H. Puterbaugh, *J. Amer. Chem. Soc.*, *75*, 1068 (1953).

57. C. A. Vanderwerf and L. F. Lemmenan, *Org. Syn.*, *Coll. Vol.*, *3*, 44 (1955).

58. C. R. Hauser and W. R. Dunnavant, *Org. Syn.*, *Coll. Vol.*, *4*, 963 (1963).

59. D. Caine, *J. Org. Chem.*, *29*, 1868 (1964).

60. P. Rosen, Ph. D. dissertation, Columbia University, 1962.

61. J. A. Marshall and D. J. Schaeffer, *J. Org. Chem.*, *30*, 3642 (1965).

62. H. D. Zook and W. L. Gumby, *J. Amer. Chem. Soc.*, *82*, 1386 (1960).

63. J. S. McConaghy and J. J. Bloomfield, *J. Org. Chem.*, *33*, 3425 (1968).

64. H. W. Whitlock, Jr., *J. Amer. Chem. Soc.*, *84*, 3412 (1962).

65. W. L. Meyer and A. S. Levinson, *J. Org. Chem.*, *28*, 2184 (1963).

66. C. A. Brown, *J. Org. Chem.*, *39*, 3913 (1974).

67. W. S. Johnson and G. H. Daub, *Org. React.*, *6*, 39 (1951).

68. H. O. House and K. A. J. Snoble, *J. Org. Chem.*, *41*, 3076 (1976).

69. H. O. House, personal communication.

70. H. Gilman and B. J. Gaj, *J. Org. Chem.*, *28*, 1725 (1963).

71. C. R. Hauser, D. S. Hoffenberg, W. H. Puterbaugh, and F. C. Frostick, *J. Org. Chem.*, *20*, 1531 (1955).

72. P. Tombouliam, *J. Org. Chem.*, *24*, 229 (1959); P. Tombouliam and K. Stehower, *J. Org. Chem.*, *33*, 1509 (1968).

73. W. B. Renfrow, Jr. and C. R. Hauser, *Org. Syn.*, *Coll. Vol.*, *2*, 607 (1943).

74. H. O. House and V. Kramar, *J. Org. Chem.*, *27*, 4146 (1962).

75. R. Levine, E. Baumgarten, and C. R. Hauser, *J. Amer. Chem. Soc.*, *66*, 1230 (1944).

76. E. J. Corey and C. W. Cantrall, *J. Amer. Chem. Soc.*, *81*, 1745 (1959).

77. E. J. Corey, M. Ohno, R. B. Mitra, and P. A. Vatakencherry, *J. Amer. Chem. Soc.*, *86*, 478 (1964).

78. R. A. Lee and W. Reusch, *Tetrahdedon Lett.*, 969 (1973).

79. R. A. Lee, C. McAndrews, K. M. Patel, and W. Reusch, *Tetrahedron Lett.*, 965 (1973).

80. H. Normant, T. Cuvigny, and D. Reisdorf, *C. R. Acad. Sci. Paris, Ser. C*, *268*, 521 (1969).

81. F. Gaudemar-Barbone and M. Gaudemar, *J. Organometal. Chem.*, *104*, 281 (1976).

82. M. W. Rathke and A. Lindert, *J. Amer. Chem. Soc.*, *93*, 2318 (1971).

83. R. A. Olofson and C. M. Dougherty, *J. Amer. Chem. Soc.*, *95*, 532 (1973).

84. J. L. Herrman, G. R. Kieczykowski, and R. H. Schlessinger, *Tetrahedron Lett.*, 2433 (1973).

85. U. Wannagat and H. Niederprüm, *Chem. Ber.*, *94*, 1540 (1961).

86. G. Stork, J. O. Gardner, R. K. Boeckmann, Jr., and K. A. Parker, *J. Amer. Chem. Soc.*, *95*, 2014 (1973).

87. C. R. Krüger and E. G. Rochow, *J. Organometal. Chem.*, *1*, 476 (1964).

88. M. Tanabe and D. F. Crowe, *Chem. Commun.*, 1498 (1969).

89. D. H. R. Barton, R. H. Hesse, G. Tarzia, and M. M. Pechet, *Chem. Commun.*, 1497 (1969).

90. D. H. R. Barton, R. H. Hesse, M. M. Pechet, and C. Wittshire, *Chem. Comm.*, 1017 (1972).

91. E. Piers, M. B. Geraghty, F. Kido, and M. Soucy, *Syn. Commun.*, *3*, 39 (1973); E. Piers, R. W. Britton, M. B. Geraghty, R. J. Keziere, and F. Kido, *Can. J. Chem.*, *53*, 2838 (1975).

92. E. J. Corey and M. Chaykovsky, *J. Amer. Chem. Soc.*, *87*, 1345 (1965).

93. J. E. McMurry and S. J. Isser, *J. Amer. Chem. Soc.*, *94*, 7132 (1972).

94. J. E. McMurry, *J. Org. Chem.*, *36*, 2826 (1971).

95. J. E. McMurry, *J. Amer. Chem. Soc.*, *90*, 6821 (1968).

96. C. H. Heathcock, R. A. Badger, and J. W. Patterson, Jr., *J. Amer. Chem. Soc.*, *89*, 4133 (1967).

97. E. Piers, W. de Waal, and R. W. Britton, *J. Amer. Chem. Soc.*, *93*, 5113 (1971).

98. J. K. Crandall, R. D. Huntington, and G. L. Brunner, *J. Org. Chem.*, *37*, 2911 (1972).

99. E. J. Corey and D. S. Watt, *J. Amer. Chem. Soc.*, *95*, 2305 (1973).

100. T. Durst, *Adv. Org. Chem.*, *6*, 285 (1969).

101. J. J. Eisch and W. C. Kasha, *J. Org. Chem.*, *27*, 3745 (1962).

102. S. Bank and S. P. Thomas, *Tetrahedron Lett.*, 305 (1973).

103. M. Vora and N. Holy, *J. Org. Chem.*, *40*, 3144 (1975).

104. R. Bourhis and E. Frainnet, *Bull. Soc. Chem. Fr.*, 3552 (1967).

105. A. J. Van der Zeenu and H. G. Gersmann, *Rec. Trav. Chim.*, *84*, 1535 (1965).

106. B. Angelo, *C. R. Acad. Sci.*, *Paris*, *Ser. C*, *273*, 1767 (1971); *276*, 293 (1973); *Bull. Soc. Chem. Fr.*, 1848 (1970).

107. S. Watanabe, K. Suga, T. Fujita, and K. Fujiyoshi, *Chem. Ind.*, 1811 (1969); *Israel J. Chem.*, *8*, 731 (1970).

108. S. Watanabe, K. Suga, T. Fujita, and K. Kujiyoshi, *Chem. Ind.*, 80 (1972).

109. K. Suga, S. Watanabe, and T. Fujita, *Aust. J. Chem.*, *25*, 2393 (1972).

110. J. Fauvarque and J.-F. Fauvarque, *Bull. Soc. Chim. Fr.*, 160 (1969).

111. J. Fauvarque and J.-F. Fauvarque, *Bull. Chem. Soc. Fr.*, 4015 (1969).

112. Y. Maroni-Barnaud, P. Marconi, and R. Cantagrel, *Bull. Soc. Chem. Fr.*, 4051 (1971).

113. Y. Koudsi and Y. Maroni-Barnaud, *Tetrahedron Lett.*, 2525 (1975).

114. H. D. Zook and T. J. Russo, *J. Amer. Chem. Soc.*, *82*, 1258 (1960).

115. H. D. Zook, T. J. Russo, E. F. Ferrand, and D. S. Stotz, *J. Org. Chem.*, *33*, 2222 (1968).

116. A. G. Pinkus, J. G. Lindberg, and A. B. Wu, *Chem. Commun.*, 1350 (1969); 859 (1970).

117. H. O. House, R. A. Auerbach, M. Gall, and N. P. Peet, *J. Org. Chem.*, *38*, 514 (1973).

118. H. O. House, A. V. Prabhu, and W. V. Phillips, *J. Org. Chem.*, *38*, 1209 (1976).

119. H. O. House and B. M. Trost, *J. Org. Chem.*, *41*, 2502 (1965).

120. H. E. Zaugg and A. D. Schaefer, *J. Amer. Chem. Soc.*, *87*, 1857 (1965).

121. H. E. Zaugg, J. F. Ratajczyk, J. E. Leonard, and A. D. Schaeffer, *J. Org. Chem.*, *37*, 2249 (1972).

122. C. Agami, *Bull. Chem. Soc. Fr.*, 39 (1968).

123. D. Martin, A. Weise, and H.-J. Niclas, *Agnew Chem.* (Int. Ed.), *6*, 318 (1967).

124. H. Normant, *Angew. Chem. (Int. Ed.)*, *6*, 1046 (1967).

125. H. Normant, *Bull. Soc. Chim. Fr.*, 791 (1968).

126. A. J. Parker, *Quart. Rev. (London)*, *16*, 163 (1962).

127. A. J. Parker, *Adv. Org. Chem.*, *5*, 1 (1965).

128. H. D. Zook and J. A. Miller, *J. Org. Chem.*, *36*, 1112 (1971).

129. T.-L. Ho, *Chem. Rev.*, *75*, 1 (1975).

130. D. C. Ayres, *Chem. Ind. (London)*, 937 (1973).

131. W. J. LeNoble, *Synthesis*, 1 (1970).

132. D. Caine, *Org. React.*, *23*, 1 (1976).

133. C. D. Ritchie and G. H. Megerle, *J. Amer. Chem. Soc.*, *89*, 1447 (1967).

134. T. J. Wallace and A. Schrishein, *Tetrahedron*, *21*, 2271 (1965).

135. K. W. Bowers, R. W. Giese, J. Grimshaw, H. O. House, N. H. Kolodny, K. Kronberger, and D. K. Roe, *J. Amer. Chem. Soc.*, *92*, 2783 (1970).

136. C. D. Ritchie, G. A. Skinner, and V. G. Badding, *J. Amer. Chem. Soc.*, *89*, 2063 (1967).

137. H. O. House, Ref. 1, pp. 349-352.

138. E. J. Corey, *J. Amer. Chem. Soc.*, *78*, 175 (1954).

139. E. J. Corey and R. A. Sneen, *J. Amer. Chem. Soc.*, *76*, 6269 (1956).

140. S. K. Malhotra and H. J. Ringold, *J. Amer. Chem. Soc.*, *87*, 3228 (1965).

141. S. K. Malhotra and H. J. Ringold, *J. Amer. Chem. Soc.*, *86*, 1997 (1964).

142. G. Subrahmanyan, S. K. Malhotra, and H. J. Ringold, *J. Amer. Chem. Soc.*, *88*, 1332 (1966).

143. F. Johnson, *Chem. Rev.*, *68*, 375 (1968).

144. J. Warkentin and O. S. Tee, *J. Amer. Chem. Soc.*, *88*, 5540 (1966).

145. D. N. Kirk and M. P. Hartshorn, *Steroid Reaction Mechanisms*, Elsevier, New York, 1968, pp. 154-163.

146. K. Iwai, H. Kosugi, and H. Uda, *Chem. Lett.*, 981 (1975).

147. D. Seebach and M. Teschner, *Tetrahedron Lett.*, 5113 (1973).

148. D. Seebach and M. Teschner, *Chem. Ber.*, *109*, 1601 (1976).

149. R. A. Rossi and J. F. Bunnett, *J. Org. Chem.*, *38*, 1407 (1973).

150. I. J. Borowitz, E. W. R. Casper, R. K. Crouch, and K. C. Yee, *J. Org. Chem.*, *37*, 3873 (1972).

151. H. O. House, D. S. Crumrine, A. Y. Teranishi, and H. D. Olmstead, *J. Amer. Chem. Soc.*, *95*, 3310 (1973).

152. H. D. Zook and W. L. Rellahan, *J. Amer. Chem. Soc.*, *79*, 881 (1957).

153. H. D. Zook, W. L. Kelly, and I. Y. Posey, *J. Org. Chem.*, *33*, 3477 (1968).

154. R. E. Ireland and A. K. Willard, *Tetrahedron Lett.*, 3975 (1975).

155. W. A. Kleschick, C. T. Buse, and C. H. Heathcock, *J. Amer. Chem. Soc.*, *99*, 247 (1977).

156. J. E. Dubois and P. Fellmann, *Tetrahedron Lett.*, 1225 (1975).

157. H. J. Ringold and S. K. Malhotra, *J. Amer. Chem. Soc.*, *84*, 3402 (1962).

158. S. K. Malhotra and F. Johnson, *J. Amer. Chem. Soc.*, *87*, 5493 (1965).

159. C. Rappe and W. H. Sachs, *Tetrahedron*, *24*, 6287 (1968).

160. H. W. Amburn, K. C. Kauffman, and H. Shechter, *J. Amer. Chem. Soc.*, *91*, 530 (1969).

161. H. C. Brown, J. H. Brewster, and H. Shechter, *J. Amer. Chem. Soc.*, *76*, 467 (1954).

162. R. D. Clark and C. H. Heathcock, *Tetrahedron Lett.*, 2027 (1974); *J. Org. Chem.*, *41*, 1396 (1976).

163. A. Anthony and T. Maloney, *J. Org. Chem.*, *37*, 1055 (1972).

164. E. J. Corey and R. A. Sneen, *J. Amer. Chem. Soc.*, *77*, 2505 (1955).

165. B. Berkoz, E. P. Chavez, and C. Djerassi, *J. Chem. Soc.*, 1323 (1962).

166. F. Sondheimer, Y. Kilbansky, Y. M. Y. Haddad, G. H. R. Summers, and W. Klyne, *J. Chem. Soc.*, 767 (1961).

167. W. W. Wells and D. H. Neiderhiser, *J. Amer. Chem. Soc.*, *79*, 6569 (1957).

168. Y. Mazur and F. Sondheimer, *J. Amer. Chem. Soc.*, *80*, 6296 (1958).

169. P. Morand, J. M. Lyall, and H. Stollar, *J. Chem. Soc. C*, 2117 (1970).

170. R. Bucourt, *Topics in Stereochemistry*, Vol. 8 (E. L. Eliel and N. L. Allinger, eds.), Wiley, New York, 1974, pp. 159-224.

171. H. J. Reich, J. M. Renga, and I. L. Reich, *J. Amer. Chem. Soc.*, *97*, 5434 (1975).

172. E. Vedejs, *J. Amer. Chem. Soc.*, *96*, 5944 (1974).

173. M. S. Newman and M. D. Farbman, *J. Amer. Chem. Soc.*, *66*, 1550 (1944).

174. H. O. House, W. F. Fischer, Jr., M. Gall, T. E. McLaughlin, and N. P. Peet, *J. Org. Chem.*, *36*, 3439 (1971).

175. S. R. Wilson, M. E. Walters, and B. Orbaugh, *J. Org. Chem.*, *41*, 378 (1976).

176. R. M. Coates, H. D. Pigott, and J. Ollinger, *Tetrahedron Lett.*, 3955 (1974).

177. B. M. Trost, K. Hiroi, and S. Kurozumi, *J. Amer. Chem. Soc.*, *97*, 438 (1975).

178. R. M. Coates, *Angew. Chem.* (Int. Ed.), *12*, 586 (1973).

179. B. M. Trost and K. Hiroi, *J. Amer. Chem. Soc.*, *97*, 6911 (1975).

180. B. M. Trost and A. J. Bridges, *J. Amer. Chem. Soc.*, *98*, 5017 (1976).

181. S. Kamata, S. Uyeo, N. Haga, and W. Nagata, *Syn. Commun.*, *3*, 265 (1973).

182. B. M. Trost and T. N. Salzmann, *J. Amer. Chem. Soc.*, *95*, 6840 (1973); B. M. Trost, T. N. Salzmann, and K. Hiroi, *J. Amer. Chem. Soc.*, *98*, 4887 (1976).

183. P. A. Grieco and J. J. Reap, *Tetrahedron Lett.*, 1097 (1964).

184. R. F. Romanet and R. H. Schlessinger, *J. Amer. Chem. Soc.*, *96*, 3701 (1974).

185. K. B. Sharpless, R. F. Lauer, and A. Y. Teranishi, *J. Amer. Chem. Soc.*, *95*, 6137 (1973).

186. P. Grieco and M. Miyashita, *J. Org. Chem.*, *39*, 120 (1974).

187. G. Stork and B. Ganem, *J. Amer. Chem. Soc.*, *95*, 6152 (1973).

188. G. Stork and J. Singh, *J. Amer. Chem. Soc.*, *96*, 6181 (1974).

189. P. Angibeaud and H. Rivière, *C. R. Acad. Sci., Paris, Ser. C*, *263*, 1076 (1966).

190. A. Spassky-Pasteur, *Bull. Soc. Chim. Fr.*, 2900 (1969).

191. H. A. Smith, B. J. L. Huff, W. J. Powers III, and D. Caine, *J. Org. Chem.*, *32*, 2851 (1967).

192. L. E. Hightower, L. R. Glasgow, K. M. Stone, D. A. Albertson, and H. A. Smith, *J. Org. Chem.*, *35*, 1881 (1970).

193. R. M. Coates and R. L. Sowerby, *J. Amer. Chem. Soc.*, *93*, 1027 (1971).

194. D. Caine, T. I. Chao, and H. A. Smith, *Org. Syn.*, *56*, 52 (1977).

195. H. L. Dryden, Jr., *Organic Reactions in Steroid Chemistry*, Vol. I (J. Fried and J. A. Edwards, eds.), Van Nostrand Reinhold, New York, 1972, p. 1.

196. M. J. Weiss, R. E. Schaub, G. R. Allen, Jr., J. F. Poletto, C. Pidacks, R. B. Conrow, and C. J. Coscia, *Tetrahedron*, *20*, 357 (1964).

197. A. J. Birch, J. E. R. Corrie, P. L. Macdonald, and G. Subba Rao, *J. Chem. Soc. Perkin I*, 1186 (1972).

198. J. E. Dubois, P. Fournier, and C. Lion, *C. R. Acad. Sci., Paris, Ser. C*, *279*, 965 (1974).

199. T. A. Spencer, R. W. Britton, and D. S. Watt, *J. Amer. Chem. Soc.*, *89*, 5727 (1967).

200. G. Stork, personal communication.

201. J. E. Dubois and J. Itzkowitch, *Tetrahedron Lett.*, 2839 (1965).

202. J. E. Dubois and P. Fellmann, *C. R. Acad. Sci., Paris, Ser. C*, *274*, 1307 (1972).

203. W. G. Dauben and R. E. Wolf, *J. Org. Chem.*, *35*, 2361 (1970).

204. G. Stork, S. Uyeo, T. Wakamatsu, P. Grieco, and J. Labovitz, *J. Amer. Chem. Soc.*, *93*, 4945 (1971).

205. P. A. Grieco, Y. Masaki, and D. Doxler, *J. Org. Chem.*, *40*, 2261 (1975).

206. J. D. McChesney and A. F. Wycpalek, *Chem. Commun.*, 542 (1971).

207. G. Stork, *Pure Appl. Chem.*, *17*, 383 (1968).

208. G. Stork, G. L. Nelson, F. Rouessac, and O. Gringore, *J. Amer. Chem. Soc.*, *93*, 3091 (1971).

209. R. A. Kretchner and W. M. Schafer, *J. Org. Chem.*, *38*, 95 (1973).

210. F. Näf and R. Decorzant, *Helv. Chim. Acta*, *57*, 1317 (1974).

211. R. A. Kretchner, E. D. Mihelich, and J. J. Waldron, *J. Org. Chem.*, *36*, 4483 (1972).

212. G. Stork and J. D'Angelo, *J. Amer. Chem. Soc.*, *96*, 7114 (1974).

213. P. Angibeaud, J.-P. Marets, and H. Rivière, *Bull. Soc. Chim. Fr.*, 1845 (1967).

214. J. B. Wiel and F. Rouessac, *Chem. Commun.*, 446 (1976).

215. H. O. House, W. L. Respess, and G. M. Whitsides, *J. Org. Chem.*, *31*, 3128 (1966).

216. J. A. Marshall and A. R. Hochstetler, *J. Amer. Chem. Soc.*, *91*, 648 (1969).

217. J. Klein, R. Levene, and E. Dunkelblum, *Tetrahedron Lett.*, 4031 (1972).

218. R. E. Ireland and G. Pfister, *Tetrahedron Lett.*, 2145 (1969).

219. D. C. Muchmore, *Org. Syn.*, *52*, 109 (1972).

220. R. K. Boeckman, Jr., *J. Amer. Chem. Soc.*, *95*, 6867 (1973).

221. R. K. Boeckman, Jr., *J. Org. Chem.*, *38*, 4450 (1973).

222. R. K. Boeckman, Jr., *J. Amer. Chem. Soc.*, *96*, 6179 (1974).

223. J. W. Patterson, Jr. and J. H. Fried, *J. Org. Chem.*, *39*, 2506 (1974).

224. P. A. Grieco and R. Finkelhor, *J. Org. Chem.*, *38*, 2100 (1973).

225. R. M. Coates and L. O. Sandefur, *J. Org. Chem.*, *39*, 275 (1974).

226. G. H. Posner, J. J. Sterling, C. E. Whitten, C. M. Lentz, and D. J. Brunelle, *J. Amer. Chem. Soc.*, *97*, 107 (1975).

227. G. H. Posner, *Org. React.*, *19*, 1 (1972).

228. E. S. Binkley and C. H. Heathcock, *J. Org. Chem.*, *40*, 2156 (1975).

229. F. Naf, R. Decorzant, and W. Thommen, *Helv. Chim. Acta*, *58*, 1808 (1975).

230. R. G. Salomon and M. F. Salomon, *J. Org. Chem.*, *40*, 1488 (1975).

231. K. K. Heng and R. A. J. Smith, *Tetrahedron Lett.*, 589 (1975).

232. T. Tanaka, S. Kurozumi, T. Toru, and M. Kobayashi, *Tetrahedron Lett.*, 1535 (1975).

233. G. Stork and M. E. Jung, *J. Amer. Chem. Soc.*, *96*, 3682 (1974).

234. P. L. Stotter and K. A. Hill, *J. Org. Chem.*, *38*, 2576 (1973).

235. D. J. Goldsmith and I. Sakano, *Tetrahedron Lett.*, 2857 (1974).

236. H. Rivière and P. W. Tang, *Bull. Soc. Chim. Fr.*, 2455 (1973).

237. A. E. Greene and P. Crabbé, *Tetrahedron Lett.*, 4867 (1976).

238. H. O. House and J. M. Wilkens, *J. Org. Chem.*, *41*, 4031 (1976).

239. H. O. House and W. F. Fisher, Jr., *J. Org. Chem.*, *34*, 3615 (1969).

240. H. O. House, *Acct. Chem. Res.*, *9*, 59 (1976).

241. J. A. Katzenellenbogen and A. L. Crumrine, *J. Amer. Chem. Soc.*, *98*, 4925 (1976).

242. D. Seebach and R. Burstinghaus, *Angew. Chem.* (Int. Ed.), *14*, (1975).

243. W. C. Still, *J. Org. Chem.*, *41*, 3063 (1976).

244. J. A. Noguez and L. A. Maldonado, *Syn. Commun.*, *6*, 39 (1976).

245. E. A. Jeffery, A. Meisters, and T. Mole, *J. Organometal. Chem.*, *74*, 365, 373 (1974).

246. M. Pereyre and J. Valade, *Bull. Soc. Chim. Fr.*, 1928 (1967).

247. M. Pereyre, G. Colin, and J. Valade, *C. R. Acad. Sci.*, *Paris*, *Ser. C*, *264*, 1204 (1967).

248. Y. Odic and M. Pereyre, *C. R. Acad. Sci.*, *Paris*, *Ser. C*, *261*, 469 (1969).

249. B. Ganem, *J. Org. Chem.*, *40*, 146 (1975).

250. J. M. Fortunate and B. Ganem, *J. Org. Chem.*, *41*, 2194 (1976).

251. E. R. H. Jones and D. A. Wilson, *J. Chem. Soc.*, 2933 (1965).

252. D. J. Pasto and P. W. Wojtkowshi, *J. Org. Chem.*, *36*, 1790 (1971).

253. G. Stork and M. Isobe, *J. Amer. Chem. Soc.*, *97*, 4745 (1975).

254. M. Pereyre and Y. Odic, *Tetrahedron Lett.*, 505 (1969).

255. E. Nakamura, T. Murofushi, M. Shimizu, and I. Kuwajima, *J. Amer. Chem. Soc.*, *98*, 2346 (1976).

256. B. M. Trost and M. J. Bogdanowicz, *J. Amer. Chem. Soc.*, *95*, 289 (1973).

257. R. M. Coates, L. O. Sandefur, and R. D. Smillie, *J. Amer. Chem. Soc.*, *97*, 1619 (1975).

258. I. Kuwajima and E. Nakamura, *J. Amer. Chem. Soc.*, *97*, 3257 (1975).

259. C. Lion and J.-E. Dubois, *Tetrahedron*, *31*, 1223 (1975).

260. R. E. Beyler, F. Hoffman, L. H. Sarett, and M. Tishler, *J. Org. Chem.*, *26*, 2426 (1961).

261. S. Binns, J. S. G. Cox, E. R. H. Jones, and B. G. Ketcheson, *J. Chem. Soc.*, 1161 (1964).

262. J.-E. Dubois and C. Lion, *Tetrahedron*, *31*, 1227 (1975).

263. J.-E. Dubois, P. Fournier, and C. Lion, *Tetrahedron Lett.*, 4263 (1975).

264. C. Wakselman and M. Mondon, *Tetrahedron Lett.*, 4285 (1973).

265. J. Valls and E. Toromanoff, *Bull. Soc. Chim. Fr.*, 758 (1961).

266. L. Velluz, J. Valls, and G. Nomine, *Angew. Chem.* (Int. Ed.), *4*, 181 (1965).

267. E. L. Eliel, N. L. Allinger, S. J. Angyal, and G. A. Morrison, *Conformational Analysis*, Interscience, New York, 1965, pp. 307-314.

268. J. W. ApSimon, *Elucidation of Organic Structures by Physical and Chemical Methods*, Vol. IV, Part III, 2nd Ed. (K. W. Bentley and G. W. Kirby, eds.), Wiley-Interscience, New York, 1972, pp. 323-334.

269. J. H. Brewster, *Elucidation of Organic Structures by Physical and Chemical Methods*, Vol. IV, Part III, 2nd Ed. (K. W. Bentley and G. W. Kirby, eds.). Wiley-Interscience, New York, 1972, pp. 79-83.

270. A. Brandstrom, *Ark. Kemi*, *6*, 155 (1954); *7*, 81 (1954); *Acta Chem. Scand.*, *7*, 233 (1953).

271. A. I. Meyers, D. R. Williams, and M. Druelinger, *J. Amer. Chem. Soc.*, *98*, 3032 (1976).

272. L. Tenud, S. Farooq, J. Seibl, and A. Eschenmoser, *Helv. Chem. Acta*, *53*, 2059 (1970).

273. P. A. Tardella, *Tetrahedron Lett.*, 1117 (1969).

274. E. J. Corey, R. Hartmann, and P. A. Vatakencherry, *J. Amer. Chem. Soc.*, *84*, 2611 (1962).

275. G. Stork, *Excerpta Medica, International Congress Series No. 219*, 101 (1970).

276. P. T. Lansbury, P. C. Briggs, T. R. Demmin, and G. E. DuBois, *J. Amer. Chem. Soc.*, *93*, 1311 (1971).

277. G. H. Douglas, J. M. H. Graves, D. Hartley, G. A. Hughes, B. J. McLoughlin, J. Siddall, and H. Smith, *J. Chem. Soc.*, 5072 (1963).

278. D. W. Theobald, *Tetrahedron*, *22*, 2869 (1966).

279. S. A. Narang and P. C. Dutta, *J. Chem. Soc.*, 2842 (1960).

280. J. M. Conia, *Bull. Soc. Chem. Fr.*, 533 (1950).

281. R. G. Pearson, *J. Chem. Educ.*, *45*, 581, 643 (1968).

282. G. Klopman, *J. Amer. Chem. Soc.*, *90*, 223 (1968).

283. G. J. Heiszwolf and H. Koosterziel, *Chem. Commun.*, 51 (1966); *Rec. Trav. Chim.*, *89*, 1153 (1970).

284. R. M. Coates and J. E. Shaw, *J. Org. Chem.*, *35*, 2601 (1970).

285. C. L. Graham and F. J. McQuillin, *J. Chem. Soc.*, 4634 (1963).

286. H. J. Reich and J. M. Renga, *Chem. Commun.*, 135 (1974).

287. P. F. Hudrlik and C. N. Wan, *J. Org. Chem.*, *40*, 2963 (1975).

288. P. Caubere, *Acc. Chem. Res.*, *7*, 301 (1974).

289. F. M. Beringer, W. J. Daniel, S. A. Galton, and G. Rubin, *J. Org. Chem.*, *31*, 4315 (1964).

290. F. M. Beringer, S. A. Galton, and S. J. Huang, *J. Amer. Chem. Soc.*, *84*, 1504 (1962).

291. R. A. Rossi and J. F. Bunnett, *J. Amer. Chem. Soc.*, *94*, 684 (1972).

292. R. A. Rossi and J. F. Bunnett, *J. Org. Chem.*, *38*, 3020 (1973).

293. J. F. Bunnett and J. E. Sundberg, *J. Org. Chem.*, *41*, 1702 (1976).

294. J. F. Bunnett, R. G. Scamehorn, and R. P. Traber, *J. Org. Chem.*, *41*, 3677 (1976).

295. J. F. Bunnett, X. Creary, and J. E. Sundberg, *J. Org. Chem.*, *41*, 1707 (1976).

296. M. F. Semmelhack, B. P. Chong, R. D. Stauffer, T. D. Rogerson, A. Chong, and L. D. Jones, *J. Amer. Chem. Soc.*, *97*, 2507 (1975).

297. M. F. Semmelhack, R. D. Stauffer, and T. D. Rogerson, *Tetrahedron Lett.*, 4519 (1973).

298. M. F. Semmelhack and H. T. Hall, *J. Amer. Chem. Soc.*, *96*, 7091 (1974).

299. C. E. Sacks and P. L. Fuchs, *J. Amer. Chem. Soc.*, *97*, 7372 (1975).

300. F. E. King, T. J. King, and J. G. Topliss, *J. Chem. Soc.*, 919 (1957).

301. H. O. House, B. A. Tefertiller, and H. O. Olmstead, *J. Org. Chem.*, *33*, 935 (1968).

302. P. L. Stotter and K. A. Hill, *J. Amer. Chem. Soc.*, *96*, 6524 (1974).

303. H. Laurent and R. Weichert, *Organic Reactions in Steroid Chem.*, Vol. 2, (J. Fried and J. A. Edwards, eds.), Reinhold, New York, 1972, pp. 86-99.

304. R. S. Matthews, P. K. Hyer, and E. A. Folkers, *Chem. Commun.*, 38 (1970).

305. R. S. Matthews, S. S. Girgenti, and E. A. Folkers, *Chem. Commun.*, 708 (1970).

306. P. Lansbury and G. E. DuBois, *Tetrahedron Lett.*, 3305 (1972).

307. G. P. Moss and S. A. Nicholaides, *J. Chem. Soc.*, *D*, 1077 (1969).

308. J. W. Cook and C. A. Lawrence, *J. Chem. Soc.*, 501 (1937).

309. H. O. House and C. J. Blankley, *J. Org. Chem.*, *33*, 47 (1968).

310. G. Stork and J. E. McMurry, *J. Amer. Chem. Soc.*, *89*, 5464 (1967).

311. R. E. Ireland, S. W. Baldwin, B. J. Dawson, M. I. Dawson, J. E. Dolfini, J. Newbould, W. S. Johnson, M. Brown, R. J. Crawford, P. F. Hudrlik, G. H. Rasmussen, and K. K. Schmiegel, *J. Amer. Chem. Soc.*, *92*, 5743 (1970).

312. R. Deghenghi, C. Revesz, and R. Gaudry, *J. Med. Chem.*, *6*, 301 (1963).

313. J. A. Marshall and A. E. Greene, *Tetrahedron*, *23*, 4183 (1967).

314. H. O. House and T. M. Bare, *J. Org. Chem.*, *33*, 943 (1968).

315. M. J. Jung, *Tetrahedron*, *32*, 3 (1976).

316. G. Stork, *Pure Appl. Chem.*, *43*, 553 (1975).

317. J. A. Marshall and W. I. Fanta, *J. Org. Chem.*, *29*, 2501 (1964).

318. H. O. House and M. J. Lusch, *J. Org. Chem.*, *42*, 183 (1977).

319. J. Bertrand, N. Cabrol, L. Gorrichon-Guigon, and Y. Maroni-Barnaud, *Tetrahedron Lett.*, 4683 (1973).

320. A. Rosen and M. Rosenblum, *J. Org. Chem.*, *40*, (1975).

321. O. Wichterle, J. Prochazka, and J. Hoffman, *Collect. Czech. Chem. Commun.*, *13*, 300 (1948).

322. D. Caine and F. N. Tuller, *J. Org. Chem.*, *34*, 222 (1969).

323. G. Stork, S. Danishefsky, and M. Ohashi, *J. Amer. Chem. Soc.*, *89*, 5459 (1967).

324. G. Stork and J. E. McMurry, *J. Amer. Chem. Soc.*, *89*, 5463 (1967).

325. J. E. McMurry, *Org. Syn.*, *53*, 70 (1973).

326. J. Pugach, Ph. D. dissertation, Columbia University, 1964; *Diss. Abstr. B*, *25* (10), 5567 (1965).

327. T. C. McKenzie, Ph. D. dissertation, Columbia, University, 1973.

328. J. W. Scott, B. L. Banner, and G. Saucy, *J. Org. Chem.*, *37*, 1664 (1972).

329. J. M. Conia and P. Briet, *Bull. Soc. Chem. Fr.*, *3881*, 3888 (1966).

330. S. Boatman, T. M. Harris, and C. R. Hauser, *Org. Syn.*, *48*, 40 (1968).

331. E. B. McCall and B. B. Willard, *J. Chem. Soc.*, 1911 (1957).

332. F. W. Swamer and C. R. Hauser, *J. Amer. Chem. Soc.*, *72*, 1352 (1950).

333. A. P. Krapcho, J. Diamanti, C. Cayen, and R. Bingham, *Org. Syn.* *47*, 20 (1967).

334. L. Ruest, G. Blouin, and P. Deslongchamps, *Syn. Commun.*, *6*, 169 (1976).

335. H. R. Snyder, L. A. Brooks, and S. H. Shapiro, *Org. Syn.*, *Coll. Vol.*, *2*, 531 (1943).

336. I. Shahak, *Tetrahedron Lett.*, 2201 (1966).

337. M. Stiles and H. L. Finkbeiner, *J. Amer. Chem. Soc.*, *81*, 505 (1959).

338. M. Stiles, *J. Amer. Chem. Soc.*, *81*, 2598 (1959).

339. S. W. Pelletier, R. L. Chappell, P. C. Parthasarathy, and N. Lewin, *J. Org. Chem.*, *31*, 1747 (1966).

340. E. J. Corey and R. H. K. Chen, *J. Org. Chem.*, *38*, 4086 (1973).

341. R. Mayer, *Newer Methods of Preparative Organic Chemistry*, Vol. 2 (W. Foerst, ed.), Academic Press, New York, 1963, pp. 101-131

342. E. Piers, R. W. Britton, and W. deWaal, *Can. J. Chem.*, *47*, 831 (1969).

343. D. A. Evans and C. L. Sims, *Tetrahedron Lett.*, 4691 (1973).

344. J. A. Marshall, G. L. Bundy, and W. I. Fanta, *J. Org. Chem.*, *33*, 3913 (1968).

345. R. W. Guthrie, W. A. Henry, H. Immer, C. M. Wong, Z. Valenta, and K. Wiesner, *Collect. Czech. Chem. Commun.*, *31*, 602 (1966).

346. R. B. Woodward, I. J. Pachter, and M. L. Scheinbaum, *Org. Syn.*, *54*, 37 (1974).

347. S. Boatman, T. M. Harris, and C. R. Hauser, *J. Amer. Chem. Soc.*, *87*, 82 (1965).

348. T. M. Bare, N. D. Hershey, H. O. House, and C. G. Swain, *J. Org. Chem.*, *37*, 997 (1972).

349. H. O. House and M. J. Umen, *J. Org. Chem.*, *38*, 1000 (1973).

350. K. Dawes, N. J. Turro, and J. M. Conia, *Tetrahedron Lett.*, 1377 (1971).

351. R. E. Ireland, P. S. Grand, R. E. Dickerson, J. Bordner, and D. R. Rydjeski, *J. Org. Chem.*, *35*, 570 (1970).

352. G. Stork, *Proceedings of 23rd International Congress of Pure and Applied Chemistry*, Vol. 2, Boston, Mass., July 1971, p. 193.

353. H. O. House and M. J. Umen, *J. Org. Chem.*, *37*, 2841 (1972).

354. J. E. Cole, W. S. Johnson, P. A. Robins, and J. Walker, *J. Chem. Soc.*, 244 (1962).

355. J. Klein, *Tetrahedron*, *30*, 3349 (1974).

356. L. H. Sarett, W. F. Johns, R. E. Beyler, R. M. Lukes, G. I. Poos, and G. E. Arth, *J. Amer. Chem. Soc.*, *75*, 2112 (1953).

357. A. Kende, *Org. React.*, *11*, 261 (1960).

358. N. J. Turro, *Acct. Chem. Res.*, *2*, 25 (1969).

359. F. G. Bordwell, *Acct. Chem. Res.*, *3*, 281 (1970).

360. R. A. Bartsch and D. M. Cook, *J. Org. Chem.*, *35*, 1714 (1970).

361. A. C. Knipe and C. J. M. Stirling, *J. Chem. Soc.*, *B*, 67 (1968).

362. A. P. Krapcho, *Synthesis*, 383 (1974).

363. R. Mayer and H. J. Schrubert, *Chem. Ber.*, *91*, 768 (1958).

364. N. A. Nelson and G. A. Mortimer, *J. Org. Chem.*, *22*, 1146 (1957).

365. G. Stork and J. Ficini, *J. Amer. Chem. Soc.*, *83*, 4678 (1961).

366. (a) C. Ho and F. T. Bond, *J. Amer. Chem. Soc.*, *96*, 7355 (1974); (b) F. T. Bond and C.-Y Bond, *J. Org. Chem.*, *41*, 1421 (1976).

367. R. B. Kelly, J. Zamecnik, and B. A. Beckett, *Can. J. Chem.*, *50*, 3455 (1972).

368. P. F. Hudrlik and M. M. Mohtady, *J. Org. Chem.*, *40*, 2692 (1975).

369. J. Meinwald, *Rec. Chem. Progr.*, *22*, 43 (1961).

370. K. B. Wiberg and G. W. Kline, *Tetrahedron Lett.*, 104 (1963).

371. E. Wenkert, P. Bakuzis, R. J. Baumgarten, C. L. Leicht, and H. P. Schenk, *J. Amer. Chem. Soc.*, *93*, 3208 (1971).

372. S. Wolff and W. C. Agosta, *Chem. Commun.*, 771 (1973).

373. S. J. Etheredge, *J. Org. Chem.*, *31*, 1990 (1966).

374. C. F. Wilcox, Jr. and G. C. Whitney, *J. Org. Chem.*, *32*, 2933 (1967).

375. C. H. Heathcock, *Tetrahedron Lett.*, 2043 (1966).

376. P. C. Mukharji and T. K. Das Gupta, *Tetrahedron*, *25*, 5275 (1969).

377. C. H. Heathcock and D. Gray, *Tetrahedron*, *27*, 1239 (1971).

378. P. C. Mukharji and A. N. Ganguly, *Tetrahedron*, *25*, 5267 (1969).

379. G. Baddeley, E. K. Baylis, B. G. Heaton, and S. W. Rasburn, *Proc. Chem. Soc.*, 451 (1961).

380. H. Christol, M. Musserson, and F. Plenat, *Bull. Soc. Chem. Fr.*, 543 (1959).

381. W. F. Johns, R. M. Lukes, and L. H. Sarett, *J. Amer. Chem. Soc.*, *76*, 5026 (1954).

382. J. L. Marshell, *Tetrahedron Lett.*, 735 (1971).

383. R. L. Cargill, D. F. Bushey, P. D. Ellis, S. Wolff, and W. C. Agosta, *J. Org. Chem.*, *39*, 573 (1974).

384. A. P. Krapcho and J. E. McCullough, *J. Org. Chem.*, *32*, 2453 (1967).

385. E. N. Marvell, D. Sturmer, and C. Powell, *Tetrahedron*, *22*, 861 (1966).

386. J. M. Conia and F. Rouessac, *Tetrahedron*, *16*, 45 (1961).

387. A. Belanger, Y. Lambert, and P. Deslongchamps, *Can. J. Chem.*, *47*, 795 (1969).

388. J. R. Wiseman, H.-F. Chan, and C. J. Ahola, *J. Amer. Chem. Soc.*, *91*, 2812 (1969).

389. I. J. Borowitz and N. Sucin, *J. Org. Chem.*, *38*, 1061 (1973).

390. H. Nozaki, H. Yamamoto, and T. Mori, *Can. J. Chem.*, *47*, 1107 (1969).

391. J. Froborg, G. Magnusson, and S. Thorén, *J. Org. Chem.*, *39*, 848 (1974).

392. G. Stork, J. M. Tabak, and J. F. Blount, *J. Amer. Chem. Soc.*, *94*, 4737 (1972).

393. K. Yoshida and Y. Yamashita, *Tetrahedron Lett.*, 693 (1966).

394. O. Touster, *Org. React.*, *7*, 327 (1953).

395. R. E. Schaub, W. Fulmor, and M. J. Weiss, *Tetrahedron*, *20*, 373 (1964).

396. H. Feuer and P. M. Pivawer, *J. Org. Chem.*, *31*, 3152 (1966).

397. H. Feuer, A. M. Hall, S. Golden, R. L. Reitz, *J. Org. Chem.*, *33*, 3622 (1968).

398. R. E. Ireland, D. C. Muchmore, and V. Hengartner, *J. Amer. Chem. Soc.*, *94*, 5098 (1972).

399. M. E. Kuehne and J. A. Nelson, *J. Org. Chem.*, *35*, 161 (1970).

400. T. A. Spencer, R. J. Friary, W. W. Schmiegal, J. F. Simeone, and D. S. Watt, *J. Org. Chem.*, *33*, 719 (1968).

401. W. M. Muir, P. D. Ritchie, and D. J. Lyman, *J. Org. Chem.*, *31*, 3790 (1966).

402. B. O. Linn and C. R. Hauser, *J. Amer. Chem. Soc.*, *78*, 6066 (1956).

403. T. A. Spencer, T. D. Weaver, R. M. Villarica, R. J. Friary, J. Posler, and M. A. Schwartz, *J. Org. Chem.*, *33*, 712 (1968).

404. M. R. Czarny, K. K. Maheshwari, J. A. Nelson, and T. A. Spencer, *J. Org. Chem.*, *40*, 2079 (1975).

405. R. A. Auerbach, D. S. Crumrine, D. L. Ellison, and H. O. House, *Org. Syn.*, *54*, 49 (1975).

406. G. Stork, G. A. Kraus, and G. A. Garcia, *J. Org. Chem.*, *39*, 3460 (1974).

407. M. Gaudemar, *C. R. Acad. Sci.*, *Paris*, *Ser. C*, 961 (1974).

408. I. Kuwajima, T. Sato, M. Arai, and N. Minami, *Tetrahedron Lett.*, 1817 (1976).

409. A. G. Schultz and Y. K. Yee, *J. Org. Chem.*, *41*, 4045 (1976).

410. J. E. Dubois and M. Dubois, *Chem. Commun.*, 1567 (1968).

411. Y. Ito, T. Konoike, and T. Saegusa, *J. Amer. Chem. Soc.*, *97*, 649 (1975).

412. T. Mukaiyama, K. Banno, and K. Narasaka, *J. Amer. Chem. Soc.*, *96*, 7503 (1974).

413. L. Fieser and M. Fieser, *Reagents of Organic Synthesis*, Vol. 5, Wiley-Interscience, New York, 1975, pp. 588-589.

414. R. H. Reuss and A. Hassner, *J. Org. Chem.*, *39*, 1785 (1974).

415. G. M. Rubottom, J. M. Gruber, and G. M. Mong, *J. Org. Chem.*, *41*, 1673 (1976).

416. S. Danishefsky, T. Kitahara, R. McKee, and P. F. Schuda, *J. Amer. Chem. Soc.*, *98*, 6715 (1976).

417. F. T. Bond, C-Y. Ho, and O. McConnell, *J. Org. Chem.*, *41*, 1416 (1976).

418. Reference 1, pp. 459-478.

419. D. Caine and A. S. Frobese, unpublished work.

420. A. G. Pinkus, W. C. Servoss, and K. K. Lum, *J. Org. Chem.*, *32*, 2649 (1967).

421. J. N. Gardner, F. E. Carlon, and O. Gnoj, *J. Org. Chem.*, *33*, 1566, 3294 (1968).

422.   G. A. Russell, E. G. Janzen, A. G. Bemis, E. J. Geels, A. J.
       Moye, S. Mak, and E. T. Strom, "Selective Oxidation Processes,"
       Advances in Chemistry, No. 51, American Chemical Society,
       Washington, D.C., 1965, pp. 112-171.

423.   J. W. Ellis, Chem. Commun., 406 (1970).

424.   J. W. Huffman, C. A. Miller, and A. R. Pinder, J. Org. Chem.,
       41, 3705 (1976).

425.   A. E. Greene, J.-C. Muller, and G. Ourisson, J. Org. Chem.,
       39, 186 (1974).

426.   J. Klein, R. Levene, and E. Dunkelblum, Tetrahedron Lett.,
       2845 (1972).

427.   Y. Ito, T. Konoike, and T. Saegusa, J. Amer. Chem. Soc., 97,
       2912 (1975).

428.   S. K. Malhotra and H. J. Ringold, J. Amer. Chem. Soc., 86, 1997
       (1964).

429.   A. J. Birch, J. Chem. Soc., 2325 (1950).

430.   K. M. Patel and W. Reusch, J. Org. Chem., 40, 1504 (1975).

431.   J. E. Telschow and W. Reusch, J. Org. Chem., 40, 862 (1975).

432.   R. A. Lee, Tetrahedron Lett., 3333 (1973).

433.   M. Tanabe and D. F. Crowe, Chem. Commun., 564 (1973).

434.   A. G. Schultz and D. S. Kashdan, J. Org. Chem., 38, 3815 (1973).

435.   A. deGroot and B. J. M. Jansen, Recl. Trav. Chim., 95, 81 (1976).

436.   G. Stork and G. A. Kraus, J. Amer. Chem. Soc., 98, 2351 (1976).

437.   G. Stork and R. L. Danheiser, J. Org. Chem., 38, 1775 (1973).

438.   A. S. Kende and R. G. Eilerman, Tetrahedron Lett., 697 (1973).

439.   C. Girard and J. M. Conia, Tetrahedron Lett., 3327 (1974).

440.   R. M. Cory and D. M. T. Chan, Tetrahedron Lett., 4441 (1975).

441.   R. B. Gammill and T. A. Bryson, Tetrahedron Lett., 2693 (1975).

442.   R. B. Gammill and T. A. Bryson, Syn. Commun., 6, 209 (1976).

443.   R. D. Clark and C. H. Heathcock, J. Org. Chem., 41, 636 (1976).

444.   P. M. Wege, R. D. Clark, and C. H. Heathcock, J. Org. Chem.,
       41, 3144 (1976).

445.   D. Caine, A. A. Boucugnani, and W. R. Pennington, J. Org. Chem.,
       41, 3632 (1976).

446.   G. Stork, R. L. Danheiser, and B. Ganem, J. Amer. Chem. Soc.,
       95, 3414 (1973).

447.   B. H. Toder, S. J. Branca, R. K. Dieter, and A. B. Smith III,
       Syn. Commun., 5, 435 (1975).

448.   L. Nedelec, J. C. Gasc, and R. Bucourt, Tetrahedron, 30, 3263
       (1974).

449. P. S. Wharton and C. E. Sundin, J. Org. Chem., 33, 4255 (1968).

450. J. M. Conia and A. LeCraz, Bull. Soc. Chem. Fr., 1934 (1960).

451. H. J. Ringold and S. K. Malhotra, Tetrahedron Lett., 669 (1962).

452. Y. Nakadavia, J. Hayashi, H. Sato, and K. Nakanishi, J. Chem. Soc. Chem. Commun., 282 (1972).

453. F. J. McQuillin and P. L. Simpson, J. Chem. Soc., 4726 (1963).

454. C. L. Graham and F. J. McQuillin, J. Chem. Soc., 4521 (1964).

455. Z. G. Hajos, R. A. Micheli, D. R. Parrish, and E. P. Oliveto, J. Org. Chem., 32, 3008 (1967).

456. T. A. Bryson, personal communication.

457. M. S. Newman, V. deVries, and R. Darlak, J. Org. Chem., 31, 2171 (1966).

458. P. S. Wharton, C. E. Sundin, D. W. Johnson, and H. C. Kluender, J. Org. Chem., 37, 34 (1972).

459. E. L. Shapiro, T. Legatt, L. Weber, and E. P. Oliveto, M. Tanabe, and D. F. Crow, Steroids, 3(2), 183 (1964).

460. G. Kruger, J. Org. Chem., 33, 1750 (1968).

461. S. Danishefsky, P. Solomon, L. Crawley, M. Sax, S. C. Yoo, E. Abola, and J. Pletcher, Tetrahedron Lett., 961 (1972).

462. H. Hart, G. M. Love, and I. C. Wang, Tetrahedron Lett., 1377 (1973).

463. A. L. Nussbaum, G. B. Topplis, T. L. Popper, and E. P. Oliveto, J. Amer. Chem. Soc., 81, 4574 (1959).

464. R. E. Schaub and M. J. Weiss, Chem. Ind. (London), 2003 (1961).

465. K. P. Dastur, Tetrahedron Lett., 4333 (1973).

466. M. Narisada and F. Watanabe, J. Org. Chem., 38, 388 (1973).

467. E. Toromanoff and R. Bucourt, Tetrahedron Lett., 3523 (1976).

468. R. L. Cargill and T. E. Jackson, J. Org. Chem., 38, 2125 (1973).

469. J. A. Edwards, M. C. Calzada, L. E. Ibanez, M. E. Cabezas Rivera, R. Urguiza, L. Carbona, J. C. Orr, and A. Bowers, J. Org. Chem., 29, 3481 (1964).

470. H. J. Ringold, E. Batres, O. Halpern, and E. Necoechea, J. Amer. Chem. Soc., 81, 427 (1959).

471. D. Caine, P. F. Brake, J. F. DeBardeleben, Jr., and J. B. Dawson, J. Org. Chem., 38, 967 (1973).

472. E. Wenkert and B. G. Jackson, J. Amer. Chem. Soc., 81, 5601 (1959).

473. K. S. Ayyer, R. C. Cookson, and K. Kagi, J. Chem. Soc. Perkin I, 1727 (1975).

474. H. E. Zimmerman, in *Molecular Rearrangements*, Part I (P. deMayo, ed.), Interscience, New York, 1963, p. 348.

475. N. W. Atwater, *J. Amer. Chem. Soc.*, *82*, 2847 (1960).

476. H. J. Ringold and G. Rosenkranz, *J. Org. Chem.*, *22*, 602 (1957).

477. F. Sondheimer and Y. Mazur, *J. Amer. Chem. Soc.*, *79*, 2906 (1957).

478. Nussbaum, T. L. Topliss, G. Brabagon, T. L. Popper, and E. P. Oliveto, *J. Amer. Chem. Soc.*, *81*, 4572 (1959).

479. M. Yanagita, M. Hirakura, and F. Seki, *J. Org. Chem.*, *23*, 841 (1958).

480. W. G. Dauben and A. C. Ashcroft, *J. Amer. Chem. Soc.*, *85*, 3673 (1963).

481. G. Stork and J. Benaim, *J. Amer. Chem. Soc.*, *93*, 5938 (1971); *Org. Syn.*, *53*, 168 (1973).

482. O. I. Fedorova, *Izv. Akad. Nauk SSSR*, 117 (1965).

483. G. H. Douglas, J. M. H. Graves, D. Hartley, G. A. Hughes, B. J. McLoughlin, J. Siddall, and H. S. Smith, *J. Chem. Soc.*, 5072 (1963); H. Smith, G. A. Hughes, and B. J. McLoughlin, *Experimentia*, *19*, 177 (163).

484. F. H. Bottom and F. J. McQuillin, *Tetrahedron Lett.*, 1975 (1967).

485. R. Zurflüh, E. N. Wall, J. B. Siddall, and J. A. Edwards, *J. Amer. Chem. Soc.*, *90*, 6224 (1968).

486. G. Stork and J. W. Schulenberg, *J. Amer. Chem. Soc.*, *84*, 284 (1962).

487. G. Amiard, R. Heymès, T. Van Thuong, and J. Mathieu, *Bull. Soc. Chim. Fr.*, 2321 (1965).

488. E. R. H. Jones, G. D. Makins, and J. S. Stephenson, *J. Chem. Soc.*, 2165 (1968).

489. F. H. Bottom and F. J. McQuillin, *Tetrahedron Lett.*, 459 (1968).

490. T. A. Bryson and R. B. Gammill, *Tetrahedron Lett.*, 3963 (1974).

491. M. Yoshimoto, N. Ishida, and T. Hiraoka, *Tetrahedron Lett.*, 39 (1973).

492. R. B. Gammill and T. A. Bryson, *Synthesis*, *6*, 401 (1976).

493. H. Christol, M. Mousseron, and R. Salle, *Bull. Soc. Chim. Fr.*, 556 (1958).

494. P. Grafen, H. J. Kabbe, O. Roos, G. D. Diana, T. Li, and R. B. Turner, *J. Amer. Chem. Soc.*, *90*, 6131 (1968).

495. C. Mercier, A. R. Addas, and P. Deslongchamps, *Can. J. Chem.*, *50*, 1882 (1972).

496. R. B. Bates, G. Büchi, T. Matsura, and A. R. Shaffer, *J. Amer. Chem. Soc.*, *82*, 2327 (1960).

497. D. Caine and J. T. Gupton III, *J. Org. Chem.*, *40*, 809 (1975); J. T. Gupton III, Ph.D. dissertation, Georgia Institute of Technology, 1975.

498. J. H. Fassnacht and N. A. Nelson, *J. Org. Chem.*, *27*, 1885 (1962).

499. A. P. Johnson, *International Symposium on Synthetic Methods and Rearrangements in Alicyclic Chemistry*, Oxford, July 22-24, 1969, Abstr., p. 13.

500. J. J. Bonet, H. Wehrli, and K. Schaffer, *Helv. Chim. Acta*, *45*, 2615 (1962).

501. P. C. Mukharji and A. N. Ganguly, *Tetrahedron*, *25*, 5281 (1969).

502. J. E. McMurry and M. G. Silvestri, *J. Org. Chem.*, *41*, 3953 (1976).

503. C. Huynh and S. Julia, *Bull. Soc. Chim. Fr.*, 1974 (1972).

504. R. A. Micheli, Z. G. Hajos, N. Cohen, D. R. Parrish, L. A. Portland, W. Sciamanna, M. A. Scott, and P. A. Wehri, *J. Org. Chem.*, *40*, 675 (1975).

505. H. O. House, W. C. Liang, and P. D. Weeks, *J. Org. Chem.*, *39*, 3102 (1974).

506. Y. Odic and M. Pereyre, *C. R. Acad. Sci.*, *Paris*, *Ser. C*, *270*, 100 (1970).

507. H. K. Dietl and K. C. Brannock, *Tetrahedron Lett.*, 1273 (1973).

508. R. E. Ireland, L. M. Mander, *Tetrahedron Lett.*, 3453 (1964); *J. Org. Chem.*, *32*, 689 (1967).

509. R. E. Ireland and L. N. Mander, *J. Org. Chem.*, *34*, 142 (1969).

510. W. Nagata, T. Sugasawa, M. Narisada, T. Wakabayashi, and Y. Hayase, *J. Amer. Chem. Soc.*, *85*, 2342 (1967); *89*, 1483 (1967).

511. S. W. Pelletier and D. L. Herald, Jr., *Chem. Commun.*, 10 (1971).

512. Y. Letourneux, G. Bijuktur, M. T. Ryzlak, A. K. Banerjee, and M. Gut, *J. Org. Chem.*, *41*, 2288 (1976).

513. S. A. G. deGraaf, P. E. R. Oosterhoff, and A. van der Gen, *Tetrahedron Lett.*, 1653 (1974).

514. G. J. Heiszwolf and H. Kloosterziel, *Rec. Trav. Chim.*, *86*, 807 (1967).

515. G. J. Heiszwolf, J. A. A. van Drunen, and H. Kloosterziel, *Rec. Trav. Chim.*, *88*, 1377 (1969).

516. R. E. Ireland and P. W. Schiess, *J. Org. Chem.*, *28*, 6 (1963).

517. R. F. Church and R. E. Ireland, *J. Org. Chem.*, *28*, 17 (1963).

518. R. E. Ireland, S. W. Baldwin, D. J. Dawson, M. I. Dawson, J. E. Dolfini, J. Newbould, W. S. Johnson, M. Brown, R. J. Crawford, P. F. Hudrlik, G. H. Rasmussen, and K. K. Schmiegel, *J. Amer. Chem. Soc.*, *92*, 5743 (1970).

519. G. Opitz and H. Mildenberger, *Ann. Chim.*, *649*, 26 (1961).

520. E. Elkik, *Bull. Soc. Chim. Fr.*, 972 (1960).

521. K. C. Brannock and R. D. Burpitt, *J. Org. Chem.*, *26*, 3576 (1961).

522. G. Opitz, H. Hellman, H. Mildenberger, and H. Suhr, *Ann. Chim.*, *649*, 36 (1961).

523. T. J. Curphey, J. C. Hung, C. C. C. Chu, *J. Org. Chem.*, *40*, 607 (1975).

524. Th. Cuvigny and H. Normant, *Bull. Soc. Chim. Fr.*, 3976 (1970).

525. G. Wittig and A. Hesse, *Org. Syn.*, *50*, 66 (1970).

526. G. Wittig and P. Suchanek, *Tetrahedron*, *Suppl. 8*, *22*, 347 (1968).

527. P. A. Grieco, C. J. Wengand, and S. S. Burke, *Chem. Commun.* 537 (1975).

528. K. Takabe, H. Fujiwara, T. Katagiri, and J. Tanaka, *Tetrahedron Lett.*, 1237 (1975).

529. H. O. House and D. J. Reif, *J. Amer. Chem. Soc.*, 77, 6525 (1955).

530. L. M. Jackman and B. C. Lange, *Tetrahedron*, *33*, 2737 (1977).

531. O. A. Reutov and A. L. Kurts, *Russ. Chem. Rev.*, *46*, 1964 (1977).

532. P. Fellmann and J.-E. Dubois, *Tetrahedron Lett.*, 247 (1977).

533. H. O. House, W. V. Phillips, T. S. B. Sayer, and C.-C. Yau, *J. Org. Chem.*, *43*, 700 (1978).

534. C. Kowalski, X. Creary, A. J. Rollin, and M. C. Burke, *J. Org. Chem.*, *43*, 2601 (1978).

535. E. Nakamura, K. Hashimoto, and I. Kuwajima, *Tetrahedron Lett.*, 2079 (1978).

536. R. E. Ireland, R. H. Mueller, and A. K. Willard, *J. Amer. Chem. Soc.*, *98*, 2868 (1976).

537. G. H. Posner and C. M. Lentz, *Tetrahedron Lett.*, 3211 (1977).

538. M. E. Jung and R. B. Blum, *Tetrahedron Lett.*, 3791 (1977).

539. R. P. Szajewski, *J. Org. Chem.*, *43*, 1819 (1978).

540. J. F. Wolfe, M. P. Moon, M. C. Sleeir, J. F. Bunnett, and R. R. Bard, *J. Org. Chem.*, *43*, 1019 (1978).

541. M. F. Semmelhack and T. M. Bargar, *J. Org. Chem.*, *42*, 1481 (1977).

542. M. W. Rathke and A. Lendert, *Syn. Commun.*, *8*, 9 (1978).

543. A. A. Millard and M. W. Rathke, *J. Org. Chem.*, *43*, 1834 (1978).

544. J. Bertrand, L. Gorrichon, and P. Maroni, *Tetrahedron Lett.*, 4207 (1977).

545. G. H. Posner and C. M. Lentz, *Tetrahedron Lett.*, 3215 (1977).

546. H. O. House, T. S. B. Sayer, and C.-C. Yau, *J. Org. Chem.*, *43*, 2153 (1978).

547. J. E. Baldwin and L. I. Kruse, *J. Chem. Soc. Chem. Commun.*, 233 (1977).

548. E. Piers and T.-W. Hall, *J. Chem. Soc. Chem. Commun.*, 880 (1977).

549. P. A. Grieco, Y. Ohfume and G. Majetich,*J. Amer. Chem. Soc .*, *99*, 7393 (1977).

550. A. K. Beck, M. S. Hoekstra, and D. Seebach, *Tetrahedron Lett.*, 1187 (1977).

551. R. A. Olofson, J. Cuomo, and B. A. Bauman, *J. Org. Chem.*, *43*, 2073 (1978).

552. T. Tamaka, S. Kurozumi, T. Toru, M. Kobayashi, S. Miura, and S. Ishimoto, *Tetrahedron*, *33*, 1105 (1977).

553. M. L. Holy and Y. F. Wang, *J. Amer. Chem. Soc.*, *99*, 944 (1977).

554. J. L. Roberts, P. S. Borromeo, and C. D. Poulter, *Tetrahedron Lett.*, 1621 (1977).

555. P. Fellmann and J.-E. Dubois, *Tetrahedron*, *34*, 1349 (1978).

556. C. T. Buse and C. H. Heathcock, *J. Amer. Chem. Soc.*, *99*, 8109 (1977).

557. K. Maruoka, S. Hashimoto, Y. Kitagawa, H. Yamamoto, and H. Nozaki, *J. Amer. Chem. Soc.*, *99*, 7707 (1977).

558. M. Samson and M. Vandewalle, *Syn. Commun.*, *8*, 231 (1978).

559. M. T. Reetz and W. F. Maier, *Angew. Chem. (Int. Ed.)*, *17*, 48 (1978).

560. J. K. Rasmussen, *Synthesis*, 91 (1977).

561. D. Caine and A. S. Frobese, *Tetrahedron Lett.*, 3107 (1977).

562. M. Samson, H. DeWilde, and M. Vandewalle, *Bull. Soc. Chim. Belg.*, *86*, 329 (1977).

563. P. G. Gassman, D. P. Gilbert, and S. M. Cole, *J. Org. Chem.*, *42*, 3233 (1977).

564. E. Vedejs, D. A. Engler, and J. E. Telschow, *J. Org. Chem.*, *43*, 188 (1978).

565. D. H. R. Barton, R. H. Hesse, C. Wilshire, and M. M. Pechet, *J. Chem. Soc.*, *Perkin I*, 1075 (1977).

566. R. E. Donaldson and P. L. Fuchs, *J. Org. Chem.*, *42*, 2032 (1977).

567. D. Spitzner, *Angew. Chem. (Int. Ed.)*, *17*, 197 (1978).

568. P. Groenewegen, H. Kallenberg, and A. van der Gen, *Tetrahedron Lett.*, 491 (1978).

569. E. J. Corey and D. Enders, *Chem. Ber.*, 111 (1978).

570. D. Enders and H. Eichenauer, *Tetrahedron Lett.*, 191 (1977).

571. D. Enders and H. Eichenauer, *Angew. Chem. Int. Ed. Eng.*, *15*, 549 (1976).

572. H. Eichenauer, E. Friedrich, W. Lutz, and D. Enders, *Angew. Chem. (Ent. Ed.)*, *17*, 206 (1978).

573. J. K. Whitesell and M. A. Whitesell, *J. Org. Chem.*, *42*, 377 (1977).

574. S. Hashimoto and K. Koga, *Tetrahedron Lett.*, 573 (1978).

575. A. I. Meyers, G. S. Poindexter, and Z. Brich, *J. Org. Chem.*, *43*, 892 (1978).

576. P. A. Wender and M. A. Eissenstat, *J. Amer. Chem. Soc.*, *100*, 292 (1978).

Chapter Three

ALKYLATIONS AND ACYLATIONS OF
PHOSPHONIUM YLIDES

H. J. Bestmann and R. Zimmermann

Institute of Organic Chemistry
The University of Erlangen-Nurenberg
Erlangen, Germany

I. INTRODUCTION

During the past few years phosphonium ylides, also called alkylidene
phosphoranes or phosphine alkylenes, have proved to be extremely
useful reagents in the field of preparative organic chemistry, with
several general reviews already published [1-9]. Being stabilized
carbanions, these ylides can react with a multitude of electrophilic
reagents to give rise to the formation of new C-C bonds. Alkylating
and acylating reagents react with phosphine alkylenes to give phosphonium salts [Eq. (1)]. Different methods make it possible to convert these salts into phosphorus-free derivatives. This discussion
is restricted to reactions with alkylenetriphenylphosphoranes
because the corresponding trialkylcompounds are not easily prepared
and handled.

$$(C_6H_5)_3P=CR^1R^2 \quad + \quad R^3X \quad \longrightarrow \quad \left[(C_6H_5)_3\overset{+}{P}-CR^1R^2R^3\right] \ X^-$$

$$\updownarrow$$

$$(C_6H_5)_3\overset{+}{P}-\overset{-}{C}R^1R^2 \tag{1}$$

As will be shown in the following sections, the alkylation or acylation of phosphonium ylides leads to the synthesis of a great variety of compounds.

II. GENERAL DATA CONCERNING THE CHEMISTRY OF PHOSPHONIUM YLIDES

Phosphonium ylides are Brönsted bases which are converted to phosphonium salts by hydrogen halides. From these salts the corresponding ylides can be regenerated by splitting off HX with a base [Eq. (2)].

$$(C_6H_5)_3P=CR^1R^2 \quad \underset{BASE}{\overset{HX}{\rightleftharpoons}} \quad \left[(C_6H_5)_3\overset{+}{P}-CHR^1R^2\right] \ X^- \tag{2}$$

The basis of many synthetic possibilities which result from reactions of ylides with alkylating or acylating compounds is the *transylidation* found in 1960 [10, 11]. This phenomenon is based on a proton transfer between the ylide and the phosphonium salt in which the ylide reacts as the base mentioned in Eq. (2). The position of the acid-base equilibrium between an ylide and a salt is determined by the groups R^1, R^2, R^3, and R^4 which influence, respectively, the acid and base strengths of the salt and the ylide [Eq. (3)]. Formation of the least basic ylide and the least acidic salt is favored.

$$\left[(C_6H_5)_3\overset{+}{P}-CHR^1R^2\right] \ X^- \quad + \quad (C_6H_5)_3P=CR^3R^4 \quad \rightleftharpoons \tag{3}$$

$$(C_6H_5)_3P=CR^1R^2 \quad + \quad \left[(C_6H_5)_3\overset{+}{P}-CHR^3R^4\right]X^-$$

Thus, the reaction of methylenetriphenylphosphorane with triphenylphenacylphosphonium bromide yields triphenylmethylphosphonium bromide and benzoylmethylenetriphenylphosphorane in almost 90% yield [Eq. (4)]

$$(C_6H_5)_3P\overset{+}{=}CH_2 \quad + \quad \left[(C_6H_5)_3\overset{+}{P}-CH_2-\underset{O}{\overset{\|}{C}}-C_6H_5\right] Br^- \longrightarrow$$

$$(4)$$

$$\left[(C_6H_5)_3\overset{+}{P}-CH_3\right] Br^- \quad + \quad (C_6H_5)_3P=CH-\underset{O}{\overset{\|}{C}}-C_6H_5$$

As already mentioned the alkylation or acylation of a phosphonium ylide leads to the formation of a phosphonium salt. In some cases it is possible to stop the reaction at this point. Often, however, the phosphonium salt spontaneously reacts with a second mole of the ylide, acting as a base, giving rise to the three types of reactions shown by Eqs. (5)-(7). Which type of reaction occurs depends on the activity of the hydrogen atoms of the salt on the α, β, or γ carbon with reference to the phosphorus atom.

α-Elimination (transylidation):

$$(C_6H_5)_3\overset{+}{P}-\overset{H}{\underset{X^-}{\overset{|}{C}}}-\overset{|}{\underset{|}{C}}-\overset{|}{\underset{|}{C}}- \quad + \quad (C_6H_5)_3P=C\overset{/}{\underset{\backslash}{}} \longrightarrow$$

$$(5)$$

$$(C_6H_5)_3P=\overset{|}{C}-\overset{|}{\underset{|}{C}}-\overset{|}{\underset{|}{C}}- \quad + \quad (C_6H_5)_3\overset{+}{P}-\overset{|}{\underset{X^-}{\overset{|}{C}}}-H$$

β-Elimination (Hofmann degradation):

$$(C_6H_5)_3\overset{+}{P}-\overset{|}{\underset{X^-}{\overset{|}{C}}}-\overset{H}{\underset{|}{\overset{|}{C}}}-\overset{|}{\underset{|}{C}}- \quad + \quad (C_6H_5)_3P=C\overset{/}{\underset{\backslash}{}} \longrightarrow$$

$$(6)$$

$$(C_6H_5)_3P \quad + \quad \overset{|}{\underset{|}{C}}=\overset{|}{C}-\overset{|}{\underset{|}{C}}- \quad + \quad (C_6H_5)_3\overset{+}{P}-\overset{|}{\underset{X^-}{\overset{|}{C}}}-H$$

γ-Elimination (Betain formation):

$$(C_6H_5)_3\overset{+}{P}-\overset{|}{\underset{X^-}{\overset{|}{C}}}-\overset{|}{\underset{|}{C}}-\overset{H}{\underset{|}{\overset{|}{C}}}- \quad + \quad (C_6H_5)_3P=C\overset{/}{\underset{\backslash}{}} \longrightarrow$$

$$(7)$$

$$(C_6H_5)_3\overset{+}{P}-\overset{|}{\underset{|}{C}}-\overset{|}{\underset{|}{C}}-\overset{|}{\underset{|}{C}}\overset{..}{\underset{}{}} \quad + \quad (C_6H_5)_3\overset{+}{P}-\overset{|}{\underset{X^-}{\overset{|}{C}}}-H$$

$$\downarrow$$

Subsequent
Reactions

While the Hofmann degradation yields a phosphorus-free compound, in the other two cases the phosphonium group must be split off by a subsequent reaction. The different ways to do so, primarily the Wittig-reaction [2, 3] or hydrolysis [4] [Eq. (8)] of the salt or

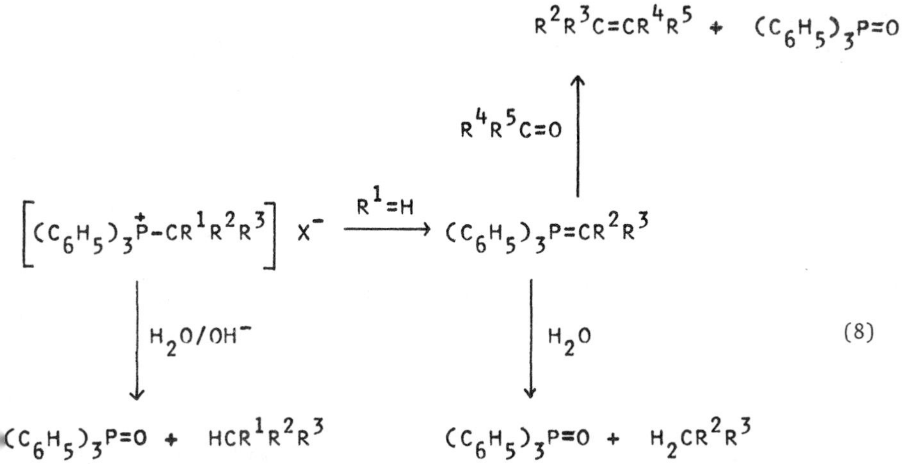

$$R^2R^3C=CR^4R^5 + (C_6H_5)_3P=O$$

$$R^4R^5C=O \uparrow$$

$$\left[(C_6H_5)_3\overset{+}{P}-CR^1R^2R^3\right] X^- \xrightarrow{R^1=H} (C_6H_5)_3P=CR^2R^3$$

$$\downarrow H_2O/OH^- \qquad\qquad \downarrow H_2O \qquad\qquad (8)$$

$$(C_6H_5)_3P=O + HCR^1R^2R^3 \qquad (C_6H_5)_3P=O + H_2CR^2R^3$$

ylide resulting from alkylation or acylation, are discussed for each special case in the following sections. When using the hydrolysis approach, one has to remember that of the four ligands on the phosphorus the one which is always split off is the one most electronegative or most stabilized by resonance. Thus, triphenylphosphine oxide is not necessarily formed in all cases.

III. ACYLATION OF PHOSPHONIUM YLIDES AND THE
RESULTING PREPARATIVE POSSIBILITIES

A. *Possibilities of Acylation*

1. Ylides and Acid Chlorides

The phosphine alkylenes *1* react with acid chlorides *2* (X = Cl) according to the scheme shown in Eq. (9) [11, 12]. The first-formed salt *3* undergoes a transylidation reaction with a second mole of the ylide *1*, to give an acylated phosphine alkylene *4* and the phosphonium salt *5a*. The stoichiometric course of the reaction thus requires 2 mol of ylide for every mole of acyl halide. The reaction is carried out with lithium salt-free solutions of the ylide *1* in an inert solvent.

1, to give an acylated phosphine alkylene *4* and the phosphonium salt
5a. The stoichiometric course of the reaction thus requires 2 mol
of ylide for every mole of acyl halide. The reaction is carried out
with lithium salt-free solutions of the ylide *1* in an inert solvent.

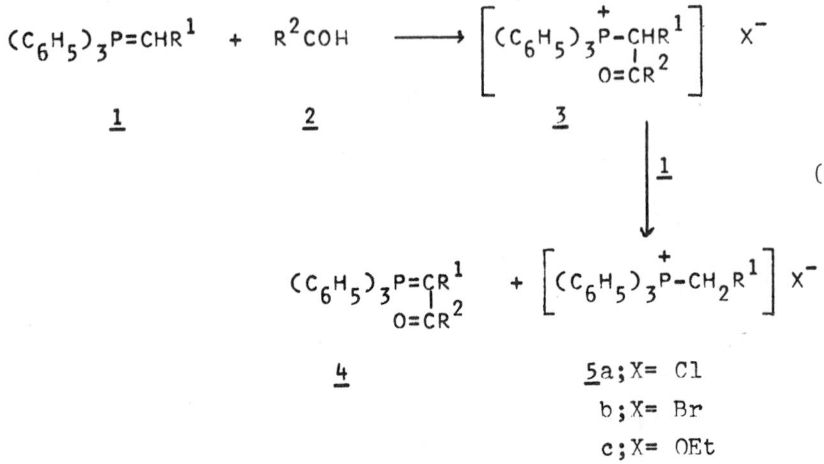

$$(C_6H_5)_3P{=}CHR^1 \;+\; R^2COH \;\longrightarrow\; \left[(C_6H_5)_3\overset{+}{P}{-}\underset{O{=}CR^2}{CHR^1}\right] X^-$$

$$\underline{1} \qquad\qquad \underline{2} \qquad\qquad\qquad \underline{3}$$

$$\Big\downarrow \underline{1} \qquad\qquad\qquad (9)$$

$$(C_6H_5)_3P{=}\underset{O{=}CR^2}{CR^1} \;+\; \left[(C_6H_5)_3\overset{+}{P}{-}CH_2R^1\right] X^-$$

$$\underline{4} \qquad\qquad\qquad\qquad \underline{5}a;X{=}\ Cl$$
$$b;X{=}\ Br$$
$$c;X{=}\ OEt$$

The most reliable method for synthesizing the necessary salt-
free ylides is to use sodium amide [12] [Eq. (10)]. All operations
are carried out under nitrogen or argon. Dry ammonia is condensed

$$\left[(C_6H_5)_3\overset{+}{P}{-}CH_2R\right] X^- \;+\; NaNH_2 \;\longrightarrow\; (C_6H_5)_3P{=}CHR \;+\; NH_3 \;+\; NaX$$

$$(10)$$

in a trap cooled with liquid air, and the required quantity of finely
chopped sodium (up to 25% excess) and a few grains of ferric nitrate
are added. When the blue sodium solution has turned gray, the
absolutely dry and finely powdered phosphonium salt is added, the
mixture stirred for a short time with a glass rod, and the ammonia
is then evaporated through a mercury valve. The evaporation can be
accelerated by heating the tube with hot air. To the residue is
added 100 ml of an inert, anhydrous solvent such as benzene, toluene,
ether, or tetrahydrofuran, and the solution is boiled for about 10
min to remove residual gas (the reflux condenser being sealed with

mercury). The solid residue is then filtered using a G3 frit.
Sodium bistrimethylsilyl amide (6), which is soluble in many organic
solvents, is also a very useful base for the generation of lithium
salt-free solutions of alkylenetriphenylphosphoranes [13] [Eq. (11)].
Because of its good solubility in many solvents and its easy handling
and exact dosing capabilities 6 is preferred in many cases to the
use of sodium amide in liquid ammonia.

$$\left[(C_6H_5)_3\overset{+}{P}\text{-}CH_2R\right] X^- + NaN\left[Si(CH_3)_3\right]_2 \longrightarrow$$
$$\underset{\underline{6}}{}$$

$$HN\left[Si(CH_3)_3\right]_2 + (C_6H_5)_3P=CHR \qquad (11)$$
$$\underset{\underline{7}}{}$$

The following synthetic variants have proved to be successful.

a. To a solution of 3.66 g (20 mmol) of sodium bistrimethyl-
silylamide (6) [14] in an anhydrous solvent such as benzene, toluene,
tetrahydrofuran, hexane, or hexamethylphosphotriamide under nitrogen
is added 20 mmol of an absolutely dry phosphonium salt. The mixture
is stirred for 30 min under room temperature and then refluxed for
1 hr. In many cases one can use the colored ylide solution immediately for further reactions. If desired, the undissolved sodium halide
can be separated by filtration under nitrogen.

b. An uncontaminated ylide solution is made as described in
method a, using 20 mmol of 6 and 20 mmol of a phosphonium salt. To
remove the hexamethyldisilazane 7, first the solvent and then 7 are
distilled at a bath temperature of approximately 100°C under high
vacuum. The residue is dissolved in the required solvent. This
ylide solution can be used immediately for further reactions.

c. To a solution of 22 mmol of 6 in 50 ml of anhydrous hexane,
20 mmol of the dry phosphonium salt are added, and this reaction
mixture is refluxed for 3 hr. It is then cooled to -20°C, whereby
the ylide crystallizes. It is sucked together with the undissolved
sodium halide and washed out of the filter cake with a solvent such

as benzene, toluene, tetrahydrofuran, or hexamethylphosphotriamide.
The ylide solution can be used for further reactions.

All the preceding operations have to be carried out under absolutely dry conditions as well as protected by nitrogen.

The reaction of the acid chloride 2 with the ylide 1 in an inert
solvent causes the salts to precipitate from the solution (they can
be used again for the preparation of 1), whereas the acylated phosphine alkylenes 4 stay in solution.

The acylations can be carried out in the following way: a
solution of 10 mmol of the acyl chloride in 50 ml of benzene is
added dropwise to a salt-free solution of 22 mmol of an alkylidenetriphenylphosphorane in 100 ml of anhydrous benzene under nitrogen
at room temperature. The addition of acyl chloride is stopped when
the ylide has been consumed, as shown by decoloration of the solution.
An excess of acyl chloride must be avoided, since this compound also
attacks the acylated ylide. When the reaction is complete, the inert
atmosphere is no longer necessary. The phosphonium chloride which
precipitates is filtered off and washed with benzene, and the solution
is concentrated in vacuum. The residue is mixed with 10 ml of ethyl
acetate, cooled for a short time in an ice-salt bath, and filtered.
A second fraction can be obtained by concentration of the mother
liquor. Evaporation of the solvent frequently leaves behind an oily
residue, but this generally dissolves in ethyl acetate and crystallizes after 24 hr in a refrigerator. Most acylalkylidenetriphenylphosphoranes can be recrystallized from ethyl acetate. The solubility
of the compounds in this solvent increases with the number of methylene groups they contain. The crystallization is frequently slow and
can be accelerated by the addition of a few milliliters of ether.
Many acylalkylidenetriphenylphosphoranes tenaciously retain the
solvent.

It is not absolutely necessary to isolate the acylated ylide
for use in further reactions. It can be hydrolyzed or made to react
with aldehydes directly after concentration of the benzene solution.
If benzene is to be used as the solvent for the reaction with

aldehydes, the acylation solution can be used directly after the phosphonium salt 5 has been filtered off.

The phosphonium chlorides 5a are formed in yields of 80 to 100% and can be used to prepare further ylide after a single reprecipitation from chloroform-ether, followed by careful drying. The chlorides are occasionally obtained in the form of oils, but these can be crystallized by trituration.

The lowest yields are obtained from the reaction of strongly basic alkylidenephosphoranes with acyl chlorides having an activated H atom close to the carbonyl group, as in phenylacetyl chloride. In these cases the reaction yields dark resinous products from which triphenylphosphine oxide can be isolated. These disadvantages are avoided if thioesters are used instaed of acyl chlorides, as described in the following discussion. By using phosphine alkylenes with strongly electron-attracting groups at the ylide carbon atom, the second mole of ylide may be replaced by triethylamine [49]. These reactions are successful, however, only if the amine is a stronger base than the ylide.

Generally acylation with acid chlorides has the disadvantage that 1 mol of the ylide 1 used is precipitated as the phosphonium salt 5a, even though compound 1 can be regenerated from 5a. An acylation procedure that avoids this disadvantage is the reaction of phosphine alkylenes with certain esters.

2. Ylides and S-Ethyl Esters of
   Thiocarboxylic Acids

A particularly useful group of acylating agents are the S-ethyl esters of thiocarboxylic acids (8) [12]. This method leads to purer products than are obtained from the acylation using acid chlorides. This reaction also occurs with transylidation. The first step is the formation of the acylalkylidenephosphorane 4 and the phosphonium ethyl thiolate 9, which precipitates [Eq. (12a)]. Like phosphonium alkoxides [15], 9 loses ethyl mercaptan on heating to form the ylide 1[Eq. (12b)]. This then reacts with further thioester (8) and is removed, in this way, from the equilibrium. This is shown by the

$$(C_6H_5)_3P=\underset{\underset{H}{|}}{C}-R^1 \quad + \quad R^2-CO-S-C_2H_5 \quad \longrightarrow$$

$$\underline{1} \qquad\qquad\qquad \underline{8}$$

(12a)

$$(C_6H_5)_3P=\underset{\underset{O=C-R^2}{|}}{C}-R^1 \quad + \quad \left[(C_6H_5)_3\overset{+}{P}-CH_2-R^1\right]SC_2H_5^-$$

$$\underline{4} \qquad\qquad\qquad\qquad\qquad \underline{9}$$

$$\left[(C_6H_5)_3\overset{+}{P}-CH_2-R^1\right]SC_2H_5^- \quad \rightleftharpoons \quad (C_6H_5)_3P=CHR^1 \quad + \quad HSC_2H_5$$

$$\underline{9} \qquad\qquad\qquad\qquad\qquad \underline{1} \qquad\qquad \underline{10}$$

(12b)

slow dissolution of the precipitated salt 9, which is frequently in the form of an oil. This method requires 1 mol of ylide less than the acyl chloride method. The reaction should be carried out in a Schlenk tube, and it is advisable to close the top of the reflux condenser with a mercury valve. Twenty millimoles of the S-ethyl thiocarboxylate are added to an absolute toluene solution of 22 mmol of an alkylidenephosphorane prepared from a phosphonium salt and sodium amide (the solution need not be filtered). The mixture is refluxed for 18 hr, during which time the oily phosphonium thiolate which initially precipitates is slowly redissolved. If an unfiltered ylide solution is used, it is necessary to remove the sodium halide by hot filtration, but a protective nitrogen atmosphere is not required for this step. Evaporation of the solution leaves the acylalkylidenetriphenylphosphorane, which can be crystallized by trituration.

Ylide solutions prepared by the use of sodium bistrimethylsilyl-amide (6) as the base are acylated in a similar manner. To an ylide solution in benzene, prepared from 20 mmol of a phosphonium salt according to method b cited in Sec. III.A.1, 20 mmol of a thiolate, 8, are added. The reaction mixture is refluxed for 48 hr, cooled, shaken three times with 25 ml portions of 0.1 N hydrochloric acid, and brought to a pH of 11-13 with sodium hydroxide. The resulting milky suspension is extracted with three 25-ml portions of benzene, and the benzene solution is dried with anhydrous Na_2SO_4. After distilling off the solvent, the residue is recrystallized from ethyl acetate/petroleum ether.

3. Further Possibilities for Acylation

Ylide solutions generally do not react with ethyl or methyl esters in the absence of salts [12, 16]. However, the ω-ethoxycarbonylalkylidenetriphenylphosphoranes (11) have been observed to undergo an intramolecular C-acylation to form the cyclic ylides 12 [17, 18]. This reaction presents new possibilities for the synthesis of cyclic carbonyl compounds.

$$(CH_2)_n \overset{COOCH_3}{\underset{CH=P(C_6H_5)_3}{}} \longrightarrow (CH_2)_n \overset{C=O}{\underset{C=P(C_6H_5)_3}{|}} + CH_3OH$$

$$\underline{11} \qquad\qquad\qquad \underline{12}$$
$$n= 3,4,5$$

Alkyl esters, however, do react with alkylidenephosphoranes in the presence of lithium salts [19, 20]. These reactions also proceed by transylidation [12]. The products obtained are the acylalkylidenetriphenylphosphoranes (4) and the triphenylalkylphosphonium bromides (5b), the latter resulting from the reaction of the initially formed alkoxides 5c with the LiBr present in the solution. Thus, the starting materials react in a molar ratio of 2:1. The yields of the acylated ylide 4 are poor.

Better results are obtained in the reaction of a salt-free solution of alkylidenephosphoranes with esters (13) having activating groups at R^2 or R^3 [12, 23, 24]. Unfortunately, acylation with esters may be accompanied by competitive reactions. It has been shown that certain activated esters give rise to the formation of enol ethers (16) by way of a Wittig reaction at the ester carbonyl group [21-24]. Methylenetriphenylphosphorane (17) reacts in dimethyl-sulfoxide with activated esters (13) in a molar ratio of 3:1 to give compounds 18, which on hydrolysis yield the corresponding isopropenyl

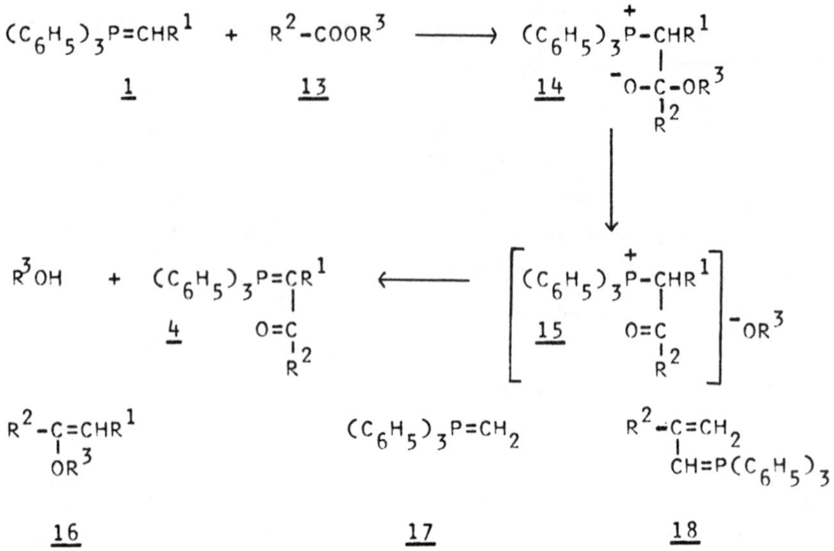

$$(C_6H_5)_3P{=}CHR^1 \ + \ R^2{-}COOR^3 \longrightarrow (C_6H_5)_3\overset{+}{P}{-}CHR^1$$

$$\underline{1} \qquad\qquad \underline{13} \qquad\qquad \underline{14} \quad {}^{-}O{-}\underset{R^2}{\overset{}{C}}{-}OR^3$$

$$R^3OH \ + \ (C_6H_5)_3P{=}\underset{O{=}\underset{R^2}{C}}{\overset{}{C}}R^1 \longleftarrow \left[(C_6H_5)_3\overset{+}{P}{-}CHR^1 \atop O{=}\underset{R^2}{C} \right]^{-}OR^3$$

$$\underline{4} \qquad\qquad\qquad \underline{15}$$

$$R^2{-}\underset{OR^3}{\overset{}{C}}{=}CHR^1 \qquad (C_6H_5)_3P{=}CH_2 \qquad R^2{-}\underset{CH{=}P(C_6H_5)_3}{\overset{}{C}}{=}CH_2$$

$$\underline{16} \qquad\qquad \underline{17} \qquad\qquad \underline{18}$$

compounds [25]. Acylimidazoles are also suitable for the acylation of the phosphorus ylides 1 [26, 27]. This reaction also proceeds by transylidation and presents in the case of R^2 = H an important method of formylating phosphonium ylides [27]. Formylation of
1 can also be achieved by dimethylformamiddichloride [28] or formic
acid ester [19, 29].

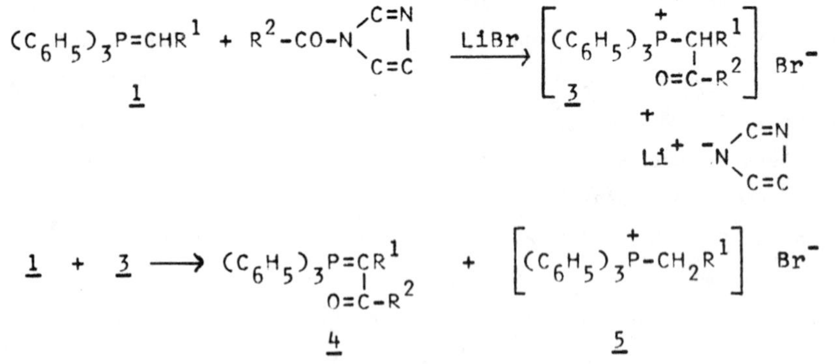

$$(C_6H_5)_3P{=}CHR^1 \ + \ R^2{-}CO{-}N{\overset{C{=}N}{\underset{C{=}C}{}}} \xrightarrow{LiBr} \left[(C_6H_5)_3\overset{+}{P}{-}CHR^1 \atop O{=}\overset{}{C}{-}R^2 \right] Br^-$$

$$\underline{1} \qquad\qquad\qquad\qquad\qquad\qquad \underline{3}$$

$$+$$

$$Li^+ \ {}^{-}N{\overset{C{=}N}{\underset{C{=}C}{}}}$$

$$\underline{1} \ + \ \underline{3} \longrightarrow (C_6H_5)_3P{=}\underset{O{=}C{-}R^2}{\overset{}{C}}R^1 \ + \ \left[(C_6H_5)_3\overset{+}{P}{-}CH_2R^1 \right] Br^-$$

$$\underline{4} \qquad\qquad\qquad \underline{5}$$

Besides acid chlorides, thioesters, activated esters, and acyl
imidazoles, anhydrides of carboxylic acids are also suitable as
acylating reagents [30].

B. *Synthesis of Carbonyl Compounds from Acylated Ylides*

   1. Ketones by Cleavage of Acylated Ylides or Acyl Phosphonium Salts

     a. Hydrolysis. The acylalkylidenetriphenylphosphoranes (*4*) are weak bases and are, therefore, stable toward cold water. On boiling with water, however, they break down to form the ketones *19a* and triphenylphosphine oxide [12, 31].

$$(C_6H_5)_3P=\underset{\underset{O=C-R^2}{|}}{C}R^1 \; + \; H_2O \; \longrightarrow \; R^1-CH_2-CO-R^2 \; + \; (C_6H_5)_3P=O$$

$$\underline{\textbf{4}} \hspace{4cm} \underline{\textbf{19}} \; a; \; R^2 = \text{alkyl or aryl}$$
$$b; \; R^2 = H$$

The acylalkylidenetriphenylphosphorane (the unpurified product of the acylation can be used) is dissolved in 50-80 ml of 80% aqueous methanol, 1-2 ml of 2 N NaOH are added, and the mixture is refluxed for 12 hr. After the addition of 3 ml of glacial acetic acid, the ketone formed is steam-distilled. The distillate is extracted with ether, the ether solution dried, and the ketone is distilled or recrystallized after the solvent has been driven off. If the product is not steam-volatile, the solution must be concentrated after the addition of the glacial acetic acid, and the ketone then separated from the triphenylphosphine oxide by treatment of the residue with petroleum ether. This separation can also be carried out conveniently by the zinc chloride method [32].

If R^1 and R^2 in *4* are strongly resonance stabilizing groups, hydrolysis may be difficult. Also the cyclic ylides *12* are hydrolyzed only on heating in o-dichlorobenzene-water-ethanol (1:3:5) in a pressure tube at $130°C$. The products are cyclic ketones and triphenylphosphine oxide [17]. In the hydrolysis of *4* with $R^2 = H$, aldehydes (*19b*) are formed [27].

     b. Reductive Cleavage. Ketones are also obtained by reductive cleavage of the acylalkylidenetriphenylphosphoranes *4* with zinc in glacial acetic acid [19]. This reaction proceeds particularly well

when the group R^2 is aromatic (yields 90%). With aliphatic groups
at R^2, the yields of ketone are much poorer than those obtained by
hydrolysis, and in some cases this method fails completely.

$$(C_6H_5)_3P=CR^1 \quad \xrightarrow{\text{Zn}} \quad R^1-CH_2-CO-R^2 \quad + \quad (C_6H_5)_3P$$
$$\underset{\underline{4}}{\overset{O=C-R^2}{}}$$

The acylalkylidenetriphenylphosphorane (5 mmol) is dissolved
in 25 ml of chloroform. Zinc powder (12 g) is added, and 50 ml of
glacial acetic acid is dropped into the boiling mixture over a period
of 1 hr. After boiling for an additional hour, the reaction solution
is steam-distilled, and the ketone is extracted from the distillate
with ether, and the ether phase is washed with 2 N NaOH and dried
over magnesium sulfate. The solvent is distilled off in vacuum,
leaving behind triphenylphosphine, which can be recrystallized from
methanol [12].

    c. Electrolysis. Another preparatively important reaction
is the electrolytic cleavage of the quaternary phosphonium salts [33].
The acylalkylidenetriphenylphosphorane *4* is dissolved in a stoichiometric quantity of mineral acid, and the resulting solution of the
salt *3* is electrolyzed in a closed vessel fitted with a reflux

$$(C_6H_5)_3P=CR^1 \quad \xrightarrow{\text{HX}} \quad \left[(C_6H_5)_3\overset{+}{P}-CHR^1 \right] X^-$$
$$\underset{\underline{4}}{\overset{O=C-R^2}{}} \qquad\qquad \underset{\underline{3}}{\overset{O=C-R^2}{}}$$

$$\Big\downarrow \text{Electrolysis}$$

$$R^1CH_2-CO-R^2 \quad + \quad (C_6H_5)_3P$$
$$\underline{19}$$

condenser, using a lead cathode which is bent around a carbon anode
a short distance away. The ketone *19* and triphenylphosphine are
obtained in yields of 60-80% [12]. This method fails with aromatic
groups at R^1, since in this case the ylides *4* are regenerated. The

same is true when R^1 is C_6H_5-CH=CH. On the other hand, this method is recommended for those compounds having aliphatic residues at R^1. Thus, the electrolytic method supplements the reductive cleavage with zinc and glacial acid, and vice versa. The advantage of these methods lies in the regeneration of triphenylphosphine.

2. α-Chloroketones by Halogenation of Acylated
   Ylides and Subsequent Hydrolysis

The phosphonium chlorides 20 are isolated from the acylalkyl-idenetriphenylphosphoranes 4 (R^1 = H) and phenyl iodide dichloride. Compounds 20 can be converted in 50-90% yield into the chloroketones 21 and triphenylphosphine oxide on treatment with a dilute Na_2CO_3 solution at $0°$-$25°C$ [34].

$$(C_6H_5)_3P=CR^1 \quad \xrightarrow{C_6H_5I-Cl_2} \quad \left[(C_6H_5)_3\overset{+}{P}-CR^1 \atop O=C-R^2 \right] \quad Cl^-$$

$$\underset{4}{} \quad \underset{O=C-R^2}{} \quad \underset{20}{}$$

$$\underset{20}{} \quad \xrightarrow[0-25°]{Na_2CO_3/H_2O} \quad R^1-\overset{Cl}{\underset{}{CH}}-CO-R^2 \quad + \quad (C_6H_5)_3P=O$$

$$\underset{21}{}$$

3. Ketones by Wittig Reactions with
   Acylated Ylides

a. *α-Branched α,β-unsaturated ketones.* The acylalkylidenetri-phenylphosphoranes 4 can react with aldehydes (22) in a Wittig reaction to form the α,β-unsaturated ketones 23 having chain branching in the α position [12, 31]. Because of the low reactivity of 4, these reactions take much longer than does the normal Wittig reaction. In order to obtain good yields it is necessary in most cases to reflux the mixture for 3 to 4 days in anhydrous benzene. Ketones

$$(C_6H_5)_3P=CR^1 \quad + \quad R^3-CHO \quad \longrightarrow \quad R^1-C=CHR^3 \quad + \quad (C_6H_5)_3P=O$$

$$\underset{4}{} \underset{O=C-R^2}{} \quad \underset{22}{} \quad \underset{O=C-R^2}{} \quad \underset{23}{}$$

do not react in this way [35]. The acylalkylidenetriphenylphosphoranes which are resistant to hydrolysis are also unable to undergo
Wittig reactions with aldehydes. The analogous reactions of the
formyl ylides *4* (R^2 = H) yield α,β-unsaturated aldehydes [29].

   b. α-Halogenated α,β-unsaturated ketones. The acyl ylides *4*
(R^1 = H) can be converted by several methods to the α-halogenated
(and pseudohalogenated) ylides, *24* [34(b) and (c), and 36-40].
Wittig reactions of these ylides lead to the formation of the α-
halogenated α,β-unsaturated ketones *25*. Since these compounds (*25*)
react with bases to form acetylenes, this reaction sequence provides
a possible synthesis of acetylenic ketones.

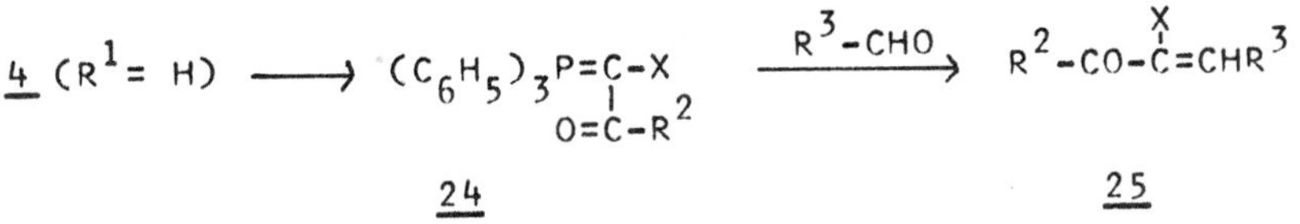

$$\underline{4}\ (R^1 = H)\ \longrightarrow\ (C_6H_5)_3\overset{P=C-X}{\underset{O=C-R^2}{}}\ \xrightarrow{R^3-CHO}\ R^2-CO-\overset{X}{\underset{}{C}}=CHR^3$$

$$\underline{24}\qquad\qquad\qquad\qquad\qquad\underline{25}$$

   c. α,β-Diketones. Acylation of methoxymethylenetriphenylphos-
phorane (*26*) with acid chlorides results in the preparation of the
ylides *27*, which undergo Wittig reaction to give the enol ethers *28*.
Acidic hydrolysis of *28* yields the 1,2-diketones *29* [29].

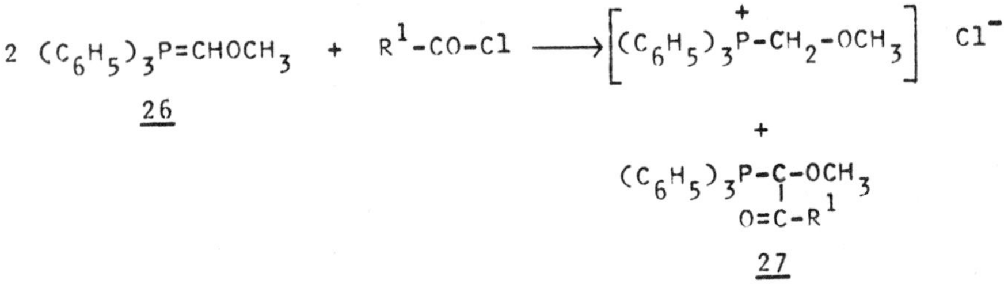

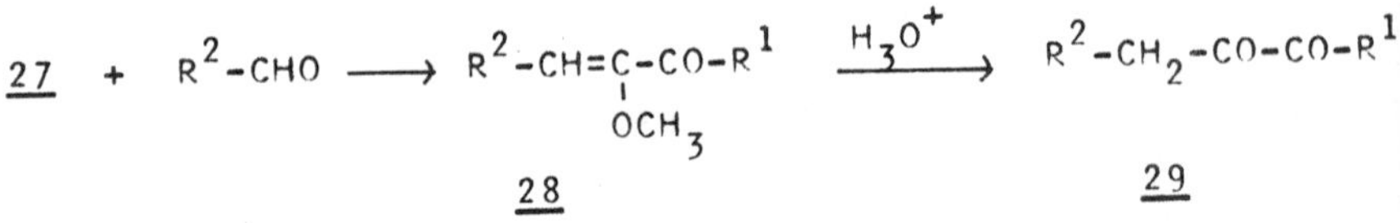

$$2\ (C_6H_5)_3P=CHOCH_3\ +\ R^1-CO-Cl\ \longrightarrow \left[(C_6H_5)_3\overset{+}{P}-CH_2-OCH_3\right]\ Cl^-$$

$$\underline{26}$$

$$+$$

$$(C_6H_5)_3\overset{P-C-OCH_3}{\underset{O=C-R^1}{}}$$

$$\underline{27}$$

$$\underline{27}\ +\ R^2-CHO\ \longrightarrow\ R^2-CH=\overset{C-CO-R^1}{\underset{OCH_3}{}}\ \xrightarrow{H_3O^+}\ R^2-CH_2-CO-CO-R^1$$

$$\underline{28}\qquad\qquad\qquad\qquad\qquad\underline{29}$$

4. 1,2-Dicarbonyl Compounds by the
   Oxidation of Acylated Ylides

   a. *α-Ketocarboxylates and α-ketothiocarboxylates.* It is
possible to oxidize the ylides *27* with lead tetraacetate or lead
dioxide to form the α-ketocarboxylic acid methyl esters *30* [41].
The analogous oxidation of the ylide *32* [prepared from 2 mol

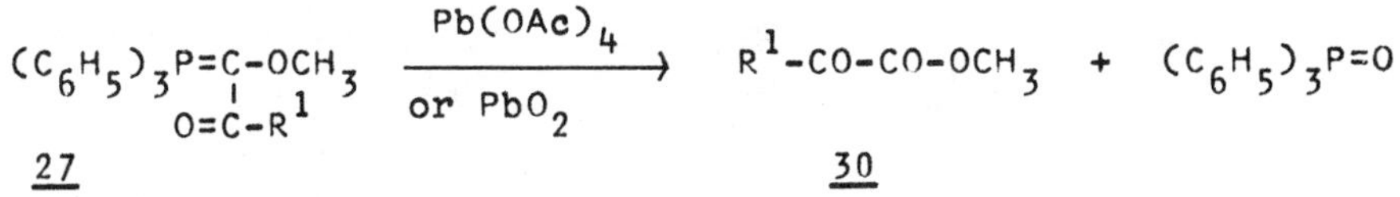

$$(C_6H_5)_3P=\underset{\substack{| \\ O=C-R^1}}{C}-OCH_3 \xrightarrow[\text{or } PbO_2]{Pb(OAc)_4} R^1-CO-CO-OCH_3 + (C_6H_5)_3P=O$$

<u>27</u>   <u>30</u>

phenylmercatomethylenetriphenylphosphorane (*31*) and 1 mol of acid
chloride] with lead tetraacetate gives the α-ketothiocarboxylic acid-
S-phenylester *33*. For the reaction mechanism see Ref. 41.

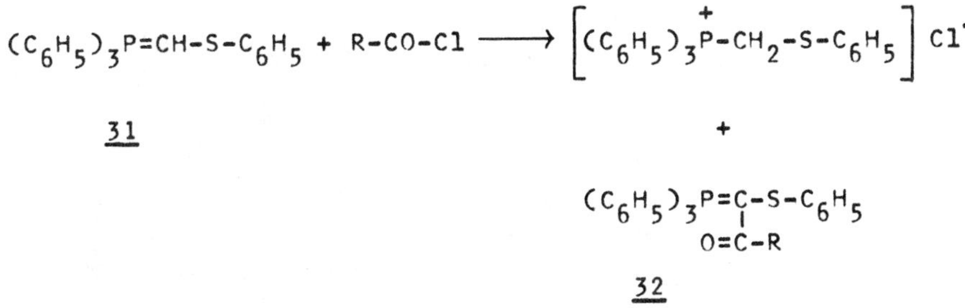

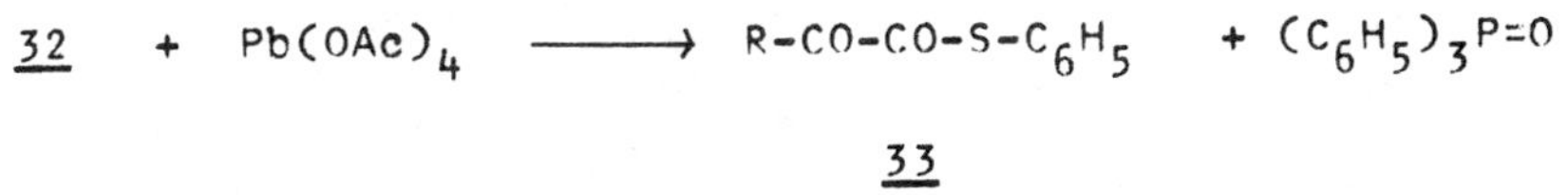

$$(C_6H_5)_3P=CH-S-C_6H_5 + R-CO-Cl \longrightarrow \left[(C_6H_5)_3\overset{+}{P}-CH_2-S-C_6H_5\right]Cl^-$$

<u>31</u>   +

$$(C_6H_5)_3P=\underset{\substack{| \\ O=C-R}}{C}-S-C_6H_5$$

<u>32</u>

$$\underline{32} + Pb(OAc)_4 \longrightarrow R-CO-CO-S-C_6H_5 + (C_6H_5)_3P=O$$

<u>33</u>

   b. *α,β-Diketones.* The various acylalkylidenetriphenylphosphoranes *4* which can be synthesized from *1* and acid chlorides are
oxidized with $KMnO_4$[42] or $NaIO_4$ [43] to give the α,β-diketones *34*.
The yields from the oxidation using $NaIO_4$ are much better than from
those in which $KMnO_4$ is used.

$$(C_6H_5)_3P=CR^1 \atop \underset{\underline{4}}{O=C-R^2} \xrightarrow[\text{or NaIO}_4]{\text{KMnO}_4} R^2-CO-CO-R^1 + (C_6H_5)_3P=O \atop \underline{34}$$

    c. 1,2-Diacylethylenes. If an aqueous solution of NaIO$_4$ is added to an aqueous solution of the phosphonium salt 5, the phosphonium periodates 35 precipitate in high yields. When these periodates (35) are heated with bases such as sodium alcoholate, sodium amide, or an organolithium, olefins (41), iodate (42), and triphenylphosphine oxide are formed [43]. The mechanism of the reaction is interpreted thus: From 35, the ylide 1 and the periodate anion 36 are formed.

$$\left[(C_6H_5)_3\overset{+}{P}-CH_2-R\right] X^- + NaIO_4 \longrightarrow \left[(C_6H_5)_3\overset{+}{P}-CH_2-R\right] IO_4^-$$
$$\underline{5} \qquad\qquad\qquad\qquad\qquad\qquad \underline{35}$$

$$\underline{35} \xrightarrow[-H^+]{B^-} (C_6H_5)_3P=CH-R + IO_4^-$$
$$\underline{1} \qquad\qquad \underline{36}$$

$$\underline{1} + \underline{36} \longrightarrow (C_6H_5)_3\overset{-}{P}\!\!-\!\!CH-R \longrightarrow (C_6H_5)_3P\!\!-\!\!CH-R$$

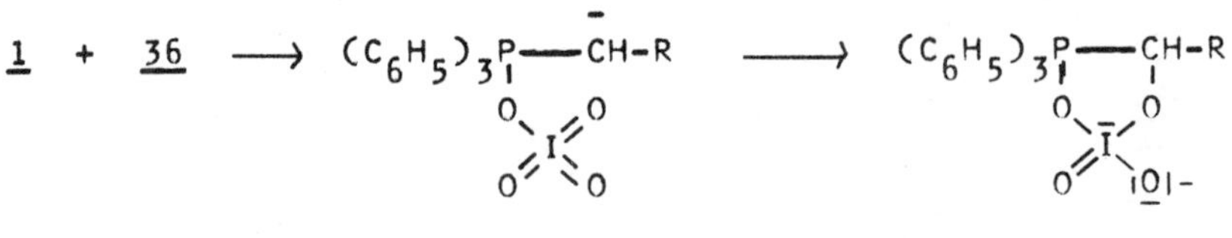

$$\underline{37} \qquad\qquad\qquad \underline{38}$$

$$\underline{38} \longrightarrow R-CHO + IO_2^- + (C_6H_5)_3P=O$$
$$\underline{39} \qquad\qquad \underline{40}$$

$$\Big\downarrow \underline{1} \qquad\qquad\qquad \Big\downarrow \underline{36}$$

$$R-CH=CH-R \qquad\qquad 2\ IO_3^-$$
$$\underline{41} \qquad\qquad\qquad \underline{42}$$

The latter attacks *1* at phosphorus leading to the intermediate *37* having a pentavalent phosphorus. From *37* a five-membered ring compound (or transition state), *38*, is formed, which decomposes into the aldehyde *39*, the ion *40*, and triphenylphosphine oxide. Compound *39* reacts with the not yet oxidized ylide *1* to give the olefin *41*, while *40* is oxidized by a second mole of *36* forming 2 mol of *42*. In agreement with the concept that a nucleophilic attack of the periodate ion takes place at the phosphorus, in particular resonance-stabilized ylides having R^1 = COR may be oxidized to 1,2-diacylethylenes.

The synthesis of α,β-diketones as well as that of 1,2-diacylethylenes by oxidation of acylalkylidenetriphenylphosphoranes may be achieved in very good yields by using ozone-phosphite adducts as oxidants [44].

C. Synthesis of Acetylene Derivatives from Acylated Ylides

The acylalkylidenetriphenylphosphoranes *4* can be represented by two resonance structures, *4a* and *4b*. The reactions with alkyl halides and acyl chlorides are often formulated as passing through *4b*, since the attack for the most part takes place on the O atom [12, 31, 45]. The thermal decomposition of the acylated ylides to form acetylenic compounds (*43*) probably also proceeds via *4b* [45, 46]. This reaction occurs only if R^1 or R^2 is an electron-withdrawing group and if neither R^1 nor R^2 is hydrogen.

$$R^2-\underset{\underset{O}{\|}}{C}-\underset{\underset{P(C_6H_5)_3}{\|}}{C}-R^1 \longleftrightarrow R^2-\underset{\underset{{}^-O}{|}}{C}=\underset{\underset{\overset{+}{P}(C_6H_5)_3}{|}}{C}-R^1 \xrightarrow{\triangle} R^2-C\equiv C-R^1$$

<u>**4a**</u>   <u>**4b**</u>   <u>**43**</u>

1. Esters of Acetylenecarboxylic Acids

The C-acylation of methoxycarbonylmethylenetriphenylphosphorane (*44*) with acyl chlorides, which proceeds by transylidation, provides

an easy route to the α-acyl-α-methoxycarbonylmethylenetriphenylphos-
phoranes 45. The thermal decomposition of these compounds leads to
very high yields of the esters of acetylenecarboxylic acids (48)
[45, 47].

$$2 \ (C_6H_5)_3P{=}CH{-}CO_2CH_3 \quad + \ R{-}CO{-}Cl \ \longrightarrow \ \left[(C_6H_5)_3\overset{+}{P}{-}CH_2{-}CO_2CH_3 \right] \ Cl$$

$$\underset{\underline{44}}{}$$

$$(C_6H_5)_3P{=}\underset{\underset{\underline{45}}{\overset{|}{O{=}C{-}R}}}{C}{-}CO_2CH_3$$

PCl$_3$ $\qquad\qquad$ $\triangle$

$$\left[(C_6H_5)_3\overset{+}{P}{-}\underset{\underset{\underline{46}}{Cl{-}C{-}R}}{\overset{\|}{C}}{-}CO_2CH_3 \right] Cl^- \ \xrightarrow{\ NaOH\ } \ R{-}\underset{\underset{\underline{47}}{Cl}}{\overset{|}{C}}{=}CH{-}CO_2CH_3 \longrightarrow \underset{\underline{48}}{R{-}C{\equiv}C{-}CO_2CH_3}$$

One gram of the ylide 45 [47] is placed in a flask with a de-
scending air condenser running to a suction tube which acts as both
an adaptor and a flask. Depending on the boiling point of the
expected ester, the pressure is reduced to between 12 and 0.05 mm
Hg, and the ylide is heated slowly to 220°-260°C. It proved to be
very successful to carry out the pyrolysis in a rotating bulb tube
[48(b)]. The greater part of the triphenylphosphine oxide as well as
the ester distills over within about 15 min at 0.05 mm Hg. The distillate is taken up in methanol and 3 ml of a concentrated sodium
hydroxide solution are added. The mixture is allowed to stand for 24
hr at room temperature, the methanol is evaporated under vacuum, and
30 ml of water are added. Triphenylphosphine oxide separated as an
oil at first, but rapidly crystallizes in the cold. The crystals are
filtered off, and the filtrate is acidified with a little concentrated
hydrochloric acid. The acid which precipitates is subsequently
recrystallized from a suitable solvent, depending on its properties.

The reaction of the ylides *45* with PCl_5 or Villsmeier reagents leads to the formation of phosphonium salts (*46*) which react with NaOH, without first being isolated, to form the esters of acetylene-carboxylic acids (*48*) via the compound *47* [48(a)]. Secondary amines may be added to *48*, and the resulting enamines may be hydrolyzed to yield β-keto esters [48(b)]. Thus, acylation of *44* by acid chlorides also provides a new route to β-keto esters.

By the thermal decomposition of acylated ylides it is also possible to obtain esters of bisacetylenecarboxylic acids [49]. The place of the ester group may also be taken by a cyano group [46].

2. Acetylenic Ketones

Reaction of the acylphosphinalkylenes *4* and the carboxylic acid anhydrides *49* gives the acylated products *50*, which are inert toward carbonyl compounds, but which lead, on thermolysis, to the acetylenic ketones *51a* or *51b* [30]. A dependence of the formation of *51a* and *51b* on R and R^1 has, up to now, not been published.

$$\underline{\mathbf{4}} \ + \ (R^2CO)_2O \ \longrightarrow \ (C_6H_5)_3P{=}C{-}CO{-}R^1 \underset{-(C_6H_5)_3P=O}{\xrightarrow{\quad \triangle \quad}}$$

$$\underset{O=C-R^2}{\qquad}$$

$$\underline{49} \qquad\qquad \underline{50}$$

$$R^1{-}C{\equiv}C{-}CO{-}R^2 \ \textit{or} \ R^1{-}CO{-}C{\equiv}C{-}R^2$$

$$\underline{51a} \qquad\qquad \underline{51b}$$

3. Diarylacetylenes

The benzoylylides *53* (obtained by reaction of the ylides *52* with benzoyl chloride) decompose on heating to the diarylacetylenes *54* and triphenylphosphine oxide [46, 50]. Strongly electron withdrawing groups on the aryl group lower the temperature of pyrolysis significantly. While the ylide *53* with Ar = C_6H_5 decomposes on heating at 300°C, the analogous ylide with Ar = $2,4\text{-}(NO_2)_2C_6H_3$ readily yields the corresponding acetylenic ketone at 60°C [50].

$$(C_6H_5)_3P=CH-Ar \xrightarrow{\quad C_6H_5COCl \quad} (C_6H_5)_3P=C-Ar$$
$$\underset{\underline{52}}{\qquad} \qquad\qquad\qquad \underset{\underline{53}}{O=C-C_6H_5}$$

$$\downarrow$$

$$(C_6H_5)_3P=O \quad + \quad Ar-C\equiv C-C_6H_5$$
$$\underset{\underline{54}}{\qquad\qquad\qquad\qquad\qquad\quad}$$

D. *Synthesis of Esters of α-Branched-β-
   Oxocarboxylic Acids and of Allenes*

    1. Esters of α-Branched-β-Oxocarboxylic
        Acids

1-Alkyl-1-alkoxycarbonylmethylenetriphenylphosphoranes (55, R^1
$\neq$ H) and acid chlorides (56) react at a temperature of 0°-20°C in a
molar ratio of 1:1 to form the acylated phosphonium salts 57 [51, 52].
The electrolysis of the salts 57 in aqueous solution using a mercury
electrode produces, besides triphenylphosphine, esters of the α-
branched-β-oxocarboxylic acids 58 [52, 53].

    The acyl chloride (0.11 mol) is added to a solution of 0.1 mol
of the α-alkyl- or α-arylalkoxycarbonylmethylenetriphenylphosphorane
in anhydrous benzene (0°-20°C). The mixture is allowed to stand for
24 hr in a refrigerator, and the benzene is then decanted from the
oily phosphonium salt. The latter is freed from solvent under vacuum
at 30°C and dissolved in a little water. The solution is extracted
with ether and the aqueous phase electrolyzed. The electrolysis is
carried out in a three-necked flask with a cooling jacket. The mid-
dle neck carries a diaphragm with a carbon anode. A pool of mercury
on the bottom of the flask is used as the cathode, to which connection
is made by a fused-in molybdenum wire. The electrolysis is carried
out at a temperature of under 35°C with a current of 1-2 A at 24 V.

    During the electrolysis the triphenylphosphine and the keto
ester separate and rise to the surface as an oil. As soon as the
current ceases to flow, the mixture is extracted with ether. The
ether phase is dried over anhydrous $MgSO_4$, and the solvent is

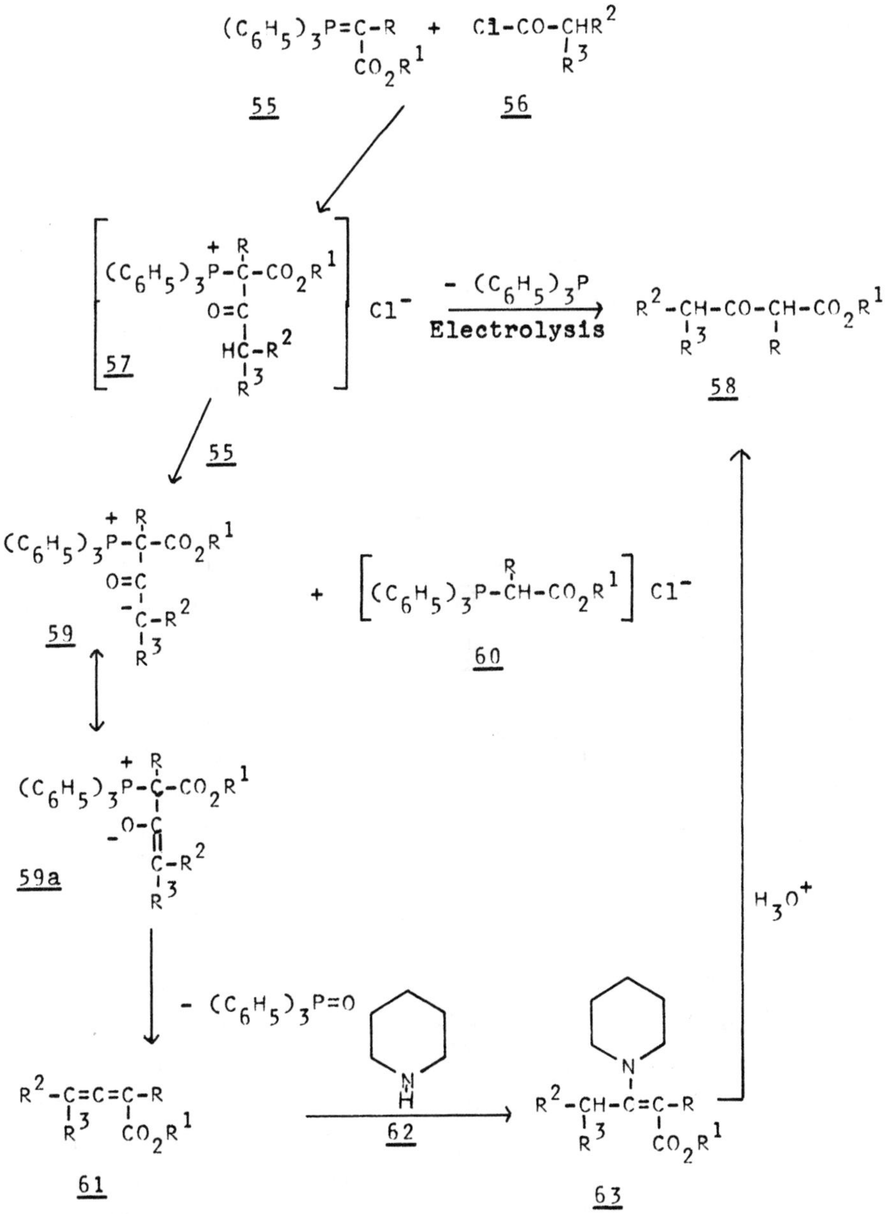

$$(C_6H_5)_3P=C-R \quad + \quad Cl-CO-CHR^2$$
$$\qquad\qquad\quad |\qquad\qquad\qquad\qquad |$$
$$\qquad\qquad CO_2R^1 \qquad\qquad\quad R^3$$

<u>55</u> <u>56</u>

$$\left[\begin{array}{c} (C_6H_5)_3P-C-CO_2R^1 \\ \quad\;\; | \\ O=C \\ \quad | \\ HC-R^2 \\ \quad | \\ R^3 \end{array} \right] Cl^-$$

<u>57</u>

$$\xrightarrow[\text{Electrolysis}]{- (C_6H_5)_3P} \quad R^2-CH-CO-CH-CO_2R^1$$
$$\qquad\qquad\qquad\qquad\qquad\quad | \qquad\qquad |$$
$$\qquad\qquad\qquad\qquad\qquad\; R^3 \qquad\qquad R$$

<u>58</u>

$$(C_6H_5)_3\overset{+}{P}-\overset{R}{\underset{|}{C}}-CO_2R^1$$
$$\qquad\quad |$$
$$\qquad O=C$$
$$\qquad\quad -\underset{|}{C}-R^2$$
$$\qquad\qquad R^3$$

<u>59</u>

$$+ \quad \left[(C_6H_5)_3\overset{+}{P}-\overset{R}{\underset{|}{C}H}-CO_2R^1 \right] Cl^-$$

<u>60</u>

$$(C_6H_5)_3\overset{+}{P}-\overset{R}{\underset{|}{C}}-CO_2R^1$$
$$\qquad\quad -O-C$$
$$\qquad\qquad\; \overset{\|}{C}-R^2$$
$$\qquad\qquad\quad R^3$$

<u>59a</u>

$$\xrightarrow{- (C_6H_5)_3P=O}$$

$$R^2-C=C=C-R$$
$$\;\; | \qquad\quad |$$
$$\;\; R^3 \quad CO_2R^1$$

<u>61</u>

$$\xrightarrow{\underset{62}{\qquad}} \quad R^2-CH-C=C-R$$
$$\qquad\qquad\qquad | \qquad\quad |$$
$$\qquad\qquad\qquad R^3 \quad CO_2R^1$$

<u>63</u>

$$H_3O^+$$

evaporated. The residue is distilled at 15 mm Hg using a rotating
bulb tube. The ester distills leaving the triphenylphosphine behind.
The latter is recrystallized from methanol, and the ester is purified
by vacuum distillation.

The yields of 57, however, become less when using the ylides
(55) in which R is larger than CH_3. In these cases large amounts of
the esters of allene carboxylic acids (61) are formed. A second
mole of the ylide 55 removes a proton from the α position with respect
to the P atom in 57 with the formation of the phosphonium salt 60 and
a betaine, 59. The latter spontaneously loses triphenylphosphine
oxide providing a simple route to the esters of the allene carboxylic
acids 61. If the latter is desired as the main product, the reaction
is carried out in a molar ratio of the ylide and the acyl chloride
of 2:1 [51].

The 1-alkylethoxycarbonylmethylidenetriphenylphosphorane 55
($R^1 = C_2H_5$) (60 mmol) is dissolved in 70 ml of anhydrous tetrahydrofuran by heating in the absence of moisture. This solution is cooled
to about $40^{\circ}C$, and 30 mmol of the acyl chloride 56, dissolved in 5-
10 ml of tetrahydrofuran, are added. After a short time the phosphonium salt 60 precipitates. The mixture is refluxed for 3-4 hr, and
the phosphonium salt 60 is filtered off and washed with 20 ml of
ether. The solvent is distilled from the filtrate through a small
column. The residue is mixed with 30-40 ml of petroleum ether,
whereupon the greater part of the triphenylphosphine oxide crystallizes. The solution is filtered, the petroleum ether is evaporated
off, and the allene carboxylate 61 is distilled in vacuum.

The difficulties in the synthesis of the β-keto esters 58 when
R is larger than CH_3 can be overcome in the following way [53]. The
allene carboxylic esters 61 are synthesized from 2 mol of 55 and 1
mol of 56. Compound 61 is reacted with piperidine (62) to give the
enamine 63, which is hydrolized by 2 N H_2SO_4 to the β-keto ester 58.
To a solution of 61 in ether is added dropwise 1 eq of piperidine
dissolved in ether. The yellow solution is refluxed for 4 hr, and
the solvent evaporated through a short column. The residue 63 is

distilled in vacuum. A solution of 63 in ether is mixed with 20 ml of 2 N H_2SO_4 and stirred for 14 hr. Fifty milliliters of ether are added, and the ethereal phase is separated and dried with Na_2SO_4. The solvent is distilled using a short column, and the residue 58 is fractionated under vacuum.

If R^1 in 55 or the acid chloride 56 is chiral, an optically active 61 yields, on hydrolysis, optically active allene carboxylic acids. Ylides containing chiral phosphorus have also been employed in this reaction [54, 55(a)]. In this connection it should be mentioned that a racemic mixture of an ylide containing a chiral phosphorus atom (available from phosphonium salts with four different ligands) with but one proton at the ylide carbon reacts with optically active acid chlorides of the type $R^1R^2R^3COCl$ (all R ≠ H) in a transylidation reaction giving rise to partial kinetic resolution. It is found that the absolute configuration of the acid chloride can be determined from the sign of the specific rotation of the precipitated acylated salt [55(b)].

2. Allenes

By the reaction of 2 mol of an ylide and 1 mol acid chloride via a phosphonium betaine, allenes are generally available [Eq. (13)]. This method is useful if R^1 is an aromatic and R^2 an aliphatic

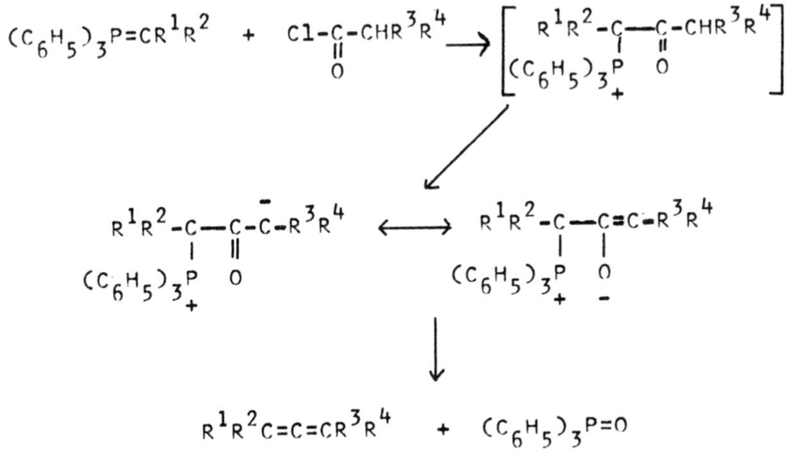

$$(C_6H_5)_3P=CR^1R^2 \;+\; Cl-\underset{\underset{O}{\|}}{C}-CHR^3R^4 \longrightarrow \left[\begin{array}{c} R^1R^2-\underset{\underset{(C_6H_5)_3\overset{+}{P}}{|}}{C}-\underset{\underset{O}{\|}}{C}-CHR^3R^4 \end{array} \right] Cl^-$$

$$R^1R^2-\underset{\underset{(C_6H_5)_3\overset{+}{P}}{|}}{C}-\underset{\underset{O}{\|}}{C}-\overset{-}{C}-R^3R^4 \longleftrightarrow R^1R^2-\underset{\underset{(C_6H_5)_3\overset{+}{P}}{|}}{C}-\underset{\underset{O^-}{|}}{C}=C-R^3R^4$$

$$R^1R^2C=C=CR^3R^4 \;+\; (C_6H_5)_3P=O \tag{13}$$

group [56(a)]. The groups R^3 and R^4 on the acid chloride can be
alkyl, aryl, or hydrogen.

E. *Preparation of α,β-Unsaturated Cyclic Ketones*
    by Acylation of Phosphonium Ylides with Enol
    Lactones

The enol lactones *64* react with the ylides *1* to give the betaines *65*.
From *65* the acylated ylides *66* are formed by a proton transfer. The
latter decompose by an intramolecular Wittig reaction yielding the
unsaturated cyclic ketones *67* and triphenylphosphine oxide. The
reaction has been specially applied in the steroid field [110].

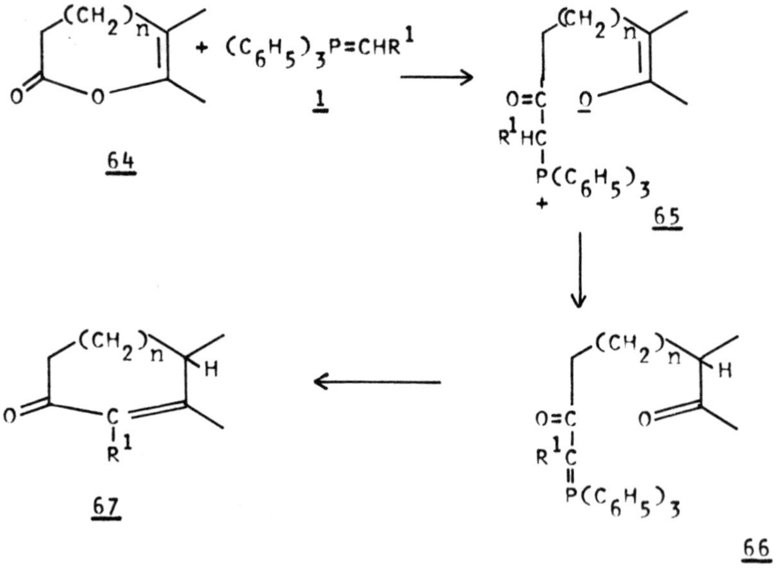

IV. ALKOXYCARBONYLATION OF PHOSPHONIUM
     YLIDES

The alkylidenephosphoranes *1* react with esters of chloroformic acid
(*68*) by transylidation. In this reaction the alkoxycarbonylation of
the ylides *1* gives the esters *55* [57]. Diesters of carboxylic acids
may also be used as alkoxycarbonylating agents [58].

$$2 \; (C_6H_5)_3P{=}CHR^1 \; + \; Cl{-}\underset{\underset{O}{\parallel}}{C}{-}OR^2 \; \longrightarrow \; \left[(C_6H_5)_3\overset{+}{P}{-}CH_2{-}R^1 \right] Cl^-$$

$$\underline{1} \qquad\qquad \underline{68} \qquad\qquad\qquad \underline{4}$$

$$+$$

$$(C_6H_5)_3P{=}\underset{\underset{CO_2R^2}{|}}{C}{-}R^1$$

$$\underline{55}$$

A. *Methoxycarbonylation of Alkylidene-*
    triphenylphosphoranes

Methylchloroformate (*68*, $R^2 = CH_3$) (0.1 mol) in 50 ml of benzene is
added dropwise, under an atmosphere of nitrogen, to a boiling salt-
free solution of 0.22 mol of an alkylidenetriphenylphosphorane in
anhydrous benzene. The phosphonium chloride *4* precipitates. The
dropwise addition is stopped immediately when the ylide has been consumed, as shown by decoloration of the red solution. An excess of
methyl chloroformate must be avoided, since this also attacks the
alkoxycarbonylated ylide *55*. The nitrogen atmosphere is unnecessary
when the reaction is completed. The phosphonium chloride is filtered
off and washed with benzene, and the solution is concentrated under
vacuum. The residue can be recrystallized from ethyl acetate or from
mixtures of ethyl acetate or benzene with petroleum ether. It is
not essential to isolate the alkoxycarbonylated ylide prior to further reactions. The benzene solution can be treated with aldehydes
directly after removal of the phosphonium chloride, or the crude
product can be hydrolyzed after the benzene has been driven off. The
phosphonium chlorides *4* are obtained in 80 to 100% yields. They
can be used for regeneration of the ylide after a single reprecipitation from chloroform with ether, followed by careful drying. Chlorides which are initially oily can be crystallized by trituration.

B. *Carboxylic Acids from Alkoxycarbonylalkyl-*
    idenetriphenylphosphoranes

The methoxycarbonylalkylidenetriphenylphosphoranes *55a*, which are
easily obtained by the method described in Sec. IV.A, can be used
for the synthesis of carboxylic acids [57, 59].

1. Hydrolysis

The alkaline hydrolysis of the ylides 55a results in the formation of the carboxylic acids 69. The ylide 55a is refluxed for 1 hr
with a 10% solution of potassium hydroxide in aqueous methanol
$(H_2O/CH_3OH, 1:1)$. After cooling, the contents of the flask are mixed

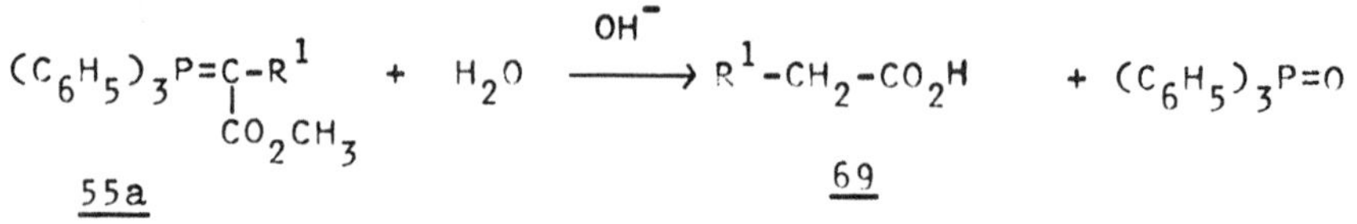

$$(C_6H_5)_3P=\underset{\underset{CO_2CH_3}{|}}{C}-R^1 \quad + \quad H_2O \quad \xrightarrow{OH^-} \quad R^1-CH_2-CO_2H \quad + \quad (C_6H_5)_3P=O$$

$$\underline{55a} \hspace{8cm} \underline{69}$$

with 10 times their volume of water, whereupon precipitation of the
greater part of the triphenylphosphine oxide occurs. The mixture is
filtered, and the filtrate is concentrated to 50-100 ml, cooled, and
brought to pH 1-3 by addition of sulfuric acid. The subsequent work-
up is carried out in accordance with the properties of the carboxylic
acid formed.

2. Wittig Reaction

The Wittig reaction of compounds 55a with aldehydes (22) leads
to the esters of the α-branched α,β-unsaturated carboxylic acids 70
[57, 59-61]. A solution containing the alkylmethoxycarbonylmethylene-
triphenylphosphorane 55a and the aldehyde 22 in benzene is heated

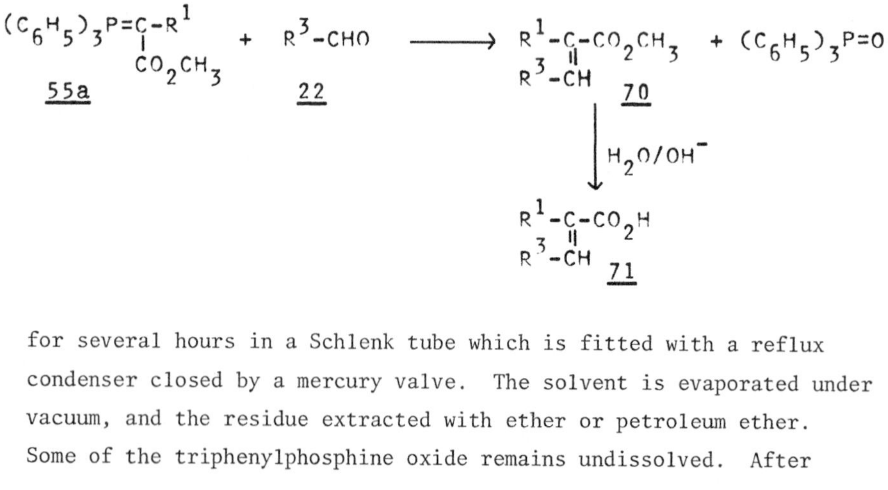

$$(C_6H_5)_3P=\underset{\underset{CO_2CH_3}{|}}{C}-R^1 \quad + \quad R^3-CHO \quad \longrightarrow \quad \underset{R^3-CH}{R^1-\overset{\|}{C}-CO_2CH_3} \quad + \quad (C_6H_5)_3P=O$$

$$\underline{55a} \hspace{4cm} \underline{22} \hspace{5cm} \underline{70}$$

$$\downarrow H_2O/OH^-$$

$$\underset{R^3-CH}{R^1-\overset{\|}{C}-CO_2H} \quad \underline{71}$$

for several hours in a Schlenk tube which is fitted with a reflux
condenser closed by a mercury valve. The solvent is evaporated under
vacuum, and the residue extracted with ether or petroleum ether.
Some of the triphenylphosphine oxide remains undissolved. After

filtration, the solvent is driven off, and the residue is hydrolyzed
as described above. Dilution of the reaction mixture with water
results in the precipitation of the remaining triphenylphosphine
oxide, and the subsequent work-up depends on the properties of the
carboxylic acid 71.

Ketones generally do not react with the ylides 55. Fodor and
Tomóskózi [62], however, were able to carry out reactions of this
type in a pressure tube at 100° to 170°C. Similar reactions can be
effected in solution by acid catalysis [35].

3. Oxidation

The alkoxycarbonylalkylidenetriphenylphosphoranes 55 can be
oxidized by KMnO$_4$ [42] or NaIO$_4$ [43] to give esters of the α-ketocarboxylic acids 72.

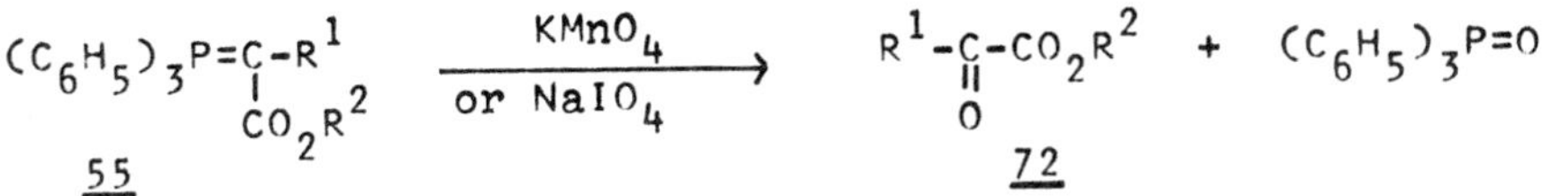

$$(C_6H_5)_3P=C-R^1 \xrightarrow[\text{or NaIO}_4]{\text{KMnO}_4} R^1-C-CO_2R^2 + (C_6H_5)_3P=O$$

<u>55</u>   <u>72</u>

C. *Reaction of Allylidenetriphenylphosphorane*
   with Esters of Chloroformic Acid: Prepara-
   tion of Esters of Polyenecarboxylic Acids

The allylidenetriphenylphosphorane (73) reacts via the resonance
form 73a with methyl chloroformate to form the salt 74 [57] from
which another molecule of the ylide 73, acting as a base, extracts a
proton from the γ position with respect to the P atom. The result
is the ylide 75, which can undergo Wittig reactions to yield esters
of the polyenecarboxylic acids 76.

V. ALKYLATION OF PHOSPHONIUM YLIDES

A. *Alkylating Reagents*

Alkyl halides, Mannich bases, activated β-chlorovinyl derivatives
(chlorinated heterocycles, β-chloro-α,β-unsaturated ketones, or
esters), and dialkylaluminumalkylideneamides can be used for the

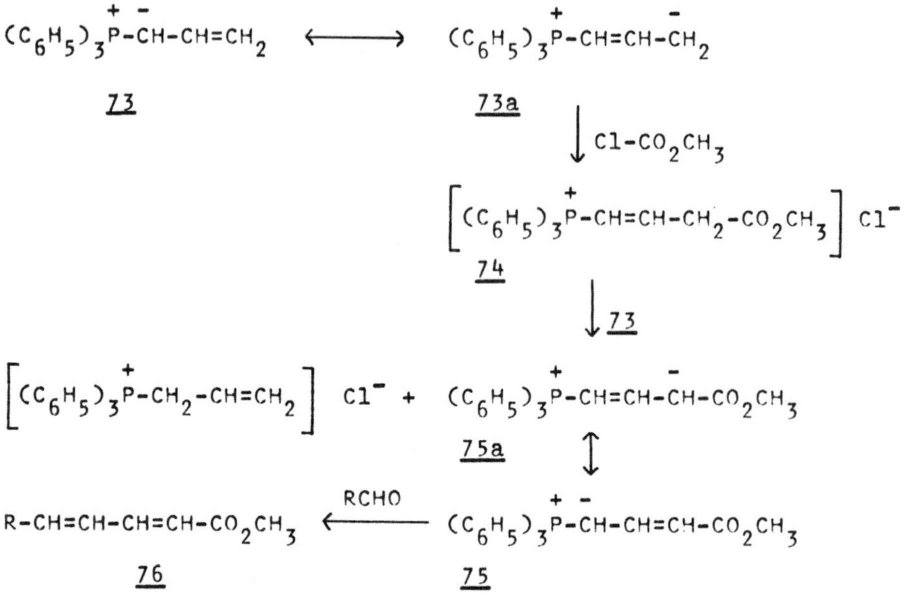

$$(C_6H_5)_3 \overset{+}{P}-\overset{-}{CH}-CH=CH_2 \quad \longleftrightarrow \quad (C_6H_5)_3 \overset{+}{P}-CH=CH-\overset{-}{CH}_2$$

$$\underline{73} \qquad\qquad\qquad\qquad\qquad \underline{73a}$$

$$\Big\downarrow Cl-CO_2CH_3$$

$$\left[(C_6H_5)_3 \overset{+}{P}-CH=CH-CH_2-CO_2CH_3 \right] Cl^-$$

$$\underline{74}$$

$$\Big\downarrow \underline{73}$$

$$\left[(C_6H_5)_3 \overset{+}{P}-CH_2-CH=CH_2 \right] Cl^- + (C_6H_5)_3 \overset{+}{P}-CH=CH-\overset{-}{CH}-CO_2CH_3$$

$$\underline{75a} \quad \Big\updownarrow$$

$$R-CH=CH-CH=CH-CO_2CH_3 \quad \overset{RCHO}{\longleftarrow} \quad (C_6H_5)_3 \overset{+}{P}-\overset{-}{CH}-CH=CH-CO_2CH_3$$

$$\underline{76} \qquad\qquad\qquad\qquad \underline{75}$$

alkylation of phosphine alkylenes. The most convenient method is the reaction with alkyl halides.

B. *Alkylation of Methoxycarbonylmethyl-*
   enetriphenylphosphorane

1. Synthesis of Carboxylic Acids

The reaction of the methoxycarbonylmethylenetriphenylphosporane *55b* with alkyl halides (*77*) leads first to the phosphonium salts *78*. If the residue R^1 is a ligand with a (-I) effect, *78* will be a strong acid and will immediately react with a second molecule of the ylide *55b* by transylidation. The resulting products are a substituted methoxycarbonylmethylenetriphenylphosphorane, *55a*, and the phosphonium salt *79* from which the starting ylide *55b* can be easily recovered by treatment with a base [59].

When *55b* is treated with alkyl iodides, the reaction in anhydrous ethyl acetate stops with the alkylated phosphonium salt *78*. This is because *78* is a weaker acid than the salt *79* which would result from transylidation owing to a (+I) effect of the aliphatic residue R^1.

$$(C_6H_5)_3P=CH-CO_2CH_3 \quad + \quad R^1X \quad \longrightarrow \quad \left[(C_6H_5)_3\overset{+}{P}-\underset{\underset{R^1}{|}}{C}H-CO_2CH_3 \right] X^-$$

$$\underline{55b} \qquad\qquad\qquad \underline{77} \qquad\qquad\qquad\qquad \underline{78}$$

$$\Big\downarrow \underline{55b}$$

$$(C_6H_5)_3P=\underset{\underset{R^1}{|}}{C}-CO_2CH_3 \quad + \quad \left[(C_6H_5)_3\overset{+}{P}-CH_2-CO_2CH_3 \right] X^-$$

$$\underline{55a} \qquad\qquad\qquad\qquad\qquad\qquad \underline{79}$$

$$H_2O \Big/ \underline{55a} \qquad\qquad \searrow \quad R^3CHO$$

$$R^1-CH_2-CO_2CH_3 \qquad\qquad\qquad\qquad R^3-CH=CR^1-CO_2CH_3$$

$$\underline{69} \qquad\qquad\qquad\qquad\qquad\qquad\qquad \underline{70}$$

However, the yield of the salt *78* decreases rapidly with increasing length of the aliphatic group R^1 owing to the lower polarity of the C-halogen bond. Since the acid-base equilibrium gives rise to partial transylidation between alkylidenephosphoranes and phosphonium salts with equal acid and base strengths, unwanted by-products can be formed. There is no doubt, however, about the course of reactions in which the inductive effect of the substituent R^1 leads to complete transylidation [59], as when R^1 is $-CH_2-CH=CH_2$, $-CH_2-CO_2CH_3$, $-CH_2-CN$, $-CH_2-C_6H_5$, or $-CH_2-CH=CH=C_6H_5$.

The alkyl halide *77* (0.2 mol) is added to a boiling solution of 0.4 mol of the methoxycarbonylmethylenetriphenylphosphorane *55b* in anhydrous ethyl acetate, and the mixture is refluxed for 2 hr. The methoxycarbonylmethyltriphenylphosphonium halide *79* which precipitates is filtered off (yield 80-95%), and the filtrate is concentrated. The residue consists of the alkylated ylide *55a*, which is often obtained as an oil, but which can generally be crystallized by trituration and recrystallized from ethyl acetate. The compounds *55a* can be used, as already mentioned in Sec. IV.B to prepare a number of different types of carboxylic acids or their corresponding esters. The alkaline hydrolysis of the salts *78* also yields the carboxylic acids *69* [59].

2. *Synthesis of Esters of β-*
    Acylacrylic Acids

The α-bromoketones *80* react with the methoxycarbonylmethylenetriphenylphosphorane *55b* to form the phosphonium salts *81*, which give
the ylide *82* and the salt *79* by transylidation with a second molecule of *55b* [63]. In the ylide *82*, the H atoms in the β position to
the P atom are activated by the adjacent carbonyl group. The ylides
82 therefore decompose to form an ester of a β-acylacrylic acid *83*,
and triphenylphosphine [57, 64].

$$(C_6H_5)_3P=CH-CO_2CH_3 \quad + \quad R-CO-CH_2Br \longrightarrow \left[(C_6H_5)_3\overset{+}{P}-\underset{CH_2-CO-R}{\overset{|}{C}H-CO_2CH_3} \right] Br^-$$

$$\underline{55b} \qquad\qquad\qquad \underline{80} \qquad\qquad\qquad\qquad \underline{81}$$

$$\underline{55b} \; + \; \underline{81} \longrightarrow (C_6H_5)_3P=\underset{CH_2-CO-R}{\overset{|}{C}}-CO_2CH_3 \quad + \quad \left[(C_6H_5)_3\overset{+}{P}-CH_2-CO_2CH_3 \right] Br^-$$

$$\underline{82} \qquad\qquad\qquad\qquad\qquad \underline{79}$$

$$\downarrow$$

$$R-CO-CH=CH-CO_2CH_3 \quad + \quad (C_6H_5)_3P$$

$$\underline{83}$$

The reaction may be looked upon as a Hofmann degradation of the
salt *81* with the ylide *55b* acting as a base. It has been shown
using compounds labeled with tritium that when an organolithium compound, sodamide, or an ylide is used for the Hofmann degradation of
phosphonium salts (*84*) in which the residue R^1 is strongly electron
attracting, the primary reaction is an α-elimination to form the
ylide *85*. Reactions with completely deuterated compounds have shown
that the alkylidenephosphorane *85* then decomposes by an intermolecular
hydrogen shift to give an olefin and triphenylphosphine [63]. It is
assumed that two molecules of the ylide *85* come together to form a
six-membered transition state which undergoes a reciprocal

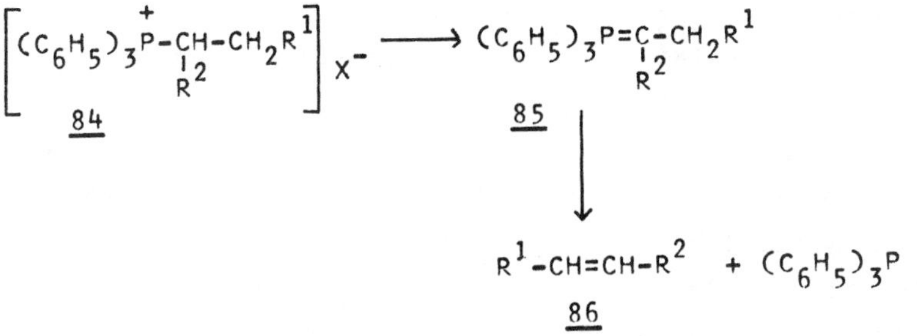

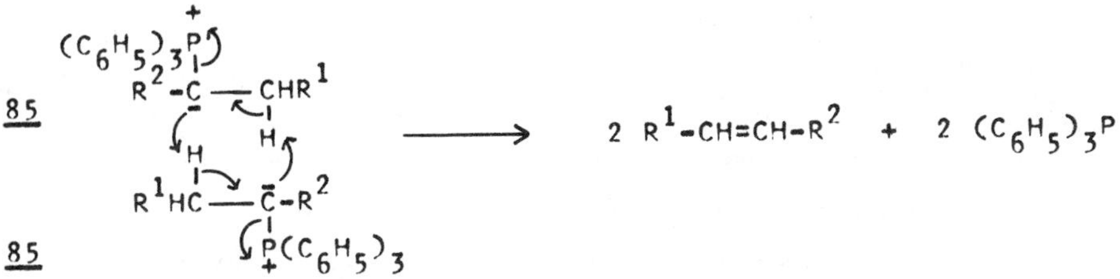

$$\left[(C_6H_5)_3 \overset{+}{P} - \underset{\underset{R^2}{|}}{C}H - CH_2R^1 \right] X^- \longrightarrow (C_6H_5)_3 P = \underset{\underset{R^2}{|}}{C} - CH_2R^1$$

$$\underline{84} \qquad\qquad \underline{85}$$

$$R^1 - CH = CH - R^2 + (C_6H_5)_3P$$
$$\underline{86}$$

$$\underline{85} \qquad\qquad \longrightarrow \qquad 2\ R^1 - CH = CH - R^2 + 2\ (C_6H_5)_3P$$

$$\underline{85}$$

β-elimination, with simultaneous loss of triphenylphosphine. This means that the Hofmann degradation of phosphonium salts, which begins with an α-elimination from the same ligand, is induced by an inter-molecular β-elimination. The ylide formation can therefore be regarded as an "interfering reaction" in the Hofmann degradation. The rate of the intermolecular hydrogen shift (reciprocal β-elimination) is determined by the residues R^1 and R^2 in the ylide *85*. Electronegative residues R^1 increase the rate by increasing the mobility of the β protons, while electronegative residues R^2 retard the reaction by decreasing the ability of the ylide to act as a proton acceptor.

This new method for the preparation of the β-aroylacrylates (R = Ar) is inferior, in a number of cases, to the reaction of maleic anhydride with aromatic compounds [65]. However, the esters of β-acylacrylic acids with aliphatic and cycloaliphatic residues R, which had previously been difficult to prepare, can now be readily obtained in high yields by a "single-pot process" [64].

To a boiling solution of 20 mmol of the methoxycarbonylmethyl-enetriphenylphosphorane *55b* in 100 ml of anhydrous benzene, 10 mmol

of the bromoketone *80* are added, and the mixture is refluxed for 2
hr. The triphenylmethoxycarbonylmethylphosphonium bromide *79* which
precipitates is filtered off (yield 90-100%), and the solution is
mixed with 10 mmol of methyl bromoacetate (possibly in slight excess)
and boiled for an additional 2 hr. The phosphonium salt *79* formed
from triphenylphosphine and methyl bromoacetate precipitates and
is filtered off (yield 70-80%). The benzene is evaporated from the
filtrate under reduced pressure. The residue is recrystallized from
methanol, cooling to between $-60°$ and $-40°C$ being preferable in many
cases. If the acrylate cannot be recrystallized from methanol, the
residue left after the benzene has been evaporated is distilled in a
rotating bulb tube at about 0.1 to 1 mm Hg. The reaction has been
applied with success in the steroid series [66]. Esters of α-branched
β-acylacrylic acids can be obtained from the alkoxycarbonylalkylidenetriphenylphosphoranes *55* and the α-bromoketones *80*. (See Secs.
IV.A and V.B.)

C. *Reactions of Phosphonium Ylides with Esters*
    of α-Halogencarboxylic Acids

The course of the reactions of alkyl halides with phosphorus ylides
is influenced not only by the polarity of the C-halogen bond but
also by the inductive effect of the halogen atom on the whole molecule. This fact is demonstrated in the reaction with the esters of
halogenated carboxylic acids. The strong (-I) effect of fluorine
in α-fluoroacetic acid ester enables nucleophilic attack of an ylide
to take place at the ester carbonyl, giving rise to the formation of
enol ethers of α-fluoroketones (Wittig reaction) [22]. With α-
chloroacetic acid esters, ylides behave as bases, abstracting a
proton from the ester. The anion thus formed gives rise, in a subsequent reaction, to the formation of esters of cyclopropane tricarboxy-
lic acid [68]. The esters of α-iodo and α-bromocarboxylic acids
react in the expected way, in a substitution reaction [69].

1. Synthesis of Esters of α,β-Unsaturated
    Carboxylic Acids

Salt-free solutions [12] of 2 mol of an alkylidenephosphorane
react with 1 mol of the α-bromo- or α-iodo-carboxylate 87, X = Br or
I, to give the α,β-unsaturated carboxylates 89, triphenylphosphine,
and the phosphonium salt 90 [69]. The polarity of the C-Br or C-I
bond enables the nucleophilic substitution of the halogen by the
ylide 86 to take place, thus giving rise to the formation of the
phosphonium salt 88. Compound 88 is activated by the inductive
effect of the ester group and reacts with a second mole of 86 in a
Hofmann degradation.

$$(C_6H_5)_3P=CR^1R^2 \;+\; R^3\underset{X}{CH}-CO_2R^4 \longrightarrow \left[(C_6H_5)_3\overset{+}{P}-CR^1R^2 \atop R^3-CH-CO_2R^4 \right] X^-$$

<u>86</u>   <u>87</u>   <u>88</u>

$$\underline{86} + \underline{88} \longrightarrow R^1R^2C=\underset{R^3}{C}-CO_2R^4 \;+$$

<u>89</u>

$$\left[(C_6H_5)_3\overset{+}{P}-CHR^1R^2 \right] X^- \;+\; (C_6H_5)_3P$$

<u>90</u>

Reactions of strongly basic ylides, 86, with α-iodo-carboxylates
give higher yields than are obtained with the corresponding bromo-
derivatives. When R^1 and $R^2 \neq H$, the use of the iodo compounds 87
causes the rate of the reaction of 87 with triphenylphosphine, formed
in the Hofmann degradation, to become more rapid than the rate of
the reaction with 86. In this case the reaction between 86 and 87
has to be carried out in a molar ratio of 1:1. The end products are
89, 90, and the phosphonium salt 91.

$$(C_6H_5)_3P \quad + \quad R^3\text{-}\underset{\underset{\underline{87}}{I}}{CH}\text{-}CO_2R^4 \quad \longrightarrow \quad \left[(C_6H_5)_3\overset{+}{P}\text{-}\underset{\underset{\underline{91}}{CO_2R^4}}{CHR^3}\right] I^-$$

This reaction may also be carried out with vinylogous halogen compounds. Two moles of the benzylidenetriphenylphosphorane 92 and 1 mol of γ-bromo-methylcrotonate (93) give in 85% yield the methyl ester of 5-phenylpentadien-(2,4) carboxylic acid (94).

$$2 \; (C_6H_5)_3P\text{=}CH\text{-}C_6H_5 \quad + \quad Br\text{-}CH_2\text{-}CH\text{=}CH\text{-}CO_2CH_3 \quad \longrightarrow$$
$$\underset{\underline{92}}{} \qquad\qquad\qquad \underset{\underline{93}}{}$$

$$(C_6H_5)_3P \quad + \quad \left[(C_6H_5)_3\overset{+}{P}\text{-}CH_2\text{-}C_6H_5\right] Br^-$$

$$+$$

$$C_6H_5\text{-}CH\text{=}CH\text{-}CH\text{=}CH\text{-}CO_2CH_3$$
$$\underline{94}$$

2. Synthesis of γ-Ketocarboxylic Acids
   and β-Acyl Acrylates

In Sec. III.C. it had been mentioned that acylalkylidenetriphenylphosphoranes (4) are described by two resonance structures, 4a and 4b. Alkylation occurs in most cases at the O atom, thus leading to the enol ether phosphonium salts 95b [26, 70, 71], but seldom at the C atom, giving the ketophosphonium salts 95a [70]. The acylmethylenetriphenylphosphoranes 96, on the other hand, are alkylated by

$$R^2\text{-}\underset{\underset{O}{\parallel}}{C}\text{---}\underset{\underset{P(C_6H_5)_3}{\parallel}}{C}\text{-}R^1 \qquad \xrightarrow{\;R^3\text{-}X\;} \qquad \left[R^2\text{-}\underset{\underset{O}{\parallel}}{C}\text{---}\underset{\underset{\overset{+}{P}(C_6H_5)_3}{\mid}}{\overset{\overset{\textstyle R^3}{\mid}}{C}}\text{-}R^1\right]X^-$$
$$\qquad\qquad \underset{\underline{4a}}{\big\uparrow} \qquad\qquad\qquad\qquad\qquad\qquad \underset{\underline{95a}}{}$$

$$R^2\text{-}\underset{\underset{\underline{O}}{\mid}}{C}\text{=}\underset{\underset{\overset{+}{P}(C_6H_5)_3}{\mid}}{C}\text{-}R^1 \qquad \xrightarrow{\;R^3\text{-}X\;} \qquad \left[R^2\text{-}\underset{\underset{R^3\text{-}O}{\mid}}{C}\text{=}\underset{\underset{\overset{+}{P}(C_6H_5)_3}{\mid}}{C}\text{-}R^1\right]X^-$$
$$\qquad\qquad \underset{\underline{4b}}{} \qquad\qquad\qquad\qquad\qquad\qquad \underset{\underline{95b}}{}$$

methyl bromoacetate (*97*) at the C atom. The salt *98* which is formed immediately undergoes a transylidation reaction with a second mole of the ylide *96*, leading to the new phosphorane *99* and the phosphonium salt *100* [72].

$$(C_6H_5)_3P=CH-CO-R \ + \ Br-CH_2CO_2CH_3 \longrightarrow \left[(C_6H_5)_3\overset{+}{P}-\underset{\underset{CH_2CO_2CH_3}{|}}{CH}-CO-R \right] Br^-$$

$$\underline{96} \qquad\qquad \underline{97} \qquad\qquad\qquad \underline{98}$$

$$\underline{96} \ + \ \underline{98} \ \longrightarrow \ (C_6H_5)_3P=\underset{\underset{CH_2CO_2CH_3}{|}}{C}-CO-R \ + \ \left[(C_6H_5)_3\overset{+}{P}-CH_2-CO-R \right] Br^-$$

$$\underline{99} \qquad\qquad\qquad\qquad \underline{100}$$

$$H_2O/OH^- \qquad\qquad C_6H_5CO_2H \qquad \Delta$$

$$R-CO-CH_2-CH_2CO_2H \qquad \left[(C_6H_5)_3\overset{+}{P}-\underset{\underset{CH_2CO_2CH_3}{|}}{CH}-CO-R \right] \longrightarrow \ R-CO-CH=CH-CO_2CH_3$$

$$\underline{101} \qquad\qquad\qquad \underline{102} \qquad\qquad\qquad\qquad \underline{83}$$

$$+$$

$$\left[C_6H_5CO_2^- \right] \qquad\qquad C_6H_5CO_2H \ + \ (C_6H_5)_3P$$

$$R = CH_3, \ C_3H_7, \ C_6H_5CHCH_3, \ cyclo-C_6H_{11}, \ C_6H_5, \ 4-CH_3O-C_6H_4$$

The ylides *99* are hydrolyzed in aqueous methanol to triphenyl-phosphine oxide, methanol, and the γ-ketocarboxylic acids *101*. The H atoms of the methylene group in the ylides *99* are activated by the neighboring ester group. On heating to 150°-180°C, *99* decomposes to triphenylphosphine and the β-acylacrylates *83*, which may, however, undergo secondary reactions at this high temperature.

The Hofmann degradation of *99* to *83* may also be achieved under more smooth conditions using benzoic acid as a catalyst. From *99* and benzoic acid the phosphonium benzoate *102* is formed. The benzoate anion attacks the phosphonium cation in the β position to the P atom leading to the formation of triphenylphosphine, β-acylacrylate (*83*), and benzoic acid. The latter again forms with *99* the corresponding salt *102*. This new synthesis for the compounds *83* complements the preparation from α-bromoketones and alkoxycarbonylmethylenetriphenylphosphoranes [64]. Since the acylated methylenetriphenylphosphoranes *96* are easily prepared from acid chlorides or thio esters and methylenetriphenylphosphoranes (c.f. Sec. III.A), the synthesis of the γ-keto acids *101* and the β-acylacrylates *83* becomes readily available as shown in Eq. (14). Compound *99* must not be purified or isolated in this procedure.

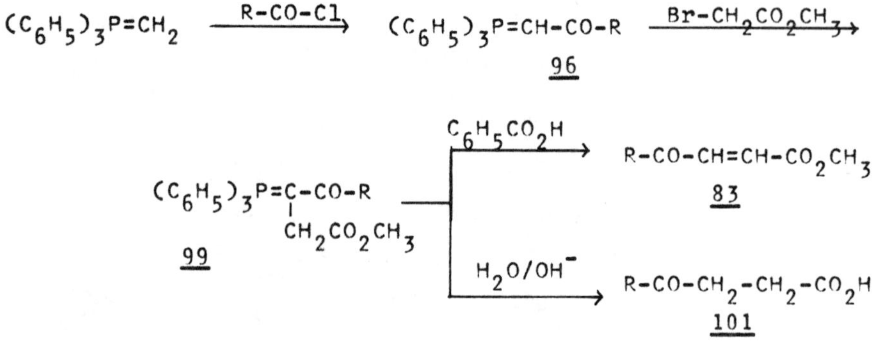

$$(C_6H_5)_3P=CH_2 \xrightarrow{R-CO-Cl} \underset{\underline{96}}{(C_6H_5)_3P=CH-CO-R} \xrightarrow{Br-CH_2CO_2CH_3}$$

$$\underset{\underline{99}}{\underset{CH_2CO_2CH_3}{(C_6H_5)_3P=C-CO-R}} \left\{ \begin{array}{l} \xrightarrow{C_6H_5CO_2H} \underset{\underline{83}}{R-CO-CH=CH-CO_2CH_3} \\ \\ \xrightarrow{H_2O/OH^-} \underset{\underline{101}}{R-CO-CH_2-CH_2-CO_2H} \end{array} \right.$$

$$(14)$$

D. *Reactions of Phosphonium Ylides with Benzyl Halides*

1. Synthesis of 1,2-Diarylethanes and Stilbenes

The benzylidene triphenylphosphoranes *103* react with benzyl-halides (*104*) in a molar ratio of 1:1 to give the phosphonium salts *105* without the occurrence of a transylidation reaction [73(a)]. From *105* the 1,2-diarylethanes *107* with substituents R and R[1] are available in two ways:

1. By electrolysis of the aqueous solution using a mercury electrode, whereby triphenylphosphine is split off (see Ref. 33).

2. By alkaline cleavage, whereby triphenylphosphine oxide is formed (see Refs. 1 and 4).

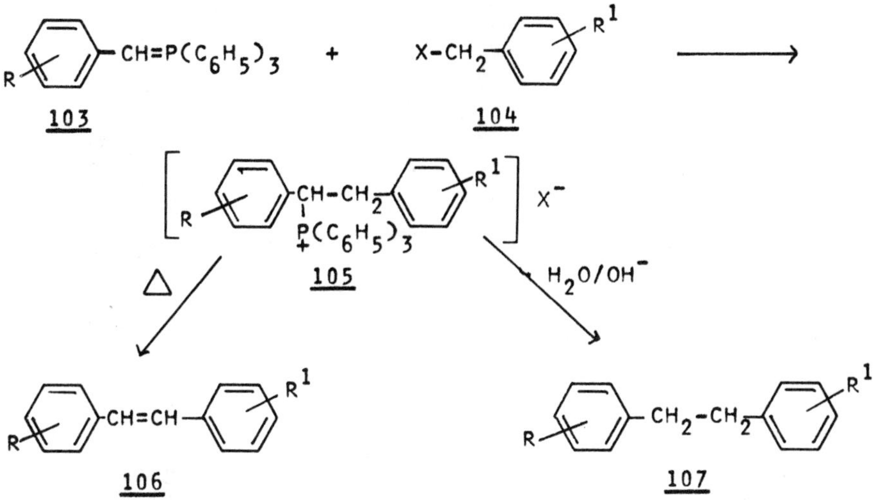

Pyrolysis of the salts *105* gives the stilbenes *106* in very good yields [73(b)].

2. Synthesis of Phenyl-Substituted Olefins

Two moles of benzyl bromide (*104*, X = Br) react with 1 mol of a strongly basic ylide, *86*, to give the phosphonium salt *108*, in which the H atoms in the β position to the phosphorus are activated, thus giving rise to a Hofmann degradation. Besides the phosphonium salts *110* and triphenylphosphine oxide, phenyl-substituted olefins *109* are isolated in 65-75% yield [73(b)].

$$(C_6H_5)_3P=CR^1R^2 \ + \ Br-CH_2-C_6H_5 \ \longrightarrow \ \left[(C_6H_5)_3\overset{+}{P}-\underset{CH_2-C_6H_5}{\overset{|}{C}R^1R^2} \right] Br^-$$

$$\underline{86} \qquad\qquad \underline{104} \qquad\qquad\qquad \underline{108}$$

$$\underline{108} \ + \ \underline{86} \ \longrightarrow \ R^1R^2C=CH-C_6H_5 \ + \ \left[(C_6H_5)_3\overset{+}{P}-CHR^1R^2 \right] Br^-$$

$$\underline{109} \qquad\qquad\qquad\qquad \underline{110}$$

$$+ \ (C_6H_5)_3P$$

a) $R^1 = H$; $R^2 = CH_3$
b) $R^1 = H$; $R^2 = $ cyclo-C_6H_{11}
c) $R^1 = R^2 = CH_3$

E. Reaction of Phosphonium Ylides with α-
    Halogenamines: Synthesis of Substi-
    tuted Allyl N,N-Dimethylamines

Two moles of an ylide, *1*, and 1 mol of the α-chloromethyldimethylamine
111 form, in a transylidation reaction, the amino-methylated phosphoranes *112*. Compounds of the *112* type, which are sensitive to air
and moisture, react with carbonyl compounds to give the allyl N,N-
dimethylamines *113* [74].

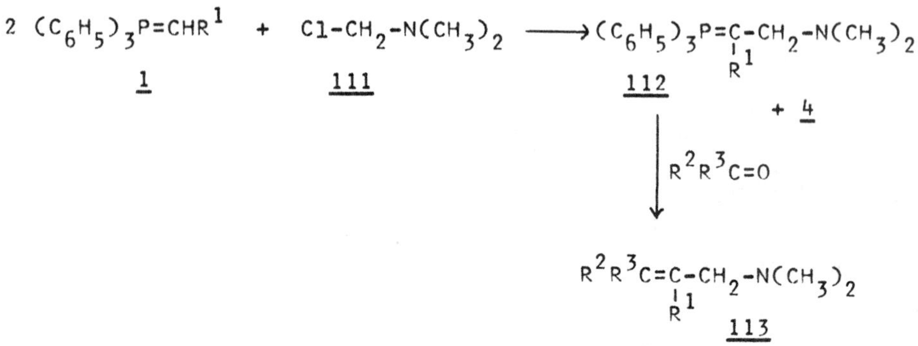

$$2 \ (C_6H_5)_3P=CHR^1 \ + \ Cl-CH_2-N(CH_3)_2 \ \longrightarrow (C_6H_5)_3P=\underset{\underset{R^1}{|}}{C}-CH_2-N(CH_3)_2$$

$$\underline{1} \qquad\qquad\qquad \underline{111} \qquad\qquad\qquad\qquad \underline{112}$$

$$+ \ \underline{4}$$

$$\Big\downarrow R^2R^3C=O$$

$$R^2R^3C=\underset{\underset{R^1}{|}}{C}-CH_2-N(CH_3)_2$$

$$\underline{113}$$

F. Alkylation of Phosphonium Ylides
    with Mannich Bases

    1. Synthesis of α-Branched α,β-
        Unsaturated Carbonyl Compounds

    The carboethoxymethylenetriphenylphosphorane *96* (R = OEt) and
the benzolymethylenetriphenylphosphorane *96* (R = C_6H_5) were found to
react readily with the Mannich bases *114* (R^1 = Ar or CH_2-CH_2-Ar) [75].
The resulting ylides *115*, which are obtained only with difficulty by

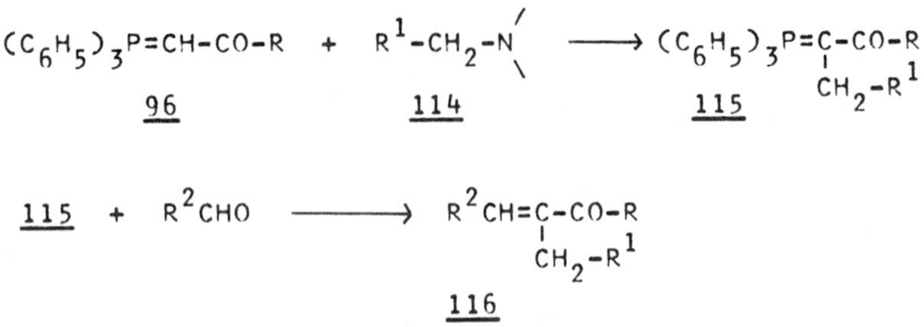

$$(C_6H_5)_3P=CH-CO-R \ + \ R^1-CH_2-N{\large\diagdown}^{\diagup} \ \longrightarrow (C_6H_5)_3P=\underset{\underset{CH_2-R^1}{|}}{C}-CO-R$$

$$\underline{96} \qquad\qquad\qquad \underline{114} \qquad\qquad\qquad\qquad \underline{115}$$

$$\underline{115} \ + \ R^2CHO \ \longrightarrow \ R^2CH=\underset{\underset{CH_2-R^1}{|}}{C}-CO-R$$

$$\underline{116}$$

other procedures, react with aldehydes to give the α-branched α,β-unsaturated carbonyl compounds *116*.

    2.   Synthesis of Benzopyran Derivatives

    a.   Some o-phenolic Mannich bases such as *117* yield products different from the "normal" ylide *115*. Owing to the participation of the phenolic hydroxyl, the alkylation may be accompanied by an internal Wittig reaction leading to benzopyrans such as *118* according to the scheme shown in Eq. (15).

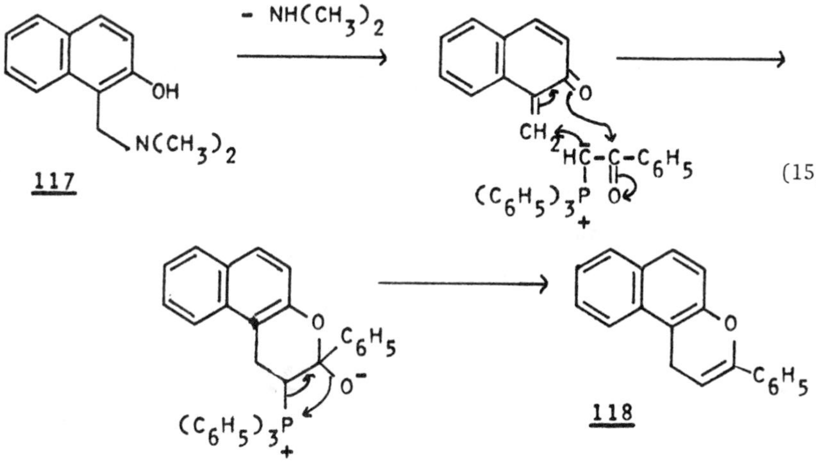

(15)

    b.   Benzopyrans are also available from the lactonization of the appropriate ylides (*115*) followed by a Wittig reaction. The Mannich base *119* reacts with *96* (R = OEt) to form the ylide *120*, which splits off 2 mol of EtOH. The resulting ylide *121* is isolated and reacts with aldehydes to give the compounds *122* [75].

The ylides *115*, resulting from the interaction of *96* and *114*, react readily with o-hydroxy benzaldehydes. The normal Wittig reaction is followed by an intramolecular condensation, in the course of which alcohol is split off and compounds such as *124* are formed [75].

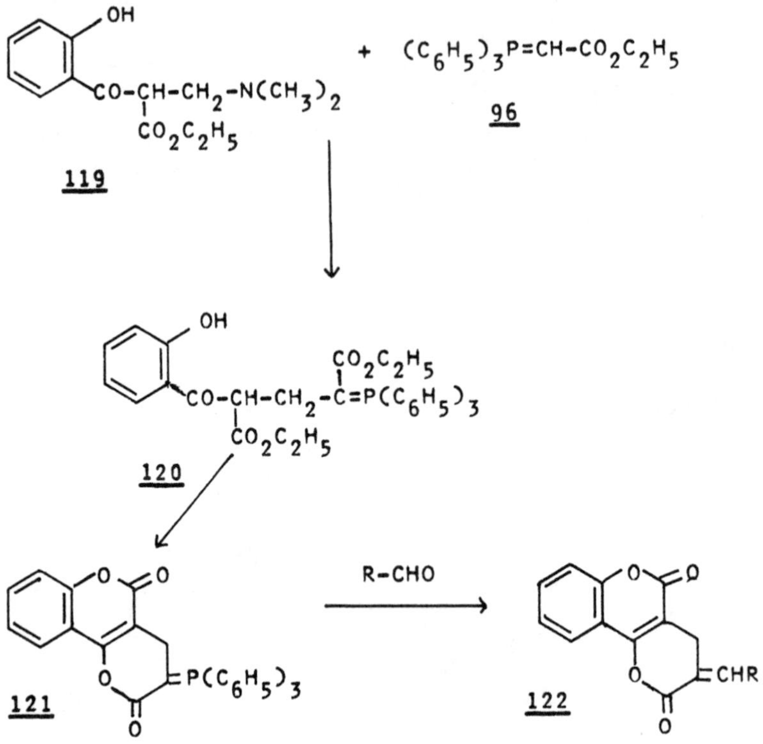
OH
CO-CH-CH2-N(CH3)2
CO2C2H5
119
+
(C6H5)3P=CH-CO2C2H5
96
OH
CO2C2H5
CO-CH-CH2-C=P(C6H5)3
CO2C2H5
120
121
=P(C6H5)3
R-CHO
122
=CHR

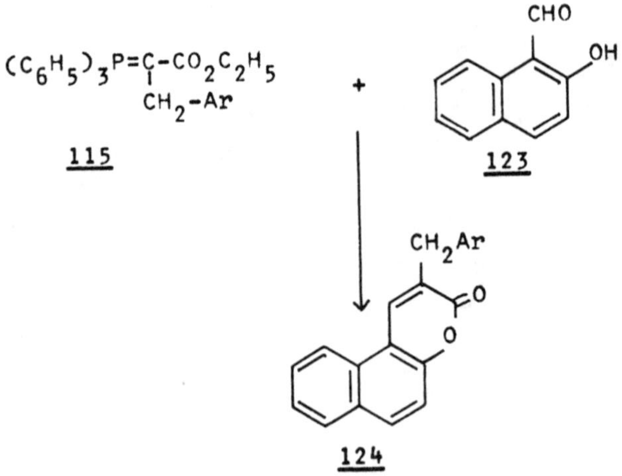
(C6H5)3P=C-CO2C2H5
CH2-Ar
115
+
CHO
OH
123
CH2Ar
124

G. Alkylation of Phosphonium Ylides by Heterocyclic Compounds

The transylidation reaction has also been applied to the synthesis of alkylated and alkenylated heterocycles [76]. The heterocycles *125* with suitable leaving groups (X = Cl, Br, SO_2CH_3, etc.) react with the alkylidene phosphoranes *126* to give the phosphonium salts *127*, which undergo transylidation by reaction with a second equivalent of the initial ylide *126*. The heterocyclic ylide *128* thus formed is then converted, in situ, either by hydrolysis to an alkyl derivative, *129*, of the heterocycle or by reaction with a carbonyl compound into an alkenyl derivative, *130* [76].

$$\text{Het-X} \;+\; \text{RCH=PR}_3^1 \;\longrightarrow\; \left[\text{Het-}\overset{+}{\underset{R}{\text{CH}}}\text{-PR}_3^1 \right] \text{X}^-$$

$$\underline{125} \qquad\qquad \underline{126} \qquad\qquad\qquad\qquad \underline{127}$$

$$\underline{126} \;+\; \underline{127} \;\longrightarrow\; \text{Het-}\underset{R}{\underset{|}{\text{C}}}\text{=PR}_3^1 \;\xrightarrow[\text{H}_2\text{O}]{\text{Na}_2\text{CO}_3}\; \text{Het-CH}_2\text{R}$$

$$\underline{128} \qquad\qquad\qquad\qquad \underline{129}$$

$$R^1 = C_6H_5 \text{ or } C_4H_9 \qquad \downarrow R^2R^3C=O$$

$$\text{Het-}\underset{R}{\underset{|}{\text{C}}}\text{=CR}^2R^3$$

$$\underline{130}$$

The preparation of heterocyclic phosphonium ylides and their subsequent reactions can be accomplished as follows: to a stirred suspension of the appropriate phosphonium salt (2.2 eq) in anhydrous 1,2-dimethoxyethane (DME) under dry nitrogen at $-30°$ to $-35°C$ is added n-butyllithium in hexane (2.2 eq). The reaction mixture is stirred for 1 hr, and the appropriate heterocycle (1 eq) in anhydrous DME is added. The mixture is allowed to warm slowly (about 1 hr) to room temperature and is stirred either under reflux or at room temperature.

The hydrolysis of heterocyclic ylides and the formation of alkyl-substituted heterocycles can be accomplished as follows:

Sodium carbonate (1 eq) in water is added to the above solution of
the heterocyclic ylide; the mixture is refluxed for 3 hr, evaporated
under reduced pressure, and then worked up by one of the following
procedures:

Method A. The mixture is suspended in chloroform or ether and
extracted with dilute aqueous hydrochloric acid, the combined aqueous
layers are made alkaline with sodium hydroxide, and the resulting
mixture is extracted with ether. The combined ether extracts are
dried and evaporated, and the product is purified by distillation
or recrystallization.

Method B. The mixture is extracted several times with hot
ether, the combined ether extracts are concentrated under reduced
pressure, and the residual material is treated with an excess of
mercuric chloride in 25% aqueous ethanol. The precipitated salt is
then collected by filtration and washed, and the heterocycle is
freed by treatment of the mercuric chloride salt with hydrogen sulfide
gas and sodium carbonate in 10% aqueous ethanol. The crude product
is purified by distillation or recrystallization.

Wittig reactions of heterocyclic ylides and the formation of
alkenyl-substituted heterocycles can be accomplished as follows:
the reaction mixture containing the "heterocyclic" ylide is treated
with an excess (about 4 eq) of an appropriate aldehyde or ketone in
anhydrous DME. In order to minimize side reactions (aldol condensations), the addition of aliphatic aldehydes and ketones to the solution of the heterocyclic ylide is performed at -30°C. The reaction
mixture is then stirred for 24 hr at room temperature and filtered,
and the excess solvent is removed under reduced pressure. The residue is treated in accordance with method B above.

The following comments can be made about this procedure for
alkylation and alkenylation of heterocycles by means of transylidation:
    1. It is applicable both to five- and six-membered ring systems
and to monocyclic as well as polycyclic heterocycles.
    2. Although the new substituent can be introduced only into a
position previously occupied by the leaving group, this restriction

confers regiospecificity to the alkylation and alkenylation reactions simultaneously. Thus, although 3-substituted quinolines cannot be prepared by this procedure, either 2- or 4-substituted derivatives can be prepared selectively, starting with an appropriately substituted precursor.

3. The requisite chloro or methylsulfonyl precursors number among the most accessible derivatives in heterocyclic chemistry. Although most of the simple heterocycles (e.g., pyridine, quinoline, and isoquinoline) are readily available in many heterocyclic systems (e.g., quinazoline), the parent substituted heterocycle is less accessible than the substituted derivatives are. The chloro compound, in fact, is often an intermediate in the formation of the unsubstituted system. Furthermore, since chloro compounds are usually prepared by the action of phosphorus oxychloride or similar reagents on the corresponding lactams, which, in turn, are often the immediate product of a ring closure reaction, the position of substitution is also assured.

4. Since a variety of leaving groups can be employed successfully, substrates which prove unreactive to displacement (i.e., 2-chloropyridine and 2-methylthioquinoline) can be converted to substrates which undergo displacement readily (i.e., 2-bromo- or 2-methylsulfonylpyridine and 2-chloro- or 2-methylsulfonylquinoline). This flexibility in the choice of the leaving group greatly extends the versatility of the procedure.

5. Flexibility is also possible in the choice of the nucleophilic alkylidenephosphorane. Methylene-, ethylidene-, and n-butylidenetriphenylphosphorane have all been used successfully. Furthermore, although 2-chloroquinoline failed to react with benzylidenetriphenylphosphorane, it reacted satisfactorily with benzylidene-(tri-n-butyl)-phosphorane, which is more nucleophilic.

6. Since the intermediate heterocyclic ylide reacts with aliphatic and aromatic aldehydes, and with ketones ranging in reactivity from acetone to benzophenone, a broad variety of alkenyl groups (and, by reduction, alkyl groups) may be selectively introduced into the heterocyclic substrate.

7. The alkenylation method appears to lead stereospecifically
to trans-olefins [77].

The versatility of this procedure, based on the transylidation
reaction, is illustrated by the synthesis of papaverine, a series
of purine derivatives, and some Cinchona alkaloids [76].

H. *Reaction of Phosphonium Ylides with*
    2-Chlorotropone

By the reaction of 2-chlorotropone (*131*) with the phosphonium ylide
1, the 2-troponyl-methylenetriphenylphosphoranes *132* are formed which
have, as indicated by x-ray analysis, a "bonding betaine" structure,
132 [78]. From *132* phosphorus-free products can be obtained either
by a Wittig reaction leading to the 2-troponylethylenes *133* [79] or
by the reaction with acetylenes, which gives rise to the formation
of the azulenes *134* with Diels-Alder products as by-products [80].

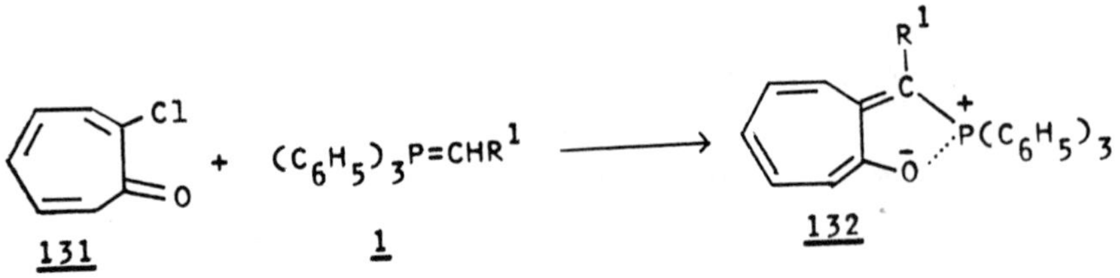

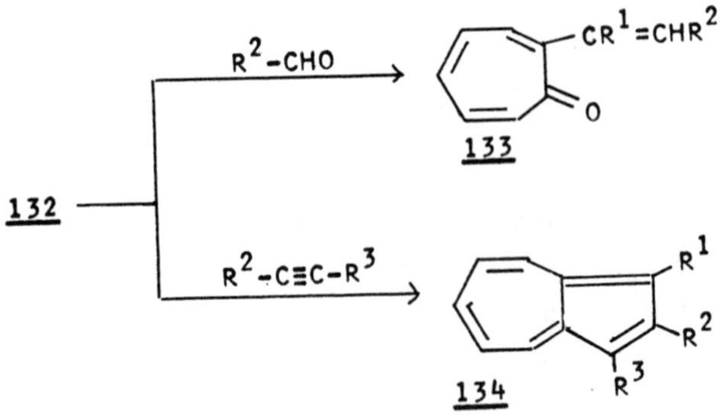

I. *Reactions of Phosphonium Ylides with β-Chloro-*
   α,β-Unsaturated Carbonyl Compounds

1. Reaction with β-Chlorovinyl Ketones

The ylides *1* react with the β-chlorovinyl ketones *135* in a transylication reaction yielding the ylides *136* and the phosphonium salts *5* [81]. The compounds *136* should be of preparative interest in the future.

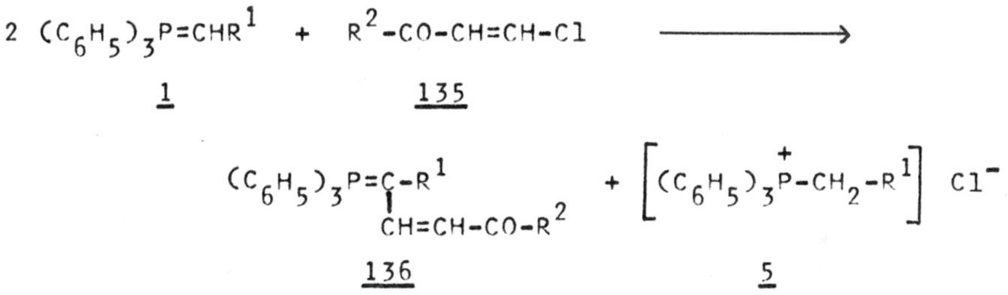

$$2 \ (C_6H_5)_3P=CHR^1 \ + \ R^2-CO-CH=CH-Cl \ \longrightarrow$$

$$\underline{1} \qquad\qquad\qquad \underline{135}$$

$$(C_6H_5)_3P=\underset{\underset{136}{CH=CH-CO-R^2}}{\overset{|}{C}}-R^1 \ + \ \left[(C_6H_5)_3\overset{+}{P}-CH_2-R^1\right] \ Cl^-$$

$$\underline{136} \qquad\qquad\qquad \underline{5}$$

2. Reaction of β-Chloroacrylates with Allylidenetriphenylphosphorane

The allylidenetriphenylphosphorane *73* may react at the γ-C atom, as was shown in the reaction with esters of chloroformic acids [57]. Compound *73* also adds to the β-chloroacrylate esters *137* and related

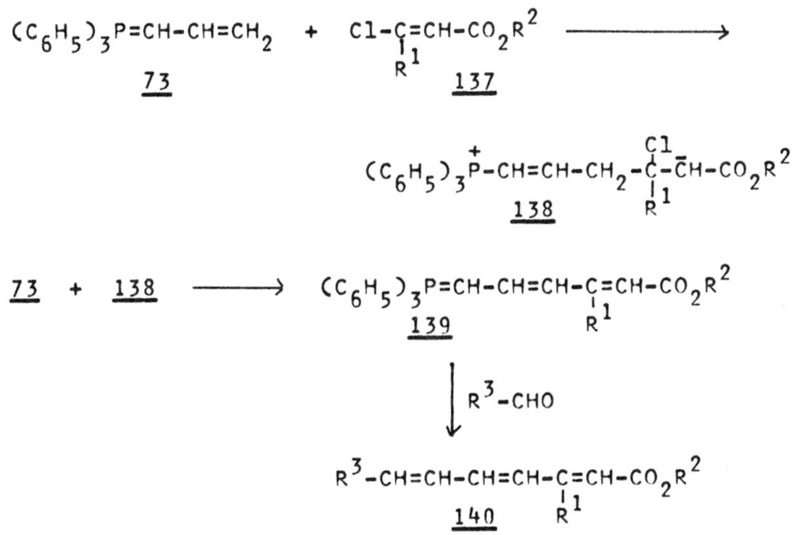

$$(C_6H_5)_3P=CH-CH=CH_2 \ + \ Cl-\underset{\underset{R^1}{|}}{C}=CH-CO_2R^2 \ \longrightarrow$$

$$\underline{73} \qquad\qquad\qquad \underline{137}$$

$$(C_6H_5)_3\overset{+}{P}-CH=CH-CH_2-\underset{\underset{R^1}{|}}{\overset{\overset{Cl}{|}}{C}}-\bar{C}H-CO_2R^2$$

$$\underline{138}$$

$$\underline{73} \ + \ \underline{138} \ \longrightarrow \ (C_6H_5)_3P=CH-CH=CH-\underset{\underset{R^1}{|}}{C}=CH-CO_2R^2$$

$$\underline{139}$$

$$\Big\downarrow R^3-CHO$$

$$R^3-CH=CH-CH=CH-\underset{\underset{R^1}{|}}{C}=CH-CO_2R^2$$

$$\underline{140}$$

Michael acceptors, presumably to give intermediates such as *138*.
Subsequent loss of chloride and deprotonation by a second mole of
73 acting as a base converts *138* into the stabilized ylide *139* [82].
Treatment of the crude product solution with a carbonyl compound
affords the expected Wittig product *140*. In this way conjugated
polyene esters or ketones are easily prepared in acceptable yield,
provided that the stabilized ylide is trapped with an aldehyde.
Ketones give only small amounts of the corresponding Wittig product.

In a typical experiment, the allyltriphenylphosphonium bromide
(0.38 g) was stirred in dry THF (10 ml) under nitrogen and treated
with lithium diisopropylamide (1.4 ml, 0.73 N in hexane-THF). After
30 min at 20°C, methyl cis-3-chloropropenoate (0.03 g) was added dropwise. The mixture was stirred for 2 hr at room temperature and
quenched with an excess of benzaldehyde (overnight). Chromatography
of the product over silica gel after aqueous work-up gave the Wittig
product in 84% yield.

The addition of the allylidenetriphenylphosphorane to methyl
β-chlorocrotonate is of special interest because the resulting ylide
139 (R^1 = CH_3) corresponds to the terminal carbons of the vitamin A
methyl ester side chain. However, *139* (R^1 = CH_3) does not appear to
be suited for stereoselective synthesis of isoprenoid polyenes.
Trapping the ylide with isobutyraldehyde affords comparable amounts
of four isomers. Apparently, elimination of chloride ion from the
zwitterionic intermediate *138* generates both E and Z isomers at
the trisubstituted double bond of *139*. Each of the E-Z isomers then
affords a pair of cis-trans isomers in the Wittig condensation step.

The net effect of the observed reactions is to replace a γ
proton of the allylidenetriphenylphosphorane by the β carbon of an
unsaturated ester or ketone. The procedure provides a facile synthesis of compounds having at least three double bonds conjugated
with a carbonyl function.

J. Alkylation of Methylenetriphenylphosphorane
by Dialkylaluminum Alkylideneamides:
Synthesis of Allylidenephosphoranes

The dialkylaluminum alkylideneamides *141* available from nitriles
and dialkylaluminumhydride in toluene, benzene, ether, or tetrahydrofuran react with methylenetriphenylphosphorane (*17*) at temperatures of 20°C or below to form the allylidenetriphenylphosphorane
142 and the dialkylaluminum amide *143* [83]. According to the proposed mechanism [83], the first step is, by analogy with the Wittig
reaction, a nucleophilic attack of the ylide carbon on the activated

$$\frac{1}{2}\left[\underset{R^2}{\overset{R^1}{>}}CH-CH=N-Al\underset{R^3}{\overset{R^3}{<}} \right]_2 \quad + \quad (C_6H_5)_3P=CH_2 \quad \longrightarrow$$

$$\underline{141} \qquad\qquad\qquad \underline{17}$$

$$\underset{R^2}{\overset{R^1}{>}}C=CH-CH=P(C_6H_5)_3 \quad + \quad H_2N-Al\underset{R^3}{\overset{R^3}{<}}$$

$$\underline{142} \qquad\qquad\qquad \underline{143}$$

R^1, R^2 = H or alkyl
R^3 = alkyl

C=N double bond of *141*, whereby the intermediate *144* is formed.
Deviating from the Wittig reaction, the betaine *144* does not decompose
to an olefin and a phosphineimine as in the reaction of ylides
with Schiff's bases [84(a)] but undergoes a 1,3-proton shift to form

$$\overset{H}{-\overset{|}{C}H-CH=N-AlR_2} \quad + \quad CH_2=P(C_6H_5)_3 \longrightarrow \quad \overset{H}{-\overset{|}{C}H-\underset{\underset{N-AlR_2}{|}}{C}H-CH_2-\overset{+}{P}(C_6H_5)_3} \longrightarrow$$

$$\underline{141} \qquad\qquad\qquad \underline{17} \qquad\qquad\qquad\qquad \underline{144}$$

$$\overset{H}{-\overset{\curvearrowright}{C}H-CH-CH=P(C_6H_5)_3} \qquad \longrightarrow \qquad -CH=CH-CH=P(C_6H_5)_3 \quad + \quad H_2N-AlR_2$$
$$\underset{H\overset{\curvearrowleft}{N}-AlR_2}{} \underline{145} \qquad\qquad\qquad\qquad \underline{142} \qquad\qquad\qquad \underline{143}$$

the 2-(dialkylaluminium amino)-alkylidenetriphenylphosphorane *145*.
The reason for the different behavior of aza-betaines compared with
that of the oxa-betaines can be seen in the stronger affinity toward
protons of the amide nitrogen.   In this connection it is of interest
that ethylidenetriphenylphosphorane reacts much slower with *141* than
methylenetriphenylphosphorane (*17*) does.   Finally, *145* decomposes in
a Hofmann degradation to *142* and *143*.   Obviously the elimination of
143 is favored by the formation of the conjugated system of *142*.   It
is of special interest that the ylides *142* of the type R^1 = alkyl
and R^2 = H, as far as investigated until now, are formed as the cis
isomers *142a*.   The reaction of *142* with carbonyl compounds gives the
conjugated dienes *146*.   In the reaction of *142* with ketones an
increase in the yield of *146* can be effected if one adds KNH_2 to
remove the dialkylaluminumamide *143* as an unsoluble complex salt.

$$R^1_{}\diagdown_{C=C}\diagup^{H}\diagdown_{C=P(C_6H_5)_3} \atop H\diagup{}^{}\diagdown H$$

142a

$$(C_6H_5)_3P=CH-CH=CR^1R^2 \ + \ R^3R^4C=O \longrightarrow R^1R^2C=CH-CH=CR^3R^4$$
$$\underline{142} \hspace{6cm} \underline{146}$$
$$+$$
$$(C_6H_5)_3P=O$$

The synthesis of conjugated dienes can be accomplished as follows
to 0.1-0.15 mol of a diisobutylaluminumalkylideneamide, *141*, in
250 ml of absolute toluene is added, dropwise over a 2-hr period, the
equivalent amount of *17* in 200 ml of toluene.   The procedure is
carried out at room temperature under an inert gas (argon), with
stirring.   After standing at room temperature for 12 hr, a 1.2- to
1.4-fold excess of the calculated amount of aldehyde or ketone is
added to the dark red toluene solution of *142* and *143* at $0°C$.   Using
aldehydes the color disappears after a few minutes at $0°C$; with ketone

higher temperatures $(70°-80°C)$ and longer reaction times (1-2 hr) are necessary. Finally, the organoaluminum is decomposed at $0°C$ by cautiously adding air-free water and 2 N H_2SO_4. The organic layer is separated, and the aqueous layer is extracted with ether. The combined extracts are washed with $NaHCO_3$, $NaHSO_4$, and water and dried with Na_2SO_4. The solvent is evaporated (at 12 Torr), and the residue is distilled under vacuum (0.2-10 Torr) to yield the conjugated dienes *146*. To separate the diisobutylaluminumamide *143*, $R^1 = R^2 = 2\text{-}C_4H_9$, KNH_2 is added at room temperature in the ratio *143*:KNH_2 = 1:2. After 12 hr the precipitated complex salt is filtered off.

The main difficulty of the described procedure is that although the ylides *142* (R^1 = alkyl, R^2 = H) are essentially cis isomers in principle, four stereoisomeric products (*146*) that have to be separated may result.

While β-chloracrylates and chloroformates (see Sec. IV.C) add to an allylidenetriphenylphosphorane at the γ position, allyl bromide (*147*) attacks the diisobutylallylidenephosphorane *148* at the α carbon with respect to the phosphorus, forming the phosphonium salt *149* [83]. Salts of the type *149*, generally available from allylic bromides and allylidenephosphoranes, yield on reduction with $Li/EtNH_2$ 1,5-dienes [84(b)].

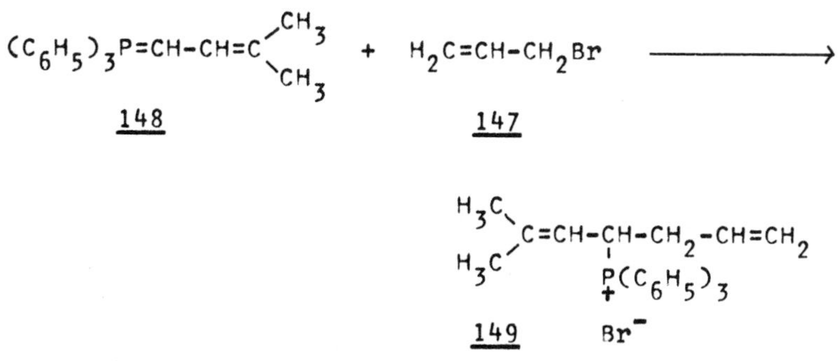

$$(C_6H_5)_3P=CH-CH=C\overset{\displaystyle CH_3}{\underset{\displaystyle CH_3}{<}} \quad + \quad H_2C=CH-CH_2Br \quad \longrightarrow$$

$$\underline{148} \qquad\qquad\qquad \underline{147}$$

$$\overset{\displaystyle H_3C}{\underset{\displaystyle H_3C}{>}}C=CH-\underset{\displaystyle \overset{|}{P(C_6H_5)_3}}{CH}-CH_2-CH=CH_2$$

$$\underline{149} \qquad Br^-$$

K. *Alkylation of Ethoxycarbonylmethylenetri-*
   phenylphosphorane by Aziridines

The acyl- and tosyl-aziridines *150* react with ethoxycarbonylmethyl-
enetriphenylphosphorane (*55*, R = C_2H_5) to give the alkylated ylides
152 [85]. It is supposed that initially a betaine, *151*, is formed,
which is transformed by a proton migration into *152*. Hydrolysis of
152 yields the N-substituted γ-amino acids *153*. Wittig reactions
give the unsaturated esters *154*.

$$R-N\underset{CH_2}{\overset{CH_2}{\big\langle}} \quad + \quad (C_6H_5)_3P=CH-CO_2C_2H_5 \quad \longrightarrow \quad R-\bar{N}-CH_2-CH_2-\underset{\overset{|}{\overset{+}{P}(C_6H_5)_3}}{CH}-CO_2C_2H_5$$

$$\underline{\textbf{150}} \qquad\qquad\qquad \underline{\textbf{55}} \qquad\qquad\qquad\qquad \underline{\textbf{151}}$$

$$\underline{\textbf{151}} \quad\longrightarrow\quad R-\overset{H}{\underset{}{N}}-CH_2-CH_2-\underset{\overset{||}{P(C_6H_5)_3}}{\overset{}{C}}-CO_2C_2H_5 \quad\xrightarrow{H_2O/OH^-}\quad R-\overset{H}{\underset{}{N}}-CH_2CH_2CH_2CO_2H$$

$$\qquad\qquad\qquad\qquad \underline{\textbf{152}} \qquad\qquad\qquad\qquad\qquad \underline{\textbf{153}}$$

$$\Big\downarrow R^1-CHO$$

$$R-\overset{H}{\underset{}{N}}-CH_2-CH_2-\underset{\overset{||}{CHR^1}}{\overset{}{C}}-CO_2C_2H_5$$

$$\underline{\textbf{154}}$$

$$R = p-NO_2-C_6H_4-CO- \quad ; \quad p-CH_3C_6H_4-SO_2-$$

L. *Ring Closure Reactions by Alkylation*
   of Phosphonium Ylides

   1. Intramolecular Ring Closure

   From an ω,ω-dihalogen compound, *155*, and triphenylphosphine the
monophosphonium salt *156* is prepared. Compound *156* is converted by
a base to the corresponding ylide *157*, which undergoes an intramolecular nucleophilic substitution in the course of which the cyclic
phosphonium salt *158* is formed [4, 86, 87(a) and (b), 88]. For
example, when Y = CH_2, compound *158* results from *156* in 91% yield

[89] if one adds to a suspension of 156 in hot water the equivalent amount of 1 N NaOH. This result shows [88] that the deprotonation 156 → 157 with subsequent alkylation, 157 → 158, is faster than the known alkaline degradation of phosphoniumhydroxides and ylides. For the usefulness of the described reaction, it is required that the reaction between 155 and triphenylphosphine yields definitely the mono salt 156. Unfortunately, in the case of Y = $(CH_2)_n$ the salts 156 are only available if n = 1 or 2 [4, 86, 87(a), 88]; for n > 2 one obtains inseparable mixtures of mono- and bisphosphonium salts [4, 88]. The Wittig reaction of 158 with Y = CH_2 or $CH(CH_3)$ gives alkylidenecyclopropanes [87(b)].

$$Y\begin{smallmatrix} CH_2-X \\ \\ CH_2-X \end{smallmatrix} + (C_6H_5)_3P \longrightarrow \left[Y\begin{smallmatrix} CH_2-\overset{+}{P}(C_6H_5)_3 \\ \\ CH_2-X \end{smallmatrix} \right] X^- \xrightarrow{\text{Base}}$$

$$\underline{\textbf{155}} \qquad\qquad\qquad \underline{\textbf{156}}$$

$$Y\begin{smallmatrix} CH=P(C_6H_5)_3 \\ \\ CH_2-X \end{smallmatrix} \longrightarrow \left[Y\begin{smallmatrix} CH-\overset{+}{P}(C_6H_5)_3 \\ | \\ CH_2 \end{smallmatrix} \right] X^-$$

$$\underline{\textbf{157}} \qquad\qquad\qquad \underline{\textbf{158}}$$

Definitive ring closure reactions by intramolecular C-alkylation of methylenephosphoranes are only possible if the starting ω,ω-bishalide has two equivalent C-halogen bonds from which only one reacts with 1 mol triphenylphosphine or if 155 has C-halogen bonds which differ in their reactivity so that only one reacts with phosphine. To the first group belong bisbromomethylaryl compounds of the type 159 with n = 0 [88, 90, 91]. In the case of n > 0 it is obvious that the bromomethyl group attached directly to the aromatic ring (Ar) is much more reactive than the bromomethyl group at the end of the aliphatic chain. As a consequence, in nonpolar solvents the monophosphonium salt 160 is formed readily. One mole of base deprotonates 160 to give the ylide, 161, which undergoes an

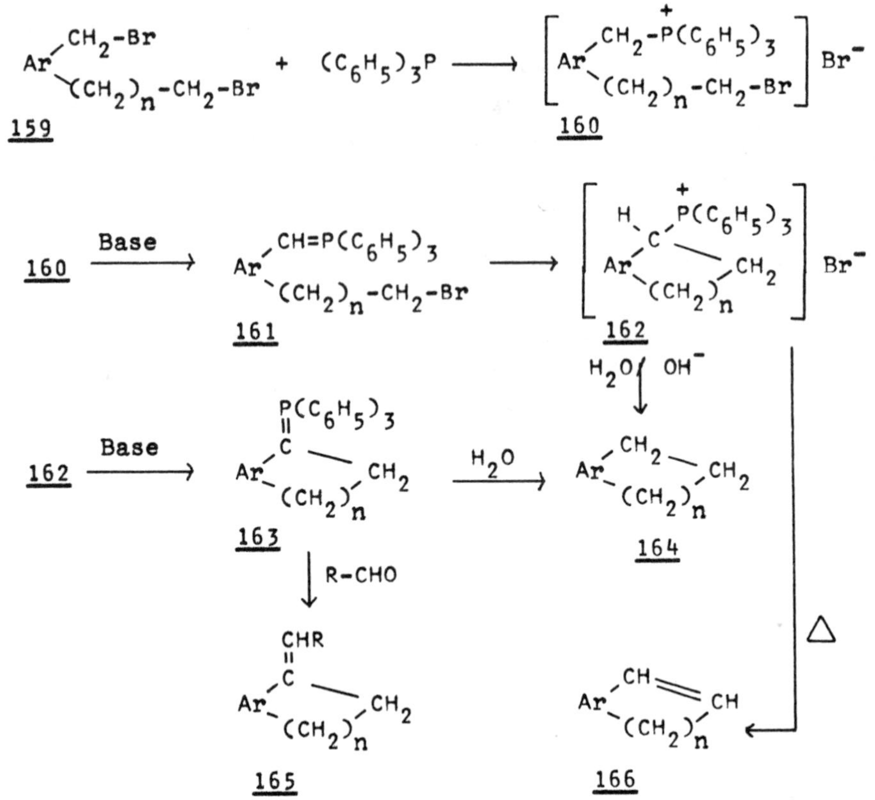

intramolecular nucleophilic substitution (C-alkylation) affording
the cyclic phosphonium salt *162*. The phosphonium salts *162* can be
deprotonated to the ylides *163* by a second mole of base.

The reaction of *163* with water yields the hydrocarbons *164* and
triphenylphosphine oxide. The same is true if *162* is hydrolyzed
with aqueous alkali. The Wittig reaction of *163* with aldehydes
leads to the compounds *165* with a semicyclic double bond. Pyrolysis
of the salts *162* gives rise to the formation of triphenylphosphine-
hydrobromide and the cyclic olefins *166*.

Because one starts with bishalogen compounds, *159*, in which the
halogen atoms are bonded to substituents of aromatic rings, the
reaction products *162-166* are polycyclic. According to this general
procedure five-, six-, and seven-membered rings have been synthesized

as the following examples show (in most cases the compounds of the type *165* and *166* have been also synthesized). Formation of five-membered rings:

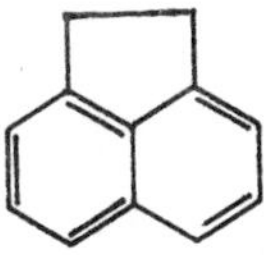

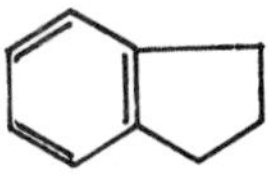

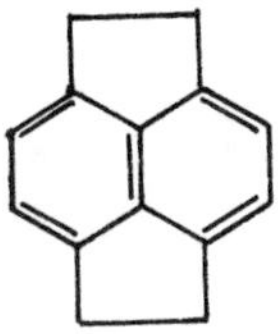

Formation of six-membered rings:

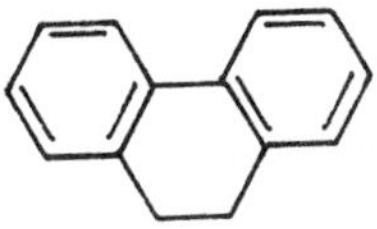

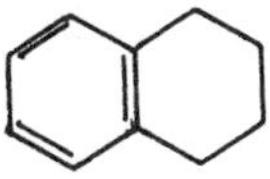

Formation of seven-membered rings:

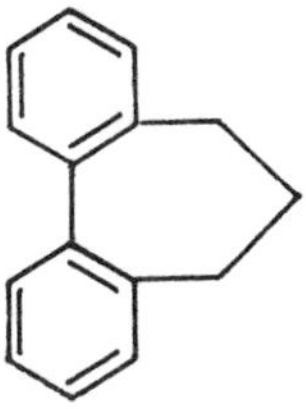

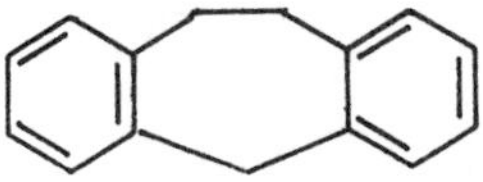

2. Combined Intra- and Intermolecular C-Alkylation

As already mentioned, the synthesis of monocyclic compounds by intramolecular C-alkylation often fails because the necessary mono-phosphonium salts with a halogen in the ω position are not readily available (with the exception of the mono salts already mentioned [*156*, $Y = (CH_2)_n$ for $n = 1, 2$]. This difficulty can be avoided by a combination of the inter- and intramolecular alkylation of ylides. The new method proceeds according to the general scheme shown in Eq. (16) [92, 93]. One mole of a dihalogen compound, *155*, reacts with 2 mol of methylenetriphenylphosphorane (*17*).

$$Y\begin{array}{c}CH_2-X\\ \diagdown\\ CH_2-X\end{array} \quad + \quad CH_2=P(C_6H_5)_3 \quad \longrightarrow \quad \left[Y\begin{array}{c}CH_2-CH_2-\overset{+}{P}(C_6H_5)_3\\ \diagdown\\ CH_2-X\end{array}\right]X^- \quad \overset{17}{\longrightarrow}$$

$$\underset{155}{} \qquad\qquad \underset{17}{} \qquad\qquad\qquad\qquad\qquad \underset{167}{}$$

$$Y\begin{array}{c}CH_2-CH=P(C_6H_5)_3\\ \diagdown\\ CH_2-X\end{array} \quad + \quad \left[(C_6H_5)_3\overset{+}{P}-CH_3\right]X^-$$

$$\underset{168}{} \qquad\qquad\qquad \underset{169}{} \qquad\qquad\qquad\qquad (16)$$

$$\Big\downarrow$$

$$\left[Y\begin{array}{c}CH_2\\ \diagup\quad\diagdown\\ CH_2\end{array}CH-\overset{+}{P}(C_6H_5)_3\right]X^- \quad \xrightarrow{\text{Base}} \quad Y\begin{array}{c}CH_2\\ \diagup\quad\diagdown\\ CH_2\end{array}C=P(C_6H_5)_3$$

$$\underset{170}{} \qquad\qquad\qquad\qquad\qquad\qquad \underset{171}{}$$

$$\overset{O_2}{\diagup} \qquad\qquad \Big\downarrow RCHO$$

$$Y\begin{array}{c}CH_2\\ \diagup\quad\diagdown\\ CH_2\end{array}C=O \qquad\qquad Y\begin{array}{c}CH_2\\ \diagup\quad\diagdown\\ CH_2\end{array}C=CHR$$

$$\underset{172}{} \qquad\qquad\qquad\qquad \underset{173}{}$$

$$+ \qquad\qquad\qquad\qquad\qquad +$$

$$(C_6H_5)_3P=O \qquad\qquad\qquad (C_6H_5)_3P=O$$

First, a phosphonium salt, *167*, is formed in an intramolecular C-
alkylation. This reacts with a second mole of the ylide *17* in a
transylidation to give methyltriphenylphosphinium halide (*169*) and
the halogenated ylide *168*. Compound *168* undergoes an intramolecular
ring closure in the course of which the cyclic phosphonium salt *170*
is formed and precipitates together with *169* from the benzene solution. As a consequence of the precipitation, the equilibrium *167* +
17 $\rightleftarrows$ *168* + *169* is permanently disturbed, so that the reaction takes
place with complete formation of *170* and *169*. The two salts can
be separated by recrystallization from water, in which *169* is readily
soluble. Often, however, this separation process can be omitted.
The use of compounds with X = Br, -I, and -O-Tos proved to be most
successful.

Bases convert *170* to the "cyclic" ylide *171*, which is the starting material for other ylide reactions. Wittig reactions with aldehydes yield compounds (*173*) with a semicyclic double bond; oxidations
with oxygen afford the ketones *172*.

In the course of this new ring closure method one C atom is
added to the C chain, which may also be interrupted by hetero atoms.
The following compounds have been synthesized according to this
method, starting from the bishalides shown [93].

Formation of four-membered rings:

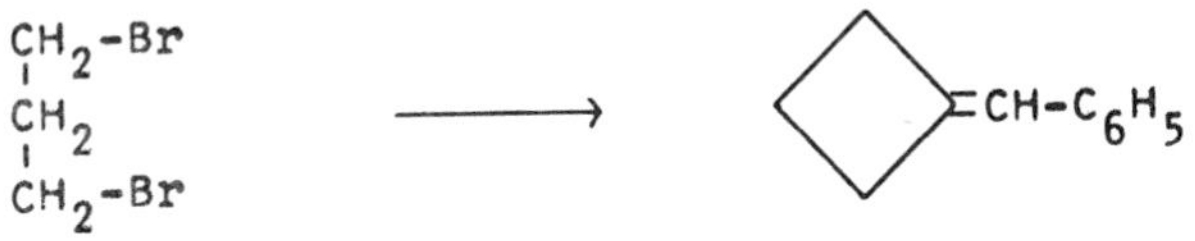

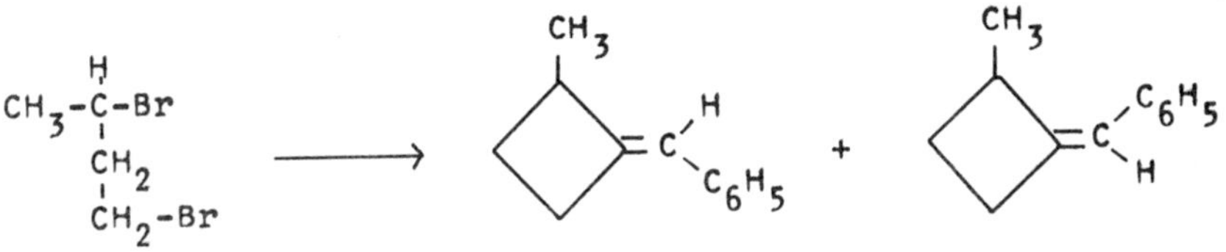

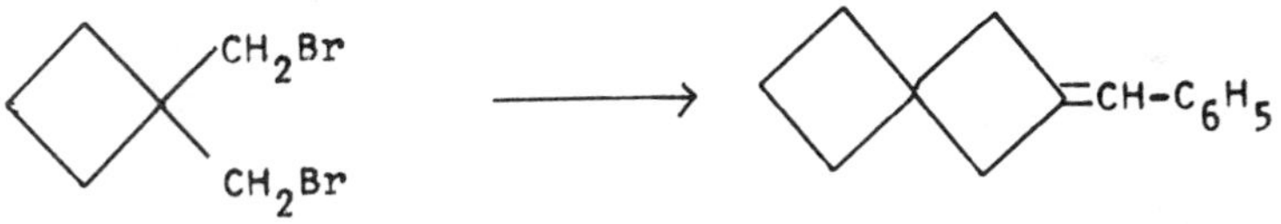

Formation of five-membered rings:

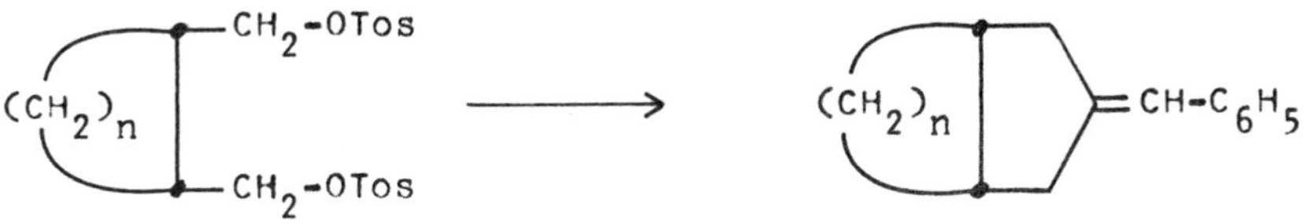

n = 2,4

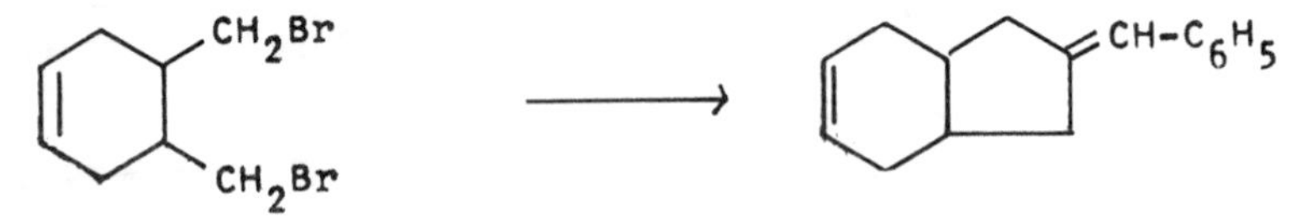

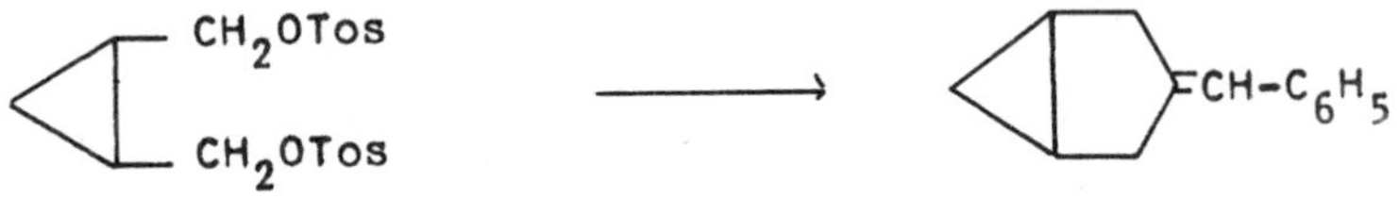

Formation of six-membered rings:

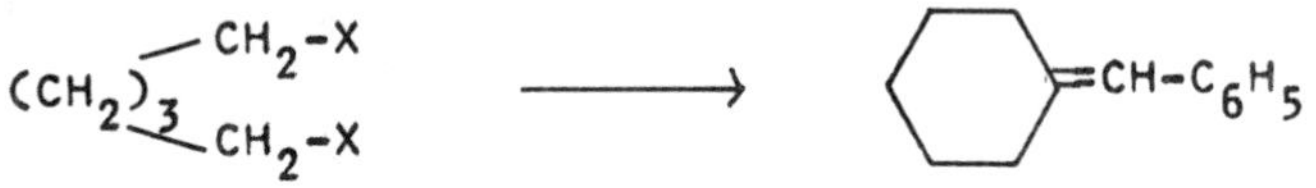

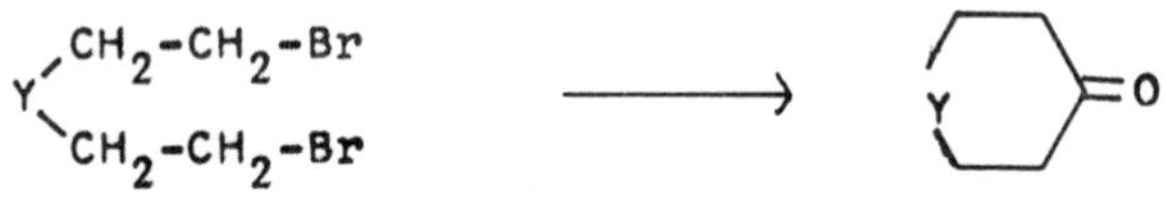

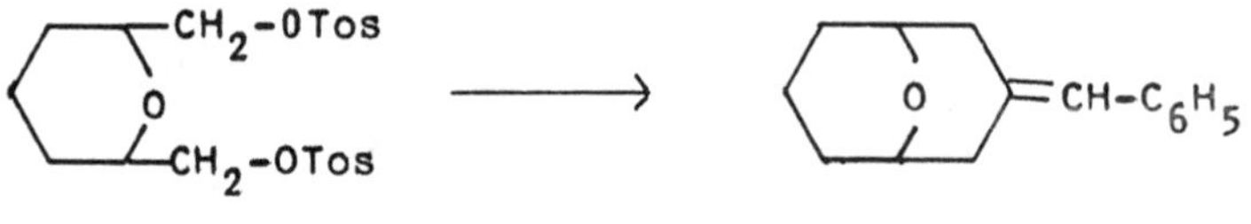

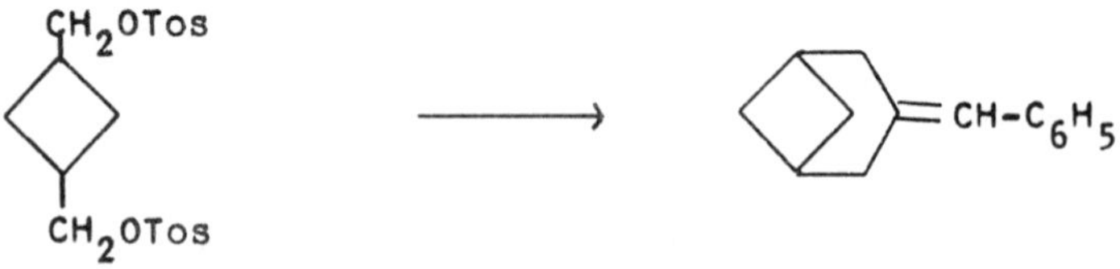

Formation of a seven-membered ring:

Further proof for the efficiency of this method is the synthesis of the helix-structured dinaptho-[2.1-a; 1.2-c]cycloheptatrien[1.3.5] (176) which can be prepared from 174 by a Hofmann degradation of the

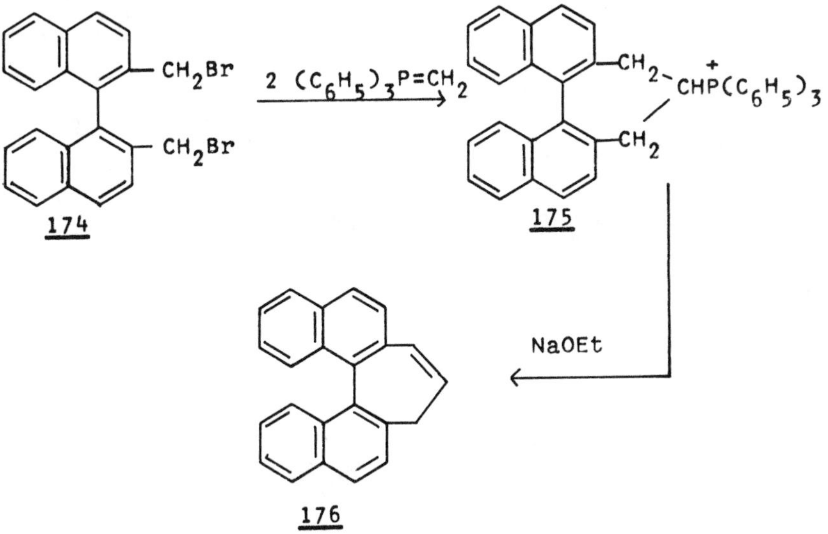

salt 175. Further, by the synthesis of the cyclic ylide 178 from 2,3,4-tribenzyl-1,5-ditosylarabitol (177), optically pure chinic acid (179) and shikimic acid (180) can be obtained [95].

Ring closure reactions of dihalides (155) with the ylide 1 yield the cyclic phosphonium salts 181, which have no more protons at the β carbon with respect to the phosphorus, so that deprotonation to an ylide is not possible [96]. In the case of $R^1 = C_6H_5$ the alkaline hydrolysis of 181 yields compounds 182. If R^1 is aliphatic, one gets substituted cycloolefins (183) on pyrolysis of 181. The pyrolysis of 181 may also give rise to the formation of semicyclic double bonds at the aliphatic ring.

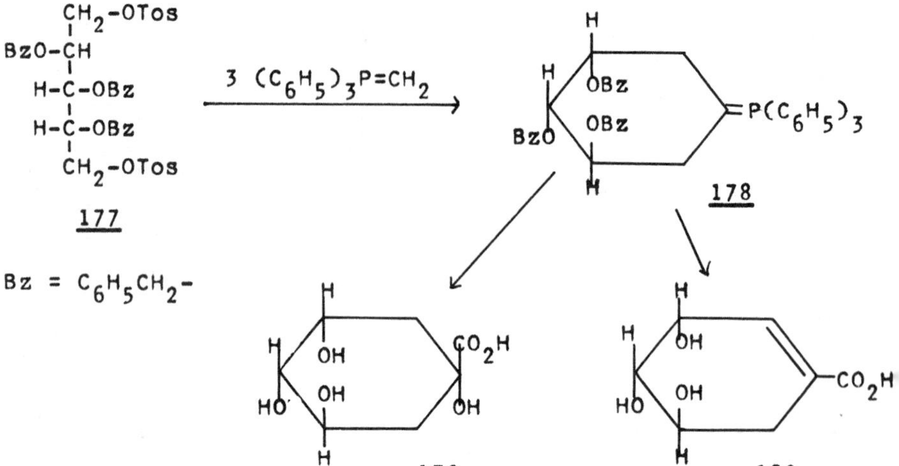

$$CH_2-OTos$$
$$BzO-CH$$
$$H-C-OBz$$
$$H-C-OBz$$
$$CH_2-OTos$$

__177__

$$\xrightarrow{3\ (C_6H_5)_3P=CH_2}$$

__178__ $=P(C_6H_5)_3$

$Bz = C_6H_5CH_2-$

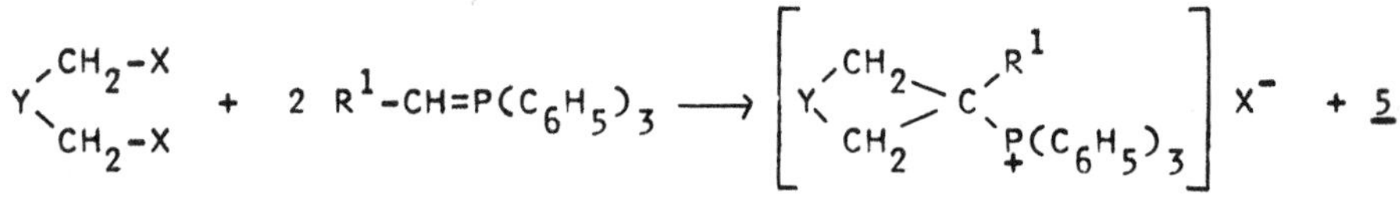

__179__

__180__

$$Y\overset{CH_2-X}{\underset{CH_2-X}{\big\langle}} \quad + \quad 2\ R^1-CH=P(C_6H_5)_3 \quad \longrightarrow \quad \left[Y\overset{CH_2}{\underset{CH_2}{\big\langle}} \overset{R^1}{\underset{\overset{+}{P}(C_6H_5)_3}{C}} \right] X^- \quad + \quad \underline{5}$$

__155__   __1__   __181__

$$Y\overset{CH_2}{\underset{CH_2}{\big\rangle}} CH-C_6H_5$$

__182__

$$Y\overset{CH}{\underset{CH_2}{\big\rangle}} C-R^1$$

__183__

3. Twofold Intermolecular C-Alkylation

The twofold intermolecular C-alkylation of the bisylides __184__
with the bishalides __155__ in the course of which the cyclic phosphonium

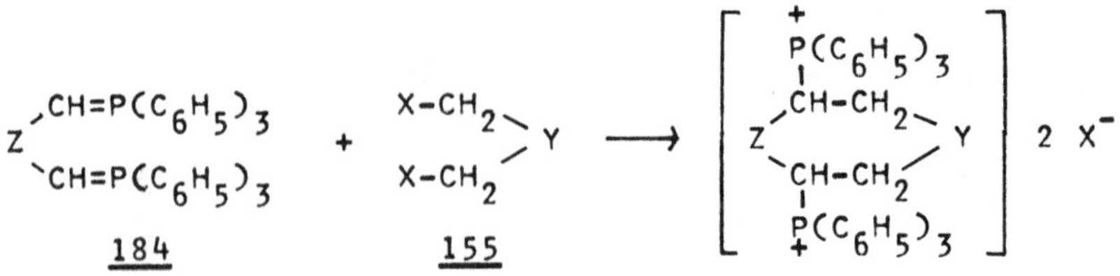

$$Z\overset{CH=P(C_6H_5)_3}{\underset{CH=P(C_6H_5)_3}{\big\langle}} \quad + \quad \overset{X-CH_2}{\underset{X-CH_2}{\big\rangle} Y} \quad \longrightarrow \quad \left[Z\overset{\overset{+}{P}(C_6H_5)_3}{\underset{\underset{\overset{+}{P}(C_6H_5)_3}{CH-CH_2}}{CH-CH_2}} Y \right] 2\ X^-$$

__184__   __155__   __185__

salts *185* are formed, represents another ring closure method [97].
This synthesis has been used successfully in the preparation of benzo-
cycloalkenes [97]. Reaction of the bisylide *186* and o-xylylenedibro-
mide (*187*) gives the bisphosphonium salt *188*, which can be hydrolyzed
to triphenylphosphine oxide and 9,10,15,16-tetrahydrotribenzo[a,c,g]-
cyclodecan (*189*) (44% yield with respect to *186*). By an analogous
series of reactions the compounds *190* and *191* have been synthesized.

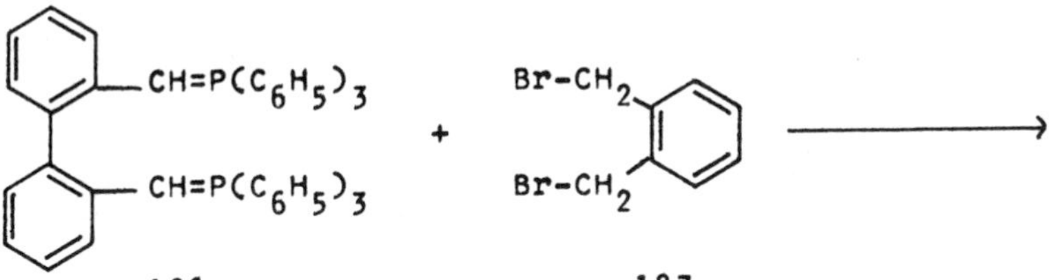

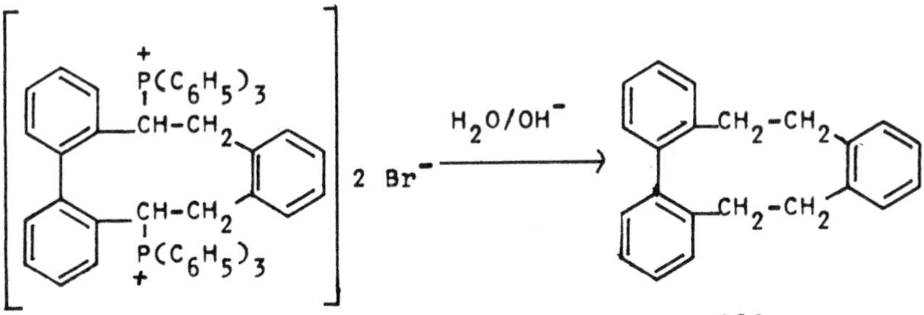

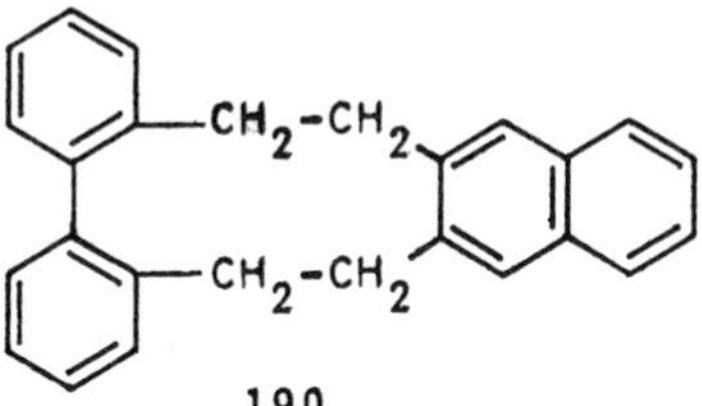

The alkaline hydrolysis of *193*, which is formed from the bisylide
192, and *187* is accompanied by a transannular reaction. Triphenylphosphine and triphenylphosphine oxide are split off giving the

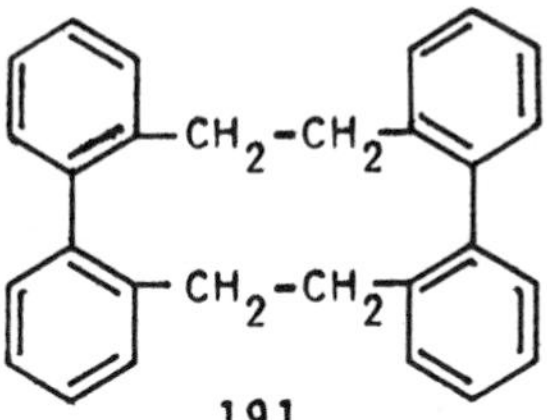

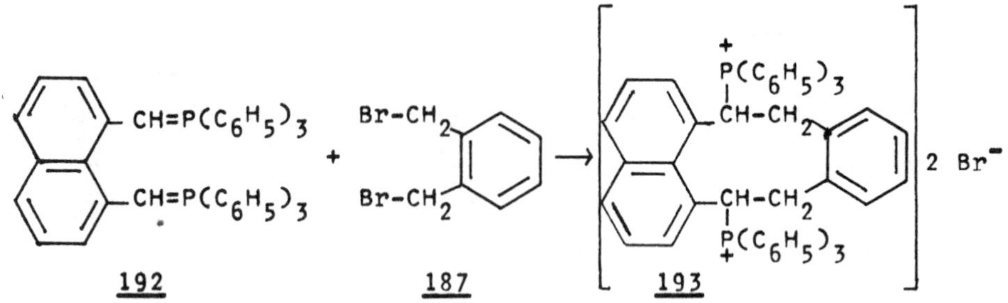

192   187   193

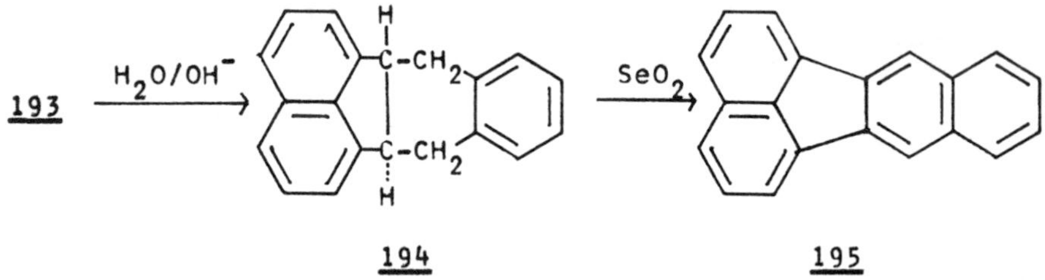

194   195

196   197

198

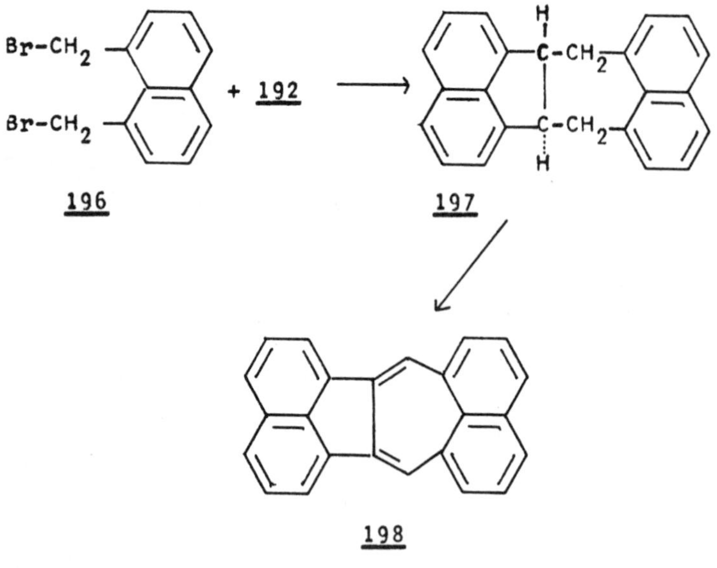

hydrocarbon *194*. Compound *194* can be dehydrogenated into benzo[k]-
fluoranthene (*195*). The analogous reaction of *192* with the dibro-
mide *196* leads, via the corresponding bisphosphonium salt, to the
polycyclic hydrocarbon *197*, from which the dinaphthoazulene (*198*) is
available by dehydrogenation with diyano-dichlorobenzoquinone.

In addition to the described two-fold C-alkylation of bisylides
with bishalides, two competing types of reactions have been observed
when aceton-1,3-diphosphorane (*199*) was alkylated with bishalides
[98].

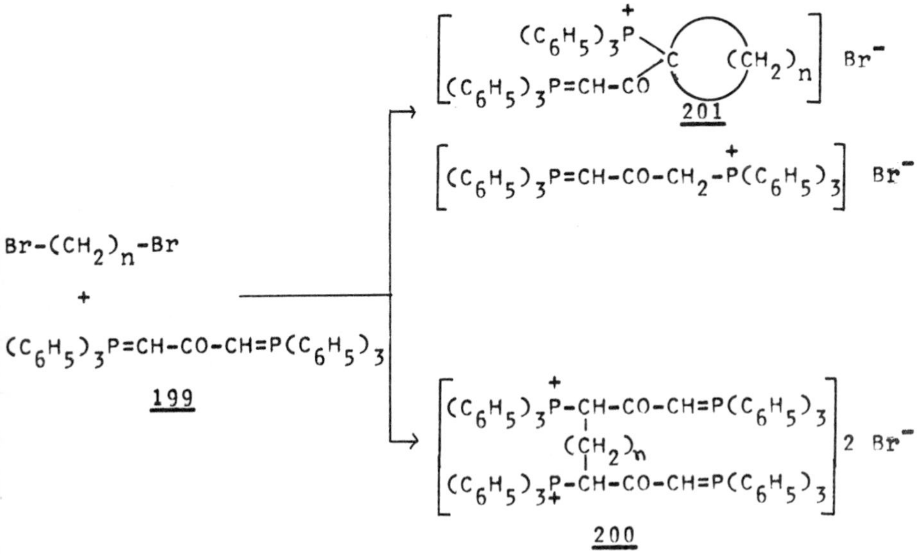

$$\left[\begin{array}{c} (C_6H_5)_3\overset{+}{P} \\ (C_6H_5)_3P=CH-C\overset{|}{O} \end{array} \underset{\underline{201}}{>\!\!C\!\!<}(CH_2)_n \right] Br^-$$

$$\left[(C_6H_5)_3P=CH-CO-CH_2-\overset{+}{P}(C_6H_5)_3 \right] Br^-$$

$$Br-(CH_2)_n-Br$$

$$+$$

$$(C_6H_5)_3P=CH-CO-CH=P(C_6H_5)_3$$

$$\underline{199}$$

$$\left[\begin{array}{c} (C_6H_5)_3\overset{+}{P}-CH-CO-CH=P(C_6H_5)_3 \\ (CH_2)_n \\ (C_6H_5)_3\overset{+}{P}-CH-CO-CH=P(C_6H_5)_3 \end{array} \right] 2\ Br^-$$

$$\underline{200}$$

1. Formation of a bis phosphonium salt, *200*, from 1 mol of
   halide and 2 mol of *199*

2. Ring closure reaction accompanied by a transylidation
   leading to the cyclic product *201*. Compound *200* as well
   as *201* can be transformed into phosphorus-free products
   by the Wittig reaction or hydrolysis.

M. Alkylation of Metalated Phosphonium Ylides

The acetylmethylenetriphenylphosphorane *202* is deprotonated by lithium
diisopropylamide or butyllithium to the lithium derivative *203*, which

$$CH_3-CO-CH=P(C_6H_5)_3 \xrightarrow{\text{LiR}} Li-CH_2-CO-CH=P(C_6H_5)_3 \xrightarrow{\text{R-X}}$$

$$\underline{202} \qquad\qquad\qquad\qquad \underline{203}$$

$$R-CH_2-CO-CH=P(C_6H_5)_3 \xrightarrow{H_2O} R-CH_2-CO-CH_3 \quad + \quad (C_6H_5)_3P=O$$

$$\underline{204} \qquad\qquad\qquad \underline{205}$$

can be alkylated to the products *204*. Hydrolysis of *204* gives the
methyl alkyl ketones *205* [99].

VI.   CYANYLATION OF PHOSPHONIUM YLIDES

The alkylidenetriphenylphosphoranes *1* react with p-tolylcyanate
(*206*) to give the 1-alkyl-1-cyano-triphenylphosphoranes *207* [11, 101].
Compounds *207* undergo Wittig reactions with aldehydes to form α-
branched α,β-unsaturated nitriles (*208*) [100, 102]. Compounds

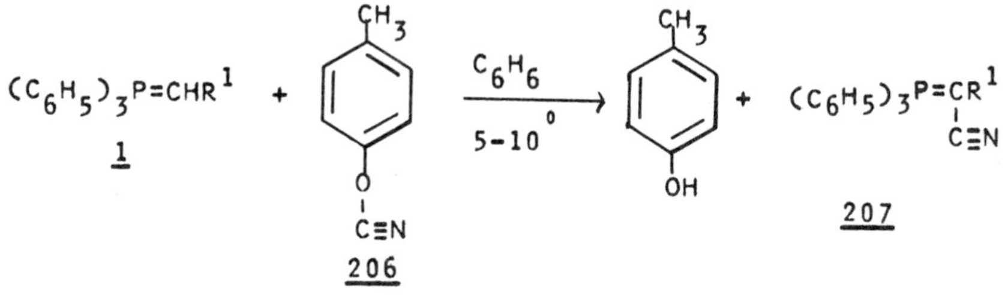

$$\underline{207} \quad + \quad R^2-CHO \quad \longrightarrow \quad \underset{\underset{\underline{208}}{CN}}{R^1-C=CHR^2}$$

207 show ambident reactivity and can be alkylated on the ylide C atom,
as well as on the N atom, by alkylating reagents [100] [Eq. (17)].
The alkylation of *207* (R^1 = H) with methylbromoacetate (*209*, R = OCH_3)
gives, in a transylidation reaction, the ylide *210* and the salt *211*.
The ylide *210* can be pyrolyzed to triphenylphosine and a Z-E mixture
of methyl 3-cyano acrylate (*212*, R = OCH_3) [100]. Bromoacetone
(*209*, R = CH_3) reacts with *207* (R = H) in an analogous way yielding
the E-3-acetyl acrylonitrile *212* (R = CH_3) [100].

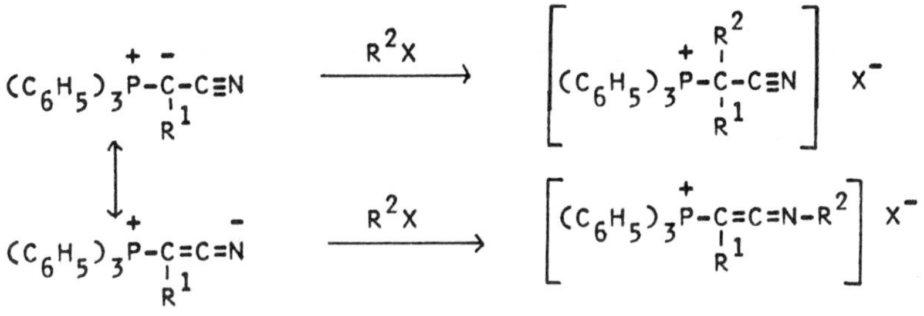

$$(C_6H_5)_3\overset{+}{P}-\underset{\underset{R^1}{|}}{\overset{-}{C}}-C\equiv N \quad \xrightarrow{\ R^2X\ } \quad \left[(C_6H_5)_3\overset{+}{P}-\underset{\underset{R^1}{|}}{\overset{\overset{R^2}{|}}{C}}-C\equiv N\right] X^-$$

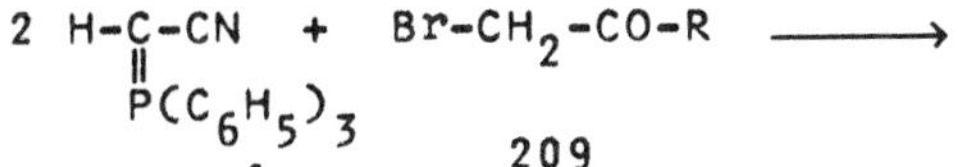

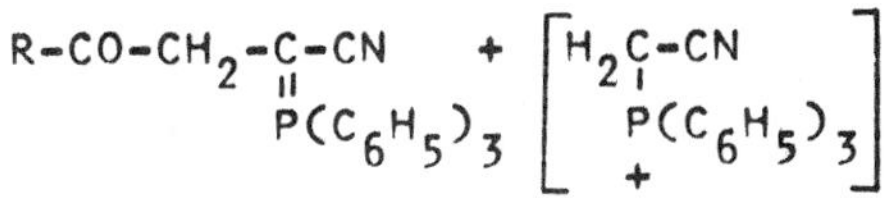

$$\updownarrow$$

$$(C_6H_5)_3\overset{+}{P}-\underset{\underset{R^1}{|}}{C}=C=\overset{-}{N} \quad \xrightarrow{\ R^2X\ } \quad \left[(C_6H_5)_3\overset{+}{P}-\underset{\underset{R^1}{|}}{C}=C=N-R^2\right] X^- \tag{17}$$

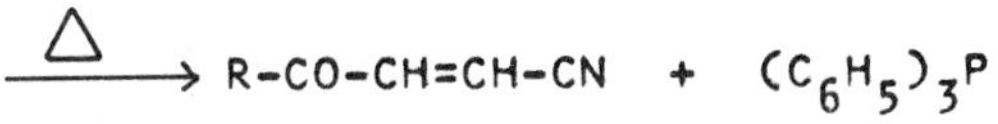

207

$$2 \ \underset{\underset{207, R^1=H}{\overset{\overset{P(C_6H_5)_3}{||}}{}}}{H-C-CN} \ + \ \underset{209}{Br-CH_2-CO-R} \ \longrightarrow$$

$$\underset{210}{R-CO-CH_2-\underset{\overset{||}{P(C_6H_5)_3}}{C}-CN} \ + \ \left[\underset{211}{\underset{\overset{|}{\underset{+}{P(C_6H_5)_3}}}{H_2C-CN}}\right]$$

$$\underset{210}{210} \ \xrightarrow{\ \triangle\ } \ \underset{212}{R-CO-CH=CH-CN} \ + \ (C_6H_5)_3P$$

VII. CARBOXYLATION OF PHOSPHONIUM YLIDES AND RELATED REACTIONS

A. *Reaction with Carbon Dioxide*

The alkylidenetriphenylphosphoranes *86* are carboxylated by CO_2 to give the betaines *213*, which decompose on alkaline hydrolysis to the carboxylic acids *214* and triphenylphosphine oxide [103]. Thermolysis of the betaines *213* with R^1 and $R^2 \neq H$ affords the allenes *215*.

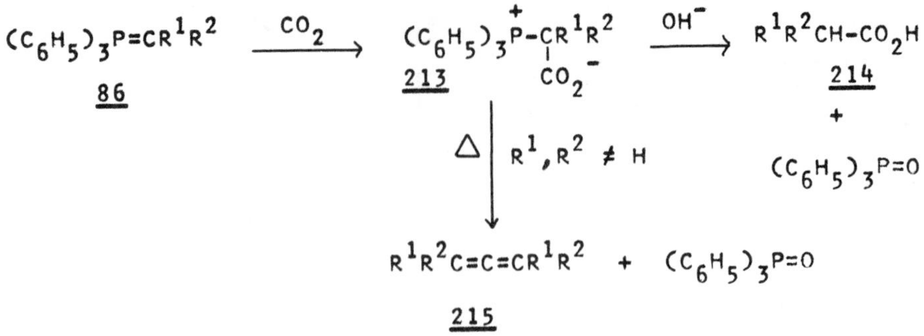

$$(C_6H_5)_3P=CR^1R^2 \xrightarrow{CO_2} (C_6H_5)_3\overset{+}{P}-\underset{\underset{CO_2^-}{|}}{C}R^1R^2 \xrightarrow{OH^-} R^1R^2CH-CO_2H$$

$$\underline{86} \qquad\qquad \underline{213} \qquad\qquad\qquad \underline{214}$$

$$+$$

$$\triangle \Big| \; R^1,R^2 \neq H \qquad\qquad (C_6H_5)_3P=O$$

$$R^1R^2C=C=CR^1R^2 \quad + \quad (C_6H_5)_3P=O$$

$$\underline{215}$$

$$R^1 = H, \; CH_3, \; CH_2-C_6H_5, \; C_3H_7, \; C_2H_5$$

$$R^2 = H, \; CH_3$$

$$R^1-R^2 = -(CH_2)_n-; \; n = 2,3,5,6$$

B. Reaction with Carbon Oxysulfide

The ylides 1 and carbon oxysulfide (216) give the betaines 217, which can be alkylated to the phosphonium salts 218. Wittig reaction of the corresponding ylides 219 affords S-esters of the α,β-unsaturated thiocarboxylic acids 220 [104].

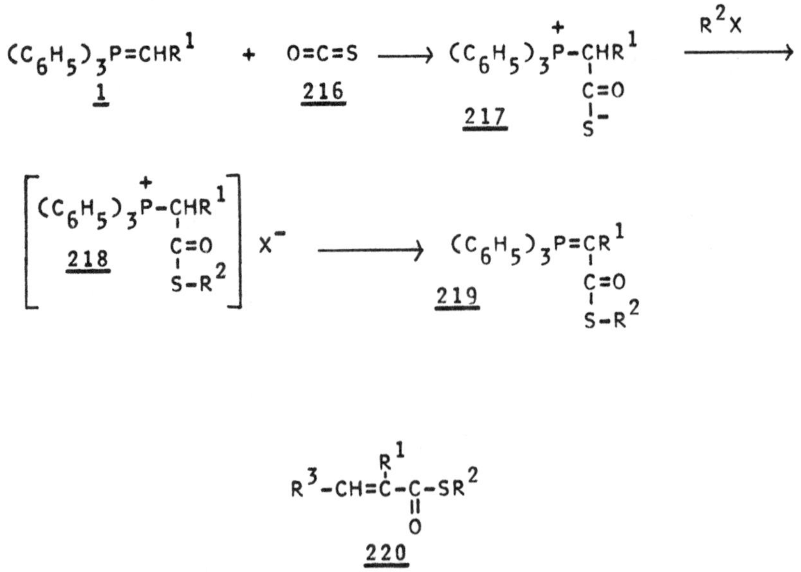

$$(C_6H_5)_3P=CHR^1 \quad + \quad O=C=S \longrightarrow (C_6H_5)_3\overset{+}{P}-\underset{\underset{\underset{S^-}{|}}{\overset{|}{C=O}}}{C}HR^1 \xrightarrow{R^2X}$$

$$\underline{1} \qquad\qquad \underline{216} \qquad\qquad\qquad \underline{217}$$

$$\left[(C_6H_5)_3\overset{+}{P}-\underset{\underset{\underset{S-R^2}{|}}{\overset{|}{C=O}}}{C}HR^1 \right] X^- \longrightarrow (C_6H_5)_3P=\underset{\underset{\underset{S-R^2}{|}}{\overset{|}{C=O}}}{C}R^1$$

$$\underline{218} \qquad\qquad\qquad\qquad \underline{219}$$

$$R^3-CH=\underset{\underset{O}{\overset{\overset{R^1}{|}}{\underset{\|}{C}}}-C-SR^2$$

$$\underline{220}$$

C. *Reaction with Carbon Disulfide*

    1. Synthesis of Ketene Mercaptals

Salt-free solutions of *1* in benzene react with CS_2 in a molar ratio of 2:1 to give the phosphonium salts *222* of α-triphenylphosphoranylidenedithiocarboxylic acids [105]. From the compounds *222* one obtains on reaction with alkyl halides (R^2X) the benzene-soluble ylides *223* and the phosphonium salts *5* which precipitate from the benzene solution. Compounds *223* do not undergo the Wittig reaction but can be alkylated by a second mole of a halide, R^3X. The resulting phosphonium salts *224* can be hydrolyzed to the ketene mercaptals *225* [105], which can be the starting material for further synthetic conversions (see, for instance, Ref. 106).

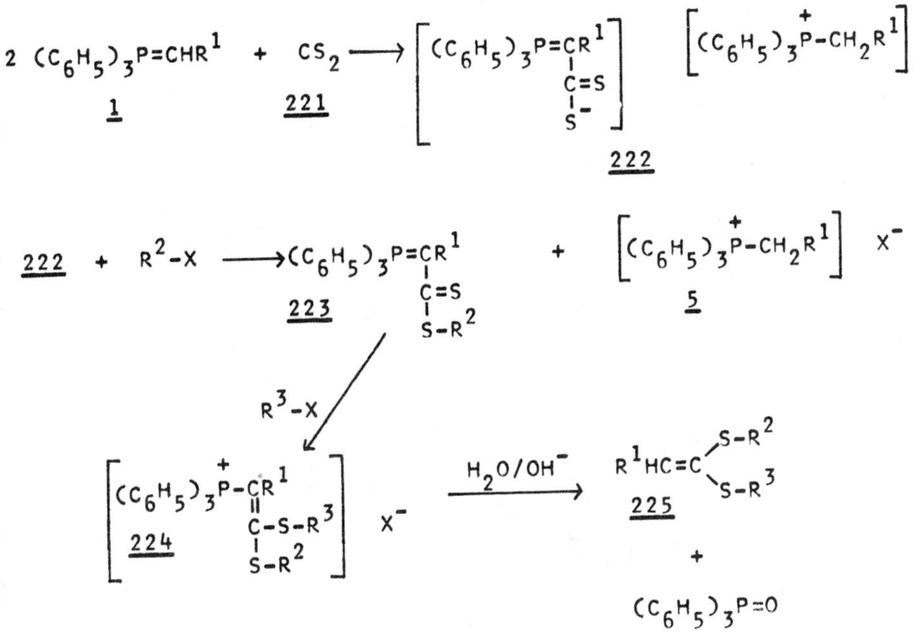

$$2\ (C_6H_5)_3P=CHR^1 \ +\ CS_2 \longrightarrow \left[\begin{array}{c} (C_6H_5)_3P=CR^1 \\ C=S \\ S^- \end{array} \right] \quad \left[(C_6H_5)_3\overset{+}{P}-CH_2R^1 \right]$$

$$\underline{1} \qquad\qquad \underline{221}$$

$$\underline{222}$$

$$\underline{222}\ +\ R^2-X \longrightarrow (C_6H_5)_3P=CR^1 \qquad +\quad \left[(C_6H_5)_3\overset{+}{P}-CH_2R^1 \right]\ X^-$$

$$\underline{223}\quad \begin{array}{c} C=S \\ S-R^2 \end{array} \qquad\qquad \underline{5}$$

$$R^3-X \Big\downarrow$$

$$\left[\begin{array}{c} (C_6H_5)_3\overset{+}{P}-CR^1 \\ C-S-R^3 \\ S-R^2 \end{array} \right]\ X^- \qquad \xrightarrow{\ H_2O/OH^-\ } \quad R^1HC=C\overset{S-R^2}{\underset{S-R^3}{\diagdown}}$$

$$\underline{224} \qquad\qquad\qquad\qquad \underline{225}$$

$$+$$

$$(C_6H_5)_3P=O$$

    2. Synthesis of Dithiocarboxylates

The reaction of the ylides *86* (R^1 and $R^2 \neq H$) with CS_2 (*221*) gives the betaines *226*. Alkylation of *226* yields the phosphonium

$$(C_6H_5)_3P=CR^1R^2 \quad + \quad CS_2 \quad \longrightarrow \quad (C_6H_5)_3\overset{+}{P}-\underset{\underset{S^-}{\overset{|}{C=S}}}{\overset{|}{C}}R^1R^2 \quad \xrightarrow{R^3-X}$$

$$\underline{86} \qquad\qquad\qquad \underline{221} \qquad\qquad\qquad \underline{226}$$

$$\left[(C_6H_5)_3\overset{+}{P}-\underset{\underset{S-R^3}{\overset{|}{C=S}}}{\overset{|}{C}}R^1R^2 \right] X^- \quad \xrightarrow[-\ (C_6H_5)_3P]{\text{Electrolysis}} \quad R^1R^2CH-\underset{S}{\overset{\parallel}{C}}-SR^3$$

$$\underline{227} \qquad\qquad\qquad\qquad\qquad\qquad\qquad\qquad \underline{228}$$

salts 227, which can be transformed into the dithiocarboxylates 228
by electrolysis [107].

D. Reaction with Isothiocyanates

The isothiocyanates 229 and the ylides 1 form the betaines 230, which
rearrange, when R = H or a group with a (-I) effect, to give the
ylides 231. The compounds 231 undergo Wittig reactions to yield the

$$(C_6H_5)_3P=CHR^1 \quad + \quad R^2-N=C=S \quad \longrightarrow (C_6H_5)_3\overset{+}{P}-\underset{\underset{S^-}{\overset{|}{C=N-R^2}}}{\overset{|}{C}}HR^1 \quad \longrightarrow$$

$$\underline{1} \qquad\qquad\qquad \underline{229} \qquad\qquad\qquad \underline{230}$$

$$(C_6H_5)_3P=\underset{\underset{HN-R^2}{\overset{|}{C=S}}}{\overset{|}{C}}R^1 \qquad \longrightarrow \qquad R^3-CH=\overset{\overset{R^1}{|}}{C}-\underset{S}{\overset{\parallel}{C}}-NH-R^2$$

$$\underline{231} \qquad\qquad\qquad\qquad\qquad\qquad \underline{232}$$

α,β-unsaturated-N-substituted thiocarboxylic amides 232 [108]. Both
230 and 231 are methylated by methyl iodide at the S atom yielding
the phosphonium salt 233, which can be deprotonated to give the
ylides 234. Hydrolysis of 234 gives the thiocarboxylic amide derivatives 235. Wittig reaction yields the corresponding α,β-unsaturated
compounds 236. Apparently, α-thiocarbonyl-stabilized ylides are also
alkylated at the S atom [109].

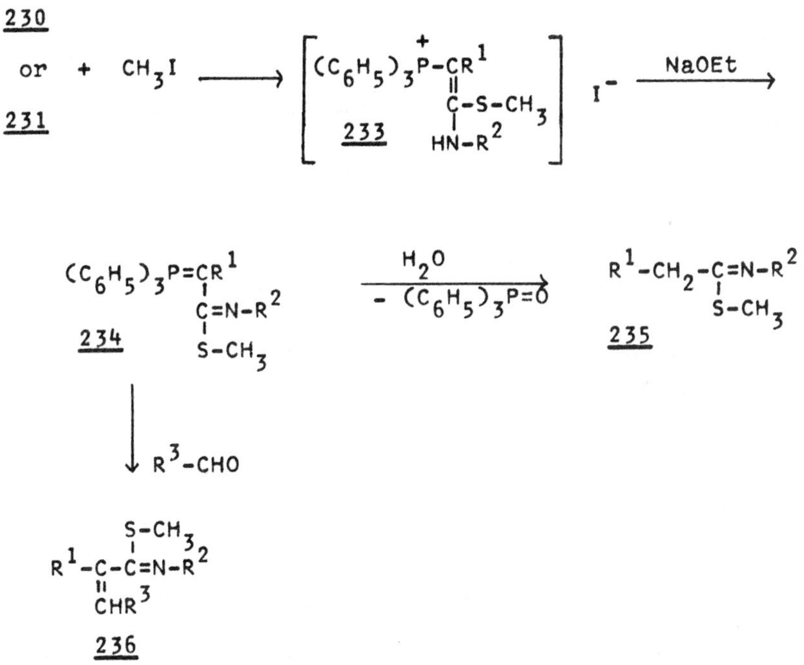

230

or

231

$+ \ CH_3I \longrightarrow \left[(C_6H_5)_3 \overset{+}{P}-\underset{\underset{HN-R^2}{\overset{\overset{\displaystyle \ \ }{C-S-CH_3}}{|}}}{\underset{\mathbf{233}}{\overset{\displaystyle \|}{C}R^1}} \right] I^- \xrightarrow{\ NaOEt\ }$

$(C_6H_5)_3 P = \underset{\underset{S-CH_3}{\overset{\overset{\displaystyle C=N-R^2}{|}}{|}}}{\overset{\displaystyle \ }{C}R^1}$

234

$\xrightarrow[- \ (C_6H_5)_3P=O]{H_2O}$

$R^1-CH_2-\underset{\overset{\displaystyle |}{S-CH_3}}{C=N-R^2}$

235

$\downarrow R^3-CHO$

$R^1-\underset{\overset{\displaystyle \|}{CHR^3}}{\overset{\overset{\displaystyle S-CH_3}{|}}{C}}-C=N-R^2$

236

REFERENCES

1. S. Trippett, *Quart. Rev.*, *17*, 406 (1963).

2. A. Maercker, *Org. Reactions*, *14*, 270 (1965).

3. A. W. Johnson, *Ylid Chemistry*, Academic Press, New York, 1966.

4. H. J. Bestmann, *Angew. Chem.*, *77*, 609, 651, 850 (1965); *Angew. Chem. (Int. Ed.)*, *4*, 583, 645, 830 (1965); published together in *Neuere Methoden der Präparativen Organischen Chemie*, Vol. V., Weinheim Bergstr., Verlag Chemie, 1967.

5. S. Trippett, in *Organphosphorus Chemistry*, Vols. I-VI, Chemical Society, London, 1970-1975.

6. H. J. Bestmann and R. Zimmermann, *Fortschr. Chem. Forsch.*, *20*, 1 (1971).

7. H. J. Bestmann and R. Zimmermann, in *Organic Phosphorus Compounds*, Vol. 3 (G. M. Kosolapoff and L. Maier, eds.), Wiley, New York, 1972.

8. H. J. Bestmann and R. Zimmermann, *Chemikerzeitung*, *96*, 649 (1972).

9. E. Zbiral, *Synthesis*, 775 (1974).

10. H. J. Bestmann, *Chem. Ber.*, *95*, 58 (1962).

11. H. J. Bestmann, *Tetrahedron Lett.*, 7 (1960).

12. H. J. Bestmann and B. Arnason, *Chem. Ber.*, *95*, 1513 (1962).

13. H. J. Bestmann, W. Stransky, and O. Vostrowsky, *Chem. Ber.*, *109*, 1694 (1976).

14. U. Wannagat and H. Niederprüm, *Chem. Ber.*, *94*, 1540 (1961).

15. G. Wittig and A. Haag, *Chem. Ber.*, *88*, 1654 (1955).

16. H. Behringer and K. Falkenberg, *Chem. Ber.*, *96*, 1428 (1963).

17. H. O. House and H. Babad, *J. Org. Chem.*, *28*, 90 (1963).

18. L. D. Bergelson, V. A. Vaver, L. I. Barsukov, and M. M. Schemyakin, *Izv. Akad. Nauk SSSR. Otd. Khim. Nauk*, 1134 (1963); *Chem. Abstr.*, *59*, 8607 (1963).

19. S. Trippett and D. M. Walker, *J. Chem. Soc.*, 1266 (1961).

20. G. Wittig and U. Schöllkopf, *Chem. Ber.*, *87*, 1318 (1954).

21. W. Grell and H. Machleidt, *Liebigs Ann. Chem.*, *693*, 134 (1966).

22. H. J. Bestmann and H. Dornauer, *Chem. Ber.*, *103*, 2011 (1970).

23. M. Le Corre, *Bull. Soc. Chim. Fr.*, 1951 (1974).

24. M. Le Corre, *Bull. Soc. Chim. Fr.*, 2005 (1974).

25. A. P. Uijttewaal, F. L. Jonkers, and A. van der Gen, *Tetrahedron Lett.*, 1439 (1975).

26. H. J. Bestmann, N. Sommer, and H. A. Staab, *Angew. Chem.*, *74*, 293 (1962); *Angew. Chem.* (Int. Ed.), *1*, 270 (1962).

27. H. A. Staab and N. Sommer, *Angew. Chem.*, *74*, 294 (1962); *Angew. Chem.* (Int. Ed.), *1*, 270 (1962).

28. G. Märkl, *Tetrahedron Lett.*, 1027 (1962).

29. E. Zbiral, *Tetrahedron Lett.*, 1483 (1965).

30. P. A. Chopard, R. J. G. Searle, and F. H. Devitt, *J. Org. Chem.*, *30*, 1015 (1965); P. A. Chopard, R. F. Hudson, and R. J. G. Searle, *Tetrahedron Lett.*, 2357 (1965).

31. F. Ramirez and S. Dershowitz, *J. Org. Chem.*, *22*, 41 (1957).

32. H. J. Bestmann, H. Buckschewski, and H. Leube, *Chem. Ber.*, *92*, 1345 (1959).

33. L. Horner and A. Mentrup, *Liebigs Ann. Chem.*, *646*, 65 (1961).

34. (a) E. Zbiral and M. Rasberger, *Tetrahedron*, *25*, 1871 (1969); (b) E. Zbiral, *Monatsh. Chem.*, *97*, 180 (1966); (c) E. Zbiral and H. Hengstberger, *Monatsh. Chem.*, *99*, 429 (1968).

35. For the acid-catalysed reactions of the stable alkoxycarbonyl-methylidenetriphenylphosphoranes see; Ch. Rüchardt, S. Eichler, and P. Panse, *Angew. Chem.*, *75*, 858 (1963); *Angew. Chem* (Int. Ed.), *2*, 619 (1963); *Chem. Ber.*, *100*, 1144 (1967); S. Fliszar, R. F. Hudson, and G. Salvadori, *Helv. Chim. Acta*, *47*, 159 (1964).

36. G. Märkl, *Chem. Ber.*, *94*, 2996 (1961).

37. G. Märkl, *Chem. Ber.*, *95*, 3003 (1962).

38. D. B. Denney and St. T. Ross, *J. Org. Chem.*, *27*, 998 (1962).

39. See also A. J. Speziale and C. C. Tung, *J. Org. Chem.*, *28*, 1353 (1963).

40. H. J. Bestmann and R. Armsen, *Synthesis*, 59 (1970).

41. E. Zbiral and E. Werner, *Monatsh. Chem.*, *97*, 1797 (1966).

42. E. Zbiral and M. Rasberger, *Tetrahedron*, *24*, 2419 (1968).

43. H. J. Bestmann, R. Armsen, and H. Wagner, *Chem. Ber.*, *102*, 2259 (1969).

44. H. J. Bestmann and L. Kisielowski, *Angew. Chem.*, *88*, 297 (1976); *Angew. Chem. (Int. Ed.)* *15*, 298 (1976).

45. S. T. D. Gough and S. Trippett, *J. Chem. Soc.*, 2333 (1962).

46. See also S. Trippett and D. M. Walker, *J. Chem. Soc.*, 3874 (1959).

47. G. Märkl, *Chem. Ber.*, *94*, 3005 (1961).

48. (a) G. Märkl, *Angew. Chem.*, *74*, 217 (1962); *Angew. Chem. (Int. Ed.)* *1*, 160 (1962); (b) H. J. Bestmann and C. Geismann, *Liebigs Ann., Liebigs Ann. Chem. 276* (1977).

49. S. T. D. Gough and S. Trippett, *J. Chem. Soc.*, 543 (1964).

50. J. J. Pappas and E. Gaucher, *J. Org. Chem.*, *31*, 1287 (1967).

51. H. J. Bestmann and H. Hartung, *Chem. Ber.*, *99*, 1198 (1966).

52. H. J. Bestmann, G. Graf, and H. Hartung, *Angew. Chem.*, *77*, 620 (1965); *Angew. Chem. (Int. Ed.)* *4*, 596 (1965).

53. H. J. Bestmann, G. Graf, H. Hartung, S. Kolewa, and E. Vilsmaier, *Chem. Ber.*, *103*, 2794 (1970).

54. I. Tömösközi and H. J. Bestmann, *Tetrahedron Lett.*, 1293 (1964). The absolute configuration of the allenic carboxylates derived therein has to be revised.

55. (a) H. J. Bestmann and I. Tömösközi, *Tetrahedron*, *24*, 3299 (1968); (b) H. J. Bestmann, H. Scholz, and E. Kranz, *Angew. Chem.*, *82*, 808 (1970); *Angew. Chem. (Int. Ed.)* *9*, 796 (1970).

56. (a) H. J. Bestmann, H. G. Liberda, and H. Salbaum, unpublished data.

57. H. J. Bestmann and H. Schulz, *Angew. Chem.*, *73*, 27 (1961); *Liebigs Ann. Chem.*, *674*, 11 (1964).

58. M. Le Corre, *C. R. Acad. Sci.*, Paris, Ser. C, *276*, 963 (1973).

59. H. J. Bestmann and H. Schulz, *Chem. Ber.*, *95*, 2921 (1962).

60. O. Isler, H. Gutmann, M. Montavon, R. Ruegg, G. Ryser, and P. Zeller, *Helv. Chim. Acta*, *40*, 1242 (1957).

61. H. O. House and G. Rasmusson, *J. Org. Chem.*, *26*, 4278 (1961).

62. G. Fodor and I. Tomoskozi, *Tetrahedron Lett.*, 579 (1961).

63. H. J. Bestmann, H. Häberlein, and I. Pils, *Tetrahedron*, *20*, 2079 (1964).

64. H. J. Bestmann, F. Seng, and H. Schulz, *Chem. Ber.*, *96*, 465 (1963).

65. H. G. Oddy, *J. Amer. Chem. Soc.*, *45*, 2156 (1923).

66. G. R. Pettit, B. Green, A. K. DasGupta, P. A. Whitehouse, and J. P. Yardley, *J. Org. Chem.*, *35*, 1381 (1970).

67. H. J. Bestmann, K. Rostock, and H. Dornauer, *Angew. Chem.*, *78*, 335 (1966); *Angew. Chem.* (*Int. Ed.*) *5*, 308 (1966).

68. H. J. Bestmann, K. Rostock, and H. Dornauer, *Liebigs Ann. Chem.*, *735*, 52, (1970).

69. H. J. Bestmann, H. Dornauer, and K. Rostock, *Chem. Ber.*, *103*, 685 (1970).

70. A. A. Grigorenko, M. J. Shewtshuk, and A. W. Dombrowski, *J. Allgem. Chem.*, *36*, 506 (1966); *Chem. Abstr.*, *65*, 737 (1966).

71. F. Ramirez, O. P. Madan, and C. P. Smith, *Tetrahedron*, *22*, 567 (1966).

72. H. J. Bestmann, G. Graf, and H. Hartung, *Liebigs Ann. Chem.*, *706*, 68 (1967).

73. (a) H. J. Bestmann, E. Vilsmaier, and G. Graf, *Liebigs Ann. Chem.*, *704*, 109 (1967); (b) H. J. Bestmann and E. Vilsmaier, unpublished results.

74. H. J. Bestmann, J. Popp, and G. Schmid, unpublished results.

75. M. P. Strandtmann, C. Cohen, C. Puchalski, and J. Shavel, *J. Org. Chem.*, *33*, 4306 (1968).

76. E. C. Taylor and S. F. Martin, *J. Amer. Chem. Soc.*, *96*, 8095 (1974).

77. For the trans-olefination of stabilized ylides see M. Schlosser, *Top. Stereochem.*, *5*, 1 (1970).

78. I. Kawamoto, T. Hata, Y. Kishida, and C. Tamura, *Tetrahedron Lett.*, 2417 (1971); 1611 (1972).

79. I. Kawamoto, Y. Sugimura, and Y. Kishida, *Tetrahedron Lett.*, 877 (1973).

80. I. Kawamoto, Y. Sugimura, N. Soma, and Y. Kishida, *Chem. Lett.*, 931 (1972).

81. E. Werner and E. Zbiral, *Angew. Chem.*, *79*, 899 (1967); *Angew. Chem.* (*Int. Ed.*) *6*, 877 (1967).

82. E. Vedejs and J. P. Bershas, *Tetrahedron Lett.*, 1359 (1975).

83. B. Bogdanovic and S. Konstantinovic, *Synthesis*, 481 (1972).

84. (a) For the reaction of phosphorus ylides with Schiff's bases see H. J. Bestmann and P. Seng, *Tetrahedron*, *21*, 1373 (1965); (b) E. H. Axelrod, G. M. Milne, and E. E. van Tamelen, *J. Amer. Chem. Soc.*, *92*, 2139 (1970); K. E. Harding and K. A. Parker, *Tetrahedron Lett.*, 1633 (1971).

85. H. W. Heine, G. B. Lowrie, and K. C. Irving, *J. Org. Chem.*, *35*, 444 (1970).

86. E. E. Schweizer, C. J. Berninger, and J. G. Thompson, *J. Org. Chem.*, *33*, 336 (1968).

87. (a) A. Mondon, *Liebigs Ann. Chem.*, *603*, 115 (1957); (b) K. Utimoto, M. Tamura, and K. Sisido, *Tetrahedron*, *29*, 1169 (1973).

88. H. J. Bestmann, R. Härtl, and H. Häberlein, *Liebigs Ann. Chem.*, *718*, 33 (1968).

89. H. J. Bestmann and E. Kranz, *Chem. Ber.*, *105*, 2098 (1972).

90. H. J. Bestmann and J. Häberlein, *Z. Naturforsch.*, *17b*, 787 (1962).

91. H. J. Bestmann, H. Häberlein, and W. Eisele, *Chem. Ber.*, *99*, 28 (1966).

92. H. J. Bestmann and E. Kranz, *Angew. Chem.*, *79*, 95 (1967); *Angew. Chem. (Int. Ed.) 6*, 81 (1967).

93. H. J. Bestmann and E. Kranz, *Chem. Ber.*, *102*, 1803 (1969).

94. H. J. Bestmann and W. Both, *Chem. Ber.*, *107*, 2926 (1974).

95. H. J. Bestmann and H. A. Heid, *Angew. Chem.*, *83*, 329 (1971); *Angew. Chem. (Int. Ed.) 10*, 336 (1971).

96. H. J. Bestmann, G. Hofmann, and E. Kranz, unpublished results.

97. H. J. Bestmann and D. Ruppert, *Angew. Chem.*, *80*, 668 (1968); *Angew. Chem. (Int. Ed.) 7*, 637 (1968).

98. A. Hercouet and M. Le Corre, *Tetrahedron Lett.*, 2491 (1974).

99. M. P. Cooke, *J. Org. Chem.*, *38*, 4082 (1973); see also M. P. Cook and R. Gosvani, *J. Amer. Chem. Soc.*, *75*, 7891 (1973) and J. D. Taylor, *Chem. Commun.*, 876 (1972).

100. H. J. Bestmann and S. Pfohl, *Liebigs Ann. Chem.*, 1688 (1974).

101. D. Martin and H. J. Niclas, *Chem. Ber.*, *100*, 187 (1967).

102. G. Schiemenz and H. Engelhard, *Chem. Ber.*, *94*, 578 (1961).

103. H. J. Bestmann, Th. Denzel, and H. Salbaum, *Tetrahedron Lett.*, 1275 (1974).

104. H. J. Bestmann and H. Salbaum, unpublished results.

105. H. J. Bestmann, R. Engler, and H. Hartung, *Angew. Chem.*, *78*, 1100 (1966); *Angew. Chem. (Int. Ed.) 5*, 1040 (1966); *Chem. Ber* 1978, in press.

106. D. Seebach and R. Bürstinghaus, *Synthesis*, 461 (1975); B. T. Gröbel, R. Bürstinghaus, and D. Seebach, *Synthesis*, 121 (1976).

107. H. J. Bestmann and E. Vilsmaier, unpublished results.

108. H. J. Bestmann and S. Pfohl, *Angew. Chem.*, *81*, 750 (1969).

109. H. Yoshida, H. Matsuura, T. Ogata, and S. Inokawa, *Bull. Chem. Soc. Jap.*, *48*, 2907 (1975); *Chem. Lett.*, 1065 (1974).

110. C. A. Henrick, E. Böhme, J. A. Edwards and J. H. Fried, *J. Amer. Chem. Soc.*, *90*, 5926 (1968).

Numbers in parentheses are reference numbers and indicate that an author's work is referred to although his name may not be cited in the text. Underlined numbers give the page on which the complete reference is listed.

A

C

Cabezas Rivera, M. E., 297(469),
    347
Cabrol, N., 201(319), 269(319),
    341
Cain, P., 52(88), 58(88), 83
Caine, D., 50(84), 83. 94(48,
    49), 96(59), 97(48, 49),
    101(48), 103(132), 108
    (48, 49), 117(48), 122(49),
    130(132), 131(191, 194),
    147(48, 49), 159(59), 160
    (48, 59), 161(48, 191),
    168(59, 191), 169(191),
    170(191), 173(194), 177
    (191), 180(149), 190(132),
    196(191), 198(48), 199
    (194), 204(322), 205(322),
    221-223(49, 194), 233(132),
    251(132), 252(132), 258
    (132), 259(132), 262(132),
    278(419), 285(445), 286
    (419, 445), 297(471), 298
    (419, 445), 299(419, 445),
    312(497), 315(471), 328
    (561), 330, 331, 334, 336,
    342, 345-347, 349, 351
Calbo, L. P., Jr., 73(112), 84
Calzada, M. C., 297(469), 347
Cane, D. E., 56(90), 83
Cantagrel, R., 100(112, 267(112),
    333
Cantrall, C. W., 98(76), 331
Carbona, L., 297(469), 347
Cardwell, H. M. E., 90, 91, 96
    (28), 192(28), 212(28),
    251(28), 258(28), 330
Cargill, R. L., 244(383), 296
    (468), 344, 347
Carlon, F. E., 282(421), 345
Carlson, R. M., 34(57), 81
Casper, E. W. R., 106(150),
    137(150), 141(150), 168
    (150), 170(150), 185(150),
    334
Caubere, P., 153(288), 340
Cayen, C., 211(333), 342
Cetenko, W. A., 73(114), 75(114),
    84
Chan, D. M. T., 285(440), 286
    (440), 297(440), 346

Chan, H.-F., 248(388), 344
Chan, T. I., 131(194), 173(194),
    199(194), 221(194), 336
Chappell, R. L., 212(339), 261
    (339), 342
Chassin, C., 28(47), 81
Chavez, E. P., 123(165), 335
Chaykovsky, M., 100(92), 104(92),
    332
Chen, R. H. K., 212(340), 260(340),
    342
Chiu, C., 35(58), 81
Chong, A., 155(296), 340
Chong, B. P., 155(296), 340
Chopard, P. A., 364(30), 373(30),
    422
Christol, H., 242(380), 311(493),
    344, 348
Chu, C. C. C., 321(523), 350
Church, R. F., 319(517), 349
Claisen, L., 8(6), 16(28), 58(93),
    79, 80, 83
Claparède, A., 16(28), 80
Clark, R. D., 46(76), 82, 112(162),
    276(162), 285(443, 444), 286
    (443), 292(444), 293(444),
    294(443), 335, 346
Clement, R. A., 39(65), 51(65), 82
Coates, R. M., 126(176, 178), 131
    (193), 132(176, 193), 133
    (176), 135(225), 136(193,
    225), 148(176, 193), 150(284),
    164(193), 167(193), 169(176,
    193), 171(193), 172(193), 173
    (193), 174(225), 177(225),
    187(193), 190(225), 196(225),
    197(193), 199(193, 225), 213
    (176), 215(193), 251(193),
    255(225), 258-260(193), 279
    (176), 336, 338, 340
Cohen, C., 392(75), 393(75), 424
Cohen, N., 315(504), 349
Cole, J. E., 228(354), 343
Cole, S. M., 328(563), 351
Colin, G., 137(247), 338
Colonge, J., 36(61), 82
Conant, J. B., 10(16, 17), 80
Conia, J. M., 87(3), 90, 96(3),
    97(3), 106(3), 115(3), 149
    (3, 280), 150(3), 164(3),
    184(3), 197(3), 210(329), 211
    (329), 223(329, 350), 245(386),